Location-Based Management for Construction

Location-Based Management for Construction

Planning, scheduling and control

Russell Kenley and Olli Seppänen

Spon Press
an imprint of Taylor & Francis
LONDON AND NEW YORK

First published 2010 by Spon Press
2 Park Square, Milton Park, Abingdon, Oxon OX14 4RN

Simultaneously published in the USA and Canada by Spon Press
270 Madison Avenue, New York, NY 10016, USA

Spon Press is an imprint of the Taylor & Francis Group, an informa business

© 2010 Russell Kenley and Olli Seppänen

Typeset in Times New Roman by
Russell Kenley using Corel VENTURA Publisher V10
Printed and bound in Great Britain by
TJ International Ltd, Padstow, Cornwall

British Library Cataloguing in Publication Data
A catalogue record for this book is available from the British Library

Library of Congress Cataloging in Publication Data
  Kenley, Russell.
  Location-based management system for construction : improving productivity using flowline / Russell Kenley and Olli Seppänen.
     p. cm.
  Includes bibliographical references and index.
  1. Building—Superintendence—Data processing. 2. Production scheduling—Data processing. 3. Location-based services. 4. Flowline. I. Seppänen, Olli. II. Title.
  TH438.4.K46 2009
  690.68 5—dc22
  2009011124

ISBN13: 978-0-415-37050-9 (hardback)
ISBN 13: 978-0-203-03041-7 (ebook)

ISBN 10: 0-415-37050-7 (hardback)
ISBN 10: 0-203-03041-9 (ebook)

Contents

Section Three   *Location-based control*

Section Five   *Case studies*

15   Case study 1: Opus Business Park   493

Figures

Tables

Preface

This book introduces a new way of thinking about the management of construction projects. As such it is expected to challenge existing beliefs, strategies and practices. The concepts presented here deliberately argue for a better way to manage, and therefore implicitly appear to criticise the way construction projects are managed now. As such it will be confrontational to many, and unfortunately it will be ignored by some.

It is difficult to write a book about the location-based management system without appearing to attack, or at least dismiss, the value of existing activity-based management systems. The problem is that most practitioners feel very sensitive about, and a degree of loyalty to, the way they currently do things. Some readers may therefore take umbrage with and reject our arguments. This is understandable given their years of hard work spent learning and developing proficiency in current techniques.

It has certainly not been our intention to cast a poor light upon existing systems. Indeed, the dominant activity-based management systems of the construction industry are well developed and extremely valuable, adding enormously to the success of the industry and dealing with a considerable amount of complexity and confusion. It is merely that, by highlighting the advantages of the new approach, it is unavoidable to point out faults in the existing approaches. If we are saying we have a better way, then we are also saying that the old way is not as good.

We would therefore like to take this opportunity to apologise to those who take our message of change as a personal affront. We target the system not the individual.

While we advocate a new location-based management system, this has much in common with existing practices. Indeed, there is much that will be familiar to practitioners, despite key differences. Our request is that you bear with us and consider the arguments. Do not be in haste to dismiss. If, after you have considered all the arguments you remain to be convinced, then we will understand. However, if you can see the value in the message, then welcome aboard. The challenge of mastering a whole new management system should not be underestimated. At the same time, it can be great fun!

A book such as this should never be seen as the work of just the authors, but rather the culmination of the continuing work of many people over a long period. While the location-based management system for construction is presented here in its entirety for the first time, many of the components will be familiar to those who have explored beyond the critical path methodology (CPM). For example, the underlying management philosophy may be understood to be derived from the lean construction community and others operating at the leading edge of construction thinking.

Location-based techniques include methods which would be more familiar to most under the terms 'line-of-balance', 'repetitive scheduling method', 'linear scheduling' or 'flowline', etc. Construction management educators have ordinarily covered such topics within their curriculum. As a rule of thumb, anyone educated before 1980 will have applied such techniques to manually schedule projects as part of their training. Further, they will frequently have used them on real, usually repetitive, projects at some time in the intervening years. Yet this knowledge and these efforts will, even by them, be dismissed as an experiment. The reliance on manual techniques, physically drawing lines on grid paper, has largely prevented any serious development of the techniques beyond special or academic needs.

We both encountered such techniques at university, and must acknowledge the thought leadership provided. Key influences include Walter Mohr and Professor Kankainen.

Walter Mohr, the first influence, was a practitioner teaching at the University of Melbourne in 1980 and one of his students was Russell Kenley. Mohr emphasised flowline scheduling as the best method for scheduling construction. In his manual for construction management, he wrote:

> Since the inception of network analysis techniques in the late 1950s it has been becoming increasingly clear that an ad-hoc approach to scheduling, even though using detailed network techniques for recording and comparing information, was not able to effectively control the timing of construction projects. Such techniques have a reasonable success rate on small non-repetitive and simple projects. On larger more complex projects, the inability of these techniques to correctly schedule (or take account of) resource limitations have severely impaired their effectiveness.
>
> The motor car industry uses techniques for balancing resources on their assembly lines. Flowline is a technique that has developed out of similar approaches. It is backed by research efforts that show historically that the better projects fell naturally into a construction pattern with similar balanced resource usage patterns. While being capable of quite sophisticated mathematical modelling, flowline is best illustrated using simple graphical techniques. One of the reasons for the success of the technique is the simplicity of this presentation.

Significantly, it is only in recent years that construction research has sought lessons from the automotive industry and actively sought to improve productivity. Current researchers may think their work is new, but it seems very little is really new in much of today's research.

The second influence, Professor Kankainen, was Professor of Construction Management and Economics at Helsinki University of Technology. He understood that CPM did not solve the problems he observed in the Finnish construction industry and commenced research into location-related methods in the 1980s. The Finnish construction industry was experiencing a boom and the economy was overheating. Subcontractors could not respond to the varying resource needs of general contractors. Many failed projects eroded the profitability of companies. CPM use was revealed to be the underlying reason and pilot studies were run to examine the applicability of flowline to construction projects. The results were dramatic: waiting time for direct labour almost totally vanished and the number of failed projects decreased, restoring profitability to those companies that adopted the new planning tools. Kankainen successfully marketed the approach to all major Finnish construction companies and it rapidly became the predominant planning method.

In the 1990s, Finland was struck by a severe economic recession. The value of real estate plummeted and construction practically stopped. An entire generation of construction engineers changed careers to IT and telecommunications (to companies like Nokia). As a consequence, CPM-based scheduling skills were largely forgotten. The alternative, flowline schedules planned by large construction companies were just drawings and were not based on quantities, resources or production rates. As a response to this trend, Kankainen commenced research aiming at improving schedule reliability and controlling production. The companies participating in the research were convinced that they needed new tools. From the drive for such tools was the location-based management system born.

Kankainen's vision was that all aspects of production could be tackled by using location-based planning and control tools. The results included location-based cost forecasts and cash flow, location-based quality management and location-driven procurement and logistics. Research was also undertaken into related implementation issues: how to change the contracts so that location-based management methods will be empowered. Also, research explored the best ways to communicate the plans and schedules to subcontractors and clients, as well as their progress status. One student working with Professor Kankainen was Olli Seppänen. He developed commercial software, DynaProject.

DynaProject was designed to replace the industry standard graphical approach, and subsequently was expanded to include all the location-based tools developed in Finland. The sum of separate tools proved to be more than the individual parts. New knowledge had to be created because the software design required much more comprehensive logic than using graphical tools in the case projects. Integrating the results of Kankainen's research projects into a comprehensive software package, and implementing it on real projects, started the thinking process that eventually led to the location based management system.

These two interests came together when at the 2003 conference of the International Group for Lean Construction (www.iglc.net). Olli demonstrated his software package, DynaProject, and it was clear that a commercial product was finally available to support the needs of flowline practitioners and researchers. We have ever since been working together to advance the supporting theory and methodology. DynaProject has evolved into Vico Software's Control.

While the development of management methods to support a location-based methodology was much further advanced in the Finnish industry, it was found that there were lessons to be learned from the international experience of construction management and CPM. It is this integration, as well as the adoption of lean construction principles, that has shaped the development of the location-based management system to its current form.

The focus of this book is to achieve a practical management system, one that is ready for immediate implementation. Lean theory tells us that there is up to 50% wasted effort in the labour component of a project, in addition poorly managed work leads to defects, rework and maintenance. Managers urgently need to recognise the current lack of efficiency as an industry problem. However, the only way to change the industry, is to start with the individual. Every construction company, every construction project and every project's clients need to find people who are willing to move beyond what they do now and adopt a new system. It is only through having willing, receptive and motivated people that an integrated management system can succeed. Our job is to provide the system with its tools and techniques. Your job is to become one of the required drivers of change.

Here you will find a complete suite of tools and methods, along with a discussion of the real world problems of implementation. This will provide a path to success for construction managers. We hope that we have managed to provide everything that a construction manager needs to prepare for changing the way they run projects in pursuit of efficiency. However, it is a new and rapidly evolving system, so we ask for patience, commitment and a spirit of experimentation. In this, you will be joining a rapidly increasing community of people learning to work in a smarter way.

Russell Kenley
Melbourne, Australia

Olli Seppänen
Miami, USA

Acknowledgements

There are so many people we would like to thank, so many who have assisted in the development of the location-based management system. Walter Mohr and Professor Kankainen have already been mentioned in the preface, but our thanks go to them for their vision and commitment to a better way. All the researchers who have worked to develop the knowledge of the field that we now call location-based planning and scheduling deserve our thanks for their insight and perseverance.

We have been assisted along the way by so many clients, contractors and consultants that there are too many to mention. So, to all those who have participated in pilot studies and have offered their projects, we wish to express our gratitude. Without real projects, this work would have remained theoretical, unproven and indeed thoroughly incomplete.

We owe an equal debt to our colleagues, those with whom we have been privileged to share the pleasure of discovery through research with rigour. These include former and current students, university colleagues, practitioners and of course work colleagues. In particular, we thank the management of Vico Software for their vision, hard work and expertise in developing the location-based technology and integrating it into a world of building information modelling (BIM) and virtual construction (VC).

We acknowledge the financial support of the School of Construction at Unitec New Zealand. They have supported the considerable time for the research, testing and writing. Thanks to Heads of School Roger Birchmore and David Nummy, and Professor Jacqueline Rowarth, former Vice President Research and Development. Similarly the Faculty of Business and Enterprise at Swinburne University of Technology has continued this support, and we express our gratitude to Professor David Hayward. The research and development which has underpinned this work was only made possible with the financial support of Finland's technology fund, Tekes.

Of course we thank our families, particularly for understanding that two weeks really means two years!

Permissions

This work would not have been possible without the kind permission to use copyright material. This has been provided by:

Taylor and Francis, for permission to use material from Construction Management and Economics. http://www.tandf.co.uk

ASCE, for permission to use material from the Journal of the Construction Division.

Elsevier, for permission to use material from the International Journal of Project Management.

Extensive use has also been made of the published literature that forms the basis of the location-based management system for construction. This includes many diagrams and tables which have been reconstructed in this work from the base equations provided. Similarly much use has been made of information that has been reformatted to include new data, or consolidated to better follow the argument of this book. Every attempt has been made to acknowledge the source of the information used in this way.

Section One

Introduction to planning and control

SECTION ONE—INTRODUCTION TO PLANNING AND CONTROL

In the following four chapters, Section One establishes the theoretical context for the location-based management system for construction (LBMS). This is where the founding literature that has influenced us in our thinking is gathered in a developmental structure. While the section is somewhat lengthy, it provides an important background to the location-based management system.

Those readers who are mainly interested in the practice of LBMS could safely skip over this section in its entirety, as it is not needed by those merely wishing to explore or implement the LBMS. However, it is at least useful to develop an understanding of the basics of the activity-based methodology (primarily CPM), of which location-based planning and scheduling is an extension and super-set. Activity-based scheduling can be reduced from location-based scheduling by isolating only one modelled location, in which case only a single logic layer (task external, Layer 1 logic) applies. Thus it is necessary to understand the origin and mathematics of CPM in order to understand the inner workings of the location-based methodology. Similarly, understanding the historical development of location-based management provides valuable context for working with the method and an appreciation of its developmental story.

There is considerable technical detail in these chapters. To aid reading, we have indicated those sections that will generally be only of academic interest, such as some of the mathematics. Such sections are indicated in the text as being safe to skip over.

The following points summarise the contents of the chapters in this section.

- Chapter 1 introduces the book and provides a brief overview of the location-based management system. This chapter also provides a walk-through of the structure of the book.
- Chapter 2 contains an analysis of the development of activity-based planning methods, including CPM and PERT. This includes some discussion of the mathematics of CPM. The chapter is introductory as this is not a book about activity-based methods. Thus it stops short of examining the recent developments of CPM in practice and the highly technical papers which extend CPM into sophisticated specific-case optimisation (such as cash flow optimisation). The purpose is to set a foundation for the method for later use.
- Chapter 3 presents an analysis of the development of location-based planning methods, in particular the many manual and analytical methods which have developed over the years including line-of-balance and flowline. Here the purpose is to establish the long history of location-based approaches and to show the evolution of methods, from graphical techniques based on simple linear mathematics towards complex integrated methods based on sophisticated algorithms.
- Chapter 4 provides an analysis of the development of methods of planning control for construction as well as an introduction to lean construction and associated control methods. It is clear that the literature relating to controlling projects is relatively weak, with the bulk of attention being derived from the lean literature. However, this is not an in-depth analysis of lean approaches to managing construction, that would be a book in itself, but rather an attempt to highlight those lean components which are adopted by the controlling mechanisms of the LBMS. Lean thinking establishes the underlying philosophy of the LBMS, but the empirical work exploring control in practice in a location-based methodology provides the practical learning and development.

Chapter 1

Introduction

This book presents a better way to manage construction projects: the location-based management system for construction. This new system is designed to reduce risk, reduce production cost, increase site harmony, improve subcontractor performance, reduce material waste, improve the quality of construction and deliver more certain outcomes. The location-based management system for construction is a comprehensive new production control system for construction, with an emphasis on the planning, scheduling and control of projects and including—in its implementation—time, cost and quality. Essentially, it is an integrated system designed to improve the profitability for all participants from clients to subcontractors. The system does, however, require a new way of thinking.

As this book is completed, the construction industry faces difficult times amid a growing world recession, even crisis. Gone are the boom times that generally held sway during the ten years of the book's research and development. Suddenly, investment funds and the construction projects they support are in short supply. Those fortunate enough to have funds are increasingly looking at their processes and asking if they can extend those funds further. The good news is there is a better way.

This book does not have all the answers, nor will it be for all practitioners. For example, it may not be for those who already have perfect systems, nor those for whom build quality presents no challenge. It may not be for those fortunate enough to always complete projects on time and or who have never had to rush to complete for a deadline or squander resources to do so, nor those who can already efficiently manage their resources such that there is no room for improvement. This book is written for the rest, the majority who are prepared to adopt change and are willing to accept there might be room for improvement.

The foundation for this better way is to challenge the dominant methodologies for the organisation and management of work. This is not the first time such a change has been proposed. In the 1960s, a new critical path methodology encountered resistance:

> It had long been recognised that the bar graph was not an ideal scheduling instrument. However, those who sought to improve on scheduling techniques took the bar graph as given, and added various signs and symbols in an effort to overcome the most obvious deficiencies. The bar graph was an ingrained part of the thinking of any project manager. It was as hard for him to think of scheduling by any other technique as it must have been for Romans to think of numbers except in terms of their own Roman Numeral System (Mauchly, 1962).

The critical path method has since become the dominant method, widely used if, increasingly, not well understood. It is now as hard for current day practitioners to think of scheduling in any way other than CPM as it was in 1962 for practitioners to replace the bar chart. Yet it is now time to do exactly that. There is little more to be achieved by continuing this emphasis on CPM. A more powerful, super-set methodology is now available.

This new methodology shifts the focus from individual discrete activities to managing the progress of tasks (work crews) as they move through a building, completing all their work location by location. This simple shift is sufficient to enable the planning and control of those crews to achieve efficient production, it does however force a shift from focusing on a 'critical path' to focusing instead on protecting production efficiency.

CONTEXT

While there have been several major advances in management systems for construction over the second half of the twentieth century, there is also a considerable body of work which remains largely ignored. Parallel to the development of CPM, a body of work has maintained a stubborn resistance to the CPM approach, claiming that it simply does not work in construction. This, it has been argued, is due to the failure of CPM to recognise the repetitive nature of construction work, and the need to manage production efficiency. In this book, this effort is aggregated under the title: location-based.

Location-based planning and scheduling provides planning for efficiency and productivity in most construction projects. At the same time it delivers a mechanism for project control. This is achieved through the use of location as the unit of analysis for scheduling work; using tasks as the unit of control. It is a mechanism for allocating resources; organising the logistics of resources; monitoring progress, cost and quality; and finally reporting all of these progressively in time to take control actions as required.

New theories which harness both the rigour of traditional scheduling methods and the power of alternative scheduling methods are integrated here into the location-based management system. The new theories are of location-based scheduling and location-based control and are combined into a management system which is, in its entirety, powerfully new. As such, it represents a revolution in project management for construction.

The planning, scheduling and controlling of construction projects should, by now, be a mature and well developed activity, with easy implementation and consistent results. Unfortunately, instead of driving project performance, we find project schedules are often produced only for the purpose of satisfying client demands or contractual requirements. The result is that the schedules frequently become mere wallpaper for site offices, they are ignored by site staff and project status data is often neglected.

Dedicated construction planners generally do a great job and are often frustrated with the failure of construction sites to follow their schedules. They often experience a disconnection between their plans and the site activities as crews deviate from the schedule. Those who think beyond the basics find the dominant critical path methodology for planning construction sites is insufficient for effective planning and control.

Almost from its inception, there has been an awareness that CPM is not a suitable planning system for construction. Birrell (1980) suggested "that the basic critical path network technique is neither a true model nor the best approximate model of the construction process". Furthermore, when it comes to controlling construction, the critical path technique on its own provides little, beyond status reports against master schedules, to assist the management of construction within projects. Continuous updating of the schedules during production tells a lot about the poor feasibility of many CPM schedules—in reality no one actually expects to implement them as planned. Controlling with CPM presents an overly optimistic view of the future: the schedules are updated but information about the actual production rates is not taken into account. Implicitly, there is always an assumption that the new schedule will somehow be implemented—until the next updating round. The end result is an overwhelming hurry in the final weeks of a project. CPM can be used to manage projects, but reliance on it as the only tool is like steering a ship without a rudder. A better method is required which enables the future to be anticipated and a course steered to achieve the planned outcome.

The location-based management system uses tasks rather than activities and concentrates on their movement through locations. It is difficult to steer a project using only disconnected activities. In contrast, having tasks to follow, monitor, forecast and steer, makes the process of steering a project intuitive and transparent. It really is that simple and this book is intended to show how to do this in practice.

A COMPARISON OF TWO PLANNING SYSTEMS

There are two main methodologies for scheduling work: activity-based scheduling and location-based scheduling—although describing them in those terms is new. These two methodologies in turn have many methods and techniques, but are principally associated with two principal scheduling methods, each designed to achieve the same purposes in different ways: CPM and either line-of-balance or flowline. While the development of these methodologies is covered in greater detail in Chapters 2 and 3, the following discussion presents an overview.

A dominant (activity-based) methodology

Activity-based scheduling is the current dominant scheduling technique and was first developed in the 1950s. The company E.I. du Pont de Nemours (DuPont) initiated research in 1956 which led to the seminal work of Kelley and Walker at the Univac Applications Research Centre in 1957. Papers by Kelley and Walker illustrated the model. They coined the term *critical path method*, the name selected... "because of the central position that critical activities in a project play in the method" (Kelley and Walker, 1959). The technique relies on the construction of a logical network of activities with four levels of complexity: deterministic (for example, CPM), probabilistic (for example, PERT), generalised activity networks and, arguably, the more recent critical chain method. Common to all these is the underlying logical structure of the model. It is a topological map of discrete activities joined by logical relationships. Each individual activity is considered free to move in time as long as it maintains its logical relationship with its predecessors and successors. Such a model suits very well any project where activities are completely discrete and have no correlation.

Consider the typical NASA or military project. These generally revolve around single location assembly (such as a missile or space vehicle) of many complex and pre-assembled components, with assembly organised sequentially but with parallel execution—a context most suitably planned with activity-based techniques and yielding a critical path. The character of an ideal activity-based schedule for projects may be described as:

- Dominated by discrete locations
- Involving much prefabrication of components
- Complex assembly of prefabricated components, involving discrete activities
- Highly sequential, in that long-duration activities are not running simultaneously
- One of many critical paths may be identified
- Resource management is a time/resource optimisation problem.

Unfortunately, this list does not describe much commercial construction at all well.

This suggests that activity-based scheduling based on CPM does not well match the character of construction projects, which consist of large amounts of on-site fabrication involving continuous or repetitive work, and in which the concept and reality of a critical path compares uncomfortably with the reality on site.

An alternative (location-based) methodology

Some of these characteristics of commercial construction align more closely with an alternative methodology based on tracking the continuity of crews working through a building. This methodology was originally based on graphical techniques, used as early as 1929 on such innovative projects as the Empire State Building, further developed by the Goodyear Company in the 1940s and expanded by the US Navy in the 1950s. This older suite of techniques found strong support in continuous general production systems (and is more typical in engineering construction) but has only found limited support in commercial construction, despite a large amount of research being undertaken in the 1960s and 1970s.

While the term 'location-based scheduling' is used in this book, elsewhere more specific names have been adopted, such as:

- Harmonograms
- Line-of-balance
- Flowline or flow line
- Repetitive scheduling method
- Vertical production method
- Time-location matrix model
- Time space scheduling method
- Disturbance scheduling
- Horizontal and vertical logic scheduling for multistory projects
- Horizontal and vertical scheduling
- Multiple repetitive construction process
- Representing construction
- Linear scheduling
- Time versus distance diagrams (T-D charts)
- Linear balance charts
- Velocity diagrams.

All these methods involve repetitive activities and for this, Harris and Ioannou (1998) suggested the generic term *repetitive scheduling method (RSM)*. However, the methods are more concerned with movement of resources through locations or places, and there really is no need for repetition, thus the term *location-based scheduling* has been coined by the authors (Kenley, 2004) as more appropriate. The character of an ideal location-based schedule for projects may be described as:

- Multiple locations (or more accurately, multiple work places)
- On-site and continuous assembly of components (including prefabricated work)
- Complex assembly involving repetitive but variable activities (work which repeats in different locations, but in which the amount or context changes)
- Equally parallel and sequential paths
- Resource management is a flow-optimisation problem, designed to achieve smooth flow and continuity in the use of resources.

DRIVERS FOR LOCATION-BASED PLANNING

The desire for the efficient management of resources is the critical driver for location-based management. The activity-based methodology sacrifices efficiency in favour of earliest completion. Even when cost driven, or with resource optimisation, there is no focus on continuity. The reasons become clear in the historical context:

The concept of CPM and PERT was created in the military/industrial environment where United States national security rightly put a low weighting on cost control and efficient use of resources. The specific project being planned and controlled was of greater importance than the efficient use of individual resources. In construction, the normal project is not of national importance, and each subcontractor is very interested in the efficient use of his resources on all projects he is working on. This creates a different situation and environment for planning than the origins of CPM and PERT. Put simply, for those who control the construction resources, minimum consumption of resources on a normal construction project is more desirable than minimum calendar duration of the whole project. At the very least, minimising cost is to be considered as much an objective as is minimising overall durations (Birrell, 1980).

This origin is of more than passing interest for those interested in construction, as it suggests one reason why planners report difficulty gaining acceptance of their schedules on site, and instead rely heavily on controlling mechanisms for work reliability such as Last Planner (see page 110).

An exemplar: the efficient production of the Empire State Building

Much has been written about the construction of the Empire State Building. Its innovations were, with the benefit of 80 years hindsight, simply astonishing. One delightful read is *Building the Empire State* by Willis (1998) which presents the history of development of the building with a rare sense of understanding of the process of construction management:

> Contrary to popular conception, the principal function of the general contractor is not to erect steel, brick, or concrete, but to provide a skilful, centralised management for coordinating the various trades, timing their installations and synchronising their work according to a predetermined plan, a highly specialised function the success of which depends on the personal skill and direction of capable executives (Starrett, cited in Willis 1998).

The project achieved milestones which would not be achievable today. These included completing a 102 level building, from sketch designs to opening for trade, in 18 months; achieving (aligned) floor cycles of one floor per day; having a high safety record (for its day) and completing under budget. Some of the innovations were:

- Guaranteed maximum price:
  A fixed price for the project was provided on the basis of sketch drawings and square metres.
- Fast-track design and construction:
  Design continued throughout construction. For example the structural steel for the upper floors was designed well after the construction of the lower floors.
- Just-in-time logistics:
  Due to the site constraints (there was no storage room), materials were delivered as required according to a tight schedule. They were moved in a single operation direct to location on the required floor within three days prior to their use.
- Learning from manufacturing to reduce cycle time:
  The production was run like an assembly line, with great emphasis on achieving continuous and aligned production.

- Location-based planning:
  The planning system involved the calculations of quantities for locations (floors and zones) and location-based planning tools—flowline.
- Location-based control:
  There was a great emphasis on controlling the work. First, actual quantities placed in locations were monitored daily. Second, the work crews were checked to ensure they were working in the correct location three times per day.

In 1929, the builders Starrett Brothers and Eken were using some of today's latest methods and proving their effectiveness—all without the aid of advanced technology. The indicators of a location-based approach are unmistakable.

Recent application in Finland

CPM has never taken hold in Finland. A large proportion of Finnish projects use location-based principles, although not to the extent presented in this book. The result is that only 1% of Finnish projects exceed schedule and, for example, Skanska has 200 projects a year and only one catastrophic project every two years. The industry's average annual profit level is now 4–6%.

The Finnish approach to project management is based on a risk management strategy rather than a CPM strategy. These issues are discussed in Chapter 7. The result is that their location-based schedules are less risky. This strategy actually delivers better performance due to reduced interference on site.

The need to support the dominant scheduling method, previously reliant on graphical techniques, led to the development of software to integrate CPM logic with location-based schedules. This was the foundation for the location-based management system described in this book, as the enabling software moved on to the international market.

Recent international developments

Location-based scheduling is no longer an intellectual curiosity. It has been adopted as the underlying methodology of one of the mainstream players in the world of building information modelling (BIM) and integrated construction systems. The scheduling planning and control system has been refined and tested on many projects. Contractors are now discovering the benefits of the LBMS in the USA, Finland, Sweden, Denmark, Hungary, the UK, Australia, New Zealand, the UAE, China and Singapore, to name just a few. While many of these are experimental, most are moving forward into mainstream application. Location-based planning and control is now a proven technique. When combined with BIM and location-based measurement and estimating, it is set to move into a dominant position. As one of the author's recently informed Vico Software, it is "the way the world plans to build"!

STRUCTURE OF THE BOOK

This book has five main sections: Introduction to planning and control; Location-based planning; Location-based control; The location-based management system; and Case studies.

Section One: *Introduction to planning and control*, introduces the background to planning and control of construction, and consists of this introduction, a review of the development of activity-based planning systems (Chapter 2), a review of the development of location-based planning systems (Chapter 3) and a review of the development of control systems including lean construction and location-based control (Chapter 4). In the course of Section One, it will become clear the extent to which this work rests on the shoulders of previous researchers. The process of writing this book has revealed just how much work has been done, by so many, in all corners of the globe. While it is exciting to see such a wealth of common belief in the new location-based approach, it is frustrating that the overwhelming sense of belief in the epistemology (and rejection of the activity-based epistemology) has made so little difference to mainstream practice. Indeed, discussions with other researchers who were involved in the 1960s has revealed that much of the primary research has been lost, literally thrown away, and must now be rediscovered.

Section Two: *Location-based planning*, presents a new methodology for the pre-construction planning of construction projects based on locations. Chapter 5 presents a new location-based model for scheduling a construction project. This model is based on five layers of CPM logic, and shows that location-based planning is a super-set for CPM— which may be interpreted as location-based planning with only a single location defined. Much of the emphasis here is in presenting the logical structures, showing their derivation in CPM and illustrating the development of the model into a much more complex method-ology. The associated suite of methods for planning projects is presented in Chapter 6, which expands the theoretical discussion of location-based planning by introducing real-world simulation methods such as unit and production system cost, production risk, procurement, design schedule, quality and learning processes (improving rates of produc-tion) within the context of a location-based schedule. It discusses measurement of quanti-ties, product-resource modelling, conventional cost loading of the schedule and a production system cost model, production system risk, and the use of buffers to mitigate production system risk. Other methods are introduced, such as procurement, pull sched-uling, design schedules, and using location-based planning to achieve well-managed hand-over of locations to drive quality and safety. Chapter 7 discusses the implementation of location-based planning: how to use the knowledge to build effective schedules for a project and to plan for both production efficiency and confidence. The discussion concentrates on techniques to build a 'good' plan, to minimise risk and to maximise feasibility.

This overall structure is replicated for the three chapters in Section Three: *Location-based control*. Chapter 8 presents a new theory for location-based control of construction, designed to enable management to take better informed control action decisions and to be proactive in maintaining the original plan. The model includes four levels of planning: base-line schedule planning, detail schedule planning, control action planning and weekly plan-ning. The chapter discusses how progress data is compared to planned values and used to calculate forecasts, from which a reactive control action planning process can be triggered to prevent future interference. Chapter 9 explores the methods required to use location-based control. It expands the theoretical discussion of location-based control by adding functional methods for improving production control, such as controlling cost, risk, procurement and quality. Tools to visualise progress are described and methods to control and forecast costs are presented. Chapter 10 discusses how to implement location-based control, how to use the knowledge to effectively manage schedules for a project and to control for both production efficiency and reliability. Location-based control processes include monitoring current status, accurate planning of implementation, forecasting prog-ress, planning control actions, prioritising tasks, ensuring prerequisites of production, and executing the plan through good assignments and communication.

Having dealt with the underlying theory, Section Four: *The location-based manage-ment system* presents the new management system holistically, adopting the new theory as well as absorbing many of the ideas developed by others over the years. Chapter 11 describes the hard components of the LBMS, providing a picture of the functions which need to be implemented. As implementation is not simple, Chapter 12 discusses the softer people issues such as system design, change management and team building, along with the harder issues such as contracts, payments systems, incentives (pain and gain) and exploiting the advantages of the system in real project reporting. The chapter suggests the methods to use to overcome resistance and to successfully implement the LBMS. These are the soft components of the LBMS. Chapter 13 discusses some of the various project types which may be managed using LBMS and considers the different strategies that apply to project types such as residential, office, retail and health projects. The chapter also discusses special project types such as civil engineering projects and maintenance work and shows how they may be managed using the LBMS. Chapter 14 looks at the special case of linear projects. These are projects for which the location is intrinsic and linear, such as road, rail, tunnelling and pipe projects. The particular case of mass haul optimisation is considered in this context.

Section Five, *Case studies*, presents practical empirical results in the form of case studies of various projects, case studies that have been instrumental in the development of the location-based management system. Chapter 15 presents case study 1: Opus Business Park—Stage 3, a 14,500 m^2 office building in eastern Helsinki. The general contractor of the project, NCC Construction Ltd, is one of the largest Nordic construction firms. The Finnish subsidiary of the company uses LBMS in all its projects. Opus was its first case study for the location-based controlling system and also incorporated location-based contracts. The LBMS provided good results in this project. The project showed that subcontractors and other stakeholders of the project can be taught to understand flowline and control diagrams. Client reporting was done based on LBMS results. Chapter 16 presents case study 2: St Joseph's NE Tower addition, a 9,476 M^2 addition to an existing hospital in Eureka, California. St Joseph Health System is an innovative hospital owner that has been at the front line of implementing 3D model-based trade coordination and change management processes. This project was selected as its first pilot project for implementing the location-based management system using a 3D-model to generate the quantities. Chapter 17 presents multiple case study components, all of which have been influential in the development of the location-based planning system in some way.

It would be quite acceptable for a reader to commence reading with this chapter, and then to move directly to Section Four—Implementation of the LBMS. However, experience indicates that implementation will rely heavily on knowledge of the underlying theory and the weight of precedent as a major reason change. It is therefore advisable for Chapter 5 at least to be understood, as this provides the rigour to the methods and the management system. Of course, the best results will come from reading the whole book but the authors recognise that is a large task. This book presents a rich picture of the LBMS, so please take your time to accept and adopt it, the result is very rewarding.

REFERENCES

Birrell, G.S. (1980). "Construction Planning—Beyond the Critical Path". *Journal of the Construction Division, American Society of Civil Engineers* **106**(3): 389–407.

Harris, R.B. and Ioannou, P.G. (1998). "Scheduling projects with repeating activities". *Journal of Construction Engineering and Management* **124**(4): 269–278.

Kelley, J.E. and Walker M.R. (1959). "Critical-Path Planning and Scheduling". *Proceedings of the Eastern Joint Computer Conference*. 160–173.

Kenley, R. (2004). "Project micromanagement: Practical site planning and management of work flow". Proceedings of the 12th *International conference of Lean Construction (IGLC12)*, Helsingør, Denmark, 194–205.

Mauchly, J.W. (1962). "Critical-Path Scheduling". *Chemical Engineering* 1962 (April): 139–154.

Willis, C., Ed. (1998). *Building the Empire State*, W.W. Norton and Company with the Skyscraper Museum. New York.

Chapter 2

The development of activity-based planning and scheduling systems

INTRODUCTION

As long as there have been projects, there have always been processes of planning and controlling construction projects, from simple notes, through bar charts and basic reactive techniques, to the relatively sophisticated graphical methods exemplified by the Empire State Building project (see page 7). However, it is only within the last fifty years that quantitative methods for project analysis and scheduling have been developed and documented, and particularly those supported with computerisation.

In this chapter, we explore the development of the traditional techniques that will be familiar to anyone who is involved in construction management. These are termed activity-based methods in response to their emphasis on networks of discrete activities or work packages. Understanding the development and underlying theory of such methods is critical to understanding what is the same with and what is different from location-based techniques—the parallel development of which is discussed in Chapter 3.

Henry L. Gantt and Frederick W. Taylor developed the now universal bar chart format in the early 1900s (O'Brien and Plotnick, 1999) but at the time this was a largely graphical technique limited by its lack of an underlying analytical method. Even so, Gantt charts (as they became known) were an excellent graphical form for representation of the organisation of production work and were soon adopted by the construction industry, where they remain the dominant form of communication of a time schedule today. While the complexity of methods and underlying theory has developed enormously, most remain dependent on this simple form of representation for the communication of the project plan. The essential characteristic of a Gantt chart is that it represents production as a series of activities on a chart with a timescale.

The analytical development of project planning methods was inevitable, given the growing complexity of construction projects and the increased emphasis on management as a discipline. Once computers were able to automate complex analysis, companies, governments and academia were in the market for tools to help with the task. In the 1950s computers became available that could be programmed to solve complex problems. The need for tight control of large and complex military projects, particularly relating to the early stages of the Cold War, led mathematicians and computer programmers to get together and develop solutions. Thus was the critical path method (referred in this book as CPM) borne. CPM has ever since been the dominant methodology for scheduling construction.

A concentration on the critical path method in the early publication record, and a corresponding dominance in commercial software applications indicates, on the face of it, a remarkable degree of success for CPM. However, right from the beginning an alternative methodology, based on location, has been developed and published, and the publication record has trended in the last fifteen years towards alternative, repetition- or location-based approaches (see Chapter 3). Nevertheless, the success of the CPM community in transferring their method into quantitative tools capable of automation has differentiated the two methodologies in practice: a significant success indeed for CPM. Location-based methods have, largely without computer support, suffered from the stigma of remaining fundamentally a graphical technique and have often been dismissed by the practitioner.

CPM is therefore the dominant methodology for planning, scheduling and controlling construction work to this day. CPM uses an activity-based methodology, a term first proposed by Kenley (2004), to differentiate between the emphasis on activities and locations in planning. Its dominance can be attributed to the following factors:

- The early publication and dissemination of a simple but rigorous method.
- The absence of any early alternative of similar rigour.
- The early requirement for enormously expensive mainframe computers, sold by the software developers of CPM, which restricted competition through access.
- The acceptance of CPM (or PERT) as a requirement for US government projects.
- The acceptance of improved project management measured through case studies.
- The incorporation of CPM into the accepted curricula for engineers.
- The courts accepting CPM as a valid method for time-related claims.
- Clients defining CPM as a contractual requirement.

This situation is now, however, changing rapidly for various reasons:

- There is increased frustration at the apparent shortcomings of CPM for the particular nature of construction.
- Growing recognition that CPM does not effectively manage or control construction.
- Sophisticated software tools that support a location-based methodology have now emerged.
- Computer processing power has become cheap and affordable.
- The incorporation of CPM techniques into a location-based methodology has revealed a more powerful methodology for scheduling construction projects (as described in Section Two of this book).

These concerns have led to the development of hybrid systems, also discussed in Chapter 3. The market is now moving towards an integrative method, based on underlying CPM analysis. It is now clear that there are two methodologies for planning, scheduling and controlling construction work. These are activity-based and location-based methodologies. Each has a long history and, while it is debatable which came first, there is no doubt that activity-based systems are the dominant form and deserve to introduce this discussion of planning and scheduling systems.[1]

This discussion will commence with a discussion of the concept of activity-based methods and then will move to the progressive development of a suite of methods which focus on the activity as the unit of analysis. These are the systems which trace back to CPM or PERT and whose methodology is activity (or event) oriented (Collins, 1964: 5), these are activity-based systems.

WARNING: TECHNICAL MATERIAL FOLLOWS

The following sections are more technical and can be safely skipped by those wanting only to learn about the location-based methodology. The discussion of the practical application of CPM, including the forward and backward pass, is important for those interested in the

---

[1] Kelley and Walker (1959) deliberately differentiate between planning and scheduling, with planning being defined as the act of stating what activities must occur in a project and in what order (technology and sequence), whereas scheduling required prior planning, and was the act of producing timetables in consideration of the plan and costs.

mathematics of flowline, because this is the basis of the standard forward and backward pass—which is used extensively in location-based planning and the calculation of task and activity timing (starting on page 32). Those wishing to skip the technical detail entirely could skip to page 38 or even to Chapter 3.

WHAT ARE ACTIVITY-BASED METHODS

Activity-based systems are planning, scheduling and control methods which concentrate on the unit of work to be done. Work is treated as a series of packages which have a time-based relationship (only) to each other. Each work package is considered discrete and the method does not explicitly take into account the physical location and its relationship to surrounding locations—there is no location-based relationship between activities.

The development of activity-based methods

Fifty years ago, the Integrated Engineering Control Group of du Pont de Nemours & Co. established a research group to investigate the prospects for applying electronic computers as an aid to coping with the complexities of managing engineering projects (Kelley and Walker, 1959). A year later Remington Rand joined the effort (in support of its clients) and introduced the resources of the Univac Applications Research Centre. This combined effort brought together James E. Kelley and Morgan Walker and they developed in 1957 a new technique called the Kelley-Walker method (O'Brien and Plotnick, 1999) and the first commercial program RAMPS (Resource Analysis and Multi-Project Scheduling) (Gordon and Tulip, 1997). They developed the method now commonly known as critical path method or CPM. The basics of this method were so successful that, according to O'Brien and Plotnick (1999) in their definitive reference to CPM, "it is interesting that no fundamental changes in this first work have been made".

Commercial applications developed rapidly. Gordon and Tulip summarise the early commercial applications which evolved from CPM as PMS from IBM, Pert200 from Honeywell, 1900PERT from ICL, GSP from K&H and an Atlas system from English Electric[2]. So successful were the basic techniques, that those original programs were largely differentiated only by their method of handling resource constraints.

The rapid dissemination of commercial applications based on the Kelley and Walker method (and similar methods discussed later) arose due to the willingness of the founding partners to disseminate their knowledge to their clients. It also arose from the early publication of the method in scientific papers (particularly by Kelley and Walker). Kelley and Walker subsequently went into private practice and continued to conduct seminars and develop the system with industry. Fifty years later, the dominant commercial CPM applications today can generally trace their origins back to these seminal commercial programs, and all rely on the same basic underlying CPM algorithms.

The exact origin of activity-based methods is a little uncertain, and there were competitors laying claim to the development of CPM. Uher (2003) suggested that the method actually originated in the UK in the mid 1950s on the construction of an electricity-generating complex, with its full potential being realised later by DuPont. This was possibly the Atlas system from English Electric.

---

[2] Some of these were PERT applications rather than specifically using the Kelley and Walker method of CPM.

Arditi (1983), citing the work of Battersby (1970), drew attention to work being undertaken in Europe in the 1950s:

> Similar research was being carried out in Europe at about the same time. In England, Andrew was using the *Controlling Sequence Duration* for scheduling maintenance as early as 1955 and the Central Electricity Generating Board had devised its *Minimum Irreducible Sequence* method in 1957. Roy in France had started work on his *Methods of Potentials* in 1958 and had perfected a working method by 1960. Finally, Wille described the construction under network control of a military airbase in 1965.

Similarly, the USA Navy developed a method specifically aimed at reviewing progress against project plans called PERT (discussed later, refer page 23) first published in the late 1950s (Malcolm et al., 1959).

Whatever the origin, Kelley and Walker have largely been attributed with the development of the CPM methodology.

The founding CPM model

By 1960 the common underlying CPM approach had been established as the method described by Kelley and Walker (1959) with the underlying mathematics detailed in Kelley (1961). The CPM technique had its origins in linear programming and, accordingly, adopted the style of that technique whereby the activities were treated as paths between nodes or connecting points. The graphical representation of this has become known as activity on the arrow, or activity diagram method (ADM). Interestingly, Kelley and Walker's 1959 paper largely separated the mathematics from the logic of the approach, reflecting the importance of the logic to acceptance of the method by industry.

The basis of method is simple and has often been described. The following description under the heading *Project Structure* (Kelley and Walker, 1959:161) is included here because of its historical significance:

> Fundamental to the critical path method is the basic representation of a project. It is characteristic of all projects that all work must be performed in some well-defined order[3]. For example, in construction work, forms must be built before concrete can be poured; in R & D work and product planning, specs must be determined before drawings can be made; in advertising, artwork must be made before layouts can be done, etc.
>
> These relations of order can be shown graphically. Each job in the project is represented by an arrow which depicts (1) the existence of the job, and (2) the direction of time-flows (from the tail to the head of the arrow). The arrows then are interconnected to show graphically the sequence in which the jobs in the project must be performed. The result is a topological representation of a project. Figure 2.1 typifies the graphical form of a project.
>
> Several things should be noted. It is tacitly assumed that each job in a project is defined so that it is fully completed before any of its successors can begin. This is always possible to do. The junctions where arrows meet are called events. These are points in time when certain jobs are completed and others

---

[3] Perhaps a misconception in construction, as experience shows that many activities seem able to be constructed out of sequence with impunity despite the existence of a defined sequence in a CPM network.

must begin. In particular there are two distinguished events, origin and terminus, respectively, with the property that origin precedes and terminus follows every event in the project.

Associated with each event, as a label, is a non-negative integer. It is always possible to label events such that the event at the head of an arrow always has a larger label than the event at the tail. We assume that events are always labelled in this fashion. For a project, P, of $n + 1$ events, origin is given the label 0 and terminus is given the label n.

The event labels are used to designate jobs as follows: if an arrow connects event i to event j, then the associated job is called job (i,j).

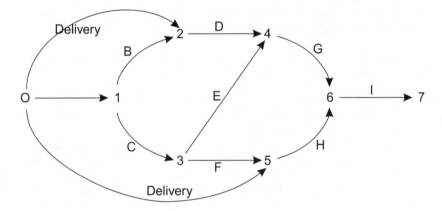

Figure 2.1 Typical project diagram (after Kelley and Walker, 1959)

During the course of constructing a project diagram, it is necessary to take into account a number of things pertaining to the definition of each job. Depending upon such factors as the purpose for making the project analysis, the nature of the project, and how much information is available, any given job may be defined in precise or very broad terms. Thus, a job may consist of simply typing a report, or it might encompass all the development work leading up to the report plus the typing. Someone concerned with planning the development work should be interested in including the typing as a job in the project while those concerned with integrating many small development projects would probably consider each such project as an individual job.

Further, in order to prepare for the scheduling aspects of project work, it is necessary to consider the environment of each job. For example, on the surface it may be entirely feasible to put 10 workers on a certain job. However, there may only be enough working space for five workers at a time. This condition must be included in the job's definition. Again, it may technically be possible to perform two jobs concurrently. However, one job may place a safety hazard on the other. In consequence, the first job must be forced to follow the second.

Finally, the initiation of some jobs may depend on the delivery of certain items—materials, plans, authorisation of funds, etc. Delivery restraints are considered jobs, and they must be included in the project diagram. A similar situation occurs when certain jobs must be completed by a certain time. Completion conditions on certain jobs also may be handled, but in a more complicated fashion, by introducing arrows in the project diagram.

Project diagrams of large projects, although quite complicated, can be constructed in a rather simple fashion. A diagram is built up by sections. Within each section the task is accomplished one arrow at a time by asking and answering the following questions for each job:

1 What immediately precedes this job?
2 What immediately follows this job?
3 What can be concurrent with this job?

By continually back-checking, the chance of making omissions is small. The individual sections then are connected to form the complete project diagram. In this way, projects involving up to 1600 jobs have been handled with relative ease (Kelley and Walker, 1959:161).

This quote described the project structure. It must be remembered, when reading these descriptions, that computers of the day were huge and slow, and the human interface was batch entry by punch card, ticker tape or magnetic tape. The aim of the mathematicians was to develop simple topological models to describe large complex project models[4]. It is this simplicity which gave the method its power. The next step was to derive durations, and then to calculate activity (job) and project durations.

Durations may be considered deterministic (non-variable) or non-deterministic (variable with a probability distribution). This early consideration of non-deterministic durations in scheduling is important, as it reflects an early awareness of the need to handle variation in the modelling. This is of interest because the practical application of CPM schedules on construction projects effectively ignores this attribute and deterministic schedules are assumed. For this reason, this discussion will concentrate on the deterministic case.

Kelley and Walker (1959) consider a case where a project P of $n+1$ events, starts at the relative time 0:

Relative to this starting time each event in the project has an earliest time occurrence. Denote the earliest time for event i by $t_i^{(0)}$ and the duration of the job by y_{ij}. We may then compute the values of $t_i^{(0)}$ inductively as follows:

$$\begin{cases} t_o^{(0)} = 0 \\ t_j^{(0)} = \max[y_{ij} + t_i^{(0)} \mid i < j, (i,j) \in P], 1 \le j \le n \end{cases} \qquad (2.1)$$

Similarly, we may compute the latest time at which each event in the project may occur relative to a fixed project completion time. Denote the latest time for event i by $t_i^{(1)}$ If λ is the project completion time (where $\lambda \ge t_n^{(0)}$) we obtain

$$\begin{cases} t_n^{(1)} = \lambda \\ t_i^{(1)} = \min[t_j^{(1)} - y_{ij} \mid i < j, (i,j) \in P], 0 \le i \le n-1 \end{cases} \qquad (2.2)$$

Having the earliest and latest event times we may compute the following important quantities for each job, (i, j), in the project:

Earliest start time $= t_i^{(0)}$ (2.3)

---

[4] The early programs could only handle between 739 jobs, 239 events and 3000 jobs, 1000 events. Modern computers have no such limits, although people may well have.

Earliest completion time[5] $= t_i^{(0)} + y_{ij}$ (2.4)

Latest start time $= t_j^{(1)} - y_{ij}$ (2.5)

Latest completion time[6] $= t_j^{(1)}$ (2.6)

Maximum time available $= t_j^{(1)} - t_i^{(0)}$ (2.7)

If the maximum time available for a job equals its duration the job is called *critical*. A delay in a critical job will cause a comparable delay in the project completion time. A project will contain critical jobs only when $\lambda = t_n^{(0)}$. If a project does contain critical jobs, then it also contains at least one contiguous path of critical jobs through the project diagram from origin to terminus. Such a path is called a *critical path*.

If the maximum time available for a job exceeds its duration, the job is called a *floater*. Some floaters can be displaced in time or delayed to a certain extent without interfering with other jobs or the completion of the project. Others, if displaced, will start a chain reaction of displacement downstream in the project.

It is desirable to know, in advance, the character of any floater. There are several measures of float of interest in this connection. The following measures are easily interpreted:

Total float $= t_j^{(1)} - t_i^{(0)} - y_{ij}$ (2.8)

Free float $= t_j^{(0)} - t_i^{(0)} - y_{ij}$ (2.9)

Independent float $= \max(0, t_j^{(0)} - t_i^{(0)} - y_{ij})$ (2.10)

Interfering float $= t_j^{(1)} - t_j^{(0)}$ (2.11)

Kelley and Walker introduce several seminal concepts in this description, in particular the concept of a critical path, of which there may be more than one through a project, and the concept of a floater with four types of float: total, free[7], independent and interfering float. These concepts form the basis of CPM.

Extending the basic model

The Kelley and Walker model included three basic extensions:

1. The non-deterministic case: where y_{ij} is a random variable with probability density $G_{ij}(y)$, thus providing a probability distribution for start and completion times.
2. The project cost function: where a job duration is assumed to be variable with a corresponding linear cost function.
3. The need to consider availability of resources. The core assumption underlying the basic CPM model is that there are unlimited resources for both executing and accelerating jobs.

---

[5] Equation 2.4 was originally published as $= t_i^{(0)} - y_{ij}$, but this was corrected in Kelley (1963)

[6] Equation 2.6 was originally published as $= t_i^{(0)}$, but this was corrected in Kelley (1963)

[7] O'Brien (1965: 228) noted that "when you look past the formula, the free float information loses its luster" as, in a chain of activities with free float, only the last activity shows free float.

These extensions are important, not so much for their methods, but for the clear indication that the original Kelley and Walker model explicitly included functionality for stochastic variation in project duration, and also a recognition that there is no such thing as a single fixed duration for a job in a project schedule, but rather a project schedule can have many solutions—with a relationship between the project duration and the project cost. In practice, CPM has become deterministic and resultant project schedules gain credence of a sort never intended by Kelley and Walker. This is an error, and indeed Kelley (2003) stated "...people are still falling into some of the same scheduling traps warned against during CPM's child-hood". The reality is that there are many possible schedules for any given project, all of which will be correct (although some may be more risky than others).

Kelley and Walker indicate that a schedule is incomplete without reference to the vari-ability in duration, cost and manpower requirements of activities in the schedule. While resource optimisation systems differentiate most commercial CPM applications, the rela-tively less understood cost component of the original model will become important in understanding the contribution of a location-based methodology to cost reduction. The CPM cost component provided for a job cost curve for each job (activity) in the project and allowed the selection of job durations to minimise cost. The assumption Kelley and Walker made was that accelerating a job (reducing its duration by increasing resources, calendar, etc.) will increase its cost. An optimum schedule will therefore be a function of choosing the appropriate duration for each job, when it may be allowed to vary, such that the project achieves minimum time with the least cost: involving a necessary trade-off.

One possible conclusion from this is that Kelley and Walker were very much aware of the relationship between cost and schedule and resource issues and attempted to accommo-date these needs in the basic model.

Cost is one of the major drivers for the ongoing research effort in construction plan-ning, scheduling and control, such as the new model based on neurocomputing and object technologies by Adeli and Karim (2001). It also forms a major part of the rationale for loca-tion-based methods—particularly those aspects which arise from the efficient use of resources.

The non-deterministic case

Deterministic models dominate the practical application of CPM, but some software has the option to include the non-deterministic case—particularly that which arises from PERT (see page 23). The Kelley and Walker (1959) model for CPM included the non-determin-istic case, providing a probability distribution for start and completion times. In the their model, each job had a duration with a mean and a standard deviation according to a normal distribution. Thus a project had a corresponding probability distribution for the project duration which was the result of the combined effect of all the ranged distributions along the paths to the end of the project. From Kelley and Walker (1959):

> Thus in the non-deterministic case we assume that the duration y_{ij} of an activity (i,j) is a random variable with probability density $G_{ij}(y)$. As a consequence it is clear that the time at which an event occurs is also a random variable, t_{ij} with probability density $H_i(t)$. We assume that event 0 is certain to occur at time 0. Further on the assumption that it is started as soon as possible, we see that $t_i + y_{ij} = x_{ij}$, the completion time for job (i,j), is a random variable with the probability density $S_{ij}(x)$:

$$S_{ij}(x) = \begin{cases} G_{ij}, \text{ if } i=0 \\ \int_{-\infty}^{\infty} H_i(u)G_{ij}(x-u)du, (i,j) \in P \end{cases} \tag{2.12}$$

Assuming now that an event occurs at the time of the completion of the last activity preceding it we can easily compute the probability density, $H_i(t)$ of $t_i = \max[x_{ij}|(i,j) \in P, i < j]$, where x_{ij} is taken from $S_{ij}(x)$:

$$H_i(t) = \sum_{(i,j) \in P} S_{ij}(t) \prod_{\substack{(k,j) \in P \\ k \neq i}} \int_{-\infty}^{t} S_{kj}(u)du, 1 \leq j \leq n \tag{2.13}$$

Several methods are available for approximating $S_{ij}(x)$ and $H_i(t)$. The one which suits our taste is to express $G_{ij}(y)$ in the form of a histogram with equal class intervals. The functions $S_{ij}(x)$ and $H_i(t)$ are then histograms also and are computed in the obvious way by replacing integrals by sums (Kelley and Walker, 1959).

These distributions are based on the forward pass starting from a fixed time ($t = 0$). Kelley and Walker found that a backward pass was not possible, as this would assume a variable start time. At this point they made an interesting statement which was not explained, yet which has great significance for the management of projects, and which is used extensively in the risk management methodology of location-based management:

> To proceed further we must introduce the notion of "risk" in defining the criticalness of a job. On the basis of this definition one would hope to obtain probabilistic measures for float which would be useful for setting up a system for management by exception.

It is interesting to compare this approach with the subsequent discussion of variability in PERT, which emphasises risk and measuring the likelihood of achieving a schedule. The basic CPM model here performs a forward pass only to derive early start and completion dates and then calculates the variability of all activities on all paths. However, a backward pass to calculate the latest start and completion dates is not possible.

The project cost function

There is assumed to be a relationship between the duration of a task and its cost. Kelley and Walker (1959) illustrated a typical job cost curve, and then found two reasonable boundaries for a job duration. These were the *crash limit*—at which point it becomes unreasonably expensive to accelerate the job further—and the *normal limit*—at which point the cost of slowing the job starts to cost more due to wasted effort or waiting. They then developed their job cost function with cost as a function of duration and bound by the optimum duration (the shortest time at which the cost is lowest) and the crash time (at which the cost becomes prohibitive to accelerate further) and simplified to be linear, as shown in Figure 2.2.

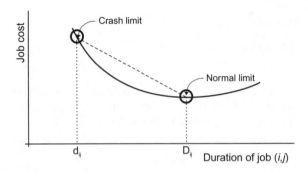

Figure 2.2 Typical job cost curve (after Kelley and Walker, 1959)

At the project level, they found that by only accelerating those activities on the critical path and solving for duration and cost using a primal-dual algorithm (Kelley, 1959; 1961), then a total project duration can be calculated and shown as a project cost relationship (Figure 2.3).

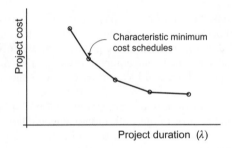

Figure 2.3 Typical project cost curve (after Kelley and Walker, 1959)

Resource optimisation

When planning a project, there is a need to consider the availability of resources. The core assumption underlying the basic CPM model is that resources are unlimited for both executing and accelerating jobs. Kelley and Walker recognised that this was often an unrealistic assumption and that, in addition to the existing ability to link jobs to force resource dependancy, there was a need to consider two resource contexts:

- Available resources are invested in *one* project
- Available resources are shared by *many* projects.

In fact, it is in this area of resource optimisation that most of the structural differences exist in the modern commercial applications of CPM.

Kelley (1963) argued that resource problems could not be solved mathematically, describing them as combinatorial. "The way the restrictions on the sequence in which jobs may be performed interacts with resource requirements and availabilities forces a solution set which is unconnected. This property exists even if we assume a solution set which is unconnected." Few such problems can be solved mathematically, Kelley advocated *ad hoc* intuitive methods, or iterative processes in which a schedule is prepared and the resource usage and profile compared to the available resource limits. The planner then manually alters the inputs until the desired profile is achieved.

Specifically, Kelley considered the tangible costs of workflow disruption on resources. However, while he felt that such a focus was commendable, he stated that "too often it is a myopic course of action". This is most likely because no method had been found to achieve continuity of resources using CPM at that time.

Resource optimisation has probably developed further under the influence of PERT, so this discussion will be continued page 26. Planning for the efficient use of resources is a key component of location-based management, and will be addressed in Section Two.

An alternative—PERT

In a development generally considered parallel with CPM, the Special Projects Office of the USA Navy Bureau of Ordnance (together with Booz, Allen & Hamilton and the Lockheed Missile Systems Division) was attempting to develop systems to control the construction of the complex Polaris missile program or Fleet Ballistic Missile (FBM) weapon system (Malcolm et al., 1959). The goal of the group was to determine whether computers could be used in planning and controlling the Polaris program, which involved 250 prime contractors and more than 9,000 subcontractors (O'Brien and Plotnick,1999). They developed the PERT system (the Program Evaluation Review Technique[8]).

PERT was very similar to CPM, but with a different strategic purpose, and it also had specific functionality for the incorporation of variation in job duration as a distribution. PERT was first implemented in 1958 and according to Mauchly (1962), a co-developer of CPM, it followed CPM, having been developed by Willard Fazar in June 1958. In contrast, Fazar (1962) stated that the project started December 1957. Even in 1962 no one was sure which came first!

The PERT system was designed to be a method to assess a particular schedule which had been developed by other techniques and which encompassed thousands of activities extending years into the future (Malcolm et al., 1959). The schedule included deadlines and milestones which were considered uncomfortable (Mauchly described them as arbitrary), and certainly not calculated from a CPM forward pass. However, it was rapidly recognised that the method had application beyond that first project and a generalised PERT model was substantially accepted within government circles in the USA, particularly by all defence and space organisations such as NASA, RCA and General Electric. PERT was particularly successful in handling multiple project situations.

The mathematics of the PERT model were not developed from first principles or from linear scheduling and were not published with the mathematical proofs, as had been provided by Kelley for CPM. Indeed, it is generally accepted that PERT adopted the mathematics of CPM from the work of DuPont, with which the developers were familiar. Nevertheless, the logic of the system was the same, involving an activity on arrow representation method and both a forward and backward pass. The different aim of the method meant that the backward pass did not take the calculated early completion date for the start of the backward pass. Rather, PERT took the planned completion date, which could be either greater or less than the calculated earliest completion date, and then calculated the probability that the date would be achieved—and indeed any milestones on the way. Under these circumstances there may not be a critical path according to the definition that slack=0 for criticality (time

---

[8] PERT was originally termed the Program Evaluation Research Task, but this usage has subsequently been replaced by the more common Program Evaluation Review Technique. For a comprehensive description of the development of PERT, including how it got its name, see Fazar (1962).

allowed is greater than the calculated earliest completion time) or times could be negative—where the completion deadlines are earlier than calculated. This concept of negative time remains a bone of contention between CPM purists to this day, who may attest that time cannot be negative. However, this is a misapprehension of the calculation which was intended to yield the likelihood of achieving the deadline. The critical path was defined as "those activities that cannot be delayed without jeopardy to the entire program" (Malcolm et al., 1959).

PERT is therefore intended to provide advice about the likelihood of achieving deadlines and a target end date. As such, it can break one of the basic rules of CPM, that the latest finish of the last activity is equal to the earliest finish. O'Brien and Plotnick (1999: 124) criticise much modern software in that the practice of adding constraints allows the same breaking of this rule. However, the different intent of PERT makes it quite acceptable. Under these circumstances, it is possible that an activity will show it must be completed (to meet the target) before it can be executed (according to the logic) in order for the target to be met. While a problem to CPM, this is in fact the design intent of PERT—to measure the probability of achieving such an aim. What the PERT system is actually saying is that it is unlikely that the activity will be completed by the target.

Kelley and Walker (1959) held that the PERT model did not include a continuous probability distribution for the range of possible job durations, rather it included three values for minimum (optimistic), most likely and maximum (pessimistic) durations. The PERT model considered that deriving a probability distribution for individual activities was unrealistic—as there was no valid way to study the majority of one-off durations for activities to develop a suitable distribution profile (Archibald and Villoria, 1967). Thus the PERT simplification adopted a common technique which is also often found in simplified Monte Carlo simulations (see the discussion of risk, Chapter 6, page 180) and has the advantage of being simple to understand and communicate.

> In obtaining raw data from engineers, it was felt that a more realistic evaluation could be made if three estimates for each activity were obtained. This practice was designed to help disassociate the engineer from his built-in knowledge of the existing schedule and to provide more information concerning the inherent difficulties and variability in the activity being estimated. Consequently, three numbers designated as the *optimistic, pessimistic* and *most likely* elapsed time estimates were developed and utilised in the interrogation process.
>
> The next task was to translate the engineers' estimates into measures descriptive of expected elapsed time t_e and the uncertainty of involved in that expectation, $\sigma(t_e)$. It was postulated that the three estimates could be used to construct a probability distribution of the time expected to perform the activity. It was felt that such a distribution would have one peak—with the most probable time estimate, m, being representative of that value. Similarly, it was assumed that there is a relatively little chance that either the optimistic or pessimistic estimates, a and b, would be realised. Hence small probabilities of a and b. No assumption was made about the position of the point m relative to a and b. It is free to take any position between the two extremes, depending entirely on the estimator's judgement.
>
> Figure 2.4 represents the situation described above. With the assumptions that the standard deviation of the distribution, $\sigma(t_e)$, could be adequately estimated as $\frac{1}{6}$ of $(b-a)$ and that the beta distribution, $f(t) = K(t-a)^{\alpha}(b-t)^{\gamma}$, is an adequate model of the distribution of an activity time, it was possible to develop equations for calculating t_e and $\sigma(t_e)$:

$$t_e = \frac{1}{3}[2m + \frac{1}{2}(a+b)], \tag{2.14}$$

$$\sigma_e(t_e) = [\frac{1}{6}(b-a)]^2 \tag{2.15}$$

With three elapsed time estimates for each activity in the plan it is possible to calculate an expected time, t_e and a measure of its potential variability $\sigma(t_e)$ for each activity (Malcolm et al., 1959).

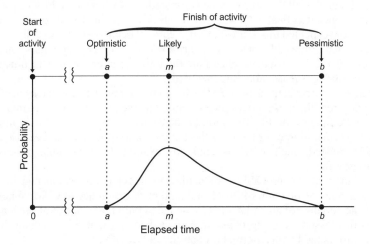

Figure 2.4 Estimating the elapsed time distribution (after Malcolm et al., 1959)

Whether an estimated duration has a normal distribution (CPM) or a beta distribution derived from three estimates is probably moot. However, Kelley and Walker (1959) took exception to the philosophy, considering that "using expected elapsed times for jobs in the computations instead of the complete probability density functions biases all the computed event times in the direction of the project start time". This distortion arises because of *merge event bias*, which occurs when several paths converge on a single node. The PERT critical path does not take variability into account when calculating paths through a network, rather it selects the longest path into a node. However, random variation may yield an alternative longer path should a simulation be run, therefore the PERT simplification will, on average, lead to an underestimation of network duration (Halpin and Riggs, 1992: 269).

   PERT calculates the distribution of a single event by only considering the longest path to that event and the relevant activity distributions contained on that path. Malcolm et al. (1959) were aware of this limitation:

> ...the time constraints of all paths leading up to an event are considered, and the greatest of these expected values is assigned to the event. The variance of this expected value is the sum of the variances associated with each expected value along the longest path.
>
> This simplification gives biased estimates such that the estimated expected time of events are always too small. It was felt that, considering the nature of the input data, the utility of other outputs possible from the data and the need for speedy, economical implementation, the method described above was completely satisfactory (Malcolm et al., 1959).

Malcolm et al. agree with Kelley and Walker's assessment, but deem it irrelevant. Of course, modern computers and software can remove this bias using multiple iterations.

Kelley and Walker were also concerned about the PERT method of working backwards from "some fixed completion time" used in comparative analysis. They were challenging the methodology of PERT in application, rather than the method itself. Their concern arose because PERT is a review technique, and did not require a network to be constructed at the start or indeed in its entirety, rather it was only necessary to model from a given point of time (the time of review) onwards.

Malcolm et al. (1959) described the computation of *slack* in the PERT system. Slack was the same concept as float in CPM, however slack was defined as the difference between the latest and the expected times at which an event will occur. "Slack exists in a system as a consequence of multiple path junctures that arise when two or more activities contribute to a third." They saw it as a measure of scheduling flexibility as it represented the time interval in which the event might reasonably be scheduled.

The PERT critical path, as per CPM, is found by joining the zero-slack events together to form a path from the present (as a review technique the present was more important than the start) to the final event. Identification of the critical path was considered important as this enabled acceleration efforts not on the critical path to be rejected because they were unprofitable. Indeed, such activities could be examined for possible performance or resource trade-offs.

Calculation of the probability of achieving target dates established in an extant schedule was a major function of PERT and a point of difference to CPM. Whereas CPM calculated the variance of all paths, PERT calculated only the variance of the longest path, and it was then assumed, using the central limit theorem, that this resultant distribution could be approximated with a normal probability distribution.

These concepts of accepting variation in durations, and therefore schedules, are very important. The concept of risk in a schedule and even designing a schedule such that it has a high probability of being achieved are little understood in today's predominantly deterministic scheduling world. Kelley (1961) also attempted to consider the implications of risk for CPM.

Risk management in scheduling and control is a powerful feature of the location-based management system and challenges the CPM assumptions, properly being closer to the PERT model of establishing a feasible (less risky) schedule and then estimating the probability of achieving it. The concept of scheduling by risk management is discussed further in Chapter 6, page 180.

PERT and cost

PERT became known as PERT-Time once the development team developed routines associated with cost. PERT with cost modelling routines became known as PERT-Cost (Schoderbek and Digman, 1967).

Resource optimisation

Resource optimisation or levelling is one of the least understood components of CPM in use. It is important to understand the origin of resource levelling in CPM because, despite the low level of understanding and use, it forms a key component in making CPM schedules useful. Efficient use of resources is desirable in any scheduling system, otherwise the schedule risks being ignored. This book introduces the location-based principles of resource optimisation which are based on protecting crews through continuity, workable backlog and planned usage breaks. CPM addressed resource problems in other ways—by levelling the total count of individual resources.

Kelley (1961) established three principles for resource optimisation:

1. The logic of the schedule remains valid.
2. There are limits on available resources which are not exceeded.
3. The duration of the project is minimised.

He proposed two basic approaches: the *serial method* and the *parallel method*. These have since been expanded by others (see below).

Serial method of resource optimisation

The serial method processes tasks sequentially in some rank order (usually the earliest start date, as this ensures that all predecessors are completed). Each job's early start is calculated in the usual way. If the required resources are unavailable, the job's start would be delayed until the required resources are available.

A construct was introduced (and remains widely used today) to cover the contingency that a job would delay the project by delaying other jobs that required the use of the same resources. Kelley assumed "that jobs can be split arbitrarily into phases or parts that are an integral number of time units (say, days) long" (Kelley, 1961). The concept of *splitting* enables a job to start at its earliest time and to split (be interrupted) when the resources are demanded by other jobs, with the job finishing at least by its latest time.

Splitting is a powerful technique designed to recognise that it is common practice in construction to divert common resources amongst suitable tasks as required. However, this sensible addition to CPM has also the effect of structurally embedding interruption and poor workflow, and thus additional cost through inefficiency, into the basic CPM engine.

Kelley included splitting in the serial method. Gordon and Tulip (1997) argued it is a property of the parallel method, which they therefore considered more suitable to construction. They suggested using a serial system unless there was a high level of splitting required, when a parallel method would be more suitable.

Parallel method of resource optimisation

The parallel method allows several jobs to be scheduled at one time. Available jobs are queued and those which have sufficient available resources are commenced, or if they are delayed, then as soon as the resources are released.

Kelley (1961) also introduced the concept of planned gang sizes. This is another important concept relevant to location-based planning (see Chapter 5), which is concerned with continuity of work for work crews. Kelley established thresholds of individual resources (such as labourers) in a gang such that above that threshold the work could continue. For example, if five workers were in a gang and only four were available, a threshold of 80% would allow the work to proceed, but with a reduced efficiency—or longer duration of 125%. Once again, this is a practice which seems reasonable on the surface to a construction practitioner—as it is a common to commence work without the required resources. However, this embeds sub-optimal work practices as crews will be less efficient.

Resource optimisation methods

Over time, five basic methods have been applied to resource optimisation in CPM, and these were summarised by Gordon and Tulip (1997), who cautioned that "it should be remembered that most of the variants that were the pride of the development teams have been rendered archaic with the advent of the personal computer and the ability to model alternative project programs reasonably quickly". The five methods are:

- Aggregation
- Cumulation
- Allocation
- Smoothing
- Levelling.

Aggregation

Aggregation is the simple calculation of the total number units of resource required in each time period (day). As aggregation may be done on both the forward and backward pass, an aggregation may be calculated for both early and late starts. A small demonstration network is used to explain the calculation of resource optimisation using aggregation (Figure 2.5, with Table 2.1 showing the early and late start and completion times).

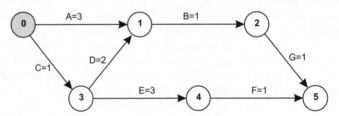

Figure 2.5 A small resource-loaded network for resource calculations

Table 2.1 Early and late starts for resource calculation for Figure 2.5

job	$y(i,j)$	ES_i	EC_j	LS_i	LC_j
A	3	0	3	1	4
C	2	0	2	0	2
D	2	2	4	2	4
B	5	4	9	4	9
E	4	2	6	6	10
F	6	6	12	10	16
G	7	9	16	9	16

Table 2.2 illustrates the early and late date aggregation of a resource. Aggregation is useful for resources which are renewable, such as labour and plant. It can be seen that the early pass is less favourable for resource utilisation, consuming a peak of 8 resource units

on day 3. This peak that might not be achievable, requiring a manager to reduce demand on resources by rescheduling. A resource histogram is a common representation (Figure 2.6).

Table 2.2 Aggregation of resources by early and late dates

Day→	1	2	3	4	5	6	7	8	9	10	11	12	13	14	15	16
Activity ↓							Early start									
A	3	3	3													
C	1	1														
D			2	2												
B					1	1	1	1	1							
E			3	3	3	3										
F							1	1	1	1	1	1				
G										1	1	1	1	1	1	1
Aggregate	4	4	8	5	4	4	2	2	2	2	2	2	1	1	1	1
Activity ↓							Late start									
A		3	3	3												
C	1	1														
D			2	2												
B					1	1	1	1	1							
E							3	3	3	3						
F											1	1	1	1	1	1
G										1	1	1	1	1	1	1
Aggregate	1	4	5	5	1	1	4	4	4	4	2	2	2	2	2	1

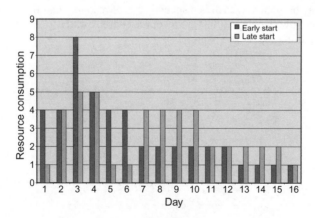

Figure 2.6 Resource histogram for the sample network

Cumulation

This involves the calculation of the cumulative total number units of resource required progressively over each time period (day). As aggregation may be done on both the forward and backward pass, a cumulation may be calculated for both early and late starts (Table 2.3).

Table 2.3 Aggregation and cumulation of resources by early and late dates

Day→	1	2	3	4	5	6	7	8	9	10	11	12	13	14	15	16
								Early dates								
Aggregate	4	4	8	5	4	4	2	2	2	2	2	2	1	1	1	1
Cumulative	4	8	16	21	25	29	31	33	35	37	39	41	42	43	44	45
								Late Dates								
Aggregate	1	4	5	5	1	1	4	4	4	4	2	2	2	2	2	1
Cumulative	1	5	10	15	16	17	21	25	29	33	35	37	39	41	43	45

Cumulative resource modelling has several purposes:

- It is useful for resource management to manage non-renewable resources or to trigger events. For example, a stock of bricks is a resource that, unless replenished, will be absorbed into the project. This is a very different type of resource from renewable resources such as labour, which is available for repetitive use. Cumulation of non-renewable resources enables the triggering for events such as "delivery of bricks" to be executed on reaching a resource threshold.
- The actual resource consumption may be compared using the resource envelope illustrated in Figure 2.7. The resource envelope is the area between the early and late cumulative curves. These form the outer bounds of the expected resource utilisation for the project and enable a manager to see if the project performance is reasonable.
- Cumulation may be used to generate *earned value* charts, whether for resources or for money. It may be seen that cash is a special type of resource and that cash flow and cumulative cash flow (Kenley, 2003) are resource charts. Similarly, location-based control adopts production graphs (see page 331).

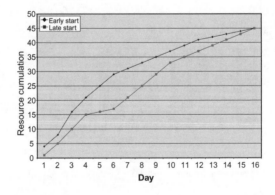

Figure 2.7 Early and late cumulative resource profiles showing the resource envelope

Allocation

Resource allocation is the process of allocating tasks by the serial or parallel methods proposed by Kelley (1961).

There are two primary categories of limitations on resource optimisation:

- Time-limited resource scheduling
- Resource-limited resource scheduling.

These are two end-points on a continuum, and solutions may be found using a mix of the two methods.

Time limited resource scheduling is the calculation of a schedule in which the scheduled activity dates are determined such that neither the project completion nor any target milestone dates are exceeded while minimising the maximum extent to which any resource availability is exceeded, and in particular using available float to provide flexibility in scheduling activities such that resource usage is levelled.

Resource-limited resource scheduling calculation of a schedule in which the scheduled activity dates are determined such that the resource constraints are considered as fixed and project completion may be delayed as necessary to avoid exceeding the maximum resource limits.

Actual resource allocation is made by Kelley's serial or parallel methods.

Smoothing

Smoothing is the process of allocating activities to yield a feasible schedule while achieving a uniform level of resource usage. For example, the latest possible schedule has removed the high resource peak of the earliest schedule in Table 2.2. This could have been smoothed further were the end date requirement relaxed.

On larger projects it is possible to allocate activities specifically to move resource peaks into times of troughs in resource demand to achieve a smoother use of resources. This is particularly useful for key items of plant, where it is not so much a limit on availability that is the issue but rather the need to avoid fluctuations in demand. Often, items of plant remain on site when not in use, and therefore charges continue to accrue.

Gordon and Tulip (1997) describe four steps in a smoothing cycle, designed to be repeated until all work is allocated:

1. Schedule any critical activities.
2. Find the most important activity yet to be scheduled.
3. Find the best place to schedule this activity, and do so.
4. Adjust the early and late times of the unscheduled activities to take account of this new scheduled activity.

Algorithms exist to enable a computer to undertake this task, using either testing of all possible options or statistical methods of least squares.

One way to visualise this process is to imagine a game of Tetris. This is the game where a sized and shaped box drops onto a pile and the user must rotate and allocate the box to fill (smooth) the base layer. The movement of the box by the player is intended to achieve a smooth filling of the layers without any gaps. In a similar way the resources are selected by their ability to both satisfy criticality and smooth the use of resources. It is not always easy or possible to fill in all the gaps.

When one considers that a project has many resources and conflicting demands, it is easy to see that smoothing could rapidly become a major problem. Whatever the solution derived, it is often necessary for the planner to make direct changes in order to level the resources further.

Levelling

Levelling is the process of further adjusting the smoothed schedule to remove any peaks and troughs in resource use. This involves complicated algorithms and solutions that are largely proprietary and do not need discussion further in this book. However, the concept of level-ling approaches the critical concept of workflow in its intent, as it provides the appearance of producing an even allocation of resources.

The concept of levelling resources is not the same as ensuring continuity of resource use. Indeed, in order to achieve levelling, it is often necessary to increase the disruption of work crews, particularly where activity splitting is employed. Levelling does not consider the continuous use of resources at all, only consistent totals. For example, consistent use of resources can involve resources jumping between activities in order balance the demand. This can be extremely disruptive to work crews.

Planning for the continuous use of resources and achieving workflow is discussed in Chapter 5.

CALCULATING TIMING USING ADM

The following discussion uses the arrow diagram method (ADM) as proposed by Kelley and Walker (1959), which uses a visualisation method in which the job is represented as a directional arrow passing between two events, an event i and an event j . The arrow has properties of duration (≥ 0) and direction and other job properties.

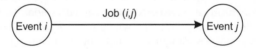

Figure 2.8 Job (i,j)

A project is then represented as a sequence of such jobs whereby the relationship between a preceding job and a successor passes through a common event (Figures 2.9 and 2.10).

O'Brien (1965: 49, 226) illustrated a matrix method for manual calculation of a network. This is very similar to the method proposed earlier by Charnes and Cooper (1962) but O'Brien focused on activities on the arrow (see page 38). Such tools were rapidly supplanted by the method of *forward pass* and *backward pass*.

An examination of Equations 2.1 and 2.2 indicate something strange on first appear-ance. In Equation 2.1, the term to express the earliest dates seems constructed from the completion date but it is in fact constructed by selecting the maximum of all possible preceding earliest completion dates plus the job duration. The process of making this calcu-lation for each job step by step, from the start to the end of the project, is known as a *forward pass*. The result is a calculation of earliest start and completion dates for each job.

$$\begin{cases} t_o^{(0)} = 0 \\ t_j^{(0)} = \max[y_{ij} + t_i^{(0)} \mid i < j, (i, j) \in P], 1 \le j \le n \end{cases} \qquad (2.16)$$

Similarly, in Equation 2.2, the term to express the latest dates is constructed by selecting the minimum of all possible latest following start dates less the job duration. The result is a calculation of latest start and completion dates for each job. This is known as a *backward pass*.

$$\begin{cases} t_n^{(1)} = \lambda \\ t_i^{(1)} = \min[t_j^{(1)} - y_{ij} \mid i < j, (i, j) \in P], 0 \le i \le n - 1 \end{cases} \qquad (2.17)$$

Each activity therefore has both earliest start dates and latest completion dates arising from the forward and backward passes respectively. O'Brien (1965) described the process as working forward and working backward through the network.

The method of performing a forward and backward pass seems to have emerged in about 1961. Until then, the calculations were done using the matrix method. O'Brien described an intuitive manual computation method in 1965 and observed that the change from the matrix method to the intuitive method occurred about 1961. It was his belief that mathematicians were not able to see the "logical" solution—it took engineers to do that. Interestingly, matrix methods are an extremely powerful means of performing these sorts of equations rapidly given the memory and processing power of modern computers.

It is easy to calculate both the forward and backward passes manually and thus this technique is a powerful teaching method as well as being easily programmable. The following sections outline the method for activity on the arrow.

Forward pass

The forward pass involves a step-wise consideration of all possible paths to each event on the schedule, commencing with the start of the project and progressing to the end.

When Event i is the project start (Event 0), then time at event i is 0. Time at Event 1 in Figure 2.9 is therefore the result of the addition of 0 (which is the time at Event i for Job A) plus the duration of Job A.

Figure 2.9 A simple two Job (i,j) network

This pattern repeats for Job B, however Event 1 represents both j for the Job A, as well as Event i for Job B. The time for Event 2 is therefore the time for Event j for Job A plus the duration of Job B.

The early finish or completion (EC_j) for a general activity in such a chain is:

$$EC_j = ES_i + y_{(i,j)} \qquad (2.18)$$

As can be seen from Figure 2.10, a network may have many paths to an event. Thus at each node, only the latest early completion (EC) date may be selected as the early start (ES) date for the following job. The earliest start date for B is the latest of the completion dates for either A or D. Technically this should actually be the later of the earliest start dates for A plus the duration of A and the earliest start dates for D plus the duration of D. Similarly the earliest start date for H is the latest of the completion dates for either F or G.

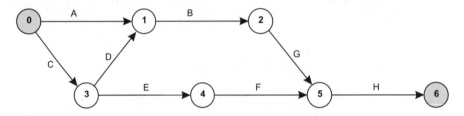

Figure 2.10 A small network showing multiple paths

Therefore the general case is:

$$ES_i = \max(ES_i + y_{(i,j)}) \tag{2.19}$$

Which is equivalent to Equation 2.1. More commonly, the general case is given as:

$$ES_i = \max(EC_j) \tag{2.20}$$

In the above example (Figure 2.10) the forward pass may be calculated by progressing forward through the network event by event. Table 2.4 provides the results of the forward pass for the network.

Table 2.4 Forward pass results

Job	Node i	Node j	ES_i	$y_{(i,j)}$	EC_j
A	0	1	0	3	3
C	0	3	0	2	2
D	3	1	2	2	4
B	1	2	4	5	9
E	3	4	2	4	6
F	4	5	6	6	12
G	2	5	9	7	16
H	5	6	16	2	18

Backward pass

The backward pass involves a step-wise consideration of all possible paths to each event on the schedule commencing with the end of the project and progressing backwards to the start.

When Event j is the project end (Figure 2.8), then time at Event j is λ (the total time derived from the forward pass). In Figure 2.9, time at Event 1 is therefore λ (time at Event j for Job B) less the duration of Job B.

The late completion (LC_j) for a general activity in such a chain is therefore:

$$LS_i = LC_j - y_{(i,j)} \qquad (2.21)$$

Once again using Figure 2.10, at each node only the earliest late start (LS) date may be selected as the late completion (LC) date for the preceding job. For example, the latest completion date for C is the earliest of the start dates for either D or E.

Therefore the general case is:

$$ES_i = \max(ES_i + y_{(i,j)}) \qquad (2.22)$$

Which is equivalent to Equation 2.2. More commonly, the general case is given as:

$$LC_j = \min(LS_i) \qquad (2.23)$$

In the above example (Figure 2.10) the backward pass may be calculated by progressing backward through the network event by event (results in Table 2.5).

It is immediately clear that here is a different set of results for each job than was achieved by the forward pass. This is because of the different durations of the multiple paths through the schedule.

Care must also be used in interpreting the start dates, as when work is scheduled in days, it actually starts on the next day. For example, while $t_0{}^0 = 0$, the first day of work for the activity following the start event is day 1, or $t_0{}^0 + 1$.

Table 2.5 Backward pass results

Job	Node i	Node j	LS_i	$y_{(I,j)}$	LC_j
H	5	6	16	2	18
G	2	5	9	7	16
F	4	5	10	6	16
E	3	4	6	4	10
B	1	2	4	5	9
D	3	1	2	2	4
C	0	3	0	2	2
A	0	1	1	3	4

Float and criticality

There are multiple paths through a complex schedule, but one or more of them will follow a line where the early and late start and finish dates will be the same (providing the project end date is the same as the earliest end date). This path is known as the critical path, as the jobs on the critical path cannot be delayed without effecting the end date of the project. Activities on the critical path are critical. Activities not on the critical path are floaters, that is they have float as defined in Equations 2.8 to 2.11. There are four definitions of job float provided by Kelley and Walker (1959):

- Total float (*TF*)

$$= t_j^{(1)} - t_i^{(0)} - y_{ij} \equiv LC_j - ES_i - y_{ij} \tag{2.24}$$

Total float is the latest completion time less the earliest start time less the job duration. This is the difference between the maximum time available and the job duration.

- Free float (*FF*)

$$= t_j^{(0)} - t_i^{(0)} - y_{ij} \equiv EC_j - ES_i - y_{ij} \tag{2.25}$$

Free float is the earliest completion time less the earliest start time less the job duration. This is difference between the minimum time available given an early start and the job duration and will never interfere with the progress of a succeeding job.

- Independent float (*IDF*)

$$= \max(0, t_j^{(0)} - t_i^{(0)} - y_{ij}) \equiv \max(0, EC_j - LS_i - y_{ij}) \tag{2.26}$$

Independent float is the greater of 0 and the earliest completion time less the latest start time less the job duration. This is the difference between the minimum time available (late start) and the job duration.

- Interfering float (*INF*)

$$= t_j^{(1)} - t_j^{(0)} \equiv LC_j - EC_j \tag{2.27}$$

Interfering float is the time between the early and late completion of the job. This time will always interfere with the progress of some succeeding job. This is also the difference between the total float (Equation 2.24) and the free float (Equation 2.25), as can be proven:

$$INF = TF - FF = (LC_j - ES_i - y_{ij}) - (EC_j - ES_i - y_{ij}) = LC_j - EC_j \tag{2.28}$$

It can be seen from Table 2.6 that all jobs in the project are critical except A, E and F, which have 1, 4 and 4 days of interfering float respectively. This form of calculation of the forward and backward pass is easily computerised.

Float is an important concept in scheduling, and has become very significant in the application in management. A schedule without float has no capacity to absorb delay. While the schedule may be correct, modern managers often rely on float to absorb the impact of delay on site activities. Constructing logic diagrams that contain float would therefore seem sensible, however float should not be used as a management tool in this way. It will be shown in Section Two that an alternative concept of a buffer may have more value for this situation. Uher (2003) argues that float may be viewed as a time contingency, giving the planner flexibility to schedule the start of a particular activity or activities within the maximum time available. This is erroneous, as contingency is used during the control phase, while scheduling occurs occur prior to construction. Using float in this way is more likely to be a function of resource management than contingency. However, it is the same as the interpretation placed upon the equivalent concept of *slack* in PERT (see page 26).

Table 2.6 Backward pass results

job	Node *I*	Node *j*	y(*i,j*)	ES*i*	EC*j*	LS*i*	LC*j*	TF	FF	IDF	INF
A	0	1	3	0	3	1	4	1	0	0	1
C	0	3	2	0	2	0	2	0	0	0	0
D	3	1	2	2	4	2	4	0	0	0	0
B	1	2	5	3	8	4	9	0	0	0	0
E	3	4	4	2	6	6	10	4	0	0	4
F	4	5	6	6	12	10	16	4	0	0	4
G	2	5	7	9	16	9	16	0	0	0	0
H	5	6	2	16	18	16	18	0	0	0	0

Logic of ADM

The ADM is very powerful for constructing construction schedules, but it must be remembered that the diagrammatic representation is merely a communication device representing a mathematical concept. In fact O'Brien and Plotnick (1999) report Kelley, who was a mathematician, as saying the CPM method was envisioned mathematically, with the logic diagram only being used to explain the approach to management. Nevertheless, without ADM and later PDM, CPM would most likely never have emerged as the powerful force it is today. Without a method to visualise the logic, it is unlikely the method would be used.

The rules for constructing ADM logic diagrams, provided before the advent of personal computers, were summarised by the AICA (1964) as follows:

1. Each activity is represented by an arrow. The length and direction of an arrow are not significant but only its position in relation to other activities which gives the time flow or job sequence.
2. Each activity is given an identification which is placed adjacent to the middle of the arrow. Sometimes double letter identification is preferred, particularly if there are more than 26 activities which is most likely to occur. Some analysts prefer to write a brief description of the activity onto the diagram.
3. Each activity begins and ends at a node—referred to as the Origin and Terminal Nodes. A node, which is indicated by a circle, represents a state of affairs in the project where one or more activities have been completed thus enabling subsequent and dependent activities to be started.
4. If one activity is dependent on another, the arrows are drawn in series.
5. If two or more activities can be carried out concurrently but cannot start until a common previous activity is completed, they are drawn with a common origin node.
6. All activities beginning at a common node are dependent on all activities which end at that node.
7. All nodes are numbered. The numbers of nodes should start at "1" (and not zero) and continue onwards in strict sequence without omitting any numbers.
8. It is usual that for every activity, the terminal node number is greater than the origin node number, but it does not necessarily have to be the next number in sequence. The more sophisticated programs now available from computers permit an almost indiscriminate numbering of nodes. However, care should be taken to determine the requirements of the proposed computer program before this general rule is broken.
9. No two activities should have identical origin and terminal node numbers, i.e. each activity must have a unique set of node numbers.

     10. Activities of no duration or cost—called Dummy Arrows—are sometimes inserted
        into the network to avoid breaking some of the above rules. To distinguish them from
        real activities, dummy arrows are drawn using a dotted line instead of a full line.
     11. There is one other type of arrow which does not represent an activity in the strict
        sense of the word. This is called a Restraint, and it represents some happening which
        becomes a controlling factor at a specific point in a network. For example: availability
        of trained personnel, or a *completion restraint* used to ensure scheduled completion of
        all or part of a project at a specific point in time (AICA, 1964).

This list was correct, although clearly dated by the subsequent advent of powerful personal
computer systems. However, there is one other key property which should have been added:

     12. Logic loops (recursive logic) are not allowed.

These logic rules are rigorous and focus attention on correct logic. For this reason, many
scheduling purists insist that ADM is the only safe way to schedule—it will avoid errors in
logic that can arise from taking the shortcuts available in an alternative method: the precedence diagram method (PDM). The logic rules are difficult to remember and understand,
particularly in building complex relationships between activities, and thus a simpler
alternative has found success in the market: the precedence diagram method.

THE DEVELOPMENT OF THE PRECEDENCE DIAGRAM METHOD (PDM)

In the early 1960s new methods for interpreting the CPM network were being developed.
There were two primary drivers: to improve the calculation methods to place less demand
on the computer, such as memory, and to make it more accessible for the layman. It was felt
that the existing methods were a barrier to acceptance.

    Charnes and Cooper (1962) demonstrated an alternative calculation system using the
theory of directed subdual algorithms. Their contribution was to focus on the precedence
relationship, as an entity, and to shift focus from the job. They developed a matrix-based
calculation method to calculate the minimum early start times. This method probably led to
the notation by O'Brien (1965: 49, 226) that a matrix method could be used for the manual
calculation of an activity on arrow CPM network. Similarly, Giffler (1963), who worked for
IBM, had recognised that:

    Underlying every scheduling problem is an order system whose elements are
    the tasks to be scheduled. The basic order relation, which connects the tasks, is
    called the *precedes* relation and is designated by the symbol $\leq$ The statement
    $a \leq b$ is taken to mean that task a must start at the same time or before b or, more
    simply, that task a precedes[9] task b. The relation $\leq$ is transitive, reflexive, and
    antisymmetric. That is, (1) $a \leq a$ for all a, (2) $(a \leq b) \wedge (b \leq c) \rightarrow a \leq c$, and (3)
    $(a \leq b) \wedge (b \leq a) \rightarrow a = b$.
        Note that the precedes relation does not exist between two tasks simply
    because one happens to start at the same time or before the other...
        The precedes relation $\leq$ includes the relation *next-precedes*, denoted by the
    symbol $\langle\langle$. We take the statement $a \langle\langle b$ to mean that task a next-precedes task b,
    or, more specifically, that there exists a transitive chain of relations $\leq$ from a to b

---

[9] Note that the word *precedes* is used as though it had precisely the same meaning as the longer phrase "*must
start before or at the same time*".

which includes no other task as intermediary. The distinction between the relations ≤ and ⟨⟨ may be clarified by the example which follows:

Order systems may be conveniently represented by directed graphs as in the following example. Figure 2.11 illustrates a system of four tasks: *a, b, c,* and *d*. The precedes relations are $a \leq a, a \leq b, a \leq c, a \leq d, b \leq b, b \leq c, b \leq d, c \leq c, c \leq d, d \leq d$. The next-precedes relations are $a \langle\langle b, a \langle\langle d, b \langle\langle c, c \langle\langle d...$

Figure 2.11 Precedes and next-precedes relations (after: Giffler, 1963: Figure 1)

Each precedes relation $a \leq b$ contains one or more chains of zero or more relations ⟨⟨. With reference to Figure 2.11, the relation $a \leq d$ consists of two chains of relations ⟨⟨, namely $a \langle\langle b \langle\langle c \langle\langle d$, and $a \langle\langle d$. The first of these is said to be of level 3, since it consists of three relations ⟨⟨; the second said to be of level 1. Each precedes relation of a task to itself is said to consist of one chain of level 0.

This approach has one major difference from previous CPM and PERT approaches, in that jobs were no longer represented as arrows between nodes, but rather as an activity on the node.

Smith (1981) attributed the development of activity-on-node (AoN) primarily to John Fondahl, an independent developer at Stanford University. The AoN concept was expanded and explained by Levy et al. (1963a) who described their network visualisation method as follows:

> First of all, each job necessary for the completion of the project is listed with a unique identification (usually a number), the time required to complete the job, and its immediate prerequisite jobs. Then each job is drawn on the graph as a circle, with its number and time appearing within the circle. Sequence relationships are indicated by arrows connecting each circle (job) with its successors, with the arrows pointing to the latter.

Levy et al. specifically noted the difference between their representation and previous representations:

> The above method of depicting a project graph differs in some respects from the representation used by Kelley and Walker...
> In the widely used Kelley-Walker form, a project graph is just the opposite of that described above: jobs are shown as arrows, and the arrows are connected by means of circles (or dots) that indicate sequence relationships. Thus all immediate predecessors of a given job connect to a circle at the tail of the job arrow. In essence then, a circle marks an event: the completion of all jobs leading into the circle. Since these jobs are the immediate prerequisites for all jobs

leading out of the circle, they must all be completed before *any* of the succeeding jobs can begin.

In order to portray accurately all predecessor relationships, "dummy jobs"[10] must be added to the project graph in the Kelley-Walker form. The method described above by [Levy et al.] avoids the necessity (and complexity) of dummy jobs, is easier to program for a computer, and seems more straightforward in explanation and application.

Levy et al. were correct in their assessment. Their diagram method, now known as precedence diagram method (PDM), has become the dominant representational form in the market. The notation employed by Levy et al. (1963a) for a job is shown in Figure 2.12, with a project schedule being formed by sets of such jobs connected by precedence links (Figure 2.13).

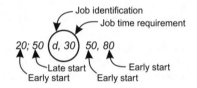

Figure 2.12 Precedes and next-precedes relations
(after: Levy et al., 1963a: Figure 4)

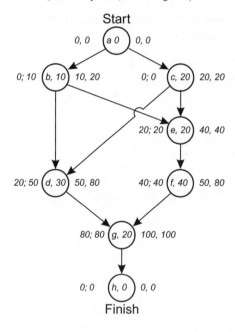

Figure 2.13 Calculation of late early and late start times for each job in a project
(after Levy et al., 1963a: Figure 4)

---

[10] Dummy jobs are more correctly constraints. However, the term dummy was used by Kelley (1961) and has become common. Constraints is the preferred term in this book.

Levy et al. (1963b) constructed several algorithms to support their work. The first calculates the early starts $ES(a)$ and early finishes $EF(a)$ for each Job a in project J, where P_a is the set of jobs where EF has already been calculated:

$$ES(a) = \max EF(x) \,|\, x \text{ in } P_a \qquad\qquad 2.29$$

$$EF(a) = ES(a) + t_a \qquad\qquad 2.30$$

The second calculates late finishes $LF(a)$ and latest starts $LS(a)$ for each Job a in project J, where S_a is the set of jobs where LS has already been calculated, assuming a known finish date (such as the calculated earliest finish date) T.

$$LF(a) = \min LS(x) \,|\, x \text{ in } S_a \qquad\qquad 2.31$$

$$LS(a) = LF(a) - t_a \qquad\qquad 2.32$$

These clearly were equivalent equations for what would later become known as the forward and backward passes. Levy et al. (1963a) clearly described this process in the early representation of two simple procedures:

> There is a simple way of computing ES and EF times by using the project graph. It proceeds as follows: (1) mark the value of S [start time for the project] to the left and to the right of Start; (2) consider any new unmarked job all of whose predecessors have been marked, and mark to the left of the new job the largest number marked to the right of any of its immediate predecessors (this is its early start time); (3) add to this number the job time and mark the result to the right of the job (early finish time); (4) continue until Finish has been reached, then stop. Thus at the conclusion of this calculation the ES time for each job will appear to the left of the circle which identifies it, and the EF time will appear to the right of the circle. The number which appears to the right of the last job, Finish, is the early finish time F for the entire project.
>
>   [Latest dates] are calculated for each job in a manner similar to previous calculations, except that we work from the end of the project to its beginning. We proceed as follows: (1) mark the value of T to the right and left of Finish; (2) consider any new unmarked job *all of whose successors have been marked*, and mark to the right of the new job the smallest number marked to the left of any of its immediate successors; (3) subtract from this number the job time and mark the result to the left of the job; (4) continue until Start has been reached, then stop. At the conclusion of this calculation the LF time for a job will appear to the right of the circle which identifies it, and the LS time for the job will appear to the left of the circle. The number which appears to the right of Start is the latest time that the entire project can be started and still finish at the target time T.

Levy et al. (1963a) refer to the Malcolm et al. (1959) concept of *slack* in calculation of the float in the PDM schedule. They therefore do not go into the same level of detail as Kelley (1959) in their treatment of float.

Constructing logical networks in PDM

The precedence diagram method (PDM) uses a visualisation method in which the activity is represented as a node connected to other activities by arrows passing between any two activities. It is also known as Activity on the Node for this reason. The node has properties of duration (≥ 0) and other activity properties such as resources. The arrow also has properties, it has a duration (≥ 0), direction and type as shown in Figure 2.14, which illustrates the simple example of the finish to start [F–S] logic link.

Figure 2.14 Two activities on the node joined by a logical relationship

PDM represents a project as a sequence of such nodes whereby the relationship between preceding activities and successors are indicated by a series of arrows. There is great emphasis on the logical representation of connections in PDM, as they are no longer simple logical relationships between activities, but rather are relationships with multiple properties.

It is much quicker to build complex relationships in PDM than ADM. However this power means it is relatively easy to make mistakes. Therefore, it is very useful to have a topological representation of the network when constructing schedules.

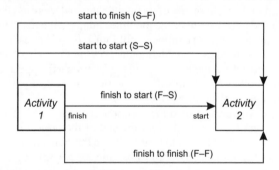

Figure 2.15 Four types of logical link

There are four types of links as illustrated in Figure 2.15, using a topological representation which displays a [S–] relationship coming from the start of the activity, a [F–] relationship coming from the end of the activity, a [–S] relationship coming into the start of the following activity and a [–F] relationship going into the end of the following activity.

Figure 2.16, panels 1 to 4, show the comparison of the different links constructed in both ADM and PDM. Note, in these examples, the link has no duration as the introduction of duration requires the use of dummies or constraints in the ADM.

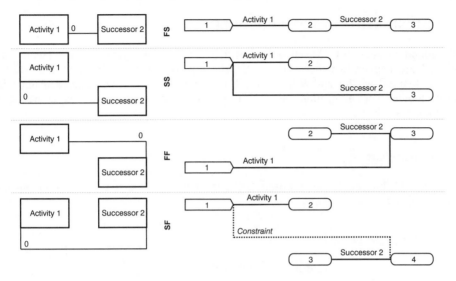

Figure 2.16 Comparison of the four types of logical links
in both PDM (left) and ADM (right)

The power of the PDM system is the ability to not only have types of relationships (which would have to be constructed by careful logical relationships and restraints in ADM) but also to allow multiple connections between the same two activities. This becomes important because of the link types. Careful examination of the comparison between PDM and ADM reveals that it is easy to leave an activity without any start or end link, and thus it will be constrained only by the end of the project. Care must be taken to make appropriate logical connections. One such example is very common and is shown in Figure 2.17.

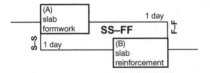

Figure 2.17 Combining S–S and F–F links between the same two tasks,
sometimes noted as SS–FF

In this case, activity B follows activity A with both [S–S] and [F–F] links (sometimes noted as [SS–FF]). An example might be *Slab Formwork* as activity A and *Slab Reinforcement* as activity B. Reinforcement might start 1 day after the start and finish 1 day after the completion of formwork. Thus, the two tasks run together but one day apart on site.

The combination of [S–S] and [F–F] links between two tasks or a group of tasks is the usual way to simulate location-based planning of repetitive activities in CPM, as long as the software supports it[11], as they can be used to form a *ladder network* or *overlap network* (see also page 81). It is common practice to establish a typical floor cycle and then to link the first and last tasks in the cycle with a lag equivalent to the duration of whichever activity will drive the repetition cycle of the work. This issue will be discussed further in Chapter 3 and Section Two.

---

[11] Microsoft Project does not allow multiple links between the same tasks.

COMMUNICATION OF CPM SCHEDULES

The communication of CPM schedules is difficult, falling back generally on a Gantt chart with a logic diagram (whether ADM or PDM) being used to convey logic for those skilled enough to use it. Mellon and Whitaker (1981) considered that the Gantt chart remained the most popular communication method in practice. There are, however, some other techniques available.

Fenced bar chart

Mellon and Whitaker (1981) proposed the addition of logic links to the Gantt diagram and termed this a *fenced bar chart*.

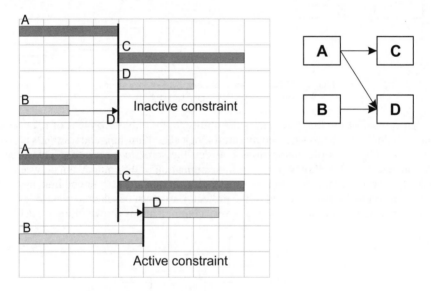

Figure 2.18 Example of a fenced bar chart
(after Mellon and Whitaker, 1981, with permission from ASCE)

The fenced bar chart included vertical lines for active logical connections and arrows to indicate float or inactive connections. This is the predecessor to today's common linked Gantt chart formats.

Timescaled arrow diagram

A timescaled arrow diagram can be used for ADM. This is a presentation of the logic diagram, with the activity on the arrow, where the arrow is shown like a bar on a Gantt chart. It is therefore timescaled. This is a powerful method, however it suffers from having many confusing lines in large relationships.

Linked Gantt chart

Even more confusing (because of the vertical compression of the visual space) is a linked Gantt chart. This is a Gantt chart with the links being shown between the activities. Except for simple projects, this can rapidly become unreadable.

DISCUSSION

The critical path method (whether CPM or PERT) is one of the most impressive success stories in modern management. It rapidly demonstrated its capacity to successfully manage complex projects time and again. It has great application for planning projects and for establishing a mechanism for control by exception. Even by 1962, it had demonstrated time reductions against manual methods of approximately 25% as well as significant cost savings (Pocock, 1962), through improved planning action, improved business orientation and improved basis for evaluation of the plan.

CPM has wide application in construction, and it now has over 50 years use in the industry. It is so successful that many ask why look further. They argue CPM is the tool for the industry and nothing else is required.

However, there is evidence of its failure too. While Pocock (1962) observed that 80% of the benefit arose merely from the process of better planning, the increased emphasis on the control and operating values of the method forecast by Pockock seems not fully realised. It consistently fails to be effectively applied on construction sites. Schedules are prepared and ignored. Projects are planned dynamically on the fly. Targets are not met and durations are exceeded. Mistakes occur and are repeated.

One problem is that CPM treats every activity as a discrete job, without any functional relationship between jobs of the same type in different locations. In reality, this does not suit construction. Another approach is needed. This is location-based planning, and it has an equally long history.

There is a wealth of further research into CPM and some very worthwhile advancements have been made, particularly in relation to modelling and optimisation using CPM. However, it is beyond the scope of this book to venture further, as it contributes little of further value to the grounding of the location-based management system.

REFERENCES

Adeli, H. and Karim, A. (2001). *Construction scheduling, cost optimisation, and management: A new model based on neurocomputing and object technologies*. Spon Press, London.

AICA (1964). "PERT, CPM and Cost Control". *The Australian Institute of Cost Accountants Bulletin*. **8**(1).

Archibald, R.D. and Villoria, R.L. (1967). *Network-based management systems (PERT/CPM)*. John Wiley & Sons, New York.

Arditi, D. (1983). "Diffusion of network planning in construction". *Journal of the Construction Division, ASCE*.

Battersby, A. (1964). *Network analysis*. 1st edition. Saint Martin's Press Inc. New York. Studies in Management series.

Battersby, A. (1970). *Network analysis for planning and scheduling*, 3rd edition Macmillan and Co. London.

Charnes, A. and Cooper, W.W. (1962). "A network interpretation and a directed subdual algorithm for Critical Path Scheduling". *The Journal of Industrial Engineering*. **13**(4): 213-219.

Collins, T.F. (1964). *Network Planning and Critical Path Scheduling*. 1st edition. Know How Publications, Berkeley, California.

Fazar, W. (1962). "The origin of PERT". *The Controller* **1959**(December): 598–621.

Giffler, B. (1963). "Schedule Algebras and their use in formulating General Systems Simulations". In *Industrial Scheduling*, edited by Muth, J.F. and Thompson, G.L., Prentice-Hall, Englewood Cliffs, NJ (4): 39-58.

Gordon, J. and Tulip, A., (1997). "Resource scheduling". *International Journal of Project Management*, **15**(6): 359–370.

Halpin, D.W. and Riggs, L.S. (1992). *Planning and Analysis of Construction Operations*. Wiley-IEEE.

Kelley, J.E. (1959). "Parametric Programming and the Primal-Dual Algorithm". *Operations Research* **7**(3): 327–334.

Kelley, J.E. (1961). "Critical-Path planning and scheduling: Mathematical basis". *Operations Research* **9**(3): 296–320.

Kelley, J.E. (1963). "The Critical-Path Method: Resources planning and scheduling". In *Industrial Scheduling*, edited by Muth, J.F. and Thompson, G.L., Prentice-Hall, Englewood Cliffs, NJ (21): 347-365.

Kelley, J.E. (2003). "The State of CPM Schedules". Opinion letter in response to "Critics can't find the logic in many of today's schedules". *Engineering News Record*, 23 June.

Kelley, J.E. and Walker, M.R. (1959). "Critical-Path Planning and Scheduling". *Proceedings of the Eastern Joint Computer Conference*. 160–173.

Kenley, R. (2003). *Financing Construction: Cash flows and cash farming*. Spon Press. London.

Kenley, R. (2004). "Project micromanagement: Practical site planning and management of work flow". Proceedings of the *12th International Conference of Lean Construction (IGLC12)*, Helsingør, Denmark: 194–205.

Levy, F.K., Thompson, G.L. and Wiest, J.D. (1963a). "Introduction to the Critical Path Method". In *Industrial Scheduling*, edited by Muth, J.F. and Thompson, G.L., Prentice-Hall, Prentice-Hall, Englewood Cliffs, NJ (20): 331–345.

Levy, F.K., Thompson, G.L. and Wiest, J.D. (1963b). "Mathematical basis of the Critical Path Method". In *Industrial Scheduling*, edited by Muth, J.F. and Thompson, G.L., Prentice-Hall, Englewood Cliffs, NJ (22): 367–387.

Malcolm, D.G., Roseboom, J.H., Clark, C.E. and Fazar, W. (1959). "Application of a technique for research and development program evaluation". *Operations Research* **7**(5): 646–669.

Mauchly, J.W. (1962). "Critical-Path Scheduling". *Chemical Engineering* 1962(April): 139-154.

Mellon, J.W. and Whitaker, B. (1981). "Fencing a bar chart". *Journal of the Construction Division, ASCE* **107**(CO3): 497–507.

O'Brien, J.J. (1965). *CPM in Construction Management: Scheduling by the Critical Path Method*. 1st edition. McGraw–Hill Book Company, New York.

O'Brien, J.J. and Plotnick, F.L. (1999). *CPM in Construction Management*. 5th edition. McGraw-Hill, Boston.

Pocock, J.W. (1962). "PERT as an analytical aid for program planning—its payoff and problems". Tenth Anniversary Meeting of the Operations Research Society of America, Washington DC, May 9: 893–903.

Schoderbek, P.P. and Digman, L.A. (1967). "Third generation, PERT/LoB". *Harvard Business Review* **45**(5): 110–110.

Smith, D.A. (1981). "Productivity engineering is 'task management'". *Civil Engineering-ASCE* **81**(8): 4–51.

Uher, T.E. (2003). *Programming and Scheduling Techniques*. UNSW Press, Sydney.

Chapter 3

The development of location-based planning and scheduling systems

INTRODUCTION

This book is about the location-based management system. So, in this chapter, we will explore the development of location-based planning theory. Location-based methods have a long history and, but for the arrival of computers and the development of commercial network planning methods, would likely be the dominant manual technique for scheduling projects. However computers did arrive, activity-based planning methods were developed (see Chapter 2) and the modern tools of CPM have come to dominate industry practice.

Activity-based methods were a wonderful development and have enhanced the construction industry enormously. Indeed, they remain the method of choice for complex projects with little or no functional activity repetition. However, despite the success of activity-based methods, there has remained an undercurrent of doubt about their suitability for real construction projects. Some researchers have realised that activity-based systems were inefficient, were failing to recognise the significance of workflow and continuity, and were essentially unreliable in their application. Over the years researchers have noted that network methods add little to solving the planning problem where there is repetition (Selinger, 1980; Reda, 1990; Russell and Wong, 1993; Arditi et al., 2002). The case is best made by Arditi et al.:

> The first problem is the sheer size of the network. In a repetitive project of *n* units, the network prepared for one unit has to be repeated *n* times and linked to the others; this results in a huge network that is difficult to manage. This may cause difficulties in communicating among the members of the construction management team. The second problem is that the CPM algorithm is designed primarily for optimising project duration rather than dealing adequately with the special resource constraints of repetitive projects. Indeed, the CPM algorithm has no capability that would ensure smooth procession of crews from unit to unit with no conflict and no idle time for workers and equipment. This leads to hiring and procurement problems in the flow of labour and material during construction (Arditi et al., 2002).

Such concerns have meant that interest in the old graphical techniques, in particular line-of-balance and flowline, never waned. At the time of the development of CPM in the early 1960s, the manual techniques that previously existed were described in both the UK (house building) and the USA (US Navy). Subsequently, in pockets around the globe, developmental work continued. Significant locations were Israel, with Selinger and Peer (early 1970s), Australia with Mohr (late 1970s), Canada with Russell (early 1980s), Finland with Kankainen (late 1980s) and several authors in the USA culminating in the enthusiasm of Arditi (2000s). Different solutions are proposed, but all are location-based.

Of this developmental work, few commercial applications have arisen. Some graphical methods were developed in Finland, but these were essentially drafting methods to assist manual planning. Russell developed a commercial application in the late 1980s, but was caught up with the recession in the property industry in the early 1990s. Some methods have been developed which draft a line-of-balance as an overlay on CPM as an extension to the dominant method.

The term location-based scheduling was first proposed in 2004 by Kenley (2004) to differentiate the emphasis on locations and activities in planning. Today, modern location-based methods for planning are available with the support of commercial software. Vico Software has emerged as an industry leader and a genuine alternative to the dominant CPM software packages, with its Control software. Control was developed by Olli Seppänen, as DynaProject, in Finland, and it arose under the guidance of Professor Kankainen.

As this new century develops, there is now a new player on the planning block. Location-based planning is gaining acceptance throughout the world at a surprisingly rapid rate. Users are recognising the advantages of protecting the work crews from poor planning. Clients are enjoying the benefits of better planning and control. Complex projects and large projects are being scheduled rapidly and with great sophistication using these new methods. The result is a better reflection of the true nature of construction work in the planning models, improved project performance and greatly increased reliability.

None of this would be possible without the work of so many people in the past. In this chapter we go back to the beginnings and explore the development of location-based methods. It is a fun journey with some genuine surprises. At the end it is easy to draw the following conclusion: location-based planning is the natural planning system for construction. It has a long history, was the earliest planning system employed on large projects, and genuinely accounts for the nature of work in construction. The authors believe most building work should be planned and controlled using location-based methods.

So, how did it develop? We now explore the detailed history of location-based methods, by whatever name they were known (and there have been many), over the past 100 years.

THE STORY OF LOCATION-BASED METHODS

Location-based planning methods utilise knowledge of location as an intrinsic component of the planning system. They are planning methods which concentrate on the relationship between the location of work and the unit of work to be done. If each work package is considered part of a location-dependent set of work packages, then progress through locations can be planned for the entire work package. When looking for the origins and development of location-based methods, we are therefore looking for methods which display (graphical) or calculate (analytical) the relationship between work and locations, and specifically treats the flow of work as important. The earliest evidence suggests such methods arose over 100 years ago.

Karol Adamiecki—the father of location-based scheduling

The nineteenth century Polish professor, Karol Adamiecki, specialised in engineering, economics and management and developed techniques which the authors believe were the foundations of location-based management. He not only developed graphical techniques, he also developed an ontology to describe production management and the complex interaction between the engineering of production and the efficiency of production.

> Karol Adamiecki was one of the most famous management researchers in Central and Eastern Europe in his times. He began his research in the Institute of Technology in St Petersburg, Russia (1884–1890), then moved to Poland. In

1896, Adamiecki invented a novel means of displaying interdependence of processes, to increase visibility of production schedules. In 1903, his theory caused a stir in Russian technical circles. He published some articles on it in *Przeglad Organizacji* (Technical Review), nos. 17, 18, 19 and 20 (1909). In 1931 he published a more widely-known article describing his diagram, which he called the harmonogram or harmonograf. Adamiecki had, however, published his works in Polish and Russian, languages little known abroad. By this time, a similar method had been popularised in the West by Henry Gantt (who had published articles on it in 1910 and 1915). With minor modifications, Adamiecki's chart is now more commonly referred to in English as the Gantt chart. Adamiecki published his first papers in management in 1898, before Frederick Winslow Taylor had popularised scientific management. (Wikipedia, 2008).

While Adamiecki was a contemporary of Gantt and Taylor[1], his interest in economics in production lead him consider efficiency in production and to develop three *laws of economy*. These, he believed, controlled or influenced the volume of effort or resources as well as the volume, size or number of useful outputs in production. Later, he published his three laws (Adamiecki, 1931):

- Law of Work Division
- Law of Concentration
- Law of Harmony in Management.

It is this body of work, together with the associated management tools and techniques, that makes Adamiecki one of the fathers of production scheduling and control (Marsh, 1975). His emphasis on the alignment of production rates, the recognition of optimum (natural) production cycles and the use of graphical tools, such as harmonograms, to plan for optimum production effectively makes Adamiecki the father of location-based planning.

While the laws of work division and concentration have much in common with location-based management systems, it is the law of harmony in management that has the greatest relevance to the LBMS and its associated methods. This law consisted of three parts:

- Harmony of choice (all production tools should be compatible with each other, with special regard to their output production speed)
- Harmony of doing (importance of time coordination, schedules and harmonograms)
- Harmony of spirit (importance of creating a good team).

Adamiecki noted somewhat dryly that his original proposal of the laws of economy was met with derision in 1903, as reviewers of the time believed that its application (and in particular the law of harmony) would result in men being converted to machines[2]. However, he was of the view that the laws were laws of nature rather than inventions of man and, as such, production which worked within the laws would lead to unprecedented economical rewards and save physical and intellectual strength. In contrast, production which ignored the laws would result in chaos and internal conflict (Adamiecki, 1931).

Adamiecki therefore expounded ideas of production theory which remain relevant to this day. His approach of developing 'natural laws' points toward the concepts of optimum

---

[1] Inventors of the bar chart or Gantt chart as it has become known.

[2] A sentiment echoed in some recent critical reviews of lean construction such as Green (1999, 2002).

production rates and optimum work crews, all concepts that we will see are components of location-based management.

One of the most significant contributions of Adamiecki was to develop his own, and more sophisticated, version of Taylor and Gantt's bar charts. The earliest versions of these were developed as early as 1896 (Marsh 1975), but generally the development was in parallel with Gantt. Adamiecki acknowledged that, while he was developing his ideas independently, his work followed theirs. Nevertheless, he took his ideas one significant step further. His diagrammatic schedules, *harmonograms*, included location as a key component for the schedule. Gantt charts do not include this feature except by coding activities.

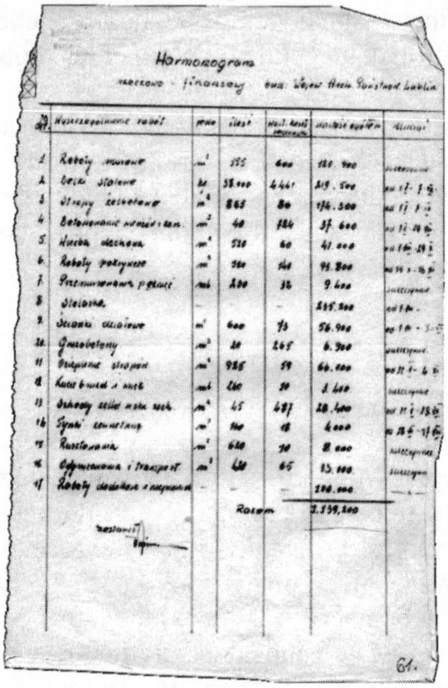

Figure 3.1 An early construction harmonogram (Archiwum Panstwowe w Lublinie, 2009)

The term *harmonogram* actually means either timetables or schedules in Polish. An early example (Figure 3.1) illustrates their use in construction, with a bill of quantities shown and including calculated durations with associated dates. While this provides the data required to prepare a chart similar to a Gantt chart, Adamiecki's charts were much more powerful. His harmonograms were significantly more complex as they included location and charted the movement of work through processes, improving efficiency by aligning production rates. The true value of harmonograms was recognised by Marsh (1975) who highlighted the network capacity of the diagrams—indicating that they already had elements integrating CPM network theory with the diagram method (also see page 54).

Figure 3.2 illustrates a harmonogram which charts time-related bars across two axis, with time on the X-axis and location on the Y-axis.
The important characteristics to observe are:

- Locations on the Y-axis
- Time on the X-axis
- The units of work are represented by bars
- The unit of production (in this case an item being manufactured) flows through locations and is represented by a line

- The production lines are not aligned
- The work at each location is discontinuous and inefficient
- The unit of production suffers discontinuities.

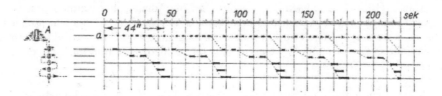

Figure 3.2 Production harmonogram showing locations (machines)
(Adamiecki, 1909: Figure 5, p76)

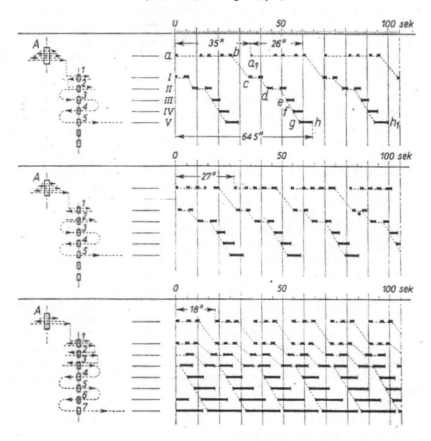

Figure 3.3 Optimisation using a harmonogram (Adamiecki, 1909: Figure 2-4, p72)

In clarification, note that this is a manufacturing production system and as such the location on the Y-axis represents an item of plant or a single process, equivalent to the crew in construction. Also, the location in construction is equivalent to the item being processed (work flows through the location) and would be approximated by the line.

Adamiecki was concerned with improving production performance. He did this by aligning the production rates, a process illustrated in Figure 3.3. Here the improvement results in a through-time cycle reducing from 35" to 18". More importantly, the production location (crew) approaches continuity.

Marsh (1975) recognised the importance of Adamiecki's harmonograms. He explained that harmonograms, which he described as graphical solutions to production problems, were one of various workflow network diagrams. Workflow networks are associated with the family of methods which include logical networks in forming the diagrams (such as PERT or CPM).

> What is particularly interesting about Adamiecki's earlier workflow network concept is that it incorporated the best features of the Gantt charts together with the network concept (Marsh, 1975).

Marsh described the technique for producing harmonograms utilising paper strips. In the method, a strip of paper represents each operation in the production process. Each strip had a sliding metal indicator representing the time to process one unit of production. A series of strips could be manipulated such that the indicators formed a continuous series between operation activities.

> Once the strips and tabs were prepared, optimising the workflow problem was simply a manual process of arranging the strips and sliding the tabs. Each strip was arranged according to the rule that all 'from' operations must be to the left of the given strip. With the strips thus arranged, the tabs were slid into the time locations which the sequence of operations dictated. This revealed the events of the critical path and resulted in an exact estimate of the production time. If necessary, further refinements in the solution obtained could be made with the use of eyeball judgment and common sense (Marsh, 1975).

Marsh's description reveals that harmonograms were very capable manual scheduling techniques. Whether or not they were truly equivalent location-based scheduling is subject to conjecture, however it is certain that workflow was a critical factor and production optimisation was the end purpose. It is true that in the factory the processes were fixed and the materials had to flow through the equipment. Thus, as always, construction presents a different problem requiring a different solution. Nevertheless, it seems clear that Karol Adamiecki provided the early thinking which has led to the development of location-based management. In particular, the concepts of alignment of production speed (production tools should be compatible) and the importance of harmonograms relate to modern day LBMS. Significantly, the importance of creating a good team to achieve harmony is only today being rediscovered through the practical application of lean theory and location-based management.

Starrett Brothers—Empire State Building

It cannot be proven that Adamiecki's ideas were directly adopted in the United States, nevertheless there appears a conceptual link between his ideas and the methods applied, with spectacular success, on the construction of the Empire State Building.

The Starrett Brothers built the Empire State Building in record time, completing 102 levels, from sketch designs to opening for trade, in 18 months achieving floor cycles of one floor per day, having a high safety record (for its day) and completing under budget. They paid special attention to production speed, they employed time coordination using schedules, and placed enormous importance on the creation of a good team. Their harmonograms included location by adopting building floors (or even zones) as the unit of control.

While many of the techniques employed are not of interest here, such as the use of specialised materials handling including rail trucks on floors (Carmody, 1930), there are several techniques which illustrate the development of a location-based management system from planning and continuing through to controlling. The project operated at the floor level of accuracy with construction zones on each floor. The planning system used a location-based measure of the work for each trade (Figure 3.4) and an harmonogram to represent the flow of work through the building (Figure 3.5).

It is likely that this is the earliest documented use of location-based scheduling in construction. The harmonogram diagram was described by Shreve:

> The interlocking schedule of the steelwork was presented diagrammatically by Balcom, using the same dates as listed on the drawing [Figure 3.4]. Reading to the right on the chart shows time passing, reading up the chart shows the building rising. The first diagonal line represents the due dates for the design; the second, the mill order; the third, shop drawings completion; the fourth, delivery of the steel to site; and the fifth, the erection of the steel. The various activities on any given date can be examined by drawing a vertical line... Similarly, the various deadlines for any given floor can be determined by drawing a horizontal line (Shreve, 1930).

The importance of an approach that uses quantities by location and represents the flow of work through the locations graphically will be shown in later chapters. However, it is important to understand the underpinning management philosophy—a philosophy driven by a desire to achieve rapid production much faster than existing norms.

To achieve a floor cycle of one day per floor, including trades such as brickwork, tight emphasis on logistics and the flow of work was required. The builder, Paul Starrett, stated:

> Our job was that of repetition—the purchase, preparation, transport to the site, and placing of the same materials in the same relationship, over and over. It was, as Shreve the architect said, like an assembly line—the assembly line of standard parts (cited in Willis, 1998).

Starrett Brothers placed a great deal of emphasis on maintaining the pace of the work, by adjusting the number of workers to keep the work on the schedule.

Interestingly, the recent concept of buffers, so much a part of location-based methodologies, can be clearly seen in the description of their approach:

> An important lesson in the pursuit of speed was to disconnect different portions of the work as much as possible. Trades move at different speeds, have special requirements and may view the same detail in entirely different ways. By eliminating as much of the contact between trades as was possible, the builders reduced the risk of cascading delays (Shreve, 1930).

They also recognised that the key to achieving the required level of performance was careful planning. There is also much in their organisation which we would these days recognise as lean production.

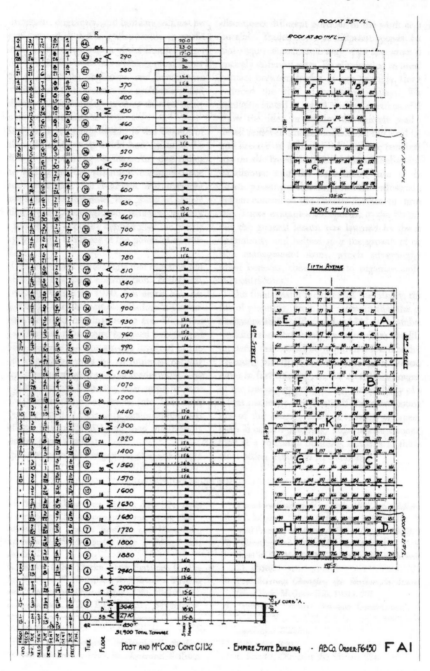

Figure 3.4 Location-based measure of structural steel and control dates per level (and zone)
on Empire State Building. Source: Shreve (1930, p772)
"Schedule for the Structural Steel for the Empire State Building, giving dates
of information and drawings required from the architects,
mill orders, shop drawings, steel delivery and steel erection."

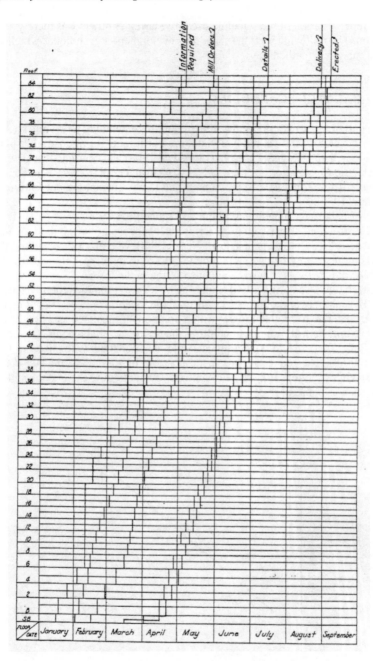

Figure 3.5 Location-based schedule for structural steel design and installation per level on the Empire State Building. Source: Shreve, (1930, p773; 1931, p346). "Chart developed from that on the opposite page [Figure 3.4] by H.G. Balcom, Consulting Engineer, working with the architects, Shreve, Lamb and Harmon, to visualise the time-co-ordination required in connection with the designing, detailing and erection of the structural steel" Shreve, (1930).

The representation of the schedule used by Starrett Brothers was merely a method of presentation. The following sections show the development of techniques which are underpinned by analytical methods.

LINE-OF-BALANCE SCHEDULING

In the first half of the twentieth century, the production scheduling technique known as line-of-balance developed into a relatively mature technique for industrial programming, commencing with the US Navy in 1942 (Lumsden, 1968) and continued particularly by the General Electric Corporation for the US Navy (Gehringer, 1958; NAVEXOS, 1962; Fink, 1965)[3] where it was used as both a planning and control technique but soon replaced by PERT (Fazar, 1959).

Line-of-balance is generally a production scheduling technique, but has found application in construction[4]. The two best descriptions of the technique applied to construction were Lumsden (1968) and the subsequent NBA (1968), who describe a method that was in the process of becoming mainstream in the UK housing industry—at the same time as the initial uptake of CPM was occurring in construction in the USA.

The technique was originally designed to be a way to handle repetitive construction, where a CPM sub-network, or other logical sub-component, could be modelling as a whole and the effective rate of production of the sub-component indicated as a line-of-balance as the repetitive units repeat. Lumsden identifies that in this way, a housing scheme ordinarily consisting of 6,000 to 12,000 activities could be modelled with 30 to 60 activities[5] (Lumsden, 1968). The reduction was made possible by modelling the sub-network as a single line, while then repeating the sub-network by continuing the line through the many locations (such as houses).

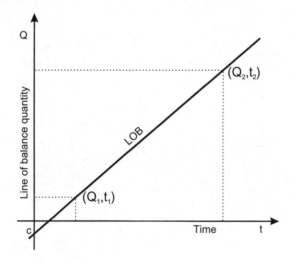

Figure 3.6 Basic line-of-balance linear relationship (after Lumsden, 1968: 1)

---

[3] For a review of the development of line-of-balance, refer Mattila and Abraham (1998); Huang and Sun (2006).

[4] Line-of-balance has important differences to the flowline method of graphical representation discussed later. This reflects its origins in production management.

[5] In LBMS these would be *tasks*.

Line-of-balance relies on the relationship between quantity of units (such as houses) delivered and the rate of unit production. This is considered a linear relationship (thus the emphasis on repetitive production where a linear relationship can hold). The relationship, illustrated in Figure 3.6, is given by the equation:

$$Q = mt + C \qquad (3.1)$$

Where: Q is the quantity delivered at time t, m is the rate of production (units per day)

This equation may also be used to find how many units are produced in a given time, and how long it will take to produce a quantity.

$$Q_1 = mt_1 + C \qquad (3.2)$$

$$Q_2 = mt_2 + C \qquad (3.3)$$

By subtraction:

$$Q_2 - Q_1 = m(t_2 - t_1) \qquad (3.4)$$

Therefore:

$$Q_2 = m(t_2 - t_1) + Q_1 \qquad (3.5)$$

Or:

$$t_2 = t_1 + (Q_2 - Q_1) / m \qquad (3.6)$$

In construction, the unit of production is a location, like a house, room, floor, etc. Typically, there are multiple activities worked in that location. For Lumsden, the line-of-balance model handled this by preparing a CPM schedule for the work to be repeated and thus deriving a duration for each line-of-balance quantity. For example, Figure 3.7 details a repetitive network with a minimum duration of 30 days, with the consequential line-of-balance network illustrated in Figure 3.8.

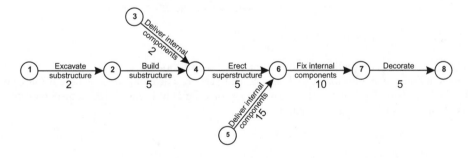

Figure 3.7 Sample unit network (after Lumsden, 1968: 14)

You may notice that Lumsden's logic uses activity on the arrow notation, thus the limits of the schedule can actually be seen on the line-of-balance to be nodes 1 and 8 (as indicated by the embedded logic diagram just visible in Figure 3.8). Lumsden then illustrates the other nodes.

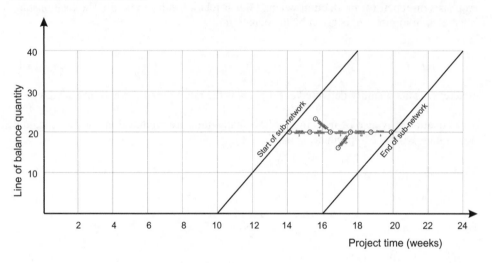

Figure 3.8 Line-of-balance limits for the sub-network (after Lumsden, 1968: 15)

Principles of line-of-balance scheduling

In this section, the principles and major characteristics of line-of-balance are discussed, as these are important for understanding the method. Line-of-balance scheduling is a combination of mathematical (production rate calculations) and graphical techniques. The principles outlined here are from Lumsden (1968) and follow the NBA (1968).

To explore the principles, it is useful to follow the example given by NBA (1968) using a simple three-activity network, Figure 3.9, with the limits of each part of the network in Figure 3.10.

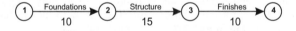

Figure 3.9 Sample sub-network for LoB (after NBA, 1968: 4)

Handover schedule or line-of-balance quantity

This is the rate of completion of production units (such as houses) and has more relevance in a production-like environment such as a housing estate. The handover schedule is mapped by the line 'end of sub-network' in Figure 3.8, or Line *D* in Figure 3.10. Lumsden (1968) used the term line-of-balance quantity.

Construction plan

This is the detailed set of activities and their logic structure (CPM) within each unit of production, such as Figure 3.7 or 3.9.

Line-of-balance schedules

NBA define the line-of-balance schedules as the Start lines for each task, these are the lines which join the start nodes for each task. These may be seen as the lines A, B, and C in Figure 3.10.

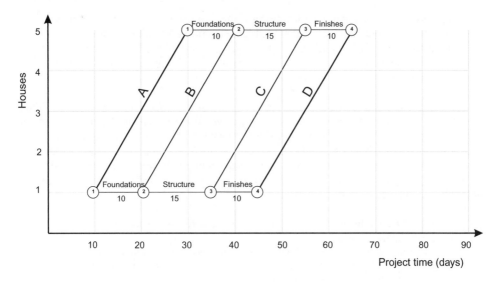

Figure 3.10 Sample sub-network showing node limits (after NBA, 1968: 4)

Time buffers

Time buffers are a critical component of line-of-balance scheduling and also of the location-based management system and arise from the need to allow for variation in production rates (see the discussion on page 144). There are many types of buffers in production management, such as inventory and stock buffers, but time-related buffers are of interest to line-of-balance.

> Time buffers provide a margin for error for and to ensure that one trade does not interfere with another, time buffers are normally inserted between the trade schedules (NBA, 1968: 4).

Lumsden (1968) defines two types: *stage* and *activity buffers*.

- Stage buffers: A time-allowance that may be made between discrete stages in the works, such as substructure, superstructure or finishes. This can allow for major impact delays such as ground conditions, weather, etc.
- Activity buffers: A time-allowance included in each activity time estimate for random differences in productivity and problems in the activity. The purpose is to delay the start of subsequent activities to minimise the effects of problems.

There is some argument that stage and activity buffers are actually the same thing. This argument flows from the fact that activity buffers are between activities in a sub-network for a project. By extension, stage buffers are between sub-networks that separate stages. In turn these represent a larger logical network, with buffers between the last activity of the earlier stage and the first activity of the following stage. Nevertheless, many practitioners will be comfortable with maintaining the division between local task-related effects and project-level impacts. Regardless of the type, buffers are not intended to allow for planning errors such as incorrect planned production rates:

> The essential function of the time buffer is to minimise the effect of disturbances between adjacent stages such that the planned smooth working of each stage is maintained and full benefits are gained from repetitive working. ...time buffers are not intended to cater for systematic slippage which implies a reduced rate of building (Lumsden, 1968).[6]

As the emphasis in using buffers is to protect the smooth working of the following activities, the buffer must be sufficient to absorb any delays. Where there is departure from the planned construction rate, Lumsden noted that substantial buffers would be required.

Lumsden noted that time buffers also create stock of completed work available for following trades. Later it will be shown that this is a location buffer, where there is stock of locations available for work should problems occur for a subsequent activity. Lumsden noted the advantages of buffers as follows:

> In general terms the advantages of building the buffer allowance into the activity duration are two-fold:
> 1. The principle can be compared with standard practice in the building industry where each job is buffered by virtue of the difference in time between the standard performance and the target performance.
> 2. It protects the project work at activity level, ie: microscopically.
> There are however, certain disadvantages which should be noted and these are as follows:
> 1. The project duration is increased but provided that the increase is consistent with a secured 5% net profit all should be well.
> 2. Consistent achievement of standard performance during the initial stages produces the temptation to bring forward subsequent work.
>   Any action in this direction should always be related to the size of buffer and the number of units outstanding.
> 3. It is difficult to cater for bad weather, material and labour shortages.
>   Of these (3) is the most significant and... stage buffers provide the best means of tackling this aspect (Lumsden, 1968).

Inserting a 5-day activity buffer (Figure 3.11) between each task highlights the line-of-balance for each task (Figure 3.12).

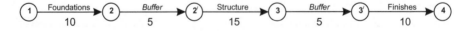

Figure 3.11 Sample sub-network for LoB (after NBA, 1968: 4)

The 5-day buffer increases the time allowed for each location by five days without impacting on the following activity. It is important to note that this only amounts to a delay to the planned project duration of five days, despite the 5-day buffer being available in each

---

[6] The significance of buffers to location-based planning is discussed in Chapter 5 and their use is discussed in Chapter 7. The significance of a reduced rate of building is discussed in Chapter 8.

location. In this case, two buffers increase the planned project duration by only 10 days for an overall construction period of 70 days.

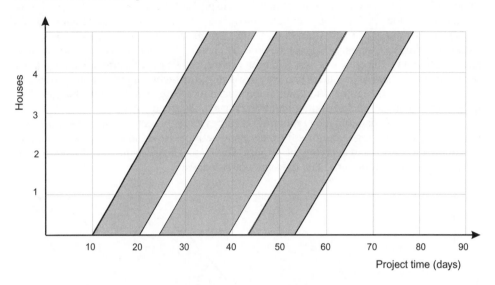

Figure 3.12 Inserting buffers between tasks (after NBA, 1968: 6)

Buffers are a risk management strategy that are designed to ensure smooth workflow according to the plan. It is important to understand their role and not remove them in either the planning or construction phases. Lumsden notes that the small cost of including a buffer is more than compensated by the increased profitability of the project.

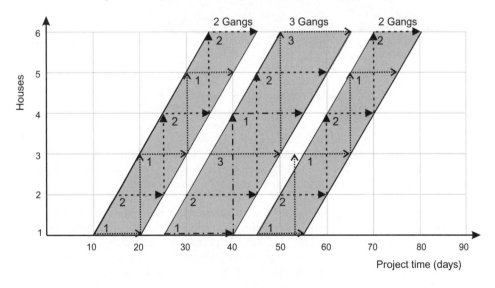

Figure 3.13 Required number of crews (after NBA, 1968: 6)

Labour resources

Figure 3.13 illustrates one of the most important aspects of the line-of-balance scheduling technique, that the graphical representation of work crews can be a form of labour resource management. For the example from Figure 3.12, it is possible to chart the number of crews working within each activity. This can be done by charting the work of a crew horizontally, and then 'moving' the crew to a new location. Figure 3.13 shows this effect. Here it can be seen that the first activity requires two crews or gangs, the second three and the third again two. This approach requires the assumption that crews or gangs can only be added to different locations instead of adding multiple crews to the same location.

The effect of changing the number of resources is easily seen. Figure 3.14 illustrates the effect of only having two crews for the second activity. Here the work slows, delaying the following activity and delaying the project. This occurs because the rate of work for the activities is no longer *aligned*. The lines are not *balanced*.

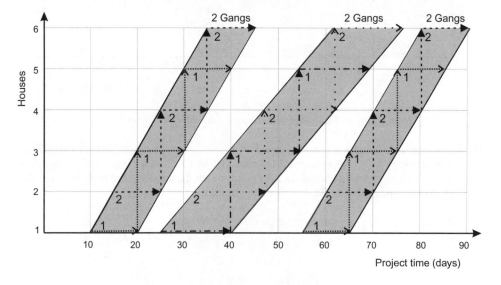

Figure 3.14 Effect of reducing crews for structure

Natural rhythm

Adamiecki referred to the law of harmony as a natural law, and reflected that going against that law would lead to chaos. Line-of-balance provides a mechanism to illustrate this and utilises the concept which Lumsden described as *the natural rhythm*.

> The idea that any repetitive process has a natural rhythm is not new, although its true significance when applied to repetitive housing is not fully appreciated by many contractors in the building industry. Any contractor embarking on a job of work does so with a view to making a profit and to this end the contractor aims to program his work such that investment in labour, materials and plant yields maximum return, i.e. minimum loss. What is frequently not understood in the housing sector of the industry is that there are optimum rates at which any house should be produced and that these rates are a function of the natural rhythm of many operations or activities involved (Lumsden, 1968).

This concept therefore implies that a work gang can produce at an inherent rate, and that production can only be sped up by adding more gangs, in multiples of the natural rate (Figure 3.15). This in turn suggests the idea of an optimum work crew for any given activity that can achieve the natural rate. Arditi (1988) later defined the natural rhythm of an activity as "the optimum rate of production that a crew of optimum size will be able to achieve".

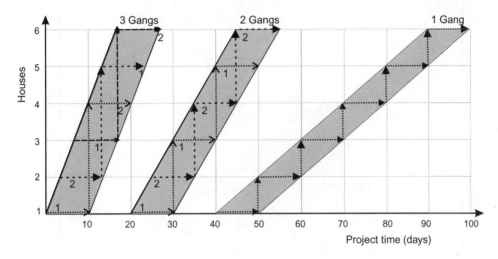

Figure 3.15 Multiples of the natural rhythm (after Lumsden, 1968: 26)

Balancing production

It follows that different activities might have different rates and that the task of the planner is to allocate resources to balance the production. The aim of line-of-balance is to balance the production rates between succeeding activities such that they can work continuously while minimising the difference in their rates to reduce overall schedule duration. The effect of a successor with a faster rate of production than its successor is illustrated in Figure 3.14. Here it can be seen that the start of the successor must be delayed to ensure continuous work. This has no effect on the project duration unless this task were to have a slower successor, which is necessarily delayed. This is termed *waiting time* (Lumsden, 1968).

Optimisation of the construction schedule therefore involves balancing the activities to avoid such delays as much as possible. The aim is to achieve the natural rate of construction for the project, or the handover schedule.

Resource histogram

Counting resources is a relatively simple matter with the line-of-balance, and is best shown as a resource histogram. This can be shown for a single task or accumulated to show project resources. A typical resource histogram may be derived as shown in Figure 3.16.

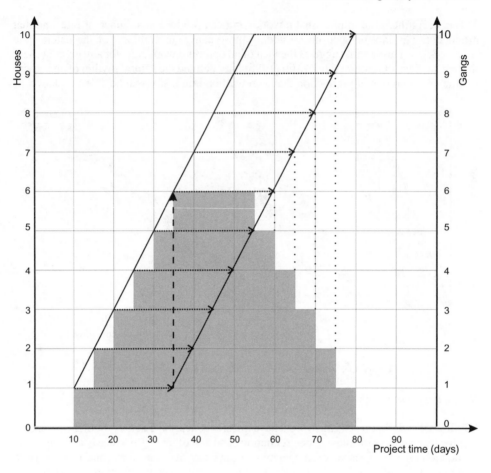

Figure 3.16 Effect of reducing crews for structure (after Lumsden, 1968: 23)

Repetitive construction

Line-of-balance, as it existed in 1968, was a largely graphical technique which relied on repetition for successful implementation.

> It is important at this point to define what is meant by repetitive working in the context that the term is used in this book and in a building environment. Essentially it means that identical operations are carried out repeatedly on successive units by the same operatives. (Lumsden, 1968: 30).

This limitation is, however, essentially a limitation of the capacity to calculate the effects of changing circumstances and differing locations rather than an inherent property of the technique itself. Nevertheless, this particular stated limitation is the one most frequently cited by authors and research students as the reason that location-based techniques do not have modern application in non-repetitive projects.

Any author today claiming the need for repetition as a reason for ignoring location-based techniques can be considered frozen in the 1960s.

Warning: technical content follows

Warning, technical content follows, feel free to skip ahead to HVLS on page 69.

Mathematics of line-of-balance

Several authors have aggregated the various formulas relating to line-of-balance technique. It may generally be agreed that these stem from the work of Selinger (1973, 1980) and his supervisor—Peer (1974). Al Sarraj (1990) foresaw the need for computerisation and standardisation of the method and drafted algorithms accordingly. The following are as published by Al Sarraj (1990). He started with the derivation of the basic line-of-balance formula $Q = mt + C$ as given in Equation 3.1, but uses the now more conventional r instead of m for the rate of production.

$$\frac{dQ}{dt} = r \tag{3.7}$$

Where Q is the quantity output in a unit of time. This yields:

$$Q = rt + C \tag{3.8}$$

Al Sarraj summarised the standardised equations as follows:

> We start with the computation of the unknown project duration D or output rate r, which are obtained using the following equations:
>
> $$r = \frac{Q - N_0}{(D - d)N} \tag{3.9}$$
>
> $$S_{kj} = F_{kj-1} + B_{kj-1} \tag{3.10}$$
>
> $$F_{kj} = S_{kj} + D_j, k = 1,\ldots,Q; j = 1,\ldots,m \tag{3.11}$$
>
> $$S_{kj} = \frac{k-1}{r} + \sum_{i=1}^{j-1}(B_i + D_i) \tag{3.12}$$
>
> $$F_{kj} = \frac{k-1}{r} + \sum_{i=1}^{j}D_i + \sum_{i=1}^{j-1}B_i \tag{3.13}$$
>
> Next we find the number of crews that work in activity A_i with the smallest duration (N_j), and the number of crews that work in activity $A_{j,}(n_j)$. This is computed using relation:
>
> $$n_i = \frac{D_i}{\left(\frac{1}{r}\right)} = rD_i \tag{3.14}$$

$$n_j = \frac{D_j}{D_i} n_i \tag{3.15}$$

Next we find the time of intersection of the lines of production of different activities, if there is an intersection. This can be found by using the equation:

$$T_j = F_{1j} + N_{uj}\left[\left(I_j - I_{j+1}\right) + I_j\right] \tag{3.16}$$

Finally, we prepare the actual schedule. This requires the starts and finishes of the unit groups and their activities. This is accomplished using the following relations:

$$S_{kj} = \frac{k-1}{r_j} + \sum_{i=1}^{j-1}(B_i + D_i) \tag{3.17}$$

$$F_{kj} = \frac{k-1}{r_j} + \sum_{i=1}^{j}D_i + \sum_{i=1}^{j-1}B_i \tag{3.18}$$

The numbers of crews, the variable activity rates in the actual schedule, and the buffers are given by:

$$n_j = \text{int}\left(rD_j\right) \tag{3.19}$$

$$r_j = \frac{n_j}{D_j}, I_j = \frac{1}{r_j} \tag{3.20}$$

$$B_j \text{ new} = B_j - \left(I_j - I_{j+1}\right)\left(G - 1\right) \tag{3.21}$$

Where:

$B(k) =$	buffer time between activity k and next activity, kH in unit network
$d, DU =$	duration of one completed unit
$D(k) =$	duration of activity k in unit network
$S(k, j) =$	start time of activity j at unit k
$F(k, j) =$	finish time of activity k in unit network
$G =$	total number of groups
$I_j - I_{j+1} =$	interval between starts of activity j and activity $j+1$
$M =$	total number of activities involved in one complete unit
$N =$	number of units to be delivered together after finishing starting units
$N(j) =$	number of groups working in activity j
$N_0 =$	number of units starting together at zero time
$Nu(j) =$	number of units to be produced
$Q =$	total number of units to be produced
$R =$	output rate (units to be produced per unit of time)
$T =$	required time to complete entire project
$T_j =$	time of intersection of two lines of production

Note: These notations above include errors and discrepancies, but in the interest of completenessare reproduced as provided in Al Sarraj (1990).

The use of such equations gives the ability to model lines-of-balance mathematically without the need for a diagram.

Horizontal and vertical logic scheduling (HVLS) method

Thabert and Beliveau (1994) attempted to retrofit the logic required for commercial construction—the vertical requirements of levels (stories) and the horizontal needs of related activities within levels. Their method, the horizontal and vertical logic scheduling (HVLS) method, applies to scheduling repetitive activities on typical floors of multistorey projects.

Thabert and Beliveau recognised that different logic structures operated, when trying to achieve continuity, within floors and between floors. They recognised that not all activities are required to be continuous between floors (and they could be intermittent). They proposed that this property could be used to incorporate continuity breaks in order to improve the schedule.

Activities contained the following parameters:

- **Work continuity**—activities fall into the categories of continuous and intermittent.
  - Continuous activities must be scheduled continuously from the first typical floor to the last.
  - Intermittent activities are scheduled in floor segments consisting of one or more floors and can be split between floors.
- **Progress rate or duration**—an activity has a normal duration and a maximum duration (based on resources) allowing scheduling flexibility.
- **Vertical workflow direction**—activities may be constructed upward or downward depending on the type of activity.

The model is not particularly sophisticated, but is notable for being the first demonstration of the line-of-balance technique (graphical method) being used in commercial construction and recognising categories of work. The method was however largely redundant due to the far greater power and modelling capacity in the integrated methods already developed (see page 84).

Thabert developed a commercial scheduling application called Space-constrained resource-constrained scheduling system (SCaRC) and described in Thabet and Beliveau (1997).

CPM/LOB

The diagrams of the formation of LoB presented here, such as Figure 3.10, show the relationship between the underlying logic (CPM) and the limits of the lines-of-balance. Suhail and Neale (1993) explored this further in their proposed combination of CPM and line-of-balance. This points the way toward integrated methodologies, however it is really a method for relaxing the demand for balancing production by utilising float available in the sub-network.

This method assumes constant repetition of sub-networks which contain float. As such, this is not an integrated approach. Nevertheless, the recognition of the role of float in line-of-balance is useful. Figure 3.17 illustrates the use of float in a simple network.

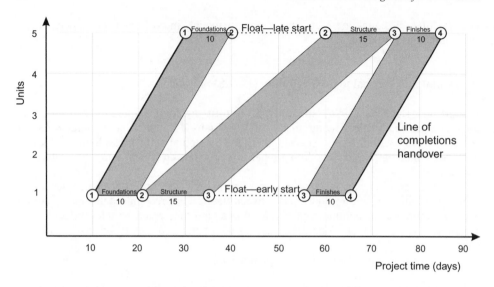

Figure 3.17 Adjusting line-of-balance for float in CPM/LOB

Multi-level LOB diagrams

Arditi is one of the most prolific researchers devoted to exploring line-of-balance in its application. Relatively recently, in the development of a software package called Chriss, two innovations were claimed. These were flexible unit networks and multi-level LoB diagrams. These were coded into a module called Lobex (Arditi et al., 2002). This work currently represents the most complete model for location-based scheduling using the line-of-balance approach.

Flexible unit networks

Flexible unit networks build on the now familiar concept of sub-networks, but specifically recognise that an activity within a network can be moved backwards or forwards in time without additional cost, within logic constraints.

Multi-level LoB diagrams

Three levels of activities are used to replace the basic sub-networks described thus far for .line-of-balance. Arditi et al. (2002) describe main activities, subactivities and sub-subactivities. This is an extremely powerful hierarchical representation of repetitive work. In this a basic network representing LoB limits may consist of a logical network of activities. One of these may represent an expanded subnetwork, and within this, one of its activities may be exploded into yet another sub-subnetwork. Thus there are three levels of hierarchical networks.

The real power of this approach is that these sub-networks may be used as libraries of standard work methods (782 sub-networks were mentioned).

Dependent and independent activities

Arditi et al. (2002) introduce the concept of dependency of networks to describe the situation where sub-networks can work at different rates subject to logical constraints, or not depending on physical constraints. Most work can operate independently—such as fit-out—but some work—such as structure—is dependent on completion of activities in other locations. This important concept was better handled by Russell and Wong (1993) and by the Layer 4 logic in the LBMS (see Chapter 5, page 139).

THE FLOWLINE METHOD

There is another way to represent the flow of work through locations. We now turn to this alternative—*flowline*. While Mohr may be attributed with naming and then documenting the flowline method (Mohr, 1979), his work was largely derived from the first to use that particular graphical representation—Selinger (1973, 1980) and his supervisor, Peer (1974)—who called them production lines. The link to line-of-balance can be seen in the work of Armon (1974), who was describing a new scheduling diagram for linear projects.

Peer found that CPM scheduling techniques did not adequately represent the use of resources in construction. He argued that:

> Limitations are imposed on the use of network analysis for planning the production process by its fundamental unrealistic assumptions of unlimited resources... and independent activities of fixed duration that can be shifted freely between earliest start and latest finish. The need for creating working continuity and balancing the while process into an integrated production system is completely neglected (Peer, 1974).

Peer's graphs specifically declare crew movements, a characteristic of flowline. He recognised that the bulk of activities , in most building projects, are repeated in a fixed sequence, performed by specialised crews and thus the plan must include for continuous production lines.

The flowline method is a graphical representation very similar to line-of-balance and for that reason is most often confused with line-of-balance. It uses very similar mathematics, yet the representation has a major difference. For that reason, the relationship of flowline to line-of-balance can be likened to the relationship between activity on arrow (AoA) and activity on node (precedence) networking in CPM.

Flowline represents the activity as a single line rather than the dual lines of line-of-balance. Thus it is similar to Precedence—with the line being the activity—whereas line-of-balance is equivalent to activity on arrow, with the lines being the nodes and constraining the activity between.

The result is that flowline is a much cleaner representation than line-of-balance. However, for the representation to be accurate, it is necessary to be explicit about the crews, as flowline otherwise does not accurately show the physical location of the crews. This becomes both possible and desirable in flowline, as detailed scheduling allows the detailed planning of work crews.

There is another, often forgotten difference: the vertical axis of line-of-balance represents the line-of-balance quantity (the cumulative production), The flowline representation presents the location on the vertical axis—in the intended sequence for construction.

It is worth noting here that in the early flowline representations, the location was treated in much the same way as the line-of-balance unit of production, and the vertical location division was of constant size. The use of a location-breakdown structure and corresponding variable vertical division was a later contribution from Finnish research.

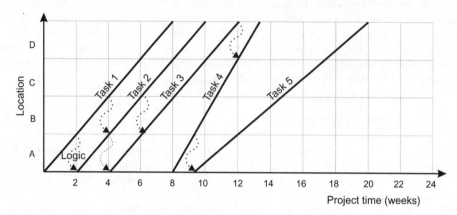

Figure 3.18 Flowlines for tasks 1–5 in locations A–D

The flowline representation is also designed to handle normal construction projects rather than the repetitive production of completed units such as housing. To do this, the project is broken down into sections of roughly equal size and content. However, instead of the horizontal line representing the production of a unit, as in ‚lin-of-balance the space occupied between the lines now represents the location. The line passes from the lower left corner (start of location, start of duration) to the upper right corner (end of location, end of duration), so its depth has meaning. In detail form, the flowline representation uses the same basic components, but represents a crew passing through a location. In Figure 3.18, the tasks 1–5 are shown passing through locations A–D. It is also easy to see that there is a logical relationship between tasks within locations.

In flowline it is possible to represent individual crews to show their physical presence in a location, or to chart a summary of the crews—a very fast way to model production.

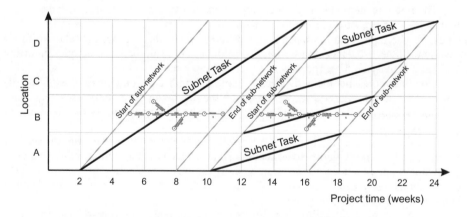

Figure 3.19 Project summary and location-detail flowlines for sub-network tasks, locations A–D

Figure 3.19 illustrates the LBMS equivalent representation of a project summary and location-level representation for a sub-network of activities. As long as activity buffers are used, either representation is practically correct. The location-level detail shows the location of work being done, whereas the project summary is a more efficient representation. The latter becomes important when building large and complex project schedules, where flow and sequence of work is most important.

Similarly, the work can be planned such that the individual sub-task crews flow. In Figure 3.19, the flow is only optimised for the entire sub-task. This requires multi-skilling, or alignment of the production rates for each sub-task. Were this possible, the flowline would be as shown in Figure 3.20.

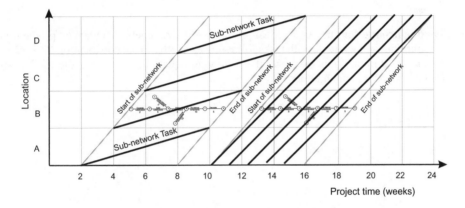

Figure 3.20 Summary and balanced sub-task flowlines for sub-network tasks, locations A–D

Warning: technical content follows

Warning, technical content follows, feel free to skip ahead to page 76.

Production sequencing

Mohr (1979) explored three types of production sequencing which are characteristic to construction scheduling. These were sequence production, parallel production and flowline production, as illustrated in Figure 3.21.

- **Sequence production**: where all works in a location are completed before commencing in the next. The project duration T is given by the following equation, where k is the module of production (the time taken for one crew to finish one location), m is the number of locations, n is the number of crews and t is the technological delay (such as curing time):

$$T_{sequence} = m \sum_{i=1}^{n} (k_i + t_i) \tag{3.22}$$

Or in the simplest case:

$$T_{sequence} = k.m.n \tag{3.23}$$

- **Parallel production**: where all activities in all locations are conducted simultaneously. The project duration T is given by the following equation, where k is the module of production, m is the number of locations, n is the number of crews and t is the technological delay:

$$T_{parallel} = \sum_{i=1}^{n}(k_i + t_i)$$ (3.24)

Or in the simplest case:

$$T_{parallel} = k.n$$ (3.25)

- **Flowline production**: this is the ideal production where repetitive balanced production flows through all locations, with tasks following each other from one location to another. The project duration T is given by the following equation, where k is the module of production and n is the number of crews:

$$T_{flowline} = k(m + n - 1)$$ (3.26)

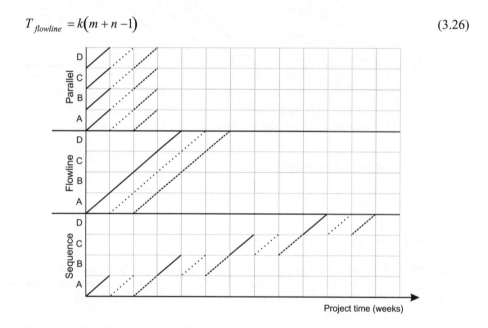

Figure 3.21 Sequence, flowline and parallel production

Earlier, Peer had developed Equation 3.26, which he described as the simplest project of a highly repetitive nature—where all activities are critical.

Peer (1974) discussed that there are four types of activities to be found on a project:

- Preparatory activities: these are the single activities which must be completed before repetitive construction can commence, and sometimes during production, and are not part of the main production.
- Main repetitive tasks: these form the production lines (flowlines) performed by specialised crews in a technological sequence.

- Interlinked activities: single or continuous activities connected only at a specific location to the rest of production (usually at the start or finish).
- External activities: single activities not part of the main production process.

Mohr's general model

Mohr (1979) developed a simplified equation for the total production time T_T (illustrated in Figure 3.22) as:

$$T_T = T_1 + T_S + T_F + T + T_U + T_D + T_R - T_A - T_C + T_E \qquad (3.27)$$

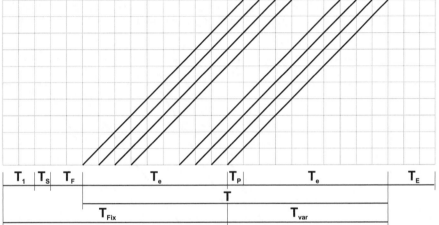

Figure 3.22 Mohr's general production equation

Where:

T_1 = Initiation (design) time.

T_S = Time to set up project before construction.

T_F = Time for foundations and footings to start of first flowline production.

T = Flowline production time using an ideal rhythmic balanced repetitive construction, where $T = T_e + T_P + T_f$

T_e = Running in time—the time to start production lines working, where

$$T_e = K(n-1) + \sum_{i=1}^{i=n} t_i$$

T_P = Time when all production lines are working.

T_f = Running out time—the time for production lines to stop working.

T_U = Additional time due to disruption arising from different quantities or rates of production in locations.

T_D = Additional time due to change in direction.

T_R = Additional time for external activities such as plastering, roof and other non-rhythmic construction—will only occur when the available time is shorter than required.

T_A = Reduction in time due to accelerating production after a change in the direction of production. This will only occur when all the following production lines speed up to match the rate of the initially accelerated production line.

T_C = The reduction in time which arises from cramming the production lines by reducing the time interval between them—will lead to interference.

T_E = Extensions of time for the project.

T_{Var} = Variable time, where $T_{Var} = T_P + T_e = K.m$

Peer's criticality

Peer highlighted the difference in approach between flowline and CPM regarding criticality. In flowline, continuity is an essential component of the criticality assumption. When production is balanced to the extreme, as in the flowline production in Figure 3.23, all work is critical.

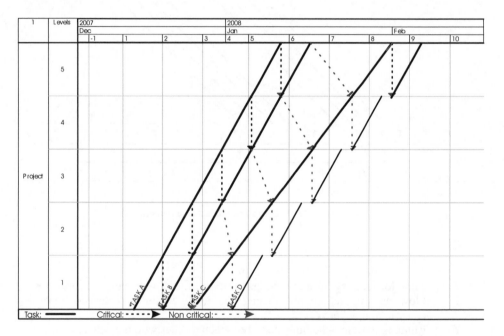

Figure 3.23 Criticality in flowline

When the process is unbalanced, construction time is determined by the slowest production line, which is then critical. The other activities are those determining the earliest possible start of this production line, and after its completion the finish of the whole project. All other activities have free floats, changing from section to section, depending on their distance from the critical path (Peer, 1974).

Peer's criticality is illustrated in Figure 3.23, where the final task is only critical in the final location due to the float which allows the activity to start earlier than optimal (for continuity) with no effect on the project duration. The criticality can be illustrated in a Gantt chart as in Figure 3.24, which shows the activities clearly in each location together with the logical relationships.

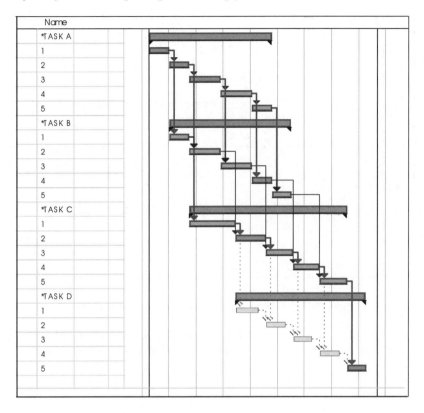

Figure 3.24 Peer's flowline criticality in a Gantt representation

Breaking production into segments

Mohr explored the relationship between time and production alignment. In the process he developed two fundamental operations to deal with a task with non-aligned production— for faster or slower production.

Non-aligned production causes interference with the following activities, thus delaying the project. The method for improving the project plan involves breaking the work into sections for both cases. In the case of slower production, the work is broken to allow multiple gangs to work simultaneously. In location-based management, this is called *splitting*. In the case of faster production, the work is broken into sections with a delay in between. In location-based management this is called non-paced or non-continuous. A third option, the default in LBMS, to increase the number of gangs in each location, was not specifically mentioned by Mohr. Figure 3.25 illustrates the two cases discussed.

Mohr noted that breaking the tasks into sections can only work if the work is carefully controlled:

> Where some of the finishing trades can work at a faster rate, but they are started too early, or their rate of production is not geared to their start date, breaks may occur to the production. When this occurs they might not return when they are required, or they might return in a fashion that does not suit following production lines (Mohr, 1979).

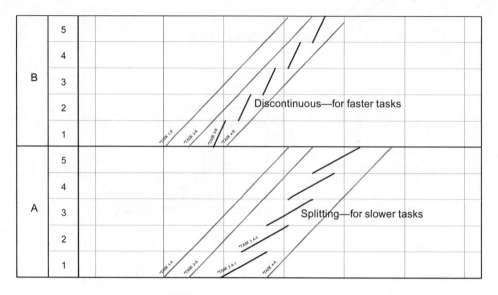

Figure 3.25 Breaking the work into sections to improve production

This effect is the first clear statement of the relationship between location-based planning and the production efficiency on site—an essential underlying principle in LBMS.

Mohr's criticality

Mohr used the concept of non-rhythmic construction to explain the concept of criticality. He defined *critical approach* (A_c) as the extra time between the finish of a production line in one section and the start of the next.

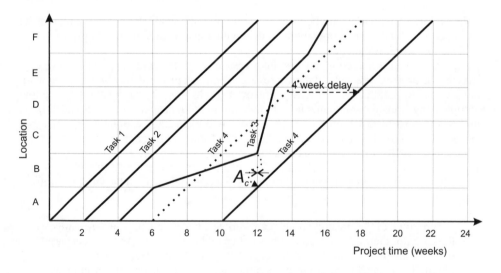

Figure 3.26 Effect of poor production control in a preceding task

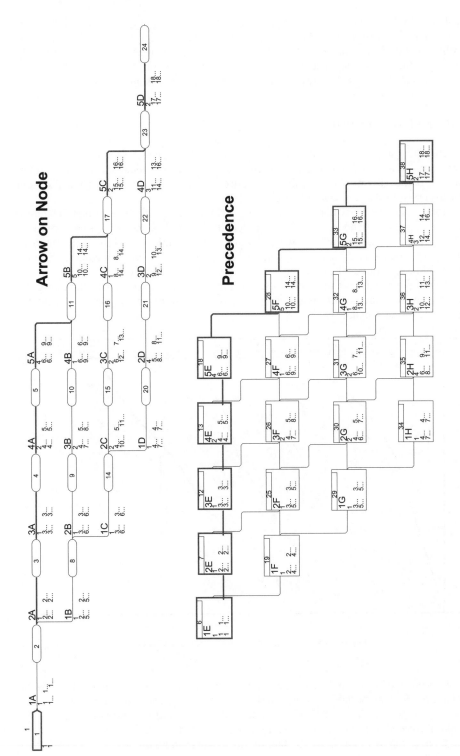

Figure 3.27 Mohr's demonstration schedule in both activity on the arrow and precedence modes

A_c cannot be negative in planning. As A_c can be calculated for each relationship in each location, it follows that "the most severe negative critical approach becomes the controlling point" in determining the location of the following task—assuming a totally rhythmic production is being resumed in the following task (Mohr, 1979: 124). The critical approach A_c may be seen delaying the planned start of the subsequent task in Figure 3.26.

Non-rhythmic construction

There are two causes of non-rhythmic construction. Work may be planned to be non-rhythmic, or it may become so during construction. Here we are concerned with the former. Mohr's use of the latter provides the first example of flowline being used as a control system and is discussed further on page 90.

Mohr recognised that not all production is rhythmic and analysed a demonstration schedule using arrow on node, precedence (Figure 3.27) and flowline formats (Figures 3.28 and 3.29). In this case, the durations were planned to be variable for tasks in different locations. This introduces a degree of complexity which illustrates the need for computerisation of the flowline technique when moving beyond rhythmic construction.

Mohr's point was that the CPM-based techniques, while capable of scheduling the network, provided neither the basis for continuous production nor an effective means for visualisation of the schedule.

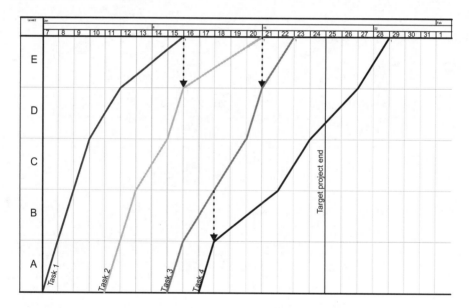

Figure 3.28 Continuous production—optimum duration

He argued that a simple bar chart could lead to assuming the project could be completed earlier than possible, if only the first activities for a task in the first location were sequenced—with each location then proceeding without regard for other locations. In the example illustrated, this would result in a duration of 15 days. A CPM analysis reveals a minimum duration of 18 days, using either of the networks in Figure 3.27. These are ladder or overlap networks. The precedence diagram most clearly shows the logical construction of a network which forces sequence in both locations and trades. This is a heavily constrained network and is labourious to build. The flowline equivalent is shown in Figure 3.29, also with a duration of 18 days. However, achieving continuous production (as is accepted by Mohr as a basic criterion of the flowline approach) requires a duration of 22 days (Figure 3.28). "Using this form of presentation it is easily seen where the work groups are discontinued to fit in with the work of other activities" (Mohr, 1979).

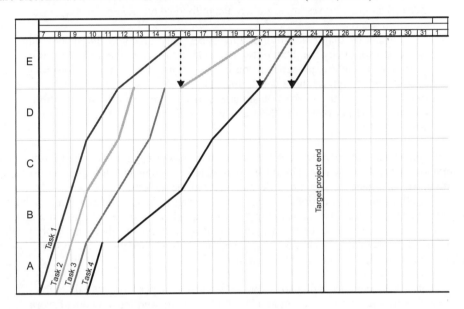

Figure 3.29 Discontinuous production—minimum duration

Buffers

Flowline relies on buffers in much the same way as line-of-balance. Actual production is likely to vary and buffers are required to reduce the incidence of negative criticality—the condition where Mohr's criticality calculation becomes negative during construction.

Work location analysis

Location is an essential component of both line-of-balance and flowline techniques. However, in the development of these tools, the emphasis was generally on repetition. Thus they were, and often remain, seen as dependent on repetition, so the location was in fact the numerical sequence of repetitive components such as housing units. Mohr clearly extended this into commercial construction and briefly discussed the selection of locations but, as with all previous work, his emphasis was on the activity or task.

The first detailed discussion of the significance of *location* as an important part of a project analysis can be found in Birrell (1980) who discussed *work location analysis*. Birrell had joined the growing number of researchers that found CPM methods unsuitable for construction planning and control, and the reasons highlight the need for study of locations. Following a description of what he saw as the failure of CPM to handle resource allocation, Birrell observed:

> Furthermore, in construction the resource allocation problem is compounded by the largely nonfungible nature of construction resources. This is a fundamental characteristic of construction resources which has been further rigidified by the construction unions. Also, productivity in construction tends to be from a squad made up from various resources rather than individual resources working in isolation from each other.
>
> The above major constraints almost certainly preclude the successful resource allocation by the [above] typical CPM approach to the topic. Put simply, the typical CPM approach to resource allocation is too simple for construction and tends to ignore these constraints. Thus CPM tends to be unsuitable to meet the needs of the construction situation. Also, there is a basic incompatibility of the CPM network with the heuristic model of the construction process (Birrell, 1980).

Birrell considered the appropriate solution was to consider each work squad (crew) as a continuous flow. All resources should be aggregated into squads, each intended to work continuously and tackling similar work in all parts of the building.

Birrell's logical assumption, and where he is clearly reflecting the extant research into repetitive construction, was that planning the work squads work to pass through the various locations in a consistent sequence was important for minimising the complexity of the construction process and the confusion of the participants. He had two aims: to simplify the planning and to improve the communication.

> This single sequence will also enable the project manager's or general contractor's site superintendent to build a "rhythm" of work and movements of work squads through the project. By this type of planning, he has the potential to create a work "pulse" for the project which can develop its own momentum to carry the work of all contractors to its completion (Birrell, 1980).

The key to Birrell's approach was the location. He found three major considerations for the construction planner when considering the physical locations in the project:

1. The vertical segmentation (Figure 3.30a).
2. The horizontal segmentation (Figure 3.30b).
3. The space available within the site, but outside the building, for material storage and handling.

The vertical and horizontal locations relate to the physical layout on the site and should not be confused with horizontal and vertical scheduling logic introduced by Thabet and Beliveau (1994) in their HVLS (page 69)—their method was not about physical but logical relationships and more closely represented the internal and external logic of LBMS.

It is interesting to note that Birrell considered the selection of appropriate locations critically important, and he also indicated the importance of having the right person make the selection. The person to undertake work location analysis had to be someone with analytical capacity and construction knowledge.[7]

---

[7] In LBMS this should be a team effort, but someone with analytic skills should be involved in the group.

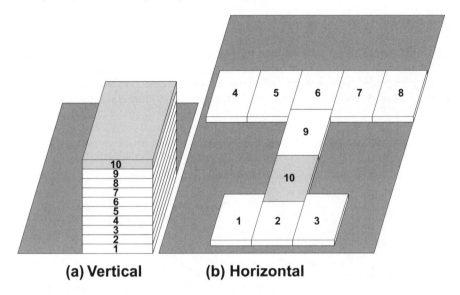

(a) Vertical   **(b) Horizontal**

Figure 3.30 Vertical and horizonal location sequencing (after: Birrell, 1980: 396)

Birrell recognised that vertical and horizontal location sequencing may also be mixed on certain projects, and that large projects are assemblies of smaller projects. From this it can be concluded that Birrell foresaw the need for a hierarchy of location sequences as large projects are divided. This hints at a location breakdown structure (as discussed in Chapter 5, page 125). He also used the term *Task* to refer to the work as it flows through locations.

Birrell's work was neither line-of-balance nor flowline, although he used the terminology of *flow lines*. In fact he dispensed with diagrammatic representation except to explain the concept of Task *work squads* 'passing through' a location (Figure 3.31).

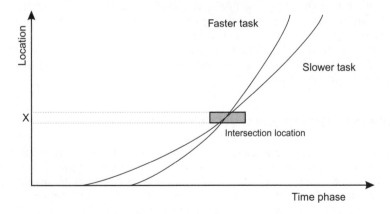

Figure 3.31 Crews 'passing through' a location (Birrell, 1980: p402)
With permission from ASCE.

Construction process matrix

Birrell (1980) constructed a matrix of work packages with work locations on the vertical axis and time periods on the horizontal axis (the sensitivity varies: half day, day or week, depending on project duration and complexity) such as Table 3.1. The work locations are sequenced as they will be built on site—"the locational analysis of the object to be constructed".

Table 3.1 Construction process matrix (after: Birrell, 1980: Figures 4, 7: Pages 398, 400)
Key: (3,9) = (Task type 3, work squad 3, in Location 9) is in time period 11

	1	2	3	4	5	6	7	8	9	10	11	12	13	14	15	16	17	18	19	20	21	22	23
10										1,10	2,10	3,10		4,10	5,10					6,10	7,10	8,10	9,10
9									1,9	2,9	3,9		4,9	5,9					6,9	7,9	8,9	9,9	
8								1,8	2,8	3,8		4,8	5,8					6,8	7,8	8,8	9,8		
7							1,7	—	3,7		4,7	5,7					6,7	7,7	8,7	9,7			
6						1,6	2,6	3,6		4,6	—					6,6	7,6	8,6	9,6				
5					1,5	2,5	3,5		4,5	5,5					6,5	—	8,5	9,5					
4				1,4	2,4	3,4		4,4	5,4					6,4	—	8,3	—						
3			1,3	2,3	3,3		4,3	5,3					6,3	7,3	8,3	9,3							
2		1,2	2,2	3,2		4,2	5,2					6,2	7,2	8,2	9,2								
1	1,1	2,1	3,1		4,1	5,1					6,1	7,1	8,1	9,1									

(In the original, cell 3,9 is circled.)

    Birrell compares his technique with the practical methods expounded by Horowitz in *Job-site management: an exercise in concurrency* (Horowitz, 1968) who reported on the construction of New York skyscrapers. Horowitz was reporting on the practical experience of HRH Construction Corporation—HRH grew from Starrett Brothers, so this is the same company that constructed the Empire State Building.

    Birrell cleverly used queueing theory to prepare the work for the construction crew. He identified that the construction process is made up of many flow lines (queues) each consisting of a work squad (a single server in queueing theory) moving through a series of locations (containers being processed by the single server) (Figure 3.32).

    Birrell's use of the term *flow line* is interesting as it is derived from the desire to emulate the manufacturing process using queueing theory. Similarly, he argued for *oscillating boundary paths* so that adjacent paths do not encroach on each other. These are clearly buffers as understood in line-of-balance or flowline. The term flow line has tended to stick and is often to be found in the location-based literature, including the integrated location-based methods which follow.

INTEGRATED LOCATION-BASED METHODS

The location-based management system requires a great deal more sophistication than the relatively simplistic models available from either line-of-balance, basic flowline or other variants of linear or repetitive scheduling. In reality, the modelling of construction projects requires the integration of CPM methods with repetitive scheduling methods.

    The integration of CPM and LoB was discussed earlier in this chapter. Of particular interest there was Arditi's work, as represented by Arditi et al. (2002). While the concept of integrating these methodologies was originally proposed by Schoderbek and Digman (1967), the concept was taken dramatically forward with the work of Alan Russell. Russell concentrated on solving the complexity of simulating real projects with logical structures. The resulting method was termed *representing construction* (with the associated software product RepCon) (Russell and Wong, 1993).

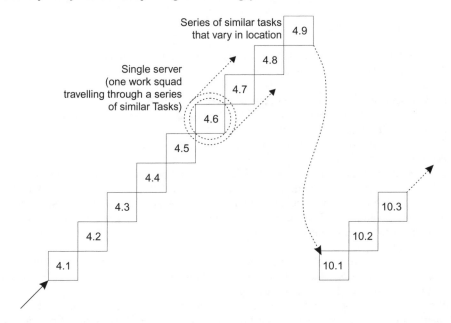

Figure 3.32 Queueing theory in construction in real world—applied to the construction process matrix. (Birrell, 1980: Figure 6, page 400, with permission from ASCE)

One of the fundamental changes involved at this time was the shift in focus from repetition to location. Stradal and Cacha (1982) referred to the space-time paradigm, clearing the way for a new emphasis on location to enable the development of better models.

Representing construction

Russell published a critically important new model for planning construction which he termed a *new generation of planning structures* and he used the title *representing construction*. This was the culmination of a decade of work. Prior to Russell's work, the general consensus of researchers was that linear methods were essentially graphical and not suitable to computerisation (Chrzanowski and Johnston, 1986). Russell and Wong (1993) changed all that by making a substantial contribution to the underlying methodology. These were intended to enable the computerisation of linear methods, but more importantly to move beyond mere repetitive construction to encapsulate non-repetitive construction.

By 1993, it was recognised that the graphical representation of the work crew was of *flow lines*. This is the terminology used by Russell, who's work may be categorised as *flowline*, although the resulting chart remained obscurely labelled as a linear planning chart, despite the projects represented being far from linear.

Apart from using work-location as the space descriptor, there were two major framework contributions made by Russell and Wong. These were:

• The definition of model attributes
• The definition of a family of planning structures.

It is safe to say that these together represent the first comprehensive location-based planning system, especially as the model attributes included a hierarchy for locations.

Definition of model attributes

Russell and Wong (1993) listed nine attributes desirable for planning, scheduling and updating (control) of construction projects.[8]

1. CPM (precedence) must be included to present functionality as a sub-set of the new model. Therefore a super-set of CPM is required, mathematically based, with two extremes—"traditional CPM and pure flow lines".
2. The terminology of CPM should be adhered to, such as *activity, precedence* or *logic relationship* and *float*.
3. All precedence relationships (FS, SS, FF, SF) should be available. Furthermore, "logic relationships among different work locations (not necessarily contiguous) should be treated (the so-called space buffer)", predecessors and successors may not share all (or any) work locations, and "the concept of precedence should be generalised so that logic linkages among the components of planning structures can be specified with a single relationship.[9]
4. Variable production rates between locations must be accommodated and the construction process need not be constant. This should allow for changing scope, complexity, site conditions and learning effects.
5. Work continuity, the delaying of commencement such that work may proceed without interruption (a requirement of most previous repetitive or linear scheduling methods including line-of-balance and flowline) should be optional.
6. The concept of work locations (now central to the model) should be broadened to include both physical and virtual (procedural) locations—for example design stages or procurement could be considered locations. A hierarchy of locations (to include both horizontal and vertical elements, should be included. This would also allow finer detail for microplanning.
7. An activity structure should have properties such as:
   • work continuity constraints
   • unlimited predecessor and successor relationships
   • location ordering
   • crewing (multiple or variable—number of locations which can be worked simultaneously)
   • pre-planned work interruptions
   • variable production rates.
8. Control (updating) should allow for changes to work location sequencing, crews and precedence relationships, while maintaining as much of the original schedule as possible. In addition repetition should be exploited.
9. More formats for communication of the schedule should be used, including linear planning chart (flowline), network diagram, bar chart and matrix chart.

The list was advanced by Russell et al. (2003) with a list of properties required by activities (Item 7), including:

• Time at each location
• Resource assignments
• Calendar

---

[8] The sequence has been altered for clarity.
[9] This requirement of generalising relationships is the key idea for creating a location-based super-set for CPM.

- Responsibility code
- Associations with product
- Cost
- Methods and as-built views of the project
- Date and float constraints
- Date and float values for early, resource levelled and late date scenarios
- Project records (for example, photos, videos, correspondence, change orders, etc.).

Definition of a family of planning structures

Even more important than the attributes, Russell's planning structures provide a ground-breaking new approach to thinking about the scheduling of activities in construction. They provide, for the first time, a structure for modelling the real flow of work and the complex inter-relationships which location-based scheduling allows. Russell and Wong (1993) expound a family of five core activity structures. These structures, together with the LBMS equivalents (see Chapter 5) are expanded below and also illustrated in Figure 3.33.

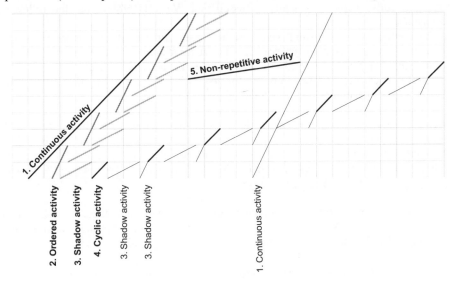

Figure 3.33 Five core activity structures

1. **Continuous activity**—work that must be executed in a specified location sequence with required continuity. [Corresponds to continuous Layer 3 logic in LBMS].
2. **Ordered activity**—work that must be executed in a specified location sequence but which can be interrupted. [Corresponds to discontinuous Layer 3 logic in LBMS].
3. **Shadow activity**—work that can be executed in any sequence, which can be interrupted and which can be executed simultaneously (it has no resource constraints). There are no implied logic links between work locations. It 'shadows' its predecessor—this is equivalent to the default assumption in CPM without resource limits. [Corresponds to workable backlog in LBMS].
4. **Cyclic activity**—work where locations of its successors are predecessors to later locations of itself. [Corresponds to Layer 4 logic in LBMS].

5. **Non-repetitive activity**—individual discrete activities. This corresponds to traditional CPM. [Corresponds to Layer 5 logic in LBMS].

Russell et al. (2003) added the following additional non-core types:

6. **Hammock activity**—an activity bridging from the start of one activity to the end of another activity and containing only time properties. [Corresponds to a summary task in LBMS].
7. **Start and finish milestones**—dummy activities which mark the start or completion of milestones.
8. **Derived activity**—for hierarchical scheduling. Hierarchical scheduling allows finer detail to be scheduled in each location within a higher level summary activity. [Corresponds with detail scheduling in LBMS].
9. **Summary activity**—to allow arbitrary groupings, independent of logic and/or hierarchical structure. [Corresponds to a summary task in LBMS].

Normal CPM precedence relationships are not sufficient to describe the relationships between repetitive activity structures. Russell and Wong (1993) define the following relationships.

For continuous, ordered and cyclic activities, precedence relationships are implied for each activity between locations in sequence. The relationships include both time and location lags (buffers). These are relationships internal to the activity structure.

Between activities, relationships are formed which are external to the activity structure. Two forms of relationship are formed:

• **Typical**—used to link between activity structures within locations, using CPM precedence links (FS, SS, FF, SF). [Corresponds to Layers 1 and 2 logic in LBMS].
• **Non-typical**—used to link from one specific location in a activity structure to a specific other location in another activity structure, or to link the same to a non-repetitive activity. [Corresponds to Layer 5 logic in LBMS].

Russell et al. (2003) note that for continuous, ordered and cyclic activities, the internal relationship $FS \geq 0$. Whereas external relationships may be $FS \pm lag$. They also noted that lags may be considered either time or spatially related.

Table 3.2 Russell's example of simplification using *representing construction*

	CPM	RepCon
Activities	720	18
Links within floors	1400	35
Links between floors	351	9

One of the most impressive features of Russell's model is that it dramatically simplifies the scheduling task using a CPM mathematical engine. For example, Russell makes this point using the example of a 40-storey commercial tower, with 18 activities per floor, 35 logic relationships connecting activities per floor and nine links between activities between floors. Using CPM this would require 720 activities and 1400 links within floors and 351 between floors. This would reduce to 18, 35 and nine respectively (Table 3.2).

Repetitive scheduling method (RSM)

While being one of the most recent integrated location-based methodologies, and with the aim to integrate previous work into a single generalised model, the *repetitive scheduling method* (RSM) from Harris and Ioannou (1998) failed to achieve its aims. The model used flow lines and develops a new approach to activity logic using two important new concepts:

- **Controlling sequence**—the chain, or sequence of activities, that establishes the minimum project duration while maintaining logic and continuity constraints. Like the critical path, this represents the shortest path through the flowline schedule, passing from activity to activity through specific locations as the logic dictates.
- **Control points**—the specific locations where the controlling sequence moves from between activities.

The method requires the construction of detailed CPM schedules for each location, such as a unit or floor, and then the construction of a flowline schedule using the logic diagram within locations (horizontal logic) and continuity constraints between locations (vertical logic). The creation of the RSM schedule is an iterative process, with the first attempt usually impractical and requiring several adjustments to logic, resources and discontinuities to achieve an optimum schedule.

The method clearly has roots in both the HVLS (see page 69) and the CPM/LoB (page 69), both line-of-balance methods, utilising the concepts of horizontal and vertical logic and using the available float in the horizontal logic schedules.

The model returns to line-of-balance approaches which predate Russell's *representing* construction model (see page 85). As such, the model does not achieve its aim of providing a generalised model. However control points and a controlling sequence is an important contribution to understanding criticality in flowline.

Harris and Ioannou's criticality

Harris and Ioannou correctly argue that the determination of a critical path does not apply in the traditional way due to the additional resource continuity requirement. They prefer the concept of controlling sequence, while recognising that this is dependent on modelling assumptions such as continuity, resources, etc. Figure 3.34 illustrates a controlling sequence through a demonstration project provided by Harris and Ioannou (1998). They note that the controlling sequence can change depending on resource decisions (such as doubling the crews to task 2 in Figure 3.34).

The controlling sequence is similar to the controlling activity path explored by Harmelink and Rowings (1998)—although their use is in true linear projects such as road or rail. As such, it is only worth noting here, however the analysis of activity types as linear, block and bar (all of which only have meaning in a linear project) is worthy of mention.

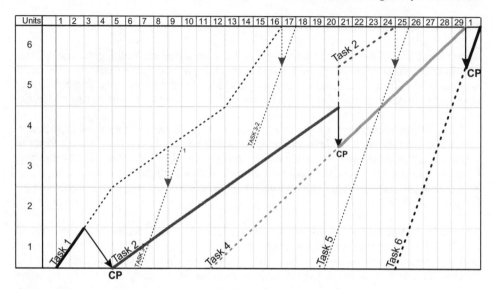

Figure 3.34 Controlling sequence passing through control points
(after Harris and Ioannou, 1998, with permission from ASCE)

DEVELOPMENT OF LOCATION-BASED PLANNING IN FINLAND

Finland has long traditions in scheduling research. Research commenced in 1970s when the Ministry of Finance started a target budgeting project, which had the goal of improving cost estimates and schedule control for building and civil engineering projects managed by the government. The end result of the project included practical guidelines for cost estimating, control and scheduling. Scheduling guidelines were close to international standards of that time: small, routine projects were planned with Gantt charts and large, complex projects were scheduled using CPM. In linear projects the recommendation was to use linear scheduling (an application of location-based planning, as discussed earlier in this chapter) and schedule control was recommended to be done based on the completion rate of tasks and by controlling schedule deviations of individual activities. Right from the start, control was seen in Finland as being integral to location-based planning (see the discussion of the development of location-based control in Finland on page 115).

In Finland, location-based planning methods have been used widely in construction since 1980s. The methods were brought to Finland and adapted to commercial construction by professors Kankainen and Kiiras from Helsinki University of Technology (Kiiras, 1989; Kankainen and Sandvik, 1993). In academic research tests it was established that the use of modified flowline planning increased productivity and decreased waiting hours for direct labour and for subcontractors (Toikkanen 1989, Venermo, 1992).

New research efforts to improve the scheduling skills of the Finnish industry were started in the end of 1990s by Professor Kankainen's research group. The results included tools such as task planning (Junnonen, 1998), project control charts, checklists to assess a schedule's feasibility (Kankainen and Kolhonen, 2005) and new contracts to support location-based control. The research results were used in a software development project to design a new software able to be used as a planning and control tool.

As a result of two decades of research, a complete schedule planning and controlling methodology based on managing schedule risk has been developed. Because of aims also shared with lean construction, namely reduced waste and interference, the Finnish results and tools have been presented at IGLC conferences (Kankainen and Seppänen, 2003; Junnonen and Seppänen, 2004; Soini et al., 2004; Seppänen and Kankainen, 2004). The features of the resulting software—DynaProject[10]—have been described in Kankainen and Seppänen (2003).

One of the most important and long term results of this effort was building productivity databases together with the largest contractors in Finland (Olenius et al., 2000). These productivity databases are continuously updated and used by all parties. They include work method descriptions and good target level productivity information for labour and material consumption information. The database is maintained by the Confederation of Finnish Construction Industries.

Research was undertaken by the Construction Economics Laboratory at Helsinki University of Technology. This work, described in more detail on page 115, arose from poor economic results on construction projects and was driven by industry. The research identified scheduling problems which arose from incorrect duration estimates, incorrect logic between tasks, multiple tasks starting at the same time, discontinuities of work arising from a lack of available work and a tendency to shift problems to the end of a project. Based on these results, companies assessed that the most important improvement area would be the planning and control system—improving both the quality and implementation of schedules.

Chapter 4 (page 115) includes the results of this research which, while driven by location-based planning, essentially found that good location-based planning was insufficient, and that location-based control was essential—thus we have included the results in the control chapter.

---

[10] DynaProject has subsequently evolved to become Control 2009 from Vico Software.

REFERENCES

Adamiecki, K. (1909). "Metoda wykreslna organizowania pracy zbiorowej w walcowniach". In Heidrich, Z. (Ed.) *O nauce organizacji: Wybor pism (1970)*. Warszawa, Panstwowe Wydawnictwo Ekonomiczne.

Adamiecki, K. (1931). "Czy Nauka Organizacji Przyczynia sie do Poglebiania Kryzysu i Bezrobocia? (L'organisation scientifique contribute-t-elle a approfondir la crise et augmenter le chomage?)". *Przeglad Organizacji*. **4**(12), Warsaw.

Al Sarraj, Z.M. (1990). "Formal development of line-of-balance technique". *Journal of Construction Engineering and Management* **116**(4): 689–704.

Archiwum Panstwowe w Lublinie (2009). *Harmonogram rzeczowo—finansowy odbudowy budynku Wojewódzkiego Archiwum*. Lublin Local Government. Retrieved May, 2009, from

http://www.lublin.ap.gov.pl/wystawy/displayimage.php?album=topn&cat=-20&pos=3

Arditi, D. (1988). "Linear scheduling methods in construction practice". *Proceedings of the Seventh Annual CMAA Conference*. Rancho Mirage, CA, CMAA.

Arditi, D., Sikangwan, P. and Tokdemir, O.B. (2002). "Scheduling system for high rise building construction". *Construction Management and Economics*. **20**(4): 353–364.

Armon, D. (1974). "Scheduling diagram: A technique for scheduling linear projects in Civil Engineering". *In the field of building*, Bulletin 215, Building Research Station, Technion–Israel Institute of Technology, Haifa.

Birrell, G.S. (1980). "Construction Planning—Beyond the Critical Path". *Journal of the Construction Division, American Society of Civil Engineers* **106**(3): 389–407.

Carmody, J.P. (1930). "The Empire State Building: X. Field Organisation and Methods". *The Architectural Forum,* LII: 495-506.

Chrzanowski, E.N. and Johnston, D.W. (1986). "Application of Linear Scheduling". *Journal of construction engineering and management* **112**(4): 476–491.

Fazar, W. (1959). "Progress reporting in the special project offices". *Navy Management Review* **1959**(April): 9–15.

Fink, N.E. (1965). "Line of balance gives the answer". *Systems & Procedures Journal*. **9**(14).

Gehringer, A.C. (1958). "Line-of-Balance". *Navy Management Review* **4**(3): 910.

Green, S.D. (1999). "The missing arguments of lean construction". *Construction Management and Economics* **17**(2): 133–137.

Green, S.D. (2002). "The human resource management implications of lean construction: critical perspectives and conceptual chasms". *Journal of Construction Research* **3**(1): 147–166.

Harmelink, D.J. and Rowings, J.E. (1998). "Linear Scheduling Model: Development of Controlling Activity Path". *Journal of Construction Engineering and Management* **124**(4): 245–341.

Harris, R.B. and Ioannou, P.G. (1998). "Scheduling projects with repeating activities". *Journal of Construction Engineering and Management* **124**(4): 269–278.

Horowitz, S. (1968). "Job-site management: an exercise in concurrency—HRH Construction Corporation". Supporting Turner, H.S. "Construction site management", in *4th CIB conference*.

Huang, R.Y. and Sun, K.S. (2006). "Non-Unit-Based Planning and Scheduling of Repetitive Construction Projects". *Journal of Construction Engineering and Management* **132**(6): 585–597.

Junnonen, J-M. (1998). *Tehtäväsuunnittelu ja laatupiiriohjattu tuotannonohjaus.* (Task planning and quality circles in production control). Licentiate Thesis, Construction

Economics and Management, Department of Civil and Environmental Engineering, Helsinki University of Technology. Espoo, Finland.

Junnonen, J-M. and Seppänen, O. (2004). "Task planning as a part of production control". *Proceedings of the 12th International Conference of Lean Construction (IGLC12)*. Elsinore, Denmark

Kankainen, J. and Kolhonen, R. (2005). *Rakennushankkeen aikataulusuunnittelu ja valvonta*. (The schedule planning and control of a construction project). Rakennustieto Oy. Helsinki, Finland.

Kankainen, J. and Sandvik, T. (1993). *Rakennushankkeen ohjaus*. (Controlling a construction project). Confederation of Finnish Construction Industries, Rakennustieto Oy, Helsinki. Finland.

Kankainen, J. and Seppänen, O. (2003). "A Line-of-Balance based schedule planning and control system". *Proceedings of the 11th International Conference of Lean Construction (IGLC11)*. Blacksburg, Virginia.

Kenley, R. (2004). "Project micromanagement: Practical site planning and management of work flow". *Proceedings of the 12th International Conference of Lean Construction (IGLC12)*, Helsingør, Denmark, 194–205.

Kiiras, J. (1989). *OPAS ja TURVA, Erityiskohteiden työnaikaista ohjausta palveleva aikataulu- ja resurssisuunnittelu*. (A schedule and resource planning system for the implementation phase of special projects). Helsinki University of Technology. Construction Economics and Management Publications.

Lumsden, P. (1968). *The Line-of-Balance method*. Pergamon Press Limited: Industrial Training Division, London, 71.

Mattila, K.G. and Abraham, D.M. (1998). "Linear scheduling: past research efforts and future directions". *Engineering Construction and Architectural Management* **5**(3): 294–303.

Marsh, E.R. (1975). "The Harmonogram of Karol Adamiecki". *The Academy of Management Journal* **18**(2): 358–364.

Mohr, W. (1979). *Project Management and Control (in the building industry)*. Department of Architecture and Building, University of Melbourne. 2nd edition, 210p.

Mohr, W. (1991). *Project Management and Control*. Department of Architecture and Building, University of Melbourne. 5th edition.

NBA (1968). *Programming House Building by Line of Balance*. The National Building Agency, London: 24.

NAVEXOS (1962). *Line of balance technology: a graphical method of industrial programming*. **P1851**. Department of the Navy, Office of Naval Material. Washington, D.C.

Olenius, A., Koskenvesa, A. and Mäki, T. (eds.) (2000). *Ratu aikataulukirja 2001*, Tampere, Finland, Confederation of Finnish Construction Industries RT and Rakennustietosäätiö.

Peer, S. (1974). "Network Analysis and Construction Planning". *Journal of the Construction Division, ASCE*, **100**(CO3): 203–210.

Reda, R.M. (1990). "RPM: repetitive project modelling". *Journal of Construction Engineering and Management* ASCE, **116**(2): 316–330.

Russell, A.D. and Wong, W. (1993). "New Generation of Planning Structures". *Journal of Construction Engineering and Management* **119**(2): 196–214.

Russell, A.D., Udaipurwala, A. and Wong, W. (2003). "A Generalised Paradigm for Planning and Scheduling". Proceedings, *Construction Research Congress*. Honolulu, Hawaii, 19–21 March. CD-Rom, 8 pages.

Schoderbek, P.P. and Digman, L.A. (1967). "Third generation, PERT/LoB". *Harvard Business Review* **45**(5): 110–110.

Selinger, S. (1973). *Method for construction process planning based on organisation requirements*. Faculty of Civil Engineering. Haifa, Technion-Israel Institute of Technology. Doctor of Science. In Hebrew with English synopsis.

Selinger, S. (1980). "Construction planning for linear projects". *Journal of the Construction Division,* ASCE **106**(CO2): 195–205.

Seppänen, O. and Kankainen, J. (2004). "An empirical research on deviations in production and current state of project control". *Proceedings of the 12th international conference of Lean Construction (IGLC12)*. Elsinore, Denmark.

Shreve, R.H. (1931). "The economic design of office buildings". *The Architectural Record.* pp339-360

Shreve, R.H. (1930). "The Empire State Building Organization". *The Architectural Forum* **LII**(6): 771–788.

Soini, M., Leskelä, I. and Seppänen, O. (2004). "Implementation of Line-of-Balance based scheduling and project control system in a large construction company". *Proceedings of the 12th International Conference of Lean Construction (IGLC12).* Elsinore, Denmark.

Stradal, O. and Cacha J. (1982). "Time space scheduling method". *Journal of the Construction Division,* ASCE **108**(3): 445–457.

Suhail, S.A. and Neale, R.H. (1993). "CPM/LoB: New methodology to integrate CPM and Line-of-Balance". *Journal of Construction Engineering and Management* **120**(3): 667–684.

Thabet, W.Y. and Beliveau, Y.J. (1994). "HVLS: Horizontal and Vertical Logic Scheduling for Multistorey Projects". *Journal of Construction Engineering and Management* **120**(4): 875–892.

Thabet, W.Y. and Beliveau, Y.J. (1997). "SCaRCSpace-constrained resource-constrained scheduling system". *Journal of Computing in Civil Engineering* **11**(1): 48–59.

Toikkanen, A. (1989). *Erityiskohteen aikataulun laadinta ja työnaikainen ohjaus.* (Scheduling and control of a Special Construction Project). Master's Thesis, Construction Economics and Management, Department of Civil and Environmental Engineering, Helsinki University of Technology, Espoo, Finland.

Venermo, T. (1992). *Pienen erityiskohteen tuotannon suunnittelu ja ohjaus.* (Control and planning of a small special construction object). Master's Thesis, Construction Economics and Management, Department of Civil and Environmental Engineering, Helsinki University of Technology, Espoo, Finland.

Wikipedia, (2008). "Karol Adamiecki". http://en.wikipedia.org/wiki/Karol_Adamiecki. accessed 15th November 2008.

Willis, C., (Ed.) (1998). *Building the Empire State*, W.W. Norton and Company with the Skyscraper Museum. New York.

Chapter 4

Approaches to planning control

INTRODUCTION

If assembling a construction project schedule prior to construction was the end of the planning process, then most construction planning managers would be contented people. Unfortunately, they often are frustrated and the reason is simple: no one follows their plan. Throughout the world, the construction industry has a problem with plans that are created, hung on the site shed wall—and then widely ignored. Sometimes they are monitored, sometimes they are updated, but rarely are they followed.

Research in Finland in the 1980s discovered that contractors were struggling to make a profit due to poor quality planning, in terms of content and quality, and poor implementation of those plans and a lack of production control. The production control problems could be seen through the incorrect use of resources, poor procurement and general failure to completely finish work. The solution is to develop comprehensive methods of control.

This chapter is about the historic development of systems which are designed from the start to create the plan to be followed, which track deviations from that plan and which forecast the consequences of deviation. Finally they allow corrective action to be taken and managed as per the original plan. These are planning control systems (as distinct from other control systems such as quality control, cost control, etc.). Along the way we will consider the activity- and location-based approaches to planning control, the way they have been implemented and highlight the intended outcome of planning control systems. We will also examine the contribution of Lean Construction theory, and in particular the Last Planner method, which has as its dominant purpose "causing events to conform to plan".

It is very likely that planning systems have developed out of a need to control construction work. Thus one view could be that this book should have first looked at the development of methods to control construction production (production control systems) and then moved to the planning systems which were developed to support that need. However, planning control systems cannot stand alone and are symbiotic with the planning system. Indeed, when considering the development of location-based methods in Finland, the authors found it extremely difficult to separate the chicken-and-egg sequence of planning and control as it emerged through research.

While it is obvious that planning systems can be used to control projects, it is less obvious that they were designed for that purpose. This is partly because, while they are designed to enable control, most of the literature concentrates on planning and leaves control almost as a natural consequence—there is perhaps a natural inclination to assume that good planning is, in itself, sufficient. There has been relatively little discussion in the literature of the control function itself, despite this being at the forefront of Kelley and Walker's (1959) thinking when they developed the critical path method—primarily as "the basis for a system for management by exception".

We will attempt here to extract the discussion of control systems from the planning and scheduling literature already covered, thus providing an opportunity to consider this vital topic independently. This focus is important if we are to properly recognise the importance of control to any management system and in particular the location-based management system. It is clear from a review of the literature that control is vital, but has received insufficient attention. This will be redressed in Section Three of this book, with the development of a location-based control system (Chapters 8 to 10).

This chapter will move through the control theory sequentially, but it is also grouped according to logical structures. It will start with a brief discussion of the theory of control, particularly as it applies to construction management. This will lead to consideration of various control systems in the dominant activity-based methodology.

Earned value performance management (EVPM) is discussed, as this is a generic control method which is independent of activity- or location-based methodologies. This powerful but imprecise tool measures deviation from planned production rates. However it is not able to force production to follow a plan. The effectiveness of CPM in achieving control is discussed, however such a discussion cannot do justice to the efforts of practitioners to make activity-based planning (CPM) successful on construction projects—this must be left to dedicated works such as Woolf (2007).

The impact of lean production theory to construction is covered as, while not strictly a control system in general management, lean has largely been adopted as a control strategy in construction. The tools and techniques, while having value for consideration during project planning, are heavily aimed at controlling the progress of the work. In particular, the Last Planner technique is a planning intervention specifically designed to ensure the effective control of projects. Lean theory applied to construction has become known as *lean construction*. This body of knowledge will be explored for relevance to planning control in construction, as the lean philosophy has an inherent emphasis on workflow and a desire to minimise production waste. These are important principles underlying the location-based management system.

The chapter concludes with the development of control systems in Finland and the evolution of location-based techniques—largely driven by the need to plan and control the rate of production to ensure successful project delivery.

What is control

Control is a very practical concept—one which every parent intuitively understands. Control is the mechanism by which a system is monitored and its behaviour corrected to ensure that performance is as planned.

There are many valid reasons for project performance to vary from the original plan, such as the plan being incorrect, the plan being misunderstood or technical problems thwarting the plan. There are also invalid reasons, such as no attempt being made to actually follow the plan, poor procurement decisions, incorrect allocation of resources or failure to complete work. Regardless of the reason, "the detection and correction of variations from the original plan are essential parts of the control process" (Woodgate, 1964: 3).

The control procedure is to measure performance, compare this with the original plan, and adjust plans, schedules and even objectives so that the project may most closely match the original plan.

Before addressing the development of control systems in construction scheduling, it is worthwhile noting that basic control theory was derived from cybernetics in 1948 (Wiener, 1996), which deals with "…an essential unity of the set of problems centring about communication, control, and statistical mechanics, whether in the machine or in living tissue".

Cybernetics comes from the Greek for steersman. The earliest form that may be identified is that of the governor in mechanical systems, described in 1868. Wiener (1996) used the analogy of the steering engines of a ship as an early and well-developed form of feedback mechanism. Cybernetics has developed into a rigorous mathematical and scientific body of knowledge to society's great benefit. It has provided the theory for much of what we take for granted, such as radio (noise removal), thermostats, automatic piloting systems, self-propelled vehicles, ultra-rapid computing machines, prosthetics and bionic ears.

The feedback mechanism of cybernetics may be found in general management theory, and in particular construction management, as the control cycle.

Control systems for construction management

Circular feedback processes are fundamental to the successful management of construction projects. The normal terminology for cycles is drawn from managerial sources: 'Shewhart's cycle' (Shewhart, 1939), 'Deming's cycle' (Deming, 1986) and terms such as PDCA (plan, do, check, act), 'control loop', 'monitoring cycle', or even 'Single Loop Learning' (Stacey, 1996). Betts and Gunner (1993) differentiate between homoeostatic systems with feedback loops and cybernetic systems with adaptive standards "where corrective action can influence both the input to a system and the process within the system itself". Whatever the terminology or the model, the process is one of setting a plan, executing performance, monitoring performance against the plan, initiating corrective action and repeating the cycle. Figure 4.1 shows the typical variant for the PDCA cycle in construction production.

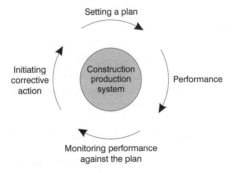

Figure 4.1 The control cycle (PDCA)

The important thing to remember with the control cycle is that it only functions in a supportive environment. When management does not initiate or maintain the cycle, feedback fails. In these circumstances, all too common with regard to project time performance management, performance is only as good as the first or latest plan, which can be based on analysis but more often emerges as the end result of intuition and discussions.

THE DEVELOPMENT OF ACTIVITY-BASED PLANNING CONTROL

Activity-based planning systems were designed with planning control in mind, but the concentration very quickly turned to and remained on the technology of planning and scheduling. It is almost as if it was assumed that covering one enabled the other. In reality, as we shall see, controlling an activity-based schedule is a complex task.

Kelley and Walker (1959) were focused on the problems of control for construction projects right from the outset. They felt that "the plan should form the basis of a system for management by exception... Under such a system, management need only act when deviations from the plan occur."

The program evaluation and review technique (PERT) was, almost by definition, a control system. Its emphasis was on forecasting the likelihood of success. In describing the early development of PERT, the US Navy reported in 1959 that "lack of adequate control information is a major handicap to the Navy Secretariat". Control was considered a major requirement for effective management systems and it may be considered almost the underlying purpose of the planning system:

> In order to discharge his responsibility effectively, the Under Secretary must be provided with a means of evaluating the performance of the technical bureaus and offices. This requires that he be furnished timely and complete basic data on the status of programs in all major functional areas (Thomsen, 1959).

There were underlying differences in the philosophies between CPM and PERT as described in Chapter 2. These were principally that PERT was a network schedule "from time now" to the end of the project and which gave an estimate of the probability of achieving the project end date and any milestones along the way. In contrast, CPM was always about a complete schedule from the start to the end—and all milestones and end dates were dictated by the schedule. This difference was directly reflected in the control strategies.

Control strategies in PERT

We start with PERT because PERT was, to a greater extent than CPM, functionally more about control and therefore effort was made to develop control strategies earlier.

Fazar (1959) identified that one of the problems with management practices, before the design of PERT, was that they failed to achieve control objectives:

> No one of these tools, singly or in combination, furnished the following required information:
> * Appraisal of the validity of existing plans and schedules for meeting program objectives;
> * Measurement of progress achieved against program objectives; and
> * Measurement of the current outlook for meeting program objectives.

The first point is a planning objective, but the other two relate to CPM and PERT objectives for controlling projects.

Fazar's team, when designing PERT, argued that time was the only practical way to control a project, as cost, resources and technical progress were too difficult to measure and relate to the progress of the work[1]. Thus PERT-Time was developed specifically to control projects to meet deadlines according to plan. In the early stages, this plan may not have been calculated by the use of PERT. Generally, later, this would be the case. The process of control consisted of recording the progress of activities and assessing the probability of achieving milestones.

Apart from the emphasis on reporting the probability of achieving milestones, the process of control in PERT and CPM were largely the same.

---

[1] Cost was incorporated in the PERT model, but was included as a property of the schedule and in no way represented the accounting cost systems. This necessary separation between cost accounting systems and the costing of planning and control remains true today.

Project progress control

There are many reasons for a project to deviate from its plan and Woodgate (1964: 187) outlined the following factors:

- Changes in technical specification of the project
- Changes in the date objectives for completion
- Changes in operating policies (eg. relative priorities)
- Revised activity time estimates
- Reassessment of resource requirements for individual activities
- Inability to utilise resources as originally planed
- Inaccurate planning of activity relationships
- unexpected technical difficulties.

The project control cycle advocated by Woodgate (Figure 4.2) describes a model which includes two loops—an evaluation of the plan and evaluation of the project performance— both operating continuously to capture project deviations.

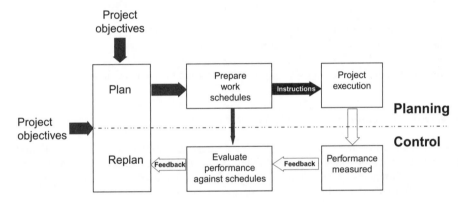

Figure 4.2 Woodgate's control cycle (after Woodgate, 1964: Figure 58)

The following progress information can be expected from the model:

1. Revision to the methods—where the methods used or the sequence of work change. This could be caused by both internal and external factors.
2. Revised durations—as the project proceeds, more is known about the duration for activities. In PERT, the original three time estimates may narrow or become one.
3. Revision of external restraints—the required milestones may change.
4. State of project progress—the list of which milestones are completed, what activities have started and finished, and which activities have started but not finished and how much work is outstanding[2].

Woodgate described two methods for calculating progress of a project:

---

[2] This is an important point, as the early practice was to record the work remaining to be completed. Recent practice has become to record the work done (except for short activities where the model is not sensitive). This makes a significant difference in modeling a project to completion. Many purists still insist on the former practice if the technology supports it (for example, Woolf, 2007: 319).

1. Modifying the network and undertaking the standard calculation.
2. Varying the calculation without changing the network.

Modifying the network and undertaking the standard calculation

The network is altered to represent the current situation. It should show the amount of work to be completed and the time elapsed since the project commenced. This is done by altering the duration of all activities to the amount of time remaining to complete the activity. In the case of a completed activity, the duration becomes zero. To take account of time passed, a dummy activity is inserted before the start of the project with a duration equal to time elapsed.

The network may now be calculated with the revised network and indicating the new expected completion and, in the case of PERT, the likelihood of achieving milestones.

If the original completion date is retained for the backward pass, then float calculations will reveal negative float for those activities which are delaying the completion of the project later than the original plan. Woodgate listed the following outputs from the control process:

1. The delay to the end of the project.
2. The critical path activities which contribute to the delay.
3. The change in the critical path from the original plan.
4. The progress position of activities (given by positive or negative float).

Varying the calculation without changing the network

Progress data effects the calculation of event earliest dates, thus enabling recalculation of the earliest project end date. If the end date is not prescribed, then the backward pass can recalculate latest dates. Where there is a set end date, then latest dates do not change. Float is effected differently depending on the end date constraints, as float is the difference between the earliest and latest dates.

In calculating the schedule, progress data is treated as follows (Woodgate 1964: 199):

- Event achieved—when an event is achieved its completion date is recorded in both earliest and latest dates.
- Activity completed—the actual date is used for the calculation. However this depends on predecessors. If a predecessor finished later then this later, date is used. If a predecessor is incomplete then this is taken into account in the later activities.
- Activity started—the start time is used as the early and late start times. The activity duration is used except where there is progress.
- Activity time outstanding—the amount of time required to complete the work from the 'time now' is used as the remaining duration and is added to the current date to give the expected completion date.

PACE—Program analysis, control and evaluation

PACE or PERT program analysis, control and evaluation, is a methodology for control of a project with uncertainty—which PERT handles well due to its probabilistic orientation (Pillai and Tiwari, 1995). PACE is primarily a structured approach to management consisting of a planning process and then a review of project performance. This method is important because it shows a connection with lean construction (discussed later in this chapter) and particularly the Last Planner methodology. Essentially, the method provides a structured process of review and discussion. The methodology is presented in Figure 4.3.

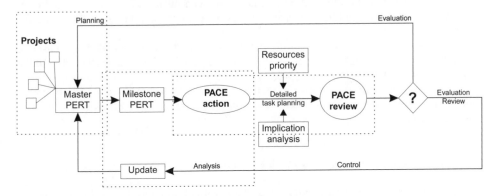

Figure 4.3 PACE methodology (after Pillai and Tiwari, 1995. With permission from Elsevier).

Control strategies in CPM

In CPM the concept of control as 'management by exception' was established by Kelley and Walker (1959). This approach differentiated CPM from PERT, in that it involved an emphasis on examining the critical path and concentrating efforts on correcting deviations where they occur on the critical path only: in other words, focusing only on those deviations which impact on the project. Glaser and Young (1961) describe the early use of CPM to control a project "at regular intervals: by measuring actual against predicted performance, management by exception is established".

> As the start of a job [activity] approaches the latest start time, or as the completion begins to come perilously close to the latest possible date, those concerned know that they have slippage on their hands of which management should be notified. Management, therefore, is truly enabled to act by exception only. (Mauchly, 1962)

One of the advantages widely perceived was the ability to give early warning of trouble, so that prompt remedial action may be taken. Thus the concept of an early warning system has been a component of control since the design of CPM.

Updating the status of a project

The process of monitoring project progress has not changed much in the 40 years since Woodgate and others first documented the steps. That which was simply called 'applying progress information to a network' is usually referred to as updating progress status—or more commonly *statusing* the project.

O'Brien and Plotnick (1999) describe a method which varies little from Woodgate's method of modifying the network and undertaking the standard calculation (page 100). To do this, all completed activities are given a zero duration. Activities in progress are given the time duration required to complete the activity. Logic is revised as necessary. In order to have a comparison against goals, the original plan is used as a baseline, and those baseline (contract) dates are used for comparison with the recalculation—providing an indication of float as positive or negative depending on the effect of the recalculation on the overall schedule. As O'Brien and Plotnick note, this method can result in late dates being earlier than early dates—a function of the early dates being realistic and the late dates being the constrained dates.

Progress updating in CPM therefore requires a contract or other schedule as a baseline schedule. Dates from milestones in this schedule are applied to the current schedule and the updated schedule then indicates progress against the baseline.

The actual progress reporting implementation varies between software programs. Indeed, different planners may apply the technology differently to suit their own organisation's needs.

Activity-based presentation techniques for control

The presentation of progress in activity-based control is limited to the use of the network (usually for activity on the arrow) or more commonly the Gantt chart—where the 'time now' line is drawn vertically through the chart, deviating to show the current completion state of individual activities (Figure 4.4). Tabular representation of task progress data may also be used.

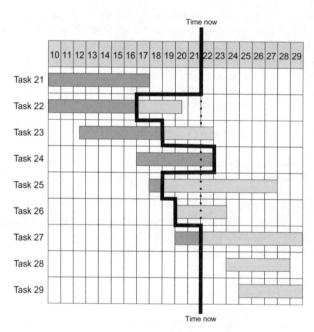

Figure 4.4 Progress shown in Gantt chart

The other way to represent progress is in aggregate. This is an attractive form of representation due to its simplicity. In this case the progress is aggregated and represented as a total percentage for the project. While this is a simple representation, it is not sensitive to the criticality of the activities on the schedule. The most common representation of aggregation is an S-curve with planned and actual represented (Figure 4.5), and this is usually based on the accumulation of expenditure (cash flow) or resources—both requiring costs or resources to be included in the model. Perhaps the most sophisticated representation of project progress is in the form of an earned-value chart.

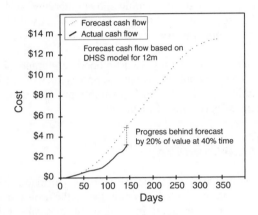

Figure 4.5 Progress monitoring using S-curves (Kenley, 2003)

Effectiveness of controlling projects using CPM

The axioms of project control identified by Woodgate are:

- The more precise the original planning the greater is the potential for accurate control and efficient performance.
- The greater the speed of feedback and response, the more accurate will be adherence to the general policies and project objectives.

Consistently, the literature presents the view that updating should occur frequently in order to ensure adequate project control. Most authors recommend updating at least every two weeks. This provides the basis for management by exception, the underlying controlling principle in CPM, however as noted by Meredith and Mantel (1995):

> ...management by exception has its flaws as well as its strengths. It is essentially an 'after-the-fact' approach to control. Variances occur, are investigated, and only then is action taken. The astute project manager is far more interested in preventing problems than curing them. Therefore the monitoring system should develop data streams that indicate variances to come.

This is problematic in an activity-based management system, as the duration of each future activity is unrelated to past activities of the same type. The CPM system does not have the capacity to make forecasts of future action without intervention by the project manager in the form of variation of future estimates of duration—perhaps based on probability and certainly based on experience.

Nevertheless, the use of CPM as a method of control clearly has positive effects. For example, Davis (1974) noted a significant difference in the performance of firms using CPM in two ways:

There are some major differences in the manner in which the *very successful* and *unsuccessful* users are employing CPM. About twice as many (proportionately) of the very successful firms employ CPM for the control of ongoing construction as do the unsuccessful firms, and many more of the successful firms employ these methods for bidding and estimating and engineering work.

Only in the area of pre-planning construction activities are the relative percentages of use about equal (successful: 96%; unsuccessful: 80%).

Of course, there are many authors who have developed very practical methods for making CPM effective on site. The most notable recent example is the work of Woolf (2007), which is an extremely practical guide to CPM planning and contains practical steps for maximising the benefits of performance recording.

EARNED VALUE ANALYSIS (EVA)

Earned value is an integrated method for measuring the performance of a project using both the schedule of work and the cost budget and monitoring the progress of work against these two plans. This method is a powerful aid in solving questions about deviation of the cash flow profile from the planned. EVA is independent of activity- or location-based methodologies.

These systems have previously been known as Cost/Schedule Control Systems Criteria (C/SCSC), as developed by the USA Department of Defence and in use since 1967, but renamed earned value analysis (EVA) in 1996. As the names suggest, they require costs to be included in the schedule model. The tool was designed by managers who recognised that increasing works program complexity and proliferating management systems demanded a reasonable degree of standardisation. Thus it is both a performance management system and a method of standardisation of the administration process.

For performance management, the earned value method is also known as 'achieved value', 'accomplished value', 'physical quantity measurement' and 'earned value performance management' (EVPM).

Fully implemented, earned value requires segmenting a project into controllable parts using a 'work breakdown structure' (WBS) which is related to the 'organisational breakdown structure' (OBS). The WBS includes all work tasks for the project. Costs are similarly broken down into a 'cost breakdown structure' (CBS). Bent and Humphreys (1996) note that the WBS is in fact a subset of the CBS—which extends to include those cost items which do not involve work.

At the heart of the breakdown, there are cost accounts consisting of either labour hours (a simplification) or total costs, and a direct relationship is established between the percentage of work done and the budget for that account. This relationship is very problematic for project control as planning systems are typically separate from accounting systems. Thus the data is either input manually, or more commonly merely the aggregation of costs or resources based on fixed estimates for each activity.

Earned value performance management utilises both mathematical and graphical relationships.

Mathematics of earned value analysis

The performance of the project is measured through a series of indicators, with the earned value (EV) being for each cost account:

$$EV = (Percentage\ complete) \times (Budget\ for\ the\ account) \tag{4.1}$$

The summing of these provides the EV for the project—which is equivalent to the work-in-progress curve. Performance measurement, however, requires this integration of time reporting and cash expended implicit in Equation 4.1. Furthermore, performance measurement requires comparison of the rate of progress with the originally planned rate of progress, or baseline scheduled progress.

Progress variance has been established to be either a variance in the time performance—schedule variance (*SV*) or a variance in the cost performance—cost variance (*CV*), or both. These indicators are:

$$SV = BCWP - BCWS \tag{4.2}$$

$$CV = BCWP - ACWP \tag{4.3}$$

Where:

$BCWP$ = Budgeted cost of work performed = *EV*
$BCWS$ = Budgeted cost of work scheduled
$ACWP$ = Actual cost of work performed

These measures provide a rapid feedback mechanism.

If *SV* is positive, then the project has completed more work than scheduled by cost (a different answer may be obtained were performance by quantity to be measured). If *SV* is negative, then the project is behind schedule.

If *CV* is positive, then the project is costing less than budgeted. If *CV* is negative then the actual cost has exceeded the budgeted cost for the work performed.

Together these two indicators inform about both schedule and cost performance. It is also possible to develop integrated measures of efficiency. These are the cost performance indicator (*CPI*) and schedule performance indicator (*SPI*):

$$CPI = BCWP\ /\ ACWP \tag{4.4}$$

$$SPI = BCWP\ /\ BCWS \tag{4.5}$$

An index value of 1.0 or greater indicates better than planned performance for either indicator. A value less than 1.0 indicates poor performance relative to the plan.

Barr (cited in Meredith and Mantel, 2001) provides a further indicator, the cost–schedule index (*CSI*) which combines *CPI* and *SPI*:

$$CSI = CPI \times SPI \tag{4.6}$$

This resolves the situation where one of the ratio indicators is less than 1.0 and the other is greater than 1.0. A problem is indicated where *CSI* < 1.0.

Applying earned value analysis

Earned value analysis requires software to be able to track the costs and progress for each individual cost centre.

There are now many software packages which will undertake this task. However, there may be a difficulty obtaining cost data in a timely manner as accounts departments usually operate on different systems and timelines, resulting in difficulty integrating this data. Therefore, many analysts use labour and worker hours to track earned value. This is a relevant approach where the bulk of the cost and the variability in cost is directly proportional to the hours expended. This may not be suitable where material consumption or waste is a significant and variable factor.

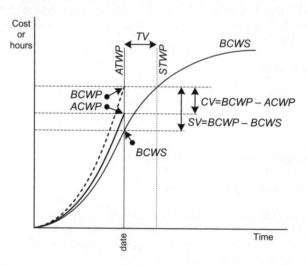

Figure 4.6 Earned value analysis—case 1 (Kenley, 2003)

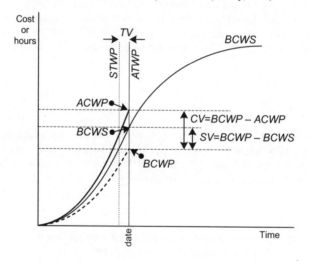

Figure 4.7 Earned value analysis—case 2 (Kenley, 2003)

The method highlights otherwise invisible problems. Figure 4.6 illustrates a case where the project appears ahead of schedule and actually is both saving money and ahead of schedule. Time variance (*TV*) is read graphically as the time difference between *BCWS* and *BCWP*, which is the difference between scheduled time of work performed (*STWP*) and the actual time of work performed (*ATWP*). In contrast, Figure 4.7 illustrates a case where the project appears ahead of schedule but is both losing money and behind schedule. This clearly demonstrates the power of EVPM.

Earned value performance management is a powerful way to provide risk management for activity-based scheduling. But it is needed because CPM is a very insensitive tool to trends in performance. The ignorance of the model to repetition means that CPM has no way to identify consistent poor performance and to forecast based on this, thus EVPM's attraction to senior management.

LEAN CONSTRUCTION

Lean theory and lean construction, one of the most important advances in construction management theory, are large topics and it is not possible to do them justice in this small section of a chapter. Therefore the focus here is to present basic lean theory and the subsequent development of lean construction and its obvious relationship to location-based planning and control[3]. In particular, this section addresses the control aspects of lean construction, which are arguably the most developed control systems for construction other than those outlined for location-based control in Chapters 8, 9 and 10.

A (very) brief overview of lean production theory

Lean production theory has developed out of studies of the Japanese car manufacturing industry, in particular of Toyota, undertaken in the 1980s and has been developing ever since. Womack et al. (1990) and Womack and Jones (2003) expounded a new production methodology which they termed *lean production*. This is sometimes known as the Toyota production system.

Lean production does more with less[4], essentially because it removes production waste from the supply chain and the production process. At the same time it focuses on delivering value to the end user, including choice. Waste, or *muda* [5] as the lean community call it, consists of the various activities or processes which do not add value. These include:

- mistakes which require rectification
- production of items no one wants resulting in inventories and remaindered goods
- processing activities or steps which are not actually needed
- unnecessary movements of employees and transport of goods from one place to another
- idle time caused by upstream activities waiting on downstream activities
- goods and services which do not meet the needs of the customer (Womack and Jones, 2003).

The principles of lean production include:

- **Value**—defined in terms of the end user, value means that the good or service must meet the needs of the end user at a specified price and time.

---

[3] Many lean construction advocates are deeply steeped in the activity-based methodology and do not accept the role which location-based planning can take in improving their systems, nevertheless flow is critical for both methodologies.

[4] Less resources—less manpower, less equipment, less time, less space.

[5] From the Japanese for waste.

- **Value stream**—an holistic concept of design (problem solving), information management and production (physical transformation) including all steps and actions required to deliver the product. The value stream can be mapped and waste eliminated.
- **Flow**—the opposite of batch and queue. The organisation of work such that the work flows rather than using a series of high-speed batch processes.
- **Pull**—the end user pulls the production such that it is only produced to suit their requirements. Each item in the production changes to suit the end requirements, thus producing choice.
- **Perfection**—the previous principles interact in a virtual circle to improve towards perfection (Womack and Jones, 2003).

The goal of lean production is to produce a reliable workflow. The application of these principles to construction is dependent on whether the interpreter is coming from an activity-based standpoint or a location-based standpoint.

The important principles which concern planning and control of construction are *flow* and *pull*, while the others apply equally. However, flow and pull may be interpreted differently depending on the planning methodology.

Activity-based interpretation of lean

In the world of activity-based planning and control, each activity in every location is considered a separate activity. Tommelein et al. (1999) referred to this as the "parade of trades", and compared "flow" as partially complete work flowing past processes on the factory production line, to "flow" as crews of various trades flowing past partially completed and stationary work on the construction site. In this view, the aim of lean construction is to:

> ...synchronize and physically align all steps in the production process, so there is little wait time for people or machines, and virtually no staging of materials or partially completed projects (Tommelein et al., 1999: 304).

The impact of variability in production is such that detailed planning is left late and work is *pulled* to be executed as ready, and as required. In this view, the specific organisation of continuous sequences of repetitive activities is considered batch processing and therefore counter to a lean production philosophy. Instead, detailed processes of late planning, including Last Planner, are used to ensure continuous flow of work.

Flow is considered the continuous supply of work-ready (prerequisites completed) activities in any sequence subject to the critical path and readiness.

Pull is considered the scheduling of work in accordance with the critical path as work is ready. "A pull technique is based on working from a target completion date backwards, which causes tasks to be defined and sequenced so that their completion releases work" (Ballard and Howell, 2003). This is arguably the most important lean concept in this view, and it is most likely that pull scheduling has arisen as a direct consequence of the prevalence of poor scheduling as a remedial response.

The main focus of lean construction in this view is making work ready so as to avoid waste due to inability to complete activities commenced due to failure of one or more prerequisites—to shield production from variation in the planning system "by making quality assignments, thereby increasing the reliability of commitment plans, such as weekly work plans" (Ballard and Howell, 1998).

So, the emphasis with an activity-based approach is to ensure that work is organised well and in particular that no work is commenced that cannot be completed. In effect, this is ensuring the flow of work as allocated. A good example can be seen in the 'project strategic planning' approach taken by Crow and Barda (2004), where the relationship between parties was considered the 'key' and the cooperative passing of work in a 'team' is used to solve the workflow reliability problem.

Location-based interpretation of lean

The parade of trades is a recognition that purely focusing on individual tasks is not realistic. Tommelein et al. (1999) note that:

> Project management schedules that use the critical-path method (CPM) describe activities with their durations and precedence relationships. The finish-to-start relationship is most often used, though it assumes sequential finality, i.e., predecessors must be 100% complete before their successors can start. This assumption certainly does not hold in the parade of trades where regular hand-offs exist between trades, and once the parade has started, all trades have to move in sync for the parade to progress at a steady pace. The CPM schedule's misrepresentation of the parade is the key reason why most superintendents use it only as a loose guide for executing work (Tommelein et al., 1999).

In the world of location-based planning and control, activities which are executed with continuous production, through many locations, are considered a single task. The concept of flow therefore requires that locations be completed sequentially, with individual requirements being pulled for that location. In the LBMS, we are therefore concerned with labour flow (continuous work, empty locations pull resources), whereas lean is primarily concerned with workflow (hand-offs between specialists, completion of one trade pulls the next). Tommelein et al. (1999) describe such as a parade of trades, referring to patterns of trade workflow sequence rather than the internal logic of individual trades. The parades they list are:

- Structural parade
- Overhead work parade
- Perimeter enclosure parade
- Interior finishes parade.

Their implication is that these parades are made up of many trades which together follow a common pattern. This is an important concept which also forms a critical component of the LBMS.

When considering continuous sequences or parades of work in LBMS, 'flow' is considered the continuous flow of resources through locations with all prerequisites completed for each location in sequence. Sequencing is according to the (flowline) plan, as location completion releases resources to the next location. Pull is considered at the task level rather than the activity level, and thus a task pulls the requirements for the commencement of a task sequence. This necessarily implies that in each location the prerequisites must be completed prior to the task arriving in that location.

The main focus of location-based lean construction is ensuring that tasks may flow without interruption due to failure of prerequisites in any location, avoiding the associated waste such as interruption, double-handling of materials, equipment or workers, or rework. The focus on flow is a clear commonality in both approaches.

Last Planner

Whether adopting an activity- or location-based methodology, the principles of one of the strongest lean construction methodologies may be applied. The Last Planner System (LPS) is an extremely powerful proprietary planning and control system developed by the Lean Construction Institute[6] and a world network of researchers, the International Group for Lean Construction[7] over the last fifteen years. It is based on lean principles but arises from an activity-based world view. Despite the contribution made to the method by the academic community, its approach will only be briefly described here and detailed analysis cannot be provided due to the copyright restrictions placed on the method.[8]

LPS is a planning and control system and these are considered as two sides of a rotating coin. Planning is the process of defining criteria for success and producing strategies, control is actively causing events to conform to plan with associated learning and re-planning.

Last Planner is designed to achieve sound production assignments by matching capacity (resource) to load. This is a social process involving discussion with site staff and planning to ensure that work is not waiting on workers and workers are not waiting on work. A typical controlling process includes evaluation by the foreman of the available resources and the available workload. The foreman notes any work remaining after all available resources have been allocated or any resource remaining after all work has been allocated. If work is left over, it goes to the workable backlog. If manpower is left over, the foreman asks for instructions from his supervisor (Ballard and Howell, 1998).

The planning system has grown in an activity-based environment and Last Planner reflects this, however it is essentially a social process rather than a technical one and planning methodology may be used. The method generally commences with a master schedule with increasing levels of detail as work approaches production. Planning moves through familiar stages: master planning, phase (pull) scheduling, look-ahead planning and weekly planning. The main contribution of LPS is in the way these are manipulated to shield production. The stages and flows of the LPS are illustrated in Figure 4.8.

Master and phase planning

Activities are planned according to the master schedule—typically a CPM schedule. The master schedule provides the overall project planning from which can be drawn procurement and resource needs and consists of milestones and long lead time items. The production is then developed together with subcontractors by those responsible for building the

---

[6] Further information and research papers may be found on the Lean Construction Institute internet site (www.leanconstruction.org).

[7] Conference publications may be found on the Lean Construction Institute internet site (www.iglc.net).

[8] LCI publish the following copyright restriction:
"We are pleased when owners, clients or their construction companies use Last Planner™ to improve their design and construction performance. We make no charge for this and place no limits on its use within a company. We do encourage companies that find it useful to become contributors to the Lean Construction Institute. We do retain a Trademark on the term and Copyright in the idea and materials to prevent people who misunderstand or misrepresent the system from using it in trade. Because we want those who design and construct with the help of the Last Planner System™ to have the best possible start, we do require that those who use the term in trade are approved by us. That is, those offering to teach, coach or apply Last Planner™ as part of a commercial offer need our approval. We also expect them to make financial and other contributions to the Institute in recognition of the financial and other benefits they are getting from the work we have put into developing Last Planner™."

phase, starting backward from the planned phase completion date. The process reveals what must be done to release work for production.

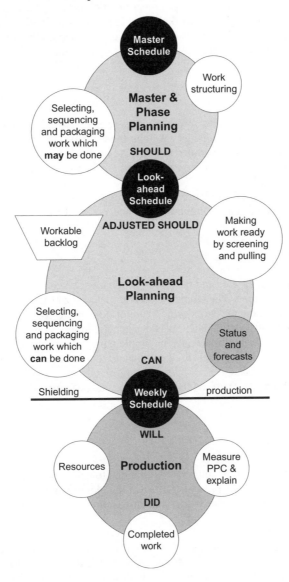

Figure 4.8 The Last Planner System illustrated

Look-ahead planning

Activities are planned in detail within a limited window, typically a four- or six-week look-ahead window. Work which cannot be made ready should be removed from the look-ahead schedule. At this stage workflow can be planned, together with sequencing and detailed methods. The work is planned on assignment level—something that can be communicated to workers.

An important concept derived from lean production is to have a supply of work available in case problems occur. This is workable backlog and is defined as work for which all prerequisites are ready but which does not require to be scheduled as yet. In principle this allows resources to maintain production when things go wrong.

Weekly planning

Detailed work schedules are derived from the look-ahead plans for each week. At this level of detail, examination of prerequisites can take place. If necessary prerequisites are not available, work should not be scheduled and instead the prerequisites should be addressed. Alternative work is then planned, possible drawn from the workable backlog. Work is pulled into the weekly plan.

LPS provides detailed guidance on selecting work to be included in the weekly planning, including definition, soundness, sequence, size and learning.

Prerequisite screening

Prerequisites are those activities and resources which must be completed or in place before an activity can be *completed*. The emphasis on completed is a critical contribution of LPS. There is no point commencing work which cannot be completed, despite this being very typical behaviour on construction sites. Waste is generated when crews start an activity but cannot complete due to lack of materials, incomplete precedent activities, faulty work, etc. Under lean principles, it is better to wait than to commence work which cannot be completed.

Only work which passes prerequisite screening can be planned at the weekly planning level.

Shielding construction

The weekly planning process is deliberately intended to shield production from poor planning. Failure of prerequisites represents failure of the planning system. LPS prevents work which cannot be completed from being scheduled, thus shielding the crews from waste generated by interruption.

> The shielding process begins with an initial screening of scheduled activities carried out in the formation of a 'look-ahead schedule', which drives a 'make ready process' that matches resources with opportunities so that production throughput is maximised. Its output is a buffer of 'sound assignments' from which to 'select assignments' for each plan period (Ballard and Howell, 1998).

Percentage planned complete

LPS provides a measure of performance in the form of a measurement of the percentage of weekly work activities planned to be completed which were able to be completed— percentage planned complete (PPC).

This performance measurement system is not without problems, for example deliberately scheduling a small series of easily completed activities can improve the PPC score, nevertheless there is a strong correlation between PPC and improved project performance (Seppänen, submitted). As such, it forms an important feedback mechanism as part of a control system.

Learning process—root cause of failure

Learning through identification of failure is a key component of the LPS. For each failed assignment, a root cause analysis is carried out to prevent the problem from happening again (Ballard, 2000). Non-completion arising from failed constraints requires improvement of the look-ahead process. Sometimes the individual planner may be at fault because of a failure to assess either capacity or risk. Whatever the cause, forecasting and alarms are included implicitly in the social process because bad forecasts will lead to failed planning which will lead to remedial action to improve forecasting and remove alarms (Seppänen, submitted).

DEVELOPMENT OF LOCATION-BASED CONTROL

Its has been recognised that location-based planning techniques provide greater opportunities for project control. For example, Schoderbek and Digman (1967) developed a model for combining PERT and LoB specifically to take advantage of the strength of line-of-balance in controlling projects in the production phase. This strength comes about because the use of production lines, whether line-of-balance or flowline, enable monitoring of trends visually.

The completion of individual tasks through units of production (line of balance) or locations (flowline) can be visually compared with the plan. The required rate of production for each trade is known and deviation from that production is clear as an actual line deviating from the planned line.

Early presentation techniques for location-based control

Line-of-balance

Line-of-balance presents many opportunities for schedules and graphs to be used to present information for control. Lumsden (1968) proposed the following main components:

1. **Activity totals by week**
   A tabular representation providing the number of units (locations) completed for each activity by week.
2. **Line-of-balance quantity chart**
   A chart showing the number of units (locations) completed for each activity in a given week.
3. **Line-of-balance progress diagram**
   A Line-of-balance diagram with the content between the limits for each task being shaded according to progress. Thus it would be shaded if completed, partly shaded if underway or blank if not started.

4. **Weekly progress report**
   A tabular representation indicating completion of a task by location.
5. **Cash flow diagram**
   A simple S-curve graph of the planned cash flow compared to the actual cash flow.

Non-rhythmic production

Mohr (1979) pointed out the use of flowline when dealing with what he described as non-rhythmic production (page 78). What he was actually discussing was the effect of poor production and how the plan would need to be changed to accommodate this problem. In the case illustrated in Figure 4.9, the following Task 4 would be delayed in starting due to the poor production of Task 3.

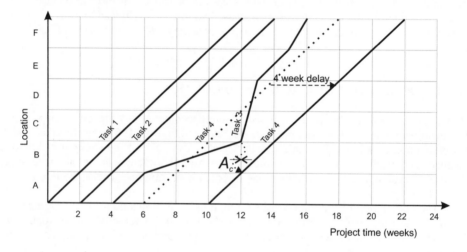

Figure 4.9 Non-rhythmic production (from Chapter 3)

The picture this presents is overly simplistic, as clearly the crew undertaking Task 4 cannot be aware that Task 3 is about to run slow at the time they are due to start. They would start as planned and then be delayed by Task 3 once on site. This requires control actions. True control action in this situation would be more like Figure 4.10, where a second crew is added to Task 4 once Task 3 is fixed. This would reduce the impact of the delay on the project to only 1 week from 4 weeks. An even better alternative would be to immediately react to the deviation of Task 3 and accelerate it by adding resources, working overtime or improving productivity. This would allow Task 4 to continue as planned. However, this optimal approach will not work if there are no buffers in the schedule. Buffers enable problems to be corrected before they affect the succeeding task.

It is easy to see how the flowline chart can be used as a control mechanism, however there is little evidence of its use in the literature. This is probably due to a combination of the effectiveness of the initial plan when location-based and, more likely, the lack of computer software to support this mechanism. A notable exception is the research that has been done in Finland during the 1990s.

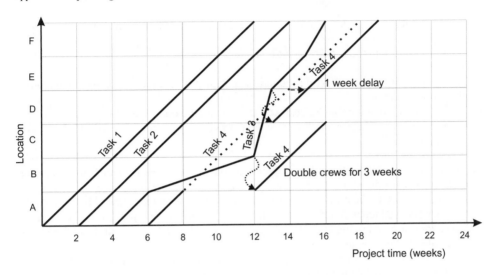

Figure 4.10 Non-rhythmic with control actions

Development of location-based control in Finland

It is not possible to discuss the development of control methods in Finland without using terminology from location-based control theory—which is discussed in Chapter 8. Therefore, some of the terms in the following section may be easier to understand once Chapter 8 is understood.

Research on location-based production control started in the 1980s and followed on from work looking at planning systems (see the discussion of the development of location-based planning in Finland on page 90). Construction costs were increasing rapidly and profit margins for general contractors were low. Three medium-sized construction companies started a large research effort which tried to explain why the economic results from projects had been so poor.

Research was undertaken by the Construction Economics Laboratory at Helsinki University of Technology. The first phase of the research involved analysing projects with both good and poor results, and discovering the causes for failure.

The main conclusion of this research was that, despite the original cost estimates being good and without major mistakes, cost overruns nevertheless occurred during the implementation phase. The causes for poor implementation were found to be poor quality planning (in terms of content and quality) and poor implementation of plans with production control being absent. The major factors were poor procurement, the incorrect use of resources and failure to completely finish work.

Planning problems arose from making incorrect duration estimates, incorrect logic between tasks, planning multiple tasks to start at the same time, or planning discontinuity into tasks (arising from a lack of available work and a tendency to shift problems to the end of a project).

Schedule control problems were associated with only controlling the work based on costs derived from company accounting systems, and therefore without knowing the total effects of any deviations.

Based on these results, companies assessed that the most important improvement area would be to improve the planning and control system—the quality and implementation of schedules.

The theoretical foundation of the research effort was goal management:

- The most important focus for production control is to make production happen according to plans
- The prerequisites for successful production control are continuous monitoring and planning good schedules
- Location-based planning is the best way to measure the quality of plans
- Location-based management allows planners to visually see the effects of deviations on other tasks.

In practice, it proved to be difficult to implement goal management principles in scheduling projects because of the attitudes of seasoned construction veterans.

Location-based planning was selected as the main scheduling method because of the good experience gained with linear scheduling in road construction projects and because of technical literature. Planning and controlling was done manually, while drafting the schedules was able to be done using computer software.

The benefits of location-based planning were seen to be visual risk evaluation of schedules, clear visualisation of sequence, visualisation of free locations and seeing the total effect of deviations based on progress data.

Action research had been the chosen research method. The first step was to define the properties of a good schedule. Then the researcher participated in controlling the schedule and planning control actions, so that the schedule could be implemented as planned. Case research was directed by two professors (Juhani Kiiras and Jouko Kankainen) and researchers were generally students who were completing their masters' theses. Over 30 case studies were carried out and each brought some improvements to the scheduling or production control techniques. Professor Juhani Kiiras wrote a white paper about planning of initial schedules, and this was used to kick off the research (Kiiras, 1989).

The main results of case studies were that:

- Production control is more important than planning
- Controlling requires good schedules, continuous monitoring and immediate reactions to deviations to decrease their effects
- The quality of plans needs to be checked to ensure that schedules are feasible and controllable and that resource requirements are the same as allowed in the cost estimate or budget
- The project should be divided into sections and zones for planning and controlling
- Continuous production is the key for good implementation of plans.

The key principles for successful scheduling were found to be that:

- Durations should be calculated based on good target productivity rates
- Buffers should be added to the schedule to protect against variability
- Production rates are synchronised by changing multiples of optimal crews
- Subcontracted work should also be planned using quantities and productivity rates
- Each location has only one task in it at a time
- Crews should be planned to have free locations to buffer against disturbances
- Tasks which are not included in the schedule should have resources allocated.

The key principles for monitoring of production were found to be the following:

- Monitoring should be done by monitoring finished locations

- Progress information is collected by the use of "control charts"
- Progress information is visualised in flowline diagrams, to show critical deviations and the effect of deviations on the total duration.

Principles for production control were divided into two groups: proactive control and reactive control. Methods for proactive control were:

- Forecasting production problems
- Risk analysis of schedule
- Detailed planning close to implementation
- Integrating procurement schedules into the production schedule to ensure starting prerequisites
- Self-commissioning.

Methods for reactive control were:

- Increasing resources by adding multiples of optimal crews
- Changing the content of a task (for example, by removing parts of the scope from a sub-contract)
- Overtime.

Case studies were documented in the students' masters' theses. Based on these, Professor Kankainen wrote a handbook of production control, which has been published in three editions (Kankainen and Sandvik, 1993, 1996; Kolhonen, Kankainen and Junnonen, 2003).

In recent years, the case study research has been continued. Some areas of interest have been ensuring prerequisites of production by task planning (Junnonen, 1998), integrating MEP and construction work schedules (Seppänen and Kankainen, 2007) and contract management in controlling of subcontracts.

DISCUSSION

Controlling construction projects has taken a huge leap forward with the work from Finland over the past thirty years. This work has substantially provided the framework for the development of the location-based management system.

However, the control methods in the location-based control system owe as much to lean thinking as to the Finnish work, and indeed to the earlier development of control systems. Therefore the LBMS is an amalgam of all the location-based planning methods considered in Chapter 3 with the control methods reviewed in this chapter.

REFERENCES

Ballard, G. (2000). The last planner system of production control. PhD thesis, Faculty of Engineering, University of Birmingham.

Ballard, G. and Howell, G. (1998). "Shielding Production: Essential step in production control". *ASCE Journal of Construction Engineering and Management*, **124**(1): 11–17.

Ballard, G. and Howell, G. (2003). "An update on last planner". *International Group for Lean Construction 11*, IGLC conference, Blacksburg, Virginia.

Bent, J.A. and Humphreys, K.K. (1996). *Effective Project Management through Applied Cost and Schedule Control*. Marcel Dekker, New York.

Betts, M. and Gunner, J. (1993). *Financial Management of Construction Projects: Cases and Theory in the Pacific Rim*. Longman, Singapore.

Crow, T. and Barda, P. (2004). "Project strategic planning: A prerequisite to lean construction". *12th Annual Lean Construction Conference*. Copenhagen, Denmark.

Davis, E.W. (1974). "CPM use in top 400 construction firms". *Journal of the Construction Division, ASCE*, **100**(CO4): 39–49.

Deming, W.E. (1986). *Out of the Crisis*. MIT Center for Advanced Engineering Study, Boston.

Fazar, W. (1959). "Progress reporting in the special project offices". *Navy Management Review* 1959(April): 9–15.

Glaser, L.B. and Young, R.M. (1961). "Critical Path Planning and Scheduling—application to engineering and construction". *Chemical Engineering Progress* **57**(11): 60–65.

Junnonen, J-M. (1998). *Tehtäväsuunnittelu ja laatupiiriohjattu tuotannonohjaus. (Task planning and quality circles in production control)*. Department of Civil and Environmental Engineering. Espoo, Finland., Helsinki University of Technology.

Kankainen, J. and Sandvik, T. (1993). *Rakennushankkeen ohjaus*. (Controlling a construction project). Confederation of Finnish Construction Industries, Rakennustieto Oy, Helsinki. Finland.

Kankainen, J. and Sandvik, T. (1996). *Rakennushankkeen ohjaus*. (Controlling a construction project). Confederation of Finnish Construction Industries, Rakennustieto Oy, Helsinki. Finland.

Kelley, J.E. and Walker, M.R. (1959). "Critical-Path Planning and Scheduling". *Proceedings of the Eastern Joint Computer Conference*.

Kenley, R. (2003). "Financing construction: Cash flows and cash farming". E&FN Spon, London, pp273.

Kiiras, J. (1989). *OPAS ja TURVA, Erityiskohteiden työnaikaista ohjausta palveleva aikataulu- ja resurssisuunnittelu.* (A schedule and resource planning system for the implementation phase of special projects). Helsinki University of Technology. Construction Economics and Management Publications.

Kolhonen, R., Kankainen, J. and Junnonen, J-M. (2003). "Rakennushankkeen ajallinen hallinta". (Time management of construction projects). Helsinki University of Technology, *Construction Economics and Management Publications* 217. Espoo, Finland.

Lumsden, P. (1968). *The Line-of-Balance method*. Pergamon Press Limited: Industrial Training Division, London.

Mauchly, J.W. (1962). "Critical-Path Scheduling". *Chemical Engineering* 1962(April): 139–154.

Meredith, J.R. and Mantel, S.J. (1995). *Project Management: A managerial approach*. John Wiley & Sons, New York.

Meredith, J.R. and Mantel, S.J. (2001). *Project Management: A Managerial Approach*. John Wiley and Sons, New York.

Mohr, W.E. (1979). *Project Management and Control (in the building industry)*. Department of Architecture and Building, University of Melbourne, Melbourne.

O'Brien, J.J. and Plotnick, F.L. (1999). *CPM in Construction Management*. McGraw-Hill, Boston.

Pillai, A.S. and Tiwari, A.K. (1995). "Enhanced PERT for programme analysis, control and evaluation: PACE". *International Journal of Project Management,* **13**(1): 39–43.

Schoderbek, P.P. and Digman, L.A. (1967). "Third generation, PERT/LoB". *Harvard Business Review* **45**(5): 110–110.

Seppänen, O. and Kankainen, J. (2007). "Skanssin loppuraportti". (Final report of Skanssi retail center). Unpublished research report.

Seppänen, O. (submitted). "Empirical research on success of production control in building construction projects". Unpublished PhD thesis submitted in fulfilment of the requirements of the degree of PhD at Helsinki University of Technology, Helsinki, Finland.

Shewhart, W.A. (1939). *Statistical Method from the Viewpoint of Quality Control*. Dover, New York.

Stacey, R.D. (1996). *Strategic management of organisational dynamics*. Pitman Publishing, London.

Thomsen, W.H. (1959). "Progress reporting for the Under Secretary of the Navy". Navy Management Review 1959(April): 7–8.

Tommelein, I.D., Riley, D.R. and Howell, G.A. (1999). "Parade Game: Impact on Work Flow Variability on Trade Performance". *ASCE Journal of Construction Engineering and Management*, **125**(5), 304–310.

Wiener, N. (1996). *Cybernetics: or control and communication in the animal and the machine*. The MIT Press, Boston.

Womack, J.P. and Jones, D.T. (2003). *Lean Thinking*. Free Press, New York.

Womack, J.P., Jones, D.T., and Roos, D., (1990). *The machine that changed the world*. Rawson Associates, New York.

Woodgate, H.S. (1964). *Planning by network: Project planning and control using network techniques*. Business Publications Limited, London.

Woolf, M.B. (2007) *Faster construction projects with CPM scheduling*. McGraw Hill, New York.

Section Two

Location-based planning

SECTION TWO—LOCATION-BASED PLANNING

The following Section Two chapters introduce and explore a new model for location-based planning and scheduling of construction projects. This is location-based planning. While the identification of the approach as location-based is new to this model, it is based on previous work as discussed in Chapter 3.

In this section, the term *planning* is used to encompass *planning*, *scheduling* and other terms which relate to the building of time-related models of construction work in order to develop a logical plan for the work. The term includes *pre-construction planning*, allowing differentiation between the upfront planning processes and the reactive planning and control processes during construction. The latter are discussed in Section Three.

The overall emphasis in location-based planning is planning for productivity. Unlike previous scheduling methodologies, location-based planning explicitly manages the continuity of work for resources and thus protects and optimises production. The tools and techniques provided in this section are designed to support the project planning team in achieving this goal.

The new model for location-based planning is rich in detail, thus the discussion is spread over three chapters. These chapters separate the theoretical discussion (Chapter 5) from the analysis of detailed methods (Chapter 6) and techniques for implementation of location-based planning (Chapter 7).

Chapter 5 presents a new theory for location-based planning of construction and reveals the power of absorbing CPM's external activity logic into the five layers of location-based task logic. All readers are encouraged to understand the generic topics, in particular locations, location breakdown structures, location-based quantities, tasks, how durations are calculated based on optimal crew sizes and location-based logic. However, the chapter includes much technical content and this may be safely skipped over by the practitioner wanting to get to applications as soon as possible.

Chapter 6 explores the methods required to use location-based planning. It expands the theoretical discussion of location-based planning by introducing real-world simulation methods such as unit and production system cost, production risk, procurement, design schedule, quality and learning processes (improving rates of production) within the context of a location-based schedule. Another level of detail is added to the measurement of quantities. Instead of just talking about finished products, for example the area of finished wall, it is also desirable to know the resources consumed in building the wall; the materials, labour, subcontracts and equipment. This is product-resource modelling. Two methods to model costs are presented. Conventional cost loading of the schedule and a production system cost model, which calculates the total labour cost of the production system including waste. This is modelled by examining the resource use and assessing the continuity of location-based tasks. This new model can be used directly to optimise the efficiency of the production system. The chapter discusses production system risk, which is strongly related to production system cost, and the use of buffers to mitigate production system risk. Other methods are introduced, such as procurement, pull scheduling, design schedules, and using location-based planning to achieve well-managed handover of locations to drive quality and safety.

Chapter 7 discusses how to implement location-based planning. It is one thing to know about location-based theory and its associated methods, it is another to know how to use that knowledge to build effective schedules for a project and to plan for both production efficiency and confidence. In this chapter the discussion concentrates on techniques to build a 'good' plan, to minimise risk and to maximise feasibility. The steps to take are outlined, and guidelines presented for executing the methods from Chapter 6. Special techniques such as schedule optimisation, cycle planning, planning buffers and assessing schedule feasibility are discussed.

Chapter 5

A new theory for location-based planning

INTRODUCTION

It is not often that a new methodology emerges in construction planning that requires the development of new theory. The location-based management system, while based on previous work which may now be termed location-based, is just such a new methodology. Understanding it requires examination of the underlying theory and the re-expression of already known concepts in accordance with that theory.

At the core of location-based management is location-based planning—everything else is based on that foundation—so the theory of location-based planning is critical to understanding location-based management systems. This chapter presents the theory, but comes with a warning of technical content. Readers will have to judge for themselves how much of the theory they need for understanding or implementation. It is recommended that practitioners should be familiarise themselves with the principles of location-based planning prior to implementing location-based management in their own organisations. Therefore, all readers should understand the generic topics, in particular locations, location-based quantities, tasks and location-based logic. The following, more technical sections, can be safely skipped by those only wanting practical understanding and less theory.

What is location-based planning

Location-based management assumes that there is value in breaking a project down into smaller *locations* and using these to plan, analyse and control work as it flows through these locations. The location provides a container for project data at a scale which is easy to monitor and analyse. Location-based planning is, in turn, concerned with the process of planning for work to protect production efficiency as work moves through locations. Specifically, the emphasis in location-based planning is to plan for productivity. Unlike previous scheduling methodologies, location-based planning explicitly manages the continuity of work for resources and thus protects and optimises production.

Location-based planning now has greater richness and analytical complexity than the previously dominant methodology, activity-based planning. However, despite its long history and ample use as a manual technique, location-based planning has generally been considered lacking in analytic rigour by the academic community. It will be shown in this chapter that this perception is erroneous. The theory of location-based planning described here is extensive and intricate. Indeed, it is the first planning system to be able to both organise activity sequence and sequence work for production efficiency.

The flow-based, cost saving, risk management and controlling-oriented location-based methods, extend basic activity-based CPM logic to yield an easy-to-use system which possesses the underlying analytic properties of CPM but specifically includes production efficiency. This is achieved by layered logic.

Layered logic is a simple process of automating the creation of a critical path network by using locations, while constraining start times and activity sequences to protect production continuity. Traditional CPM uses a single layer of logic which operates only between any two activities. Production occurring inside an activity is described only by duration, and there is no recognition of the repetition of work in multiple locations. Location-based planning introduces new layers of logic which add more detail to both the internal task production of the location-based task, and to the external links between tasks.

The location-based planning system differentiates between activities and tasks, where a task is made up of a sequence of activities in differing locations. The definition of a task is that it contains work or activities, in a sequence of locations, which can be done by a single crew or split among multiple crews. Location-based planning then uses CPM *external logic* to define the logic or connection between different activities within locations wherever they occur. However, unlike CPM, the planning system also considers a task's own *internal logic*, by calculating durations based on quantities and allowing the planner to plan the location sequence and production rate to achieve continuous production. A location-based task contains multiple CPM activities, each corresponding with discrete physical locations.

All the analytic features of CPM are preserved when examining the logic between activities within locations, as activity sequencing is driven by normal CPM algorithms with familiar concepts such as precedence and lags. The best way to understand this is to consider a project with only one location: in this project, location-based planning for the project would be exactly the same as activity-based planning using CPM. In contrast, the use of multiple locations brings many new and powerful possibilities: the planner can explicitly plan the behaviour of the crews working on the task, including when workflow may be broken or when the work should be done continuously. Because the internal logic of a task is no longer a 'black box', it is possible to take actions to change a task's production rate to achieve a better alignment of production, to cut project duration and to decrease the schedule risks of production (see Chapter 7). It ultimately provides a much improved mechanism for control of production (Section Three).

Planning is also made easier by handling multiple locations as a single planning entity—the dependency logic can be automatically copied for each location. It is not unusual for a CPM schedule to need over 20,000 activities organised into duplicated chains of connected logic. In contrast, these could be modelled with just 100 tasks across 200 locations (and thus owning 200 location-specific activities for each task). When production properties must change, the task can be changed, changing all its activities in a single step.

The combination of activity-based and location-based logic makes planning more economic, as less time is devoted to manually copying links and updating changes of logic. Instead, more time can be devoted to planning how to optimise the use of resources, time and production rates for different tasks. Automation of the task using computer software allows the solving of such complex problems without difficulty. It is no longer necessary to rely on repetition to achieve this goal.

A location-based planning system, which is based on the extension of activity-based logic with location-based logic, also offers sophisticated new tools for planning and analysis. Apart from optimisation for efficiency, Monte Carlo risk analysis can be combined with the system to highlight the risky parts of a project. Forecasts can be calculated based on the actual start dates and actual production rates, and location-based logic can be used to calculate the impact of control actions taken to recover delay from the original schedule. It is also possible to forecast when the work will need to be made discontinuous (interrupted). These are powerful tools for time-claims assessment, particularly with regard to costs. Breaking production flow costs the contractor real money and location-based flow logic allows the claim agent to show the results of any changes or deviations graphically, while still using well understood critical path logic as a theoretical foundation.

This chapter describes how locations are defined, how the activities are combined into location-based tasks by use of location-based quantities, how durations are calculated based on optimal crew sizes and how the layered dependency logic operates in location-based planning. More advanced analysis tools are described in the later chapters of Section Two, as well as the implications for controlling location-based projects (Section Three).

LOCATION BREAKDOWN STRUCTURE

Locations in a project are defined by a location breakdown structure (LBS), which has many properties in common with the work breakdown structure—WBS (PMI, 1996). It is possible for the project to be broken down in many different ways. However, locations must be hierarchical so that a higher level location logically includes all the lower level locations.

Often this breakdown is simple and obvious. However, in special projects, the breakdown structure might not be straightforward, or might include logical (non-physical) locations. Fortunately, there are some general guidelines that can be used. These guidelines apply to commercial construction and do not apply to special cases such as linear infrastructure projects (road and rail for example) which are discussed in Chapter 13.

Each of the location hierarchies has a different purpose. The highest level is used to optimise construction sequence, where the sequence and timing can be changed to optimise overall production. The structures of such sections are independent of each other, therefore it is possible to start them in any sequence or to build them simultaneously. The middle levels are used to plan production flow of structure (and often reflect physical constraints). The lowest levels are used for planning detail and finishes. The guidelines for commercial projects are:

- The highest level location hierarchies should consist of locations where it is possible to build the structure independently of other sections (for example, individual buildings or structurally independent parts of large buildings).
- Middle levels should be defined so that the flow can be planned across middle level locations (for example, floors in a residential construction project, where a floor is usually finished before moving to the next floor).
- The lowest level locations should generally be small, such that only one trade can effectively work in the area (for example, apartments, individual retail spaces, corridors). The lowest level location should be able to be accurately monitored (that is, the foreman must be able to assess whether or not the work is completed in that location).

Refurbishment projects, which are not driven by the need to erect structure, may vary from these rules (see Example 3, page 128). The highest level is, in such cases, not driven by structure but often by access to the building—particularly where existing operations continue during the construction.

Depending on the size of the project there can be from one to six hierarchy levels (six is the maximum depth that has yet been needed in the authors' experience). Interestingly, six levels is generally recommended as the limit for a work breakdown structure. This reflects the lack of value of breaking any structure into too many levels of detail as the lower levels risk becoming meaningless.

Each task is defined at (and belongs to) a hierarchy level. For example, the structure is raised one floor at a time, so the logical hierarchy level is the floor. Finishes are done one apartment at the time, so the logical hierarchy level is the apartment. It is important not to apply inappropriate detail to an activity. There is no point defining structure tasks at an apartment level, as apartments do not exist at the time of creating the structure. Creating such artificial divisions will create needless additional activity detail that would burden a management system.

To illustrate the method for designing location breakdown structures, three examples are presented. First, a simple residential construction project with three buildings. Second, a special case, a stadium where the structure and the functional spaces are built using different location breakdown structures. Finally, a hospital refurbishment project with multiple-use fit-out.

Example 1

The first example is a residential construction project of 10 risers, A–J (Figure 5.1).

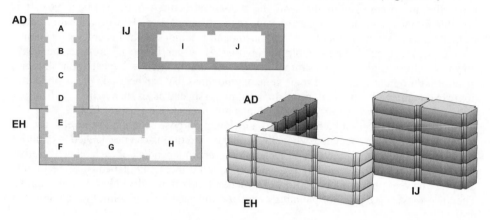

Figure 5.1 A schematic picture of a residential construction project

The highest hierarchy level has been decided on the basis of providing independence of structure. The section comprising of risers A–D can be built independently from sections E–H and I–J, and in any sequence. Notice that risers A–H are actually part of a single large residential complex, but can be built either in sequence or simultaneously with different crews (sections do not need to be whole buildings). Sections are further divided into risers A–H and the risers are divided into floors (four floors in A–H and six in I–J).

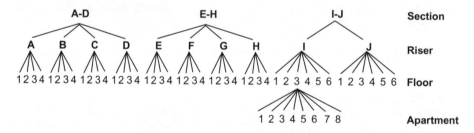

Figure 5.2 A location breakdown structure for the buildings in Figure 5.1

Risers and floors form the middle hierarchies of the LBS. Individual apartments can provide the lowest hierarchy level (5 to 8 apartments per floor per riser) and can be used to facilitate the monitoring and control of tasks which operate at that level of accuracy, such as finishes trades. The resulting location breakdown structure is shown diagrammatically in Figure 5.2.

While this vertical representation is a familiar hierarchical representation of the location breakdown structure (Figure 5.2) often used this way in a spreadsheet, it is also useful to document the breakdown horizontally (as seen in the second example, illustrated in Figure 5.3).

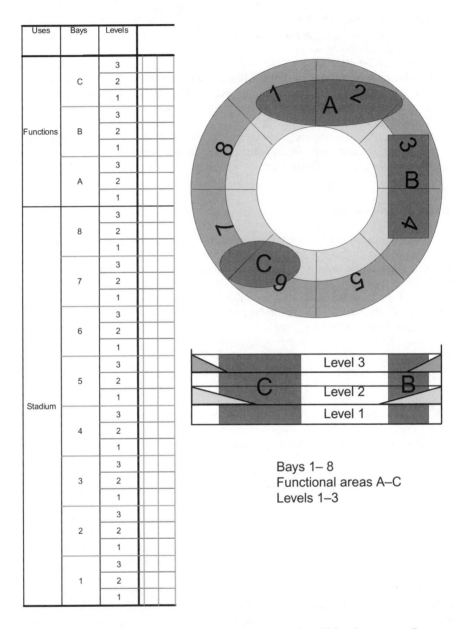

The table portion shows:

Uses	Bays	Levels			
Functions	C	3			
		2			
		1			
	B	3			
		2			
		1			
	A	3			
		2			
		1			
Stadium	8	3			
		2			
		1			
	7	3			
		2			
		1			
	6	3			
		2			
		1			
	5	3			
		2			
		1			
	4	3			
		2			
		1			
	3	3			
		2			
		1			
	2	3			
		2			
		1			
	1	3			
		2			
		1			

Bays 1– 8
Functional areas A–C
Levels 1–3

Figure 5.3 A schematic picture and location breakdown structure of a multi-function sports stadium

Example 2

The second example illustrates the LBS for a sports stadium which contains multiple occupancies or functional areas (Figure 5.3). This example illustrates a technique for using the highest level to allow disparate functional relationships within the LBS. This is a special case, where the main sports stadium with its bays and levels, forms one branch of the LBS and the bays forms the highest level structure with separable parts for the main building. The functional areas, which occupy multiple floors not linked to bays (sports museum, corporate functions, office building, etc.), form another branch and therefore the separable parts for those components.

This method effectively creates different entities (as if separate buildings) within a single building, which forms the highest hierarchy level of the LBS. Levels form the middle levels of the LBS and may be the lowest level within the building component. Similarly levels would form the middle levels and rooms would form the lowest levels for the functional areas component.

This use of the location breakdown structure to model functional or virtual hierarchies is very important for developing complex models of interaction between structures and fit-out—which often have different location breakdown requirements.

Refurbishment example 3

The third example is a hospital refurbishment project in a building with many functions and uses (Figure 5.4). In this example, structure is not relevant and therefore there is no need for the normal separable level of hierarchy. The middle levels are formed when each floor is divided into logical areas of roughly equal size, and with similar services within. Further division of the project into functional units, such as operating theatres, wards, specialist laboratories, etc., may be used for detailed fit-out items and to form the lowest level of the location breakdown structure.

This project highlights a special case, as there is no need to be driven by the structure in planning this example. Rather, as the building is to continue being used during the fit-out with progressive handover, a LBS driven by functional area is the most appropriate division.

LOCATION-BASED QUANTITIES

Quantities are an integral part of the logic of location-based scheduling, and in particular the internal logic of a task. The bill of quantities (measure) of a task defines explicitly all the work that must be completed before a location is finished and the crew may continue to the next location. For example, a tiling task can include waterproofing and floor plastering work to be done with the same crew. Thus a task's bill of quantities may include many items, even with differing units, gathered into a single task where quantities of each of the items may vary from location to location. This approach has been extensively adopted since the late 1980s in Finland (Kiiras, 1989) but due to the adoption of manual methods, their location sensitivity was generally limited to sections and floors. Such a limitation no longer applies with modern software.[1]

After a location breakdown structure is constructed, quantities may be estimated by location. The actual planning begins with a project's bill of quantities (BOQ) and the first role of the planner is to lump related BOQ items into logical packages. Items can be joined into a single package, if the work:

- Can be done with a single crew
- Has the same dependency logic outside the package
- Can be completely finished in one location before moving to the next location.

---

[1]   Such as Vico Software's Control.

Project	Functions	Levels	Zones	
	Carpark	1	1	
	Plant Room	1	1	
	Diagnostics	1	5	
			4	
			3	1 floor of diagnostic services divided into 5 logical areas
			2	
			1	
	Perioperative	1	5	
			4	
			3	1 floor of perioperative services divided into 5 logical areas
			2	
			1	
	Paediatrics - New	Level 5 Roof	1	
		Level 5, L4 Roof	2	
			1	4 floors of new paediatric services divided into 1 or 2 logical areas
		Level 4	1	
		Level 3, AHU	1	
	Paediatrics - Refurbish	Level 4	2	1 floor of refurbished paediatric services divided into 2 logical areas
			1	
	Preliminaries	1		

Stage 1 B

Figure 5.4 A LBS for a simple hospital project

The second role of the planner is to ensure that all relevant quantities are allocated to a task. This approach ensures that nothing of relevance is left out of the schedule, and enables very powerful ways of controlling a project as well as handling change orders and claims (see Section Three). Furthermore, the scheduling process can often reveal any mistakes made in estimating quantities, if the same bill of quantities is used for both planning and estimating. It also ensures that the same assumptions are used in both the cost estimate and the schedule.

Creating a bill of quantities by location

The creation of a bill of quantities is best illustrated by the following example. First, the location breakdown structure of the project is defined. In this example project, there are two buildings with four floors and a roof in each building. This can be transformed into an table (Table 5.1), where items relate to all the work to be done which either form individual tasks or which aggregate to form single tasks.

Table 5.1 Sample breakdown structure

		Project:	Project											
		Section:	Building A					Building B						
Code	Item	Consumption	Floor:	1	2	3	4	Roof	1		2	3	4	Roof	Unit

The location breakdown structure should be prepared in advance, before the quantity surveyor measures, and the schedule planner should participate in this process. There are many ways to derive quantities by location, ranging from a manual take-off to an integrated 3D-model based take-off. These methods will be discussed in Chapter 12.

The bill of quantities for an individual task might look like Table 5.2. The plasterboard wall task includes different kinds of walls, which can each be completed with the same plasterboard crew—thus they can form a single task. There are quantities in all locations except the roof. The quantities are generally larger in Building B than in Building A and the smallest quantities are in the first floor. The consumption column and the calculations will be described later.

Table 5.2 Sample BOQ for the task: plasterboard walls, including item consumption rates

		Project:	Project
		Section:	Building A
		Floor:	1
Code	Item	Consumption	man hours/unit
456100	Erect plasterboard walls between apartments	0.65	1
456226	Mount 13mm special gyproc panelling on plasterboard wall	0.16	8.6
456216	Mount 13mm gyproc paneling on plasterboard wall	0.16	8.6
456146	Mount 79mm panelling on dwelling room wall adjacent to sauna	0.46	8.6
456136	Mount 92mm panelling on washroom wall	0.46	10.7
456126	Mount 79mm panelling on washroom wall adjacent to sauna	0.46	6.4
456116	Mount 92mm panelling on dwelling room wall	0.46	57.4

2	3	4	Roof	Building B 1	2	3	4	Roof	Unit
									M2
26	26	26		31	35.4	35.4	35.4		M2
26	26	26		25.5	30	30	30		M2
26	26	26		25.5	30	30	30		M2
22.5	22.5	22.5		14.1	19.6	19.6	19.6		M2
9.5	9.5	9.5		3.1	6.2	6.2	6.2		M2
118.6	118.6	118.6		109.8	134.3	134.3	134.3		M2

The tasks quantities (with their consumption rates) result in the flowline schedule in Figure 5.5. Flowline figures show the location breakdown structure on the vertical axis and the timeline on the horizontal axis. Tasks are shown as diagonal lines. These illustrate the flow of the work through locations. In this figure, it is clear that no work is required for the roof of either building, because quantities in these locations are zero. It is also apparent that it takes a noticeably shorter time to finish the first floor (steep slope) of BuildingA , and that then the slope becomes gentler for the balance of the project. Consumption rates are constant through all locations, so the change is due to a variation in quantities in the first location.

A location-based bill of quantities enormously strengthens the planning process when using the new theory for location-based planning. If the quantities change or there is a variation (change order), the schedule may be updated by changing the quantity. The logic will then be updated automatically.

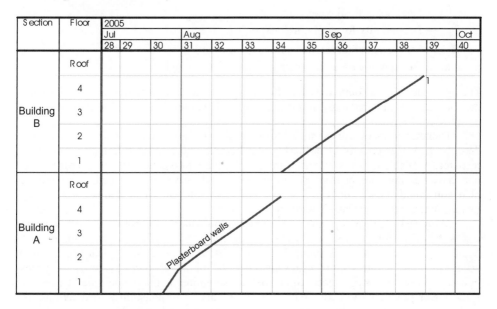

Figure 5.5 Flowline for sample task: plasterboard walls

DERIVING DURATIONS FROM RESOURCES AND PRODUCTION

In the above section, mention was made of consumption rates. This section describes how the quantities in a location may be transformed into durations by use of consumption rates, or resources and productivity data.

Labour consumption is a property of each individual BOQ item which together form the task. The consumption rate indicates the amount of worker or machine time (measured in worker or machine hours) that it takes to produce one unit of each item. It is different from the production rate (how many production units can be produced in a time unit) because it does not require information about the number of crews or shift lengths. However, the consumption value must be based on assumption of optimal crew composition because it may vary for different crew sizes.

The consumption rate may be based on historic data, which may be collected by the quantity surveyor or estimator, or it can be found in general productivity databases. When historical data is used for any future use, the particular circumstances of that project need to be taken into account.

The total quantity of worker hours needed to complete a location is the sum of the individual hours for each BOQ item in the task, which are in turn calculated by multiplying the quantity in that location by the labour consumption rate for the item.

The calculation of total worker hours per location is described in Equation 5.1:

$$h^T{}_j = \sum_{i=1}^{i=n} (Q_{i,j} \times R^C{}_{i,j}) \qquad\qquad (5.1)$$

Where:

$h^T{}_j$ = Total hours for the location j

$Q_{i,j}$ = The quantity for item i, in location j

$R^C{}_{i,j}$ = The production rate for item i, in location j

and where there are n items, i, being grouped into a single task.

The aggregation is in hours, therefore items of work can have different units of measurement and still be gathered into a single task. This calculation is required for each location of each task.

The total hours of work required for the task, whether in total or by location, doesn't describe how the work is to be performed, because there is no information about the resources available to make the calculation of duration. In practice, tasks use crews which may have varying compositions of resources, and which when combined with labour or plant consumption rates, yield an effective production rate which can then be used to calculate the duration for the task for a given location.

Many tasks have an optimal crew composition which will more efficiently perform the work. Therefore, work can best be accelerated or decelerated by increasing or decreasing in multiples of the optimal crew size (Arditi et al., 2002). It is less efficient to accelerate or decelerate work by using a less optimal mix of resources or crew size. If a less optimal mix of resources is used, the labour consumption number should be increased to show lower productivity. Ultimately, it is the planner's responsibility to determine the optimal crew size for the particular circumstances of each task in each project.

Resources may vary in their contribution to the total effort of the crew. Therefore, each resource must have a factor for its individual productivity. The default productivity factor would be 1.0, which means that the worker contributes one standardised worker hour for each hour worked. Less or more productive resources have factors other than 1.0.

It is important to note that the use of productivity factors will depend on the basis of the calculation of consumption in the underlying database of historic rates. This is because consumption may include non value-adding activities such as hauling, cleaning, etc. or may be limited to direct production. In the former case, workers performing non value-adding activities would have a production factor of 0.0. In the latter case, all resources would be based on a production factor of 1.0. Similarly, if consumption has been given in machine hours, the machine driver and any assisting labour, should have a production factor of 0.0 and the machine would have production factor of 1.0.

For special circumstances, or to allow for the effect of starting difficulties or learning processes (Chapter 6), it is possible that work in some locations may be easier or more difficult than that which formed the basis in the historic data, and therefore may vary from the project average. For this reason, locations may have difficulty factors which can be applied to the entire task for each location.

The duration (number of shifts or days) in each location may be calculated based on the quantities and crews selected using the following steps:

1. Calculate the quantity of worker hours needed to complete the location.
2. Divide the result by the sum of production factors of the selected resources to get the duration in hours.
3. Divide by the shift length to get the duration in shifts.
4. Multiply the duration (shifts) by the difficulty factor.

This duration is the optimum rhythm of the selected crew. It is described in Equation 5.2

$$D^S_{\,j} = \left(\frac{h^{\,T}_{\,j}}{\sum_{i=1}^{i=n} P_{i,j}} \right) \div h^S \times d_{\,j} \tag{5.2}$$

$$Duration = \left(\frac{Total\ hours}{\sum_{i=1}^{i=n} productivity_{\,i}} \right) \div shift\ length \times difficulty$$

Where:

$D^S_{\,j}$ = Duration in shifts for the location j

$h^{\,T}_{\,j}$ = The total hours in location j

$P_{i,j}$ = The productivity for item i, in location j

h^S = Shifts length (hours) for the task

$d_{\,j}$ = The difficulty in location j

LAYERED LOGIC: LAYERING CPM LOGIC IN LOCATION-BASED METHODS

The new theory of location-based scheduling involves far more than merely linking like activities in chains to derive resource optimisation—as sometimes suggested by CPM practitioners and some authors (for example, Suhail and Neale, 1993). Rather, it involves several layers of interactive CPM logic which combine to form a powerful location-based logic, *layered logic*, which involves the following:

1. External logical relationships between activities within locations.
2. External higher-level logical relationships between activities driven by different levels of accuracy.
3. Internal logic between activities within tasks.
4. Phased hybrid logic between tasks in related locations.
5. Standard CPM links between any tasks and different locations.

Layered logic is not hierarchical, instead all logics apply equally. Layers 1, 2, 4 and 5 logic affect the logic network directly and can be implemented using normal forward and backward passes of the CPM algorithm. Layer 3 logic includes the possibility to force tasks to be continuous—which requires augmenting the CPM algorithm to allow later locations to pull earlier locations. Therefore the resultant forward and backward pass calculations require multiple iterations. It is certainly possible to argue that location-based logic is extended CPM logic, however traditional CPM algorithms do not support the continuity requirements needed to perform location-based planning. The following logic layers are required.

Layer 1—external dependency logic between activities within locations

The *external logic* of location-based tasks controls the links between activities or tasks within locations and this forms Layer 1 logic. In location-based logic, it is assumed that in each location the logic between separate tasks is similar. This greatly simplifies the complexity, as it is only necessary to create a link between two tasks, which is then copied to each individual CPM activity in each location. Thus it is possible to consider a generic logic network defining the relationships between activities in any location. A logical connection created between two tasks is therefore automatically created and replicated for each location, regardless of any relationships existing between locations. Normal CPM calculations can be used for Layer 1 logic links, because Layer 1 is just a way of automating CPM logic creation by use of locations. Indeed, CPM is a single location model using Layer 1 logic.

Figure 5.6 presents two tasks which occur on every floor of two buildings. There is a Layer 1: F–S link between the activities, such that Task 1 must be finished on any floor before Task 2 can commence on the same floor.

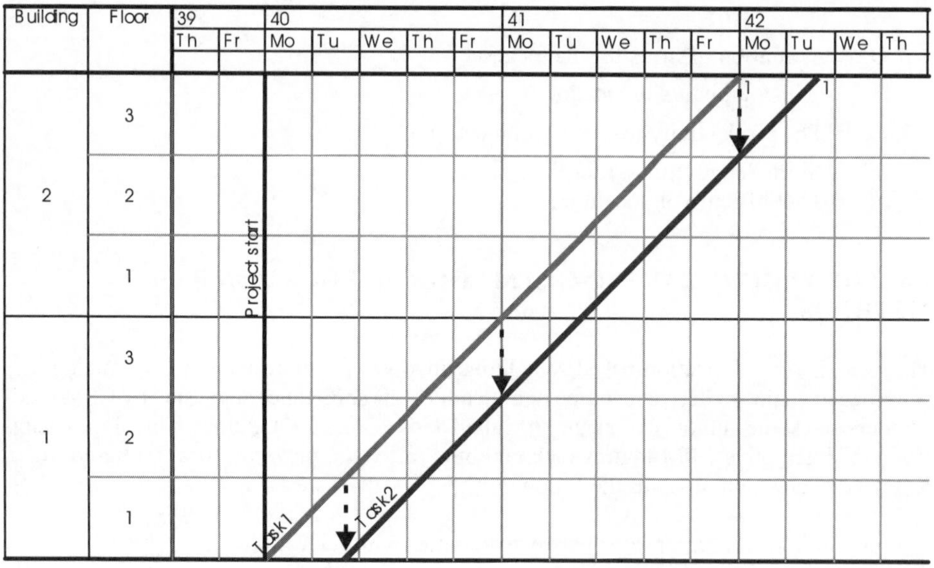

Figure 5.6 Layer 1 logic—External dependency logic between activities within locations

Layer 2—external higher-level logical relationships between activities driven by different levels of accuracy

Layer 2 logic extends Layer 1 logic to provide CPM logic where task relationships exist at different levels of accuracy. Each location-based task must be allocated an accuracy level which corresponds to a hierarchy level in the location breakdown structure. The accuracy level means the lowest level of locations which is relevant to the task. For example, the natural accuracy level for structure would be either floor or pour, depending on the project, as the structure is raised sequentially by floor. The natural accuracy level for finishes may be the individual apartment or even a room.

Just as tasks have a location-based accuracy level, similarly each link between tasks has an accuracy level. This link can be defined at any level of accuracy which is the same or higher (rougher) than the highest accuracy level of the two location-based tasks. For example (Figure 5.7), if roof work and concrete floor finishing work both have the floor level of accuracy, setting the accuracy level of link between the tasks to the building would mean that the roof must be finished in a building before the concrete floor finishing work can start in the same building.

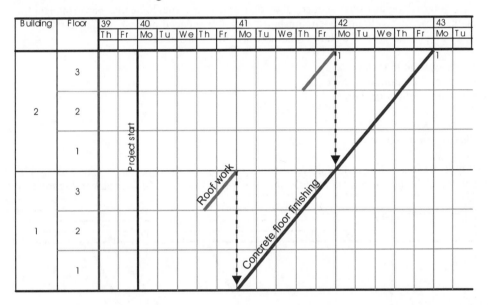

Figure 5.7 Layer 2 logic—External dependency logic between activities driven by different levels of accuracy

Layer 3—internal dependency logic between locations within activities

The third logic layer is critical to the achievement of flow of resources and uninterrupted work, and thus may be considered *flow logic*. This is the internal logic within a task between locations, and therefore between the activities in locations that form a task.

Unless deliberate decisions are made to the contrary, by using multiple work crews or splitting the task to allow parallel work, it may be assumed that work will flow between locations sequentially with finish-to-start links. The underlying assumption is that the crew completely finishes a location before moving to work in the next location. Thus, even though external CPM logic may allow work to commence early, the work cannot proceed until the resources are released from the previous location. It is not that internal logic takes precedence, but rather that both must apply.

In addition, the task has an internal location sequence which describes the sequence in which the locations will be completed for that task. The sequence for each task is independent of other tasks or any perceived sequence for the project, and it is one of the important planning decisions to be made when scheduling projects.

It is common in CPM planning for such structures to be modelled using chains of activity sequence. However, frequently such models use overlapping tasks—for example, a chain of activities with S–S and F–F links with two days lag on each. Some may consider it a shortcoming that this is not a possibility with Layer 3 logic, however this actually arises from a misunderstanding of what the overlaps mean. It either means that more than one crew is briefly engaged (or a single crew is spreading over more than one location) or it means that completing locations continues while starting up new locations. The former represents less than optimum production and should be discouraged, the latter represents difficulty in completing work in locations. Nevertheless if such a structure is required, it can be modelled by recognising that the true duration is the period from commencement in one location to the commencement in the next location, and the overlap can be modelled by use of a buffer (or an increased buffer).

Internal logic ensures sequential work, however external logic may cause interruption in the flow of work in the circumstance that the succeeding task is quicker than the preceding task. Figure 5.8 shows that there are three special cases:

- Slower continuous work
- Faster discontinuous work
- Faster continuous work.

Slower continuous work

Where the succeeding task is slower than its predecessor, the task starts as early as possible and gradually gets further behind the preceding task (Task 1—Slower task in Figure 5.8).

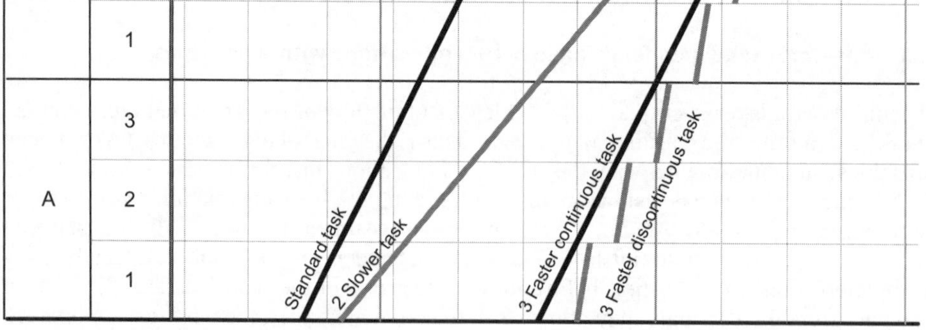

Figure 5.8 Layer 3 logic—Three cases of internal logic

Faster continuous work

Where the succeeding task is faster than its predecessor, but is to remain continuous, the task start is delayed until the work may complete as early as possible but remain continuous (Task 2—Faster continuous in Figure 5.8).

Faster discontinuous work

Where the succeeding task is faster than its predecessor and is allowed to be discontinuous, the task starts as early as possible in each location (Task 3—Faster discontinuous in Figure 5.8). An example would be the way most site managers commonly react with faster activities, by driving subcontractors to commence work they cannot execute to completion and giving the appearance of greater efficiency.

Example of the impact of making a task continuous

The third layer of internal logic provides standard CPM links to other locations for the same task, whereas the first layer provides links to other tasks within the same location. The CPM calculations could be used directly in the forward pass if everything started as soon as possible—a requirement of CPM (Figure 5.9). However, third layer internal logic breaks this requirement and allows for faster work to be forced to be continuous, thus changing the assumptions for the calculations. This is desirable, as the alternative will lead to discontinuous work which may cost money and cause disturbance on the site.

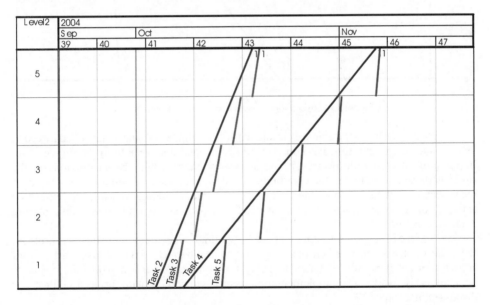

Figure 5.9 Layer 3 logic—Illustrating discontinuous faster tasks and continuous slower tasks

Many CPM planners think continuity is the same as making tasks start as late as possible, however the continuity requirement apply regardless of total float, making this actually very different.

Let us take a simple sample project with five floors and four tasks, each with F–S start links between tasks in each location and with everything starting as soon as possible (the default CPM assumption). The situation is shown in Figure 5.9. Here, Task 3 is faster than Task 2 in each location so there is a break of flow of one day on each floor. Task 5 is faster than Task 4 and its flow is broken for four days on each floor. This means that the workers of Tasks 3 and 5 will have to slow down or they will run out of work (and probably leave the site). Note the end time for the CPM solution (Friday, week 45).

Flowline requires a more flow-oriented planning with continuous work execution, which would force the tasks to be continuous—in other words, it would delay the start dates of overly fast tasks. Layer 3 location-based scheduling logic allows the user to choose the tasks which should be done continuously and the tasks where the flow can be broken.

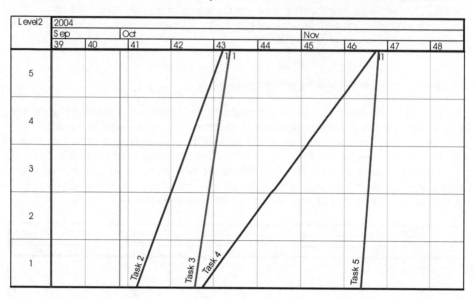

Figure 5.10 Layer 3 logic—Used to make all tasks continuous

Figure 5.10 shows the above example, but where all the tasks have now been forced to be continuous. The end time has consequentially jumped one week to Friday of week 46. Importantly, this schedule has less risk because work crews will not have cause to leave the site, which saves cost due to the continuity of production. There is still room for improvement because it would be possible to save further time by adding resources to bottleneck activities or removing resources from overly fast tasks. Such optimisation is discussed in detail in Chapter 7.

While the principle of forcing work to be continuous seems simple, there is a problem with site implementation when planning to have deliberately delayed task starts. Work supervisors are generally reluctant to delay the commencement of work, preferring to demand early work. The implementation issues concerned with project control are discussed in Chapter 12.

Example of different task sequences

Suppose a project has three hierarchy levels within its location breakdown structure. The top hierarchy level (project) has one location. The second level (section) has two locations. Each section has three floors (on hierarchy level three). This structure may be observed in Figure 5.11. Three tasks have been created. Task 1 is at the project level of accuracy, so it functions as a basic CPM activity. Task 2 is on section level of accuracy with location sequence 1, 2. Task 3 is on section level of accuracy with location sequence 2, 1. Task 4 is on floor level of accuracy with location sequence 2–1, 2–2, 2–3, 1–1, 1–2, 1–3. Task 5 is on floor level of accuracy with location sequence 1–1, 1–2, 1–3, 2–3, 2–2, 2–1. Figure 5.12 shows a Gantt chart of the project.

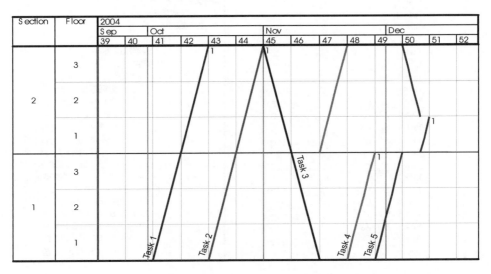

Figure 5.11 Layer 3 logic—Flowline indicating the visualisation of different sequences

Task activities, in locations, occur sequentially and are bound to other locations of the same task by finish-to-start links. The number and accuracy level of locations (and thus the number of CPM activities included) are dependent on the task's chosen accuracy level and the project's location breakdown structure. In the flowline, the location sequence for individual activities is clearly visible, as is the space between (locations where no work is in progress).

Layer 4—additional location-based logic links

Layer 1 and 2 logic links are external logic links between tasks and within locations, Layer 3 logic is internal logic between locations but within tasks but a new layer of logic, Layer 4 logic, is required to model the special case of location-lags in the sequencing of external logic.

Just as there can be a time lag between activities in CPM, there can be a location lag between tasks in location-based planning. For example, pouring in situ concrete for horizontal structures (such as slabs) interferes with the formwork of the floor above as well as with the interior works often two floors below its location due to temporary propping until the slab can bear its own weight. It does not affect the same floor, or a higher LBS level (the whole building) so this cannot be simply modelled by accuracy levels. However, this can be modelled using just one location-lag link for each case. Formwork is preceded by pouring horizontal concrete with a lag of 1 floor. Interior works are preceded by pouring horizontal concrete with a lag of –2 floors. A positive location lag of 1 means that a *lower* location must be finished before starting the succeeding task on the next floor. A negative lag of 2 floors means that two *higher* locations must be finished before starting the succeeding task on a floor. Location-based lags can work only within a group of locations, to avoid examples such as the floors of one building restricting the finishes in the next building. This

Hierarchy	Name	Duration	Start	2004														
				Sep	Oct				Nov				Dec					
				39	40	41	42	43	44	45	46	47	48	49	50	51	52	
+1	**Task 1**	10	4.10.2004															
-2	**Task 2**	10	18.10.2004															
2.1	Task 2 - S1	5	18.10.2004															
2.2	Task 2 - S2	5	25.10.2004															
-3	**Task 3**	10	1.11.2004															
3.1	Task 3 - S2	5	1.11.2004															
3.2	Task 3 - S1	5	8.11.2004															
-4	**Task 4**	10	15.11.2004															
4.1	Task 4 - S2 - F1	1.7	15.11.2004															
4.2	Task 4 - S2 - F2	1.7	16.11.2004															
4.3	Task 4 - S2 - F3	1.7	18.11.2004															
4.4	Task 4 - S1 - F1	1.7	22.11.2004															
4.5	Task 4 - S1 - F2	1.7	23.11.2004															
4.6	Task 4 - S1 - F3	1.7	25.11.2004															
-5	**Task 5**	10	29.11.2004															
5.1	Task 5 - S1 - F1	1.7	29.11.2004															
5.2	Task 5 - S1 - F2	1.7	30.11.2004															
5.3	Task 5 - S1 - F3	1.7	2.12.2004															
5.4	Task 5 - S2 - F3	1.7	6.12.2004															
5.5	Task 5 - S2 - F2	1.7	7.12.2004															
5.6	Task 5 - S2 - F1	1.7	9.12.2004															

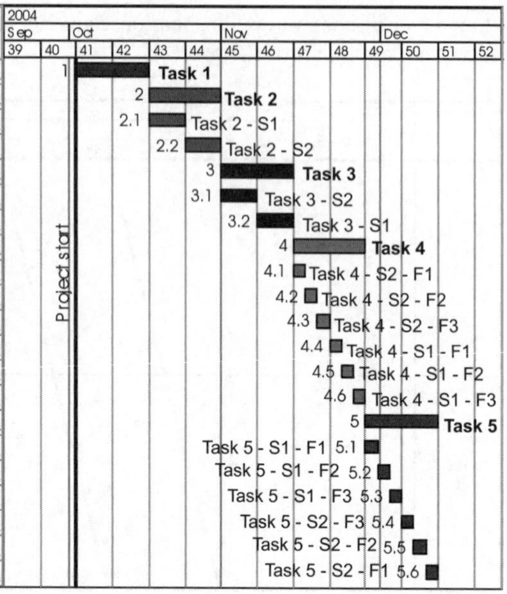

Figure 5.12 Layer 3 logic—Gantt chart illustrating sequence of locations

location grouping needs to be planned for each project based on physical links between locations (usually on basis of one location being on top of another). It is important to note that location lags are often used to create cycles, which would otherwise lead to circular logic. A typical cycle might be:

- Set out floor
- Form columns
- Reinforce columns
- Pour columns
- Slab formwork
- Slab reinforcement
- Pour slab.

In this case each task follows the other in logical sequence. Pouring the slab necessarily ultimately follows setting out the floor. However the next floor up cannot be set out without a slab to stand on, so the task 'set out floors' must follow the task 'pour slab'—but with a location lag of 1, that is the floor below (within its location group).

Figure 5.13 shows an example of two buildings of 20 and 10 floors. There are four tasks on each floor: formwork, reinforcement, concrete and interior work. The links are Layer 1 links, except that formwork is preceded by concrete with lag of 1 floor and interior work by concrete with lag of –2 floors (Layer 4 links). The sequence of locations (Layer 3 links) has been manipulated so that waiting times are minimised. The formwork crew can work in the second building when it would otherwise have had to wait in the first building. The vertical axis of the flowline figure has been adjusted, based on the sequence of formwork to show continuous line. The formwork crew can therefore work continuously by working in two buildings. However, the reinforcement and concrete tasks have waiting time because of their faster production rate. Interior work has been split into two parts to optimise continuity and duration.

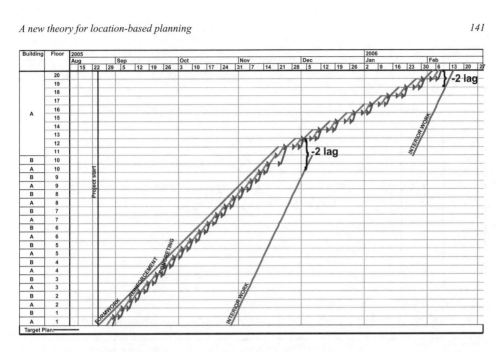

Figure 5.13 Layer 4 logic—enables complex location sequencing based on location relationships

Note that the combination of Layer 4 and Layer 1 links above prevents making concreting task continuous. This happens because knowing the finish time of concreting on Floor 1 is required before the start time of Formwork on floor 2 can be calculated. If concreting was able to be made continuous, formwork would need to be recalculated which would again affect concreting. This cascading effect would cause an infinite loop of calculations. In this case Layer 4 logic overrides Layer 3 continuity logic.

Layer 5—standard CPM links between any tasks and different locations

Layer 5 reintroduces the standard CPM logic link which can be between any task and any location. It is normally applied between tasks, but can be applied internally to a task. If applied internally to the task, the logic overrides the internal logic of the task: for example, constraining possible sequences. If created between two tasks and two locations, this link type can be used to account for special circumstances such as one part of the LBS affecting another part of the LBS. This layer is needed to model, for example, links between structure and finishes in complex location breakdown structures where the same LBS cannot be used for both structure and finishes (for example, the sports stadium, with a different LBS branch for each of the structure and functional spaces).

In Figure 5.14, Task 2's first location 2–1 must follow Task 1's location 1–2 (a single location Layer 5 link). There is also a constant Layer 1 link between the tasks such that Task 1 must succeed Task 2 on every floor. This example illustrates that multiple layers of logic must be able to be active simultaneously.

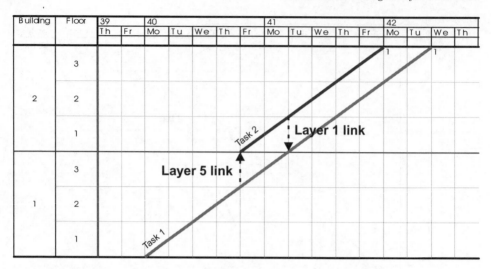

Figure 5.14 Layer 5 logic—fixed links between tasks and locations, combined with Layer 1 logic

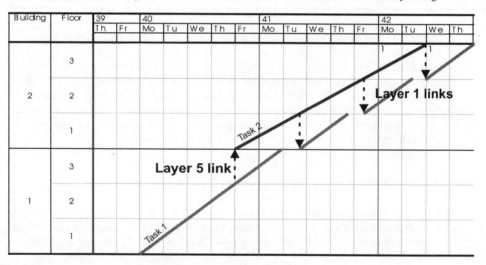

Figure 5.15 Layer 5 logic—Task 2 cannot be continuous due to conflicting Layer 5
and Layer 3 logic where the precedent task is slower

In the special case illustrated above, the relationship between the two tasks in fact becomes circular. This works for this example, however, were Task 1 to be slower than Task 2, Task 2 could not be made continuous. In which case, Layer 5 logic takes precedence over flow logic and thus Task 2 will become discontinuous (Figure 5.15). Any attempt to make Task 2 continuous would relocate the Layer 5 link in time—with a cascading effect.

Circular location-based logic

In addition to the standard CPM process of checking for circular logic between activities (in locations), circular logic also needs to be checked for at the task level (the entire location-based task). Location-based tasks can have circular logic which does not result in circular logic between location activities. As described above, this can easily happen with a combination of Layer 4 or 5 logic with other layer logic links. In such a situation, ordinary CPM calculations would not be affected, but updating start dates to achieve continuity may cause a problem with circular updating of tasks, as changing one task may break the continuity of another task, forcing another recalculation, and so on. In this case, the start dates can be shifted by the continuity heuristic only to the point where the shift does not affect the previous activity.

Conditional task logic in locations without quantities

If there are no quantities measured in a location, no work is assumed in that location and therefore the dependency logic flows through (appears to bypass) the location. Its predecessors and successors are controlled by that task only where it exists. In its absence, they will be governed only by any direct logical relationships with those earlier or later tasks. A precedence relationship cannot be assumed to be implied across the missing task in that location. It is necessary to add further precedence relationships if they are required, to allow for conditional logic (logic that depends on the actual work in each location).

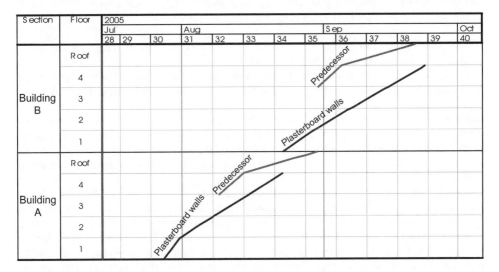

Figure 5.16 Conditional logic when tasks are without quantities in specific locations

Let us return to the example introduced in Figure 5.11, and now assume we have a task which is required only on the roof and the highest floor. In Figure 5.16 the predecessor to the plasterboard task appears to flow through the plasterboard activity where there is no plasterboard (on the roof). The plasterboard wall must be done after the task predecessor on each level, where (and only where) both activities exist. This fact can be used to optimise the sequence of work by starting the succeeding task in locations which become available earlier because of the absence of that predecessor.

The treatment of the logic relationships between tasks in the event that quantities do not exist in a given location is therefore a special case. While the most obvious assumption is that all precedence tasks to the missing task should become immediately available to the successor task, this will restrict the practical application of location-based logic, which can use the absence as a switching mechanism—conditional logic—to construct alternative logic paths depending on the circumstances within a single project, or to allow for complex relationships between activities which vary depending on the particular mix. Thus logical relationships which are always required must be specifically included in the task relationships if it is likely that a missing task will lead to incorrect logic. For example, installing plasterboard walls may be a predecessor for installing mechanical ducts in the locations where they happen together. Plasterboard walls are preceded by the roof being waterproof on the building level. If the plasterboard walls are not required in a location, the mechanical ducts should not be automatically linked to the roof, because they will not be affected by wet conditions, thus the link is invalid.

Dependency lags and buffers

Time lags and buffers are not an additional layer of logic (unlike location-lags/buffers which form Layer 4 logic), as they belong to all dependency relationships. However, they are a critical part of the layered logic of the location-based management system. Lags are a well-known component of CPM logic, however buffers will be unfamiliar to those not versed in lean construction. The following definitions apply:

- Lag: the required fixed duration of a logical connection between two activities or tasks. Examples are curing time or start-up delay.
- Buffer: the additional absorbable allowance provided to absorb any disturbance between two activities or tasks as a component of the logical connection between two tasks.

Buffers appear very similar to lags, except they are an additional time allowance meant to protect the schedule and are intended to absorb minor variations in production. The continuous flow of work needs to be protected—as will be explained in the discussion of production system risk (Chapter 6). Therefore, each dependency link has an additional attribute: a buffer.

A buffer functions technically in exactly the same way that a lag of the dependency does, however it is added on top of the lag only when calculating the forward and backward pass of the planned schedule to establish earliest and latest start and finish dates. It is also used to calculate the float of locations and location-based tasks. The buffer part of the link is ignored when forecasting future problems during the implementation phase or when doing risk analysis and analysing the probability of interference. In risk analysis and forecasting, the buffer may be absorbed by delay without impacting on the overall project duration.

RESOURCE LEVELLING

Layer 3 logic automatically levels resource consumption within the context of a single location-based task. However, if the same resources are shared across multiple tasks, additional resource levelling methods are required. Resource levelling over multiple location-based tasks is similar to Layer 3 logic. However, instead of sequencing just locations for a single task, the task–location pairs must be forced to be sequential for continuous resource consumption for a given resource type.

For example, a plasterboard wall subcontractor may perform many different activities with the same labour resource—layout and top track, framing and stud installation, installation of first board and closing the wall. Often there will not be sufficient resources to undertake all these at once and yet it is unlikely that a separate crew will be mobilised for each task. Therefore the sequence of work for plasterboard wall resources must be planned for the whole scope of work. In a project of three locations, the sequence might be as follows:

- Layout and top track 1, 2, 3
- Framing and stud installation 1, 2
- Installation of first board 1,2
- Framing and stud installation 3
- Installation of first board 3
- Closing the wall 1, 2, 3.

In the location-based management system, multiple location-based tasks which share the same resource will only be allowed simultaneous execution when resource constraints and other logic allows it. However, Layer 3 logic forces the same resources to be used within a single location-based task.

For example, even if additional resources were made available, the system would not permit layout and top track to begin in all three locations at once. Nevertheless, overlapping of framing and stud installation with layout and top track or with installation of first board would be possible if enough resources could be found. When resource constraints are exceeded, the resources are allocated using the planned sequence.

Figure 5.17 shows the example where one crew performs all the sub-tasks of plasterboard walls. The sequence has been optimised so that the electrical work can be done between the installation of first board and closing the wall.

Workable backlog

Projects always include tasks which do not require direct logical successors in the same location and may be undertaken as background to other activities. Such tasks may make use of spare resources as they become available and as long as they are not required for critical tasks. Examples include cleaning, electrical switchboard installation and installing accessories. Sometimes entire locations may be considered workable backlog, for example a parking garage in an office building, or a location on a branch of the LBS which can be completed independently and well ahead of any delivery requirements, such as the renovation of a floor in an existing building adjacent to the main project.

In the location-based planning model, it is possible to define a task as workable backlog. Workable backlog, or planned work complete, is an important concept in lean construction (Ballard, 2000; Mitropolous, 2005) and can be used to provide work when gaps appear in the sequencing of work for a crew. Such gaps may occur when a task is sequenced as early as possible, when resources are shared, or when logical constraints such as location-lags disrupt smooth production sequencing (such as a structure production cycle). Workable backlog tasks may be scheduled last, after all the schedule calculations are finished and resource needs for critical tasks are known. They can then utilise available resources which are on site but would otherwise have to leave because of a break in workflow caused by discontinuous work or finishing of some other task. Workable backlog tasks are special in that they can be undertaken discontinuously (the task can be split at the activity level), meaning that work continues only when resources are available from other tasks and where there is no penalty in disrupting the workflow (see Figure 5.18).

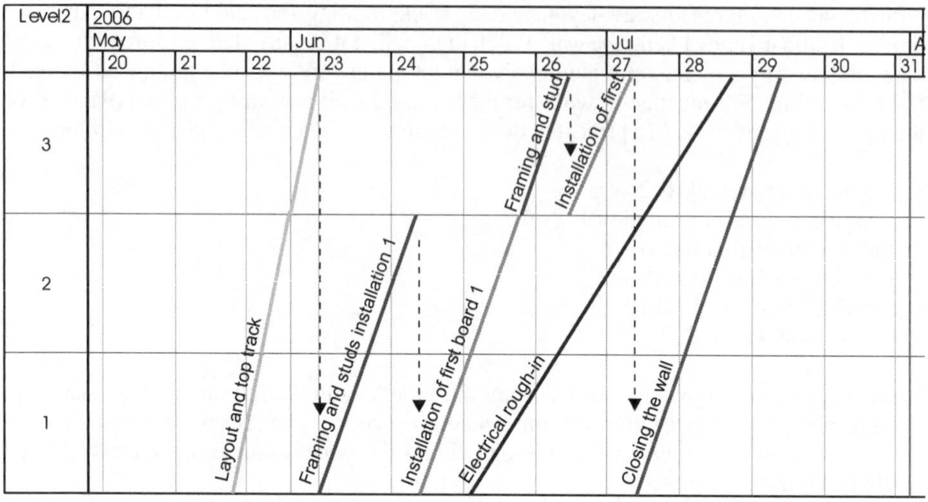

Figure 5.17 Resource levelling for shared-resource tasks

Workable backlog is the location-based equivalent to *splitting* in CPM resource levelling procedures. The difference is that it results from a deliberate planning decision to select which tasks are allowed to be split, and it is only allowed in special circumstances. This ensures production efficiency is protected, while allowing backlog tasks flexibility.

The main application of this is to have a list of tasks and/or locations for each subcontractor which are not critical in terms of the schedule but which need to be completed before the end of the contract. If, for any reason, there is a break of work in a critical production activity, the resources can work on workable backlog activities. In effect, these tasks function as a work buffer against variability.

Workable backlog tasks have normal predecessor and successor relationships (of any layer except Layer 3). They are scheduled as in normal CPM to get the earliest start and latest finish dates. The earliest start and latest finish in a location define the available time window for the workable backlog task. The resource use is examined for each day of this task window. For each day that resource use for any critical tasks is less than the available resources (the difference between the resource use of the previous day and the current day) and if the day is within the task window of one or more workable backlog tasks that use the same resource type, the available resources may be allocated to workable backlog tasks. If there are multiple eligible tasks or locations, resource allocation should be decided based on the following heuristics:

- For each workable backlog task and location calculate the remaining worker hours needed to complete the location per available time frame (from current day to latest close of the task window). This calculation defines how many worker hours should be produced in a day on average.
- Available resources are allocated to workable backlog tasks in proportion to this value, the most critical, breaking the ties in favour of the task having the largest value.
- Additional resources are required once the average daily production equals or exceeds a full shift.

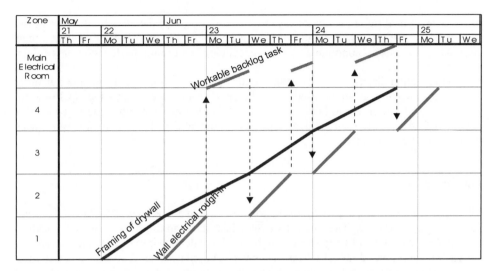

Figure 5.18 Workable backlog being used to level resource consumption for two shared-resource tasks

Planning should not allow workable backlog to become critical.

In some cases, workable backlog tasks may be mutually dependent. When this occurs, the available time frame should be calculated for each task independently—ignoring the link—and the overlap of the various available time windows should be divided in each location relative to the proportion of worker hours needed to perform the task in that location.

It is important that the scheduling of workable backlog tasks is not allowed to result in the delay of a critical task, or in the backlog task itself becoming critical. If resources are not available for the available time period, new resources must be mobilised sufficient to perform all workable backlog tasks so that they can be completed on time. If resource constraints are exceeded by this, the planner should reconsider the dependency relationships, the workable backlog status of tasks, or change the resource constraints.

Figure 5.18 shows the example of two normal tasks and a workable backlog task. In this example, framing of plasterboard wall tasks must happen before wall electrical rough-in can start on a floor. Because the electrical rough-in is a faster activity, it would become discontinuous. However, electrical work in the main electrical room is an available workable backlog task, which can proceed at any time when the wall electrical rough-in is unable to proceed. This workable backlog task can fill in the gaps and completely balance the resource flow, albeit with a planned increase in mobilisation and demobilisation.

WARNING: TECHNICAL MATERIAL FOLLOWS

The following sections are more technical and can be safely skipped by practitioners. Jump to the section on splitting on page 156.

SCHEDULE CALCULATIONS

Underlying a location-based planning methodology is the assumption of a layered CPM logic. This means that it is possible to derive a comprehensive CPM schedule from the heuristics which form the location-based model. This is important, as the CPM engine provides analytical rigour, while the layered heuristics provide the modelling power of the system. The resulting CPM schedule will include normal CPM logic links, as well as pulled start dates to ensure continuous work sequencing according to location-based task requirements.

This section discusses the algorithms underlying a location-based flowline schedule. As already discussed, location-based scheduling involves a powerful layering of CPM logic but following location-based heuristics, rather than the more common single layer CPM approach. The basic principles and methods remain the same and normal forward and backward passes are performed.

Forward pass

The main principles and objectives of the forward pass of a five-layered CPM are the same as in a single-layer CPM. The idea is to give earliest start and finish dates for each location of each task. Normal CPM algorithms can be used to calculate initial start dates. However, for each location-based task, the start dates are manipulated based on the chosen timing option using appropriate heuristics. With the exception of circular logic and resource constraints, location-based tasks are first sorted by precedence order, so that all locations for all predecessors of a location-based task (in any location) will be processed before any location of the task can be processed.

Circular location-based links can occur when Layers 4 and 5 conflict with Layers 1 or 2, with each other, or because of resource constraints. For example, the sequence *formwork–reinforcement–concreting* has a Layer 1 link on each floor but formwork must succeed concreting with a location lag of 1 floor (a Layer 4 link).

In these special cases it is not possible to calculate all locations of the predecessor before all locations of the successor. However, it is often possible to calculate many locations of the same task in sequence. The locations of the predecessor task are added first to the sort, and continue to be added until the first location that depends on a successor task's start date is encountered or a location for a task is found to be using the same resource that has been planned to be in that location earlier in the sequence. Start dates are calculated for these unencumbered locations in sequence, then the start dates are calculated until the predecessor's uncalculated start dates are needed to calculate a location's start date. Whenever there is more than one location of a location-based task in sequence, it is possible to force that part to be continuous using the same mechanism as for location-based tasks without circular links. In effect, the Layer 3 logic is split around the Layer 4 or 5 link.

The initial start dates for locations are calculated using the ordinary CPM algorithm (which has been described in Chapter 2). In case of a resource constraint, the resource availability is checked for each scheduled location in the priority sequence. If the start date has to be delayed because it has exceeded the maximum available resources, the start date is delayed until resources become available. In this case, a temporary resource link is added to the network from the location, releasing resources to the location requesting them. This link is used in the backward pass calculation to calculate total and free floats. Adding this link also breaks the continuity calculations for the task which releases the resources.

If desired, sequence optimisation can be calculated for any task at this stage. To minimise duration, the optimum sequence for any individual activity can be defined by calculating the earliest start date for each location, ignoring the internal logic links. The best sequence is then achieved by sorting the locations in ascending earliest start order. However, this crude heuristic does not work for the optimisation of the sequence of early tasks, which define the project's sequence. Additionally some of the tasks have logical constraints on sequence (like structure, which must necessarily go upwards in sequence).

Thus location optimisation works only for certain activities. It might be noted that normal CPM has no such rules for sequencing, apart from the planner's logical connection order.

After the possible location sequence optimisation, the start dates of locations are adjusted, based on the timing option selected from the following:

- As soon as possible
- Continuous
- As soon as possible and continuous
- Manual timing.

Depending which option is selected, the following steps must be taken to adjust the start dates for each location-based task or, in the event of circular logic or constraining resource constraints, for each task part.

As early as possible

This is the default CPM option, so start dates are as calculated as early as possible and not adjusted. The resulting task plan may be either continuous or discontinuous for the task, depending on the rate of its predecessors. A faster task will become discontinuous.

As early as possible and continuous

This is the default option for protecting production efficiency.

1. If the task, when calculated to be as early as possible, is found to be continuous (for all location pairs, the start date of the succeeding location equals the finish date of the preceding location), use the as soon as possible dates.
2. If the task, when calculated to be as early as possible, is not found to be continuous:
   a. Sort the locations in order of the descending finish date.
   b. For each location, shift the start date so that the finish date equals the start date of the next location.

The resulting task plan is always continuous and always as early as possible.

Continuous

When using a risk management methodology, or when planning suggests a start on a specific date, the task can have a target start date. In this case, the task will not be set as early as possible, but the task will still be planned to be continuous to protect production:

1. If the current start date of the first location is less than the target start date:
   a. Shift the start date of the first location to the target start date.
2. If the task is continuous (for all location pairs, the start date of the succeeding location equals the finish date of the preceding location), use the calculated dates.
3. If the task is not found to be continuous:
   a. Sort the locations in order of the descending finish date.
   b. For each location, shift the start date so that the finish date equals the start date of the next location.

The resulting plan is always continuous but may not be as early as possible.

Manual timing

When using a risk management methodology, or when a task must start on a specific date, the task will not be set as early as possible, and the task need not be continuous:

1. If the current start date of first location is less than the target start date: shift the start date of the first location to the target start date

The resulting plan may be continuous or discontinuous for the task.

Special calculations

The continuity calculations may cause shifts in the timing of succeeding activities. Therefore, the calculations are not the same as scheduling tasks to start as late as possible within the constraints of float—float has not even been calculated yet. Rather, productivity is optimised by making work continuous. In some cases this may lead to a longer total schedule duration. However, optimisation of production rates and being able to achieve higher productivity will, in most cases, offset this duration increase and most projects will achieve duration cuts after the schedule has been optimised. Schedule optimisation is described in detail in Chapter 6.

Finally, the start times of any workable backlog tasks are calculated based on the availability of resources. For each workable backlog task, the available time period is calculated for each location from the finish date of the last predecessor and the start date of the first successor. If there are workable backlog tasks which have links to other workable backlog tasks, the overlapping available time period is allocated for each task in relation to the total worker hours in the location. Workable backlog task locations are sorted into sequence of ascending available time period. Resources are allocated to workable backlog tasks using the following procedure:

1. Go through all the days of the available time period, then for each day:
   a. If, for the resource type required, the resource use on that day is less than the resource use on the previous day, allocate those resources to the task, reduce the number of required worker hours by the shift length times the number of resources.
2. If, after the forward check there are required worker hours left:
   a. Go backwards through all the days of the available time period, and for each day:
   b. If the resource needs of the next day are greater than the resource needs of the current day, then allocate the difference of resources to the task, reduce the number of required worker hours by shift length multiplied by the number of resources.
3. If required worker hours > 0:
   a. Calculate the average resources needed = Worker hours left per shift length per available work period.
   b. Allocate additional resources from the beginning of the time period.

The first part of the procedure tries to implement the workable backlog task by filling in the gaps of resource use, the second part checks whether the resources needed after the available time period can be mobilised earlier and the third part ensures that the work can be implemented during the available task window by mobilising additional resources as necessary. This procedure ensures that workable backlog tasks can never delay other tasks which would otherwise require additional updating rounds of the schedule.

Table 5.3 Example forward pass calculations: starting data

		Location			
	Consumption	A	B	C	Unit
Task 1	2	40	30	60	M2
Task 2	1	80	20	70	M3
Task 3	1	30	50	60	M2

Example calculation of location-based forward pass

Table 5.3 shows the starting data for a simple project with three tasks and three locations. Task 1 is succeeded by Task 2 in every location with a F–S relationship (Layer 1) and Task 2 is succeeded by Task 3 in every location with F–S relationship (Layer 1). Layer 3 links for all tasks go from Location 1 to Location 2 to Location 3.

Table 5.4 Example forward pass calculations: pass calculations

Task	Location	Duration (days)	Calculation
1	A	10	40 m^2 * 2 worker hours / m^2 / 8 hours/day / productivity 1
1	B	7.5	30 m^2 * 2 worker hours / m^2 / 8 hours/day / productivity 1
1	C	15	60 m^2 * 2 worker hours / m^2 / 8 hours/day / productivity 1
2	A	10	80 m^3 * 1 worker hour / m^3 / 8 hours / day / productivity 1
2	B	2.5	20 m^3 * 1 worker hour / m^3 / 8 hours / day / productivity 1
2	C	8.75	70 m^3 * 1 worker hour / m^3 / 8 hours / day / productivity 1
3	A	3.75	30 m^2 * 1 worker hour / m^3 / 8 hours / day / productivity 1
3	B	6.25	50 m^2 * 1 worker hour / m^3 / 8 hours / day / productivity 1
3	C	7.5	60 m^2 * 1 worker hour / m^3 / 8 hours / day / productivity 1

By assuming an optimum crew of one resource for each task, a shift length of eight hours and a standard productivity of 1.0 for each resource, the durations can be calculated for each location as shown in Table 5.4.

Tasks 1 and 3 are done by the same subcontractor with the maximum resource use of one resource each. Resources are assigned to Task 1 as a first priority and Task 3 as the second priority (this is the default assumption because of precedence). Task 2 has resource availability of one resource. Tasks 1 and 2 each have both 'as early as possible' and 'continuous' timing options set and Task 3 requires only 'as early as possible' to be set.

The forward pass starts from Task 1 because it is the first in precedence. Early start and finish dates are calculated for all locations of Task 1 before considering any location of Task 2 or Task 3 according to normal CPM algorithms. This gives the following early start and early finish dates:

Location A: ES 0.00   EF 10.00
Location B: ES 10.00   EF 17.50
Location C: ES 17.50   EF 32.50

Because Task 1 has the timing option of both 'as early as possible' and 'continuous', the successor location earliest start dates are compared to the predecessor location earliest finish dates. In this case they are equal, so no further action is taken.

The forward pass continues next to all locations of Task 2. Using ordinary CPM calculations the earliest start dates are as follows:

Location A: ES 10.00   EF 20.00
Location B: ES 20.00   EF 22.50
Location C: ES 32.50   EF 41.25

Task 2 is set to 'as early as possible' and 'continuous', so before going further, the continuity constraint needs to be taken into account. Location B has an earliest finish on day 22.5 and Location C has an earliest start on day 32.5. Therefore, Location B needs to be pulled forward by 10 days which will in turn pull Location A forward by 10 days. After modification, the earliest start and earliest finish dates of Task 2 are as follows:

Location A: ES 20.00   EF 30.00
Location B: ES 30.00   EF 32.50
Location C: ES 32.50   EF 41.25

Note that continuity calculations have been made before even considering Task 3 or before doing the backward pass and establishing float values. Task 3 uses the continuity-adjusted earliest start and finish dates of Task 2 for its calculations. The earliest start and finish dates of Task 3 (without resource constraints) would be as follows:

Location A: ES 30.00   EF 33.75
Location B: ES 33.75   EF 40.00
Location C: ES 40.00   EF 47.50

Because Task 3–Location A and Task 1–Location C together exceed the resource constraint for subcontractor 1, a temporary link must be created in the network. This will shift the start date of Location A to follow Task 1–Location C. The final earliest start and finish dates are:

Location A: ES 32.50   EF 36.25
Location B: ES 36.25   EF 42.50
Location C: ES 43.50   EF 50.00

The timing option for Task 3 was to start 'as early as possible', so continuity is not checked for this task (it happens to be continuous anyway in this example). The earliest dates can be used to draw the flowline diagram (Figure 5.19).

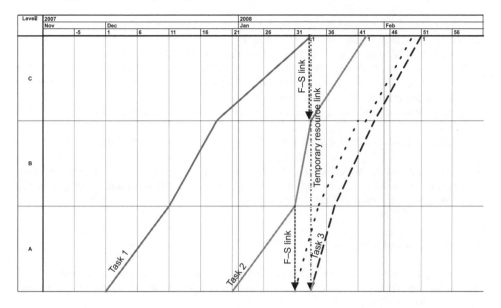

Figure 5.19 Flowline of sample forward pass

Backward pass and calculating floats for location-based tasks

Layer 3 internal logic links and continuity constraints make the backward pass different from ordinary CPM calculations. Two backward passes are required: one for calculating the location-based equivalent of total float and one for calculating the location-based equivalent of free float. There are also adjustments required to take into account the location-based heuristics which affect the calculation and interpretation of total and free float.

Backward pass for total float

Normal CPM calculations apply to location-based calculations of the backward pass, except that there are special aspects which relate to the location-based methodology, in particular the impact of continuity and resource constraints. The first is that time buffers, which are explicitly planned to protect work from interference, are used as a mandatory lag in total float calculations. The next is that all logic layers are used when calculating total float.

Location-based total float applies to the task and represents the time which the task may be moved without affecting overall project duration. This indicates whether a task is on the critical path or how close it is to being critical.

For location-based tasks which have been set to be continuous, a delay in any location will result in a delay in all the other locations (the delay pushes the succeeding locations and the continuity constraint pulls the preceding locations). Therefore, the total float of all the locations must be the same, otherwise the critical path may, in some cases, disappear. If two locations share the same resources, the same effect applies for resource constraints because a temporary resource link has been added to the network, but over multiple location-based tasks.

The procedure for calculating the latest finish dates is as follows:

1. Go through the locations in the opposite sequence of the forward pass.
2. For each location of the location-based task:
   a. Calculate the latest finish and latest start dates as in ordinary CPM.
   b. If the task has been set to be continuous, find the location with the least float. Adjust the latest finish and latest start dates of all the locations so that the float is the same as in the location with the least float.
   c. If the task does not have a continuity requirement, the CPM float can be used for the . location.
   d. The total float of the location-based task is the minimum float of the locations.

The total float can be used to estimate criticality as in ordinary CPM. The location-based task can be moved forwards without affecting overall project duration as long as it has location-based total float available for the task.

Example of backward pass for total float

Continuing the previous example, the backward pass starts in the opposite direction—from the last location of Task 3. According to normal CPM calculations, the latest start and finish dates and float for Task 3 are:

Location C: LS 42.50   LF 50.00   total float 0.00
Location B: LS 36.25   LF 42.50   total float 0.00
Location A: LS 32.50   LF 36.25   total float 0.00

The timing option is set 'as early as possible', so the dates do not need to be adjusted. For Task 2, the initial unadjusted latest start and finish dates are as follows:

Location C: LS 33.75   LF 42.50   total float 1.25
Location B: LS 31.25   LF 33.75   total float 1.25
Location A: LS 21.25   LF 31.25   total float 1.25

Task 2 is set to be 'continuous', so total float is checked. In this example, they are equal. If they were not equal, the latest start and finish dates would be adjusted to make them equal.
    Finally, the float calculations are made for Task 1. The temporary link in the network, caused by the resource link between Location C of Task 1 and Location A of Task 3, will be used in the calculations.

Location C: LS 17.50   LF 32.50   total float 0.00
Location B: LS 10.00   LF 17.50   total float 0.00
Location A: LS 0.00   LF 10.00   total float 0.00

In this example, the critical path goes through Task 1 and then transfers to Task 3 through the resource link. Task 2 has total float of 1.25 days.

Backward pass for free float

In single-layer CPM, free float exists when an activity may be delayed without affecting the following activity according to the logic. If this definition were to be strictly used in location-based logic, the free float of all except the last location of a continuous task would always be zero—hardly a useful tool—because every location is always immediately followed by the next location for the task and therefore has no free float. Therefore, in location-based planning, the concept of free float is only interesting if the internal logic links are disregarded in the float calculation. Of course, the same problem really exists in CPM, as free float can only exist for the last activity in a chain of activities because everything is as early as possible.

The concept of free float changes in location-based planning to represent how much the location can be delayed before interfering with the next activity external to the task. To calculate free float, the latest start and finish dates are calculated so that a mandatory technical lag is used instead of the buffer, and the Layer 3 logic links and resource links are disregarded. It may therefore be seen that the free float actually tells how much buffer there is in the location.

Free float is a measure of how much freedom the subcontractor for the task has to organise or re-plan production without affecting other tasks. Because the calculation does not take the continuity constraint into account, the backward pass does not need to be modified, other than ignoring resource links.

An important implication of this interpretation of free float is that if some locations have differing free floats, or if the free float is larger than the planned buffer, it means that there is inefficiency or waste in the schedule arising from poorly aligned production rates or varying quantities. The float should be removed by aligning production rates better. For example, a task can have float in later locations if the production rate of the successor is slower than the production rate of the task. To optimise production, any float greater than the buffer should be removed by realigning production—by increasing the production rate of the successor.

Location activity free floats should not be used to adjust task start dates, because if the task is continuous, only free float which exists in all locations is available to the whole task. In other words, task free float is the minimum free float of any location which may be used to adjust start dates.

Example of a backward pass for free float

Using the previous example, a backward pass of location-based free float can be illustrated. A backward pass for free float proceeds in the same sequence as for total float but all resource-based flow logic is ignored. Latest start and finish dates for Task 3 are as follows:

Location C:	LS 42.50	LF 50.00	free float 0.00
Location B:	LS 43.75	LF 50.00	free float 8.50
Location A:	LS 46.25	LF 50.00	free float 13.75

Latest start and finish dates for Task 2 are as follows:

Location C:	LS 33.75	LF 42.50	free float 1.25
Location B:	LS 33.75	LF 36.25	free float 3.75
Location A:	LS 22.50	LF 32.50	free float 2.50

Because Task 2 free float is almost equal in all locations, it is well aligned to the successor task.

The latest start and finish dates for Task 1 are as follows:

Location C: LS 17.50   LF 32.50   free float 0.00
Location B: LS 22.50   LF 30.00   free float 12.50
Location A: LS 10.00   LF 20.00   free float 10.00

Because free float for the earlier locations are larger, the schedule could be optimised by speeding up the first task so that free float would be equal in all locations.

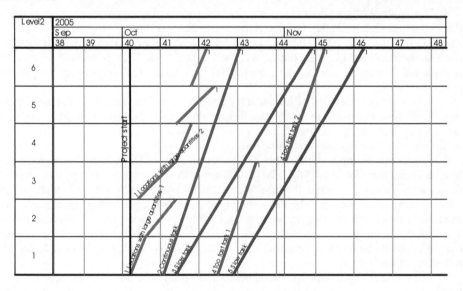

Figure 5.20 Typical examples of task splitting

SPLITTING

There are situations where some locations of a task may need to be split to form a new sub-task (as shown in Figure 5.20). Typically, this enables the following planning decisions:

- Different resources are required in a given location
- Multiple locations must work concurrently with different crews
- A planned break is required between some locations
- Some locations require a different logical relationship
- It is necessary to make one part of the task 'continuous' while another part is set 'as early as possible'.

Splitting allows the use of different internal logic for different locations of the same task. Each sub-task can have different resources and logic links. It may have durations which overlap with other sub-tasks and it will not be constrained by other sub-tasks unless explicit links are planned between them. However, split parts remain part of the same task in the sense that similar quantity items are shared between them—in other words, even though they may be different crews, they are doing the same work. Splitting is an important feature of planning, but its real value is realised during performance measurement and production control, as outlined in Chapter 8.

Splitting does not have a great effect on schedule calculations, because sub-tasks may be calculated independently of each other, as if they were separate schedule tasks.

Figure 5.20 shows some typical examples of splitting. Task 1—locations with large quantities—has been split so that one crew works in locations 3, 4 and 6 and one crew works in Locations 1, 2 and 5. This enables both of the crews to have continuous work, to not interfere with each other in the same location and to align the production rate of the whole task with the succeeding Task 2—continuous. Task 4—too fast—is between the two slower tasks, but is already working with its optimal rhythm, so instead of artificially slowing down the task or delaying the start date, a break has been planned between Locations 3 and 4. This is better than having all the locations start as soon as possible because there is only one break point instead of one between each location.

PUTTING IT ALL TOGETHER

There has been a great deal of new theory introduced in this chapter, with location-based layered logic and new approaches. Given the complexity, it is useful to summarise the assumptions and processes and put it all together. The following outlines the assumptions, the starting data and the method for calculating the model in location-based planning.

Assumptions

The described system is based on the following set of assumptions:

- All the work included in a task will be undertaken by common resources (this is a property of a location-based task)
- Resources apply the same production factors for all the items in the task that use any common resources (a simplification from real life)
- The resource production factors are independent of the item or item quantity
- There is an optimal crew composition for each task
- Resources complete all the locations included in the task in sequence (overlapping requires task splitting)
- Resources completely finish a location before moving to the next location.

These assumptions require a different methodology of planning and controlling work and new kinds of contracts to support project control. These issues are discussed in Chapters 9 and 10—the discussion of control methodology and implementation.

Starting data

The practical implementation of the location-based planning system requires the following starting data for the calculation of the schedule:

- Location breakdown structure (LBS)
- Location-based quantities estimated for locations of the LBS with the following properties:
  - Item name

- Consumption rate (hours / unit)
- Accuracy level (on the LBS)
- Quantity for each location at that accuracy level (can be zero, which means that the work does not exist in that location)
- Unit.
- Location-based schedule tasks with the following properties:
    - Name
    - Included quantity items
    - Accuracy level (on the LBS) defined by the roughest accuracy level of the included quantity items
    - Locations (defined by the locations of the included quantity items)
    - One or more sub-tasks containing locations.
- Sub-tasks with the following properties:
    - Included locations
    - Sequence of locations (Layer 3 logic sequence)
    - Optimal crew
    - Number of optimal crews
    - Task type: normal task or workable backlog task
    - Production factors of resources in the optimal crew
    - Resource constraints for each resource type
    - Location difficulty factors
    - Timing option: as early as possible; as early as possible and continuous; continuous; manual
    - Target start date—for use when the task is not as early as possible
    - Shift length (time units) and calendar (days worked, holidays and days off)
    - Logic links to predecessors with the following properties:
        - Predecessor task
        - Link type (F–S, S–S, S–F, F–F)
        - Lag (shifts or time units)
        - Buffer (shifts)
        - Location accuracy of link: Layer 1 link—if as fine (low) as the rougher of the two task accuracies; Layer 2 link—if rougher (higher); or specify that the link is between individual locations—Layer 5 link
        - Location lag: Layer 4 link if not equal to zero
        - For Layer 4 links, does the link apply only in the same branch of the LBS.

Calculating the model

The CPM model may therefore be calculated from the location-based heuristics by performing the following steps (details of these calculations have been described earlier in this chapter).

1. Calculate the durations of each location of each task based on quantities, resources and productivity rates.
2. Create the activity network from location-based links:
    a. Layer 1 links: create a link of the same type between the locations shared by both tasks.
    b. Layer 2 links: the link will affect either the first or last location within the sequence of locations at the task's own accuracy level because the task will always have a more accurate location breakdown than the accuracy level of the link:

     i. F–S: create a F–S link between the last location of the predecessor within the link's level of accuracy and the first location of the successor within the link's level of accuracy.

     ii. S–S: create a S–S link between the first location of the predecessor within the link's level of accuracy and the first location of the successor within the link's level of accuracy.

     iii. S–F: create a S–F link between the first location of the predecessor within the link's level of accuracy and the last location of the successor within the link's level of accuracy.

     iv. F–F: create a F–F link between the last location of the predecessor within the link's level of accuracy and the last location of the successor within the link's level of accuracy.

   c. Layer 3 links: create a F–S link between the location of the same sub-task in the location sequence.

   d. Layer 4 links: create a link of the same type between locations shared by both activities, offset by as many locations as indicated by the location lag. When calculating the offset, the project's location grouping is used. Any links between different location groups are ignored.

   e. Layer 5 links are already between single locations so no further action needs to be taken.

3. Calculate the forward pass of the schedule:

   a. Sort location-based tasks to precedence order so that all locations of the predecessors of a location-based task (in any location) are before any location of the task, unless Layer 4 or 5 logic creates circular location-based logic.

   b. In the case of circular location-based logic, the locations are sorted so that the Layer 1, 2 or 4 predecessor's locations are added first to the sort, until the first location that depends on a successor task's start date. Then the successor start dates are calculated until the predecessor's uncalculated start dates are needed to calculate a location's start date.

   c. For each task (or sub-task) in the sorted order:

     i. Calculate the earliest start using normal CPM calculations both using Layer 3 links and disregarding Layer 3 links.

     ii. If desired, do a sequence optimisation by changing the sequence so that locations are done in sequence of ascending earliest start order (disregarding Layer 3 links).

     iii. Check resource availability and shift the start date forward if insufficient resources are unavailable.

     iv. Adjust the start dates of locations based on the selected continuity option in order to force continuity.

   d. Record the start and finish dates for each task and location activity.

   e. Calculate the times of workable backlog activities according to resource availability.

4. Calculate the backward pass of the schedule:

   a. Use normal CPM calculations except for the following:

     i. Total float uses the buffer lag instead of the technical lag and the total float of tasks which have been planned to be continuous are adjusted so that all the locations have equal float.

     ii. Free float disregards Layer 3 links.

   b. Record total float, free float and criticality for each task and location activity.

AUTOMATION OF A LOCATION-BASED PLANNING SYSTEM

There are only minor differences between projects of the same general type (residential, retail, etc.) that mainly apply to the location breakdown structure and the quantities of work in each location. Otherwise, optimal crews, productivity rates and precedence relationships are often remarkably similar. Therefore, it is possible to automate much of the manual work which is involved in scheduling a project, to create rapid initial drafts of a schedule based on standardised packages of tasks, crews and precedence relationships which may be combined with the project-specific LBS, quantities and productivity rates.

This concept is a shift from the view that every project is individual and different and must be uniquely scheduled. In contrast, the reality is that the schedules of most projects are remarkably similar and, if approached as such, a great deal of information from past projects may be used which would otherwise be lost. A discussion of the individual characteristics of different project types can be found in Chapter 13.

Automated schedule creation

The creating of location-based tasks can be automated by defining the quantity items that are usually produced together in a standard project. This can be practically implemented by using the quantity item code (in the BOQ) or description as an identifier and lumping the quantities together in the same way as in a standardised template project. For the duration calculation, the same crews used in the template project can be adopted. The resulting schedule may be created using the following steps:

1. Go through all quantity items. For each item:
   a. If the quantity item's code matches with a quantity item code or description in the template:
      i. Assign the quantity item to the schedule task specified in the template. If the task does not yet exist, create a new task.
      ii. If the quantity item does not have a productivity rate and has the same unit as in the template, take the productivity rate from the template.
   b. If the quantity item's code does not match with a quantity item code or description from the template:
      i. Leave the quantity unscheduled.
   c. If there are more quantities, go to step 1. If all quantity items have been processed, go to step 2.
2. Go through all the schedule tasks created:
   a. Assign the same optimal crew as in the template with the same production factors.
   b. Create the same precedence links that are found in the template if both tasks exist in the current project:
      i. Layer 1 links: enter a similar Layer 1 link.
      ii. Layer 2 links: copying requires that the LBS has the same hierarchy levels, if the hierarchy levels match, enter a similar Layer 2 link.
      iii. Layer 3 links: these are not copied because they are internal to the task.
      iv. Layer 4 links: enter a similar Layer 4 link.
      v. Layer 5 links: these cannot be copied because they are project specific, and are particularly specific to the project LBS.

Use of automated schedules

It should be noted that automated schedules are just a starting point for planning. A human planner must model project-specific issues and check all constraints. If the template does not have all the items (quantities) of the planned project, the planner must assign the remaining quantities to new schedule tasks. The aim of automated scheduling is not to give the best possible schedules, but rather to remove the manual work of initiating the tasks and precedence relationships (which will be similar in most projects) of a new project based on a past template.

SUMMARY

There are many new concepts involving logic presented in this chapter. These form the underlying basis of a suite of potential tools, methods and methodologies for location-based planning. They also establish the foundation logic for the generation of a new theory for location-based control. The remainder of this section will explore the practical application of location-based planning. Section Three will explore both location-based control theory and the associated tools, methods and methodologies.

REFERENCES

Arditi, D., Tokdemir, O.B. and Suh, K. (2002). "Challenges in Line-of-Balance Scheduling". *Journal of Construction Management and Engineering*, November/December, 545–556.

Ballard, G. (2000). *The Last Planner™ system of production control*. PhD Thesis. School of Civil Engineering. Birmingham, The University of Birmingham.

Kiiras, J.M. (1989). "Erityiskohteiden työnaikaista ohjausta palveleva aikataulu- ja resurssisuunnittelu" ("A schedule and resource planning system for the implementation phase control of special projects"). *#OPAS ja TURVA*, Helsinki University of Technology Construction Economics and Management Publications. Espoo, Finland.

Mitropoulos, P.T. (2005). "'Planned work ready': A proactive metric for project control". *Proceedings of the International Group for Lean Construction, 13th annual conference*. Sydney, Australia: 235–242.

PMI (1996). *A guide to the project management body of knowledge*. Project Management Institute. Newtown Square, PA.

Suhail, S.A. and Neale, R.H. (1993). "CPM/LoB: New methodology to integrate CPM and Line-of-Balance". *Journal of Construction Engineering and Management* **120**(3): 667–684.

Chapter 6

Location-based planning methods

INTRODUCTION

Chapter 5 introduced the theory of location-based planning (or scheduling), particularly the power of combining CPM's activity logic with location-based logic, and generic topics such as locations, location breakdown structures, location-based quantities, tasks, location-based logic and calculating durations based on optimal crew sizes. While there was much technical detail supporting that theory, there is more to a management system than the theory and basic techniques—as the power comes from the ability to model complex contexts and to apply the theory to the real world. The theory provides the basis for practical methods.

This chapter expands the theoretical discussion of location-based planning by introducing real-world simulation and analytical methods such as unit and production system cost, production risk, procurement, design schedule, quality and learning processes (improving rates of production) within the context of a location-based schedule. These methods start with the components of a location-based plan: location-based quantities, the start and finish dates for locations, production rates and resources. This chapter discusses the use of these methods in pre-planning construction. Chapter 9 will then describe how the plans made with these methods can be controlled during the construction phase.

In this chapter, another level of detail is added to the measurement of quantities. Instead of just talking about finished products, for example an area of finished wall, it is also desirable to know the resources consumed in building the wall: the materials, labour, subcontracts and equipment. While the quantity of the measured item can be taken-off directly from drawings or by using a 3D-model, the resources used to produce the item can only be modelled from that measure using relative quantities (for example, square metres of board per square metres of finished wall). This is product-resource modelling. Its use will be assumed when discussing the other methods in this chapter. However, it remains possible to use measured quantities only, with some loss of power.

Two methods to model costs are presented. The first one is conventional cost loading of the schedule based on directly incurred unit-rate costs, taken from the standpoint of a single actor in the construction process (owner, GC, subcontractor). This method enables accurate modelling of cash flow using quantities and start and finish dates of locations to calculate estimated times of payment and the calculation of earned value. While this approach can effectively optimise the cash flow and overhead costs of a single actor, it is silent on the beneficial effects continuous production may have on reducing production costs or indeed on identifying the costs of production interruption. Such costs arise from the interaction of many players, and an alternative cost model is required. A production system cost model calculates the total labour cost of the production system including waste, which is modelled by examining resource use and the continuity of location-based tasks. This new model can then be used directly to optimise the efficiency of the production system.

Production system risk is strongly related to production system cost. It arises from variation between the contract and actual start dates as well as variability in the performance production rates of resources. Buffers are used to mitigate production system risk because they can absorb delay, preventing succeeding tasks from being affected. However, once crews leave a site, risk will be increased because of the uncertainty and possible delay in these crews returning. This chapter presents the components of a location-based simulation model for production system risk, allowing the analysis of cost and risk trade-offs.

Procurement activities can be planned once the location-based schedule has been finished and accepted. In addition to traditional scheduling of procurement as activities, methods for planning procurement include pull scheduling of procurement events, and calculating material handling and storage requirements (logistics) based on each location's quantities and production times. Pull scheduling (see Chapter 4) is a lean construction concept which, when applied to procurement, allows procurement activity task dates to be calculated by setting procurement activities as predecessors to the project plan with the property of being as late as possible. The production system then pulls the procurement activities as required. This change in philosophy arises from the need for continuous flow of production to be the central point of control.

The design schedule is related to the procurement schedule, however it has important differences. In many complex projects, design becomes continuously more detailed. If the production rate for design is not known, it may become a bottleneck for production tasks. Therefore the most important components of design for production purposes should be scheduled as location-based tasks just as any other production task (with the location being a virtual representation of the physical location).

Good planning is more likely to lead to a safe site with good quality control. Location-based planning uses buffers to shield production and allow time for important quality processes such as quality inspection and tests. Location-based planning and control can be used to achieve managed handover of locations to drive quality and safety.

Learning theory has its own extensive body of literature that describes the benefits to production which may be derived from repetition of an activity. Modelling learning expands the scheduling theory of Chapter 5 to allow that it is a realistic assumption that productivity can improve if the task crew can work continuously doing the same work in the same project across multiple locations. Traditional construction learning models generally follow production models and consider learning to arise from repetition of units of production. However, there are more factors to be considered, such as continuity of work, other work done by the same resources and the number of resources participating in the work. Moreover, because a location-based planning system does not assume repetition of only similar locations, locations alone do not work as the basis of a learning model for construction. Rather, this chapter presents an adapted model for learning that is founded on repetition through the quantities of a scheduled task, the number of crews and work continuity.

MODELLING CASH FLOW AND PRODUCTION SYSTEM COST

Traditional construction cost modelling takes a static approach to the construction process, and therefore struggles to represent the dynamic building production process adequately or accurately. The focus of traditional cost modelling is to represent or estimate the cost of production from a fixed viewpoint—the relationship between the client and the contractor. As such, in the traditional view, fixed estimates of costs related to unit quantities are sufficient, as these reflect the contractual agreement between the parties, and are suitable for estimating, monitoring cash flow and certifying payments. However, a true representation of the cost of construction must take into account the dynamic costs of production, including the waste. While the cost of production waste may not appear real—in the sense that it is not reflected in contractual agreements—nevertheless it represents a major component of the total system cost to the head contractor and subcontractors—and so is very real in the sense of increasing the true cost. Thus production system cost information is necessary to improve the efficiency of a project from the standpoint of all parties to the construction project. Modelling production system cost requires a different approach to the dominant static approach.

The current static approach to cost modelling is generally based on built-up rates, using historic performance data, and the cost of production is assumed to have been included through the labour component. Production is assumed not to be a variable, thus past performance is considered a suitable predictor of future performance. In this way, the efficiency of past production will be reflected in the historic data, which will then be used in the forecast costs for tenders or price estimates—thus ensuring that the forecast reflects past performance. An inefficient production system is therefore likely to be used to predict future costs which, of course, will be consistent with the originating inefficient system. This systemic inefficiency is concealed however, as no one will ever be surprised with the high production cost in the estimate, as they would be unaware of any problem in the system. The result is that most managers do not believe waste to be a true production cost. Furthermore, in practice, actual costs will track the forecasts, despite the inefficiency of production, so the entire system will be internally consistent. This leads to a false sense of confidence in outcomes and production complacency, as managers will not be held to account for faulty production systems when performance equals planned production and actual costs match forecasts.

Consider the simple case of a faulty conveyor belt in a production line. If the belt breaks down 20% of the time, then a true production rate of ten units per hour will appear as eight units in the historic data. If the historic data forecasts production of eight units, then when actual production measures as eight units management will not be aware that there is a problem with the production process. Plans will never request a production rate of the achievable ten units and rectification action will not be implemented. It is only when detailed examination of the production system reveals that production capacity is ten units per hour that an investigation may be launched, the machine repaired and ten units per hour used to forecast future production. Of course, if the machine is not repaired, the forecast will be unreliable.

Construction cost management systems assume that historic labour data is reliable. The concept of waste in the production system, while well documented (see Chapter 4), is ignored when projects are priced and planned. The inefficiency of production is concealed within the collected costs, yet the estimator can take confidence that actual costs will be correctly forecast—because actual production will be equally inefficient as past production.

Under these circumstances, an estimator has few choices for adjusting the cost estimate of a building. They may change the quantities (for example, reducing floor area), change the materials choices (removing the 'gold taps') or possibly choose a different construction system. While this appears to provide sufficient choice, it is silent on the one thing which relates to the production system: production efficiency (or doing the job well). This is somewhat surprising, as production efficiency is one thing most contractors would strongly represent to clients as their specific strength.

Management also has a conflict of interest. While there is an incentive to improve production efficiency, this is in direct conflict with most managers' implicit hubris that they are already operating at optimum production efficiency. It takes a lot of experience and nerve to admit that your methods are not currently optimum.

In the authors' experience in Finland and the USA, planned durations based on historic data are approximately 20% to 30% more than the real production rate, although some trades such as structure and claddings are much closer. The greatest variation is in the complex areas of finishes, where contractors typically lose control of production efficiency. This means that the builder plans to take possibly 30% longer than necessary, together with the associated costs, because that is the way it has worked in the past. There is clearly room for improvement by targeting that waste. In part, this current way of working is a product of planning using activity windows, a planned time allowance for an activity that includes all the disrupting influences which delay completion of an activity.

A better system, and that advocated for a location-based management system, is to directly model the production system and production system cost. This is a complex strategy and requires a substantial change in the mind-set of both the quantity surveyor or cost engineer and management. The advantage is that it makes clear that changes in the management of process on the construction site can have a direct impact on the total cost, something which is almost universally ignored in present modelling or forecast systems.

There is an old adage which states "you can't improve or manage what you can't measure". The construction industry currently does not measure and report the component costs of production, so we must doubt the ability of management to manage or improve production performance. There is therefore a critical first step to be taken by management in adopting a location-based management system: they must first accept their ability to improve performance by reducing production cost, then they must measure and report it. They will not then continue to ignore the production efficiency of their plans and schedules.

In order to understand the calculation of production system cost in location-based management, there are two main discussions.

- First, the traditional approach to cost loading a schedule is discussed, including the preparation of cash flow, which focuses on the allocation of costs to building elements reflecting the contractual pricing structures.
- Second, the components required to model production system cost are identified, so that it can be measured and reported as part of a location-based management system. This discussion focuses on the allocation of costs to resources, and estimates the true cost of production based on the actual time spent and estimates of the waste components of the production system. In particular, the inherent waste that exists in any project plan, in the form of waiting time, relocation, double-handling, mobilisation and demobilisation, are explicitly identified and expressed in the model. This enables planners to optimise their production plans to minimise production waste and the associated, very real, costs.

The following sections first revisit the quantity information used in Chapter 5 to develop location-based schedules and show how additional information can be used to develop production cost models.

Elemental- and resource-based modelling

Thus far—to make it simple—we have limited our discussion of quantity to a single-level concept that only describes the quantity of an element to be produced—the completed wall, floor, beam, etc. This is usually enough for planning and scheduling purposes because labour consumption can be evaluated at an elemental level. However, for cost loading and procurement purposes, it is beneficial to know more about how the quantities of the project will be produced. Specifically, the system should measure separately the resources which make-up the element, including equipment, direct labour and subcontracts. The resources can then be mapped to the elements.

The main assumption in mapping resources to elements is that the mapping relationship is linear. This means that there will always be a constant increment of resources corresponding to each additional unit of an element. For example, a plasterboard wall is measured in square metres and has the following resources: frames (plates and studs), wall board, insulation wool and installation labour. The quantities of resources depend on the wall type. If there is one board on each side, the quantity of board is 2 m^2 per m^2 of wall. If there is one stud every 300 mm, the quantity of frames is 4.1 m per m^2 (from the Finnish productivity database). Installation labour determines the duration of the task and is directly measured in worker hours per unit. Therefore, if quantities are modelled using two levels (elements and resources), duration calculations can use the labour resource consumption directly.

The resource level is used to separate the payments for different resource types in cash flow forecasting and to aggregate resources from the same supplier to a procurement task, thus providing a demand schedule to plan resource deliveries from all schedule tasks where the same resource is being used. It may also be used in the control part of the system to update the cost forecast based on agreed unit prices of resources (see Chapter 9).

Cost loading the location-based schedule using elemental costs

It is the traditional approach to cost load the schedule directly, using the cost estimate for a project. This is very powerful in location-based management, because quantities are included in the location-based tasks, thus providing a direct link between the estimate and the schedule—broken down by location. Every quantity element (or if the two-level quantity system is used, every resource) has an estimated unit cost. By summing the costs of resources together it is possible to calculate the cost for the entire schedule task or for any activity location of the schedule task. Cost loading a schedule is nothing new, but having location quantities provides greater power in the practical application of such estimates.

A cost loaded schedule (which includes production rates, start and finish dates, and quantities for each location) can also be used to calculate the timing of payments, progressive cash flow and earned value. (The latter is of reduced relative importance given the greatly increased confidence provided by location-based performance data when compared with other planning systems as discussed in Section Three.)

Payments can be divided into two main types: time-based or milestone-based payments. Time-based payments most commonly occur monthly, but can also occur weekly, fortnightly and bimonthly, and are calculated from the work-in-progress. Milestone-based payments apply when payments are linked to achieving a particular milestone, for example completing the whole task or a given completion rate for completing specific progressive locations. In practice the two-level quantity system becomes important at the task level, because it is possible that a scheduled task includes a mix of subcontracts, materials bought by the general contractor and direct labour to help the subcontractor. Each of these resource types has its own payment type, for example the subcontractor may be paid by milestone, the direct labour wages may be paid fortnightly and the material supplier might be paid for each delivery.

This may be illustrated by using the plasterboard wall example of earlier chapters. In this example, the plasterboard work is subcontracted, supporting logistics (carrying boards, cleaning etc.) is done by direct labour and the materials are purchased according to the company's volume purchase agreement with all materials being invoiced monthly. The costs of the task are formed by:

- Material deliveries: for wool (1 m^2 per wall m^2 at €0.80 per m^2), boards (2 m^2 per wall m^2 at €1.20 per m^2) and frames (4,1 m per wall m^2 at €0.50 per m).
- The agreed subcontract price is €22.00 per m^2 (with the productivity estimate of 0.46 worker hours per m^2, this corresponds to €48.00 per hour). There is a payment milestone after the second and fifth locations.
- The need for direct labour is estimated to be 0.15 worker hours per m^2 and the cost of direct labour (including on-costs such as mandatory social security payments) is €25.00 per hour. Direct labour is paid fortnightly.

Note that by knowing the productivity estimate and unit rate per unit quantity, it is possible to estimate how much the subcontractor has allocated for profit and wasted productivity in his bid. To get the maximum cost benefit from the location-based management system, it is worthwhile estimating these values.

Table 6.1 Schedule expense events for plasterboard

Date	Amount	Calculation
Direct labour (every two weeks)		
5/12/2005	522	4 days of work * 34.8 m² / day * 0.15 worker hours / m² * €25 / hour
19/12/2005	1,304	10 days of work * 34.8 m² / day * 0.15 worker hours / m² * €25 / hour
2/01/2006	1,304	10 days of work * 34.8 m² / day * 0.15 worker hours / m² * €25 / hour
16/01/2006	1,182	9.1 days of work * 34.8 m² / day * 0.15 worker hours / m² * €25 / hour
Materials (monthly)		
1/12/2005	365	2 days of work * 34.8 m² / day * €5.25 total material cost / m²
1/01/2006	4,017	22 days of work * 34.8 m² / day * €5.25 total material cost / m²
1/02/2006	1,655	9 days of work * 34.8 m² / day * €5.25 total material cost / m²
Subcontract (milestone)		
14/12/2005	8,800	400 m² * €22 / m²
13/01/2006	16,500	750 m² * €22 / m²

The resulting payments and calculations are shown in Table 6.1, with the corresponding expenses shown as a function of time in Figure 6.1, assuming a two-week payment delay for subcontracts and materials. The top part of the figure shows the timing of actual payments by highlighting with filled circles in the flowline diagram.

Client payments to the general contractor (cash inflow) depend on the project type. They can be tied to milestones, fixed dates or to work-in-progress. Clients wishing to maximise the benefits of applying location-based management tools on their projects will, in the future, tie the payment schedule to the completion of locations for the most critical schedule tasks. However, currently payment schedules are usually tied to overall work-in-progress (completion of the overall schedule) or fixed monthly payments. The timing of income payments can be modelled in the same way as expenses, based on completion rates of quantities in locations regardless of the payment reason (Seppänen and Kenley, 2005).

For example, the client could make a prepayment of €3,000 and then pay a lump sum of €6,000 for completing each location to approved quality standards and then an additional €7,000 two weeks after completing the entire contract. The result is the cumulative cash inflow. When the previously calculated cash outflow expenses are subtracted from the cash inflow incomes, the result is the progressive net cash flow. Figure 6.2 shows the net cash flow curve for the plasterboard wall example. In this case, despite the initial prepayment, the general contractor needs to self-finance the project for a few weeks toward the end of the contract.

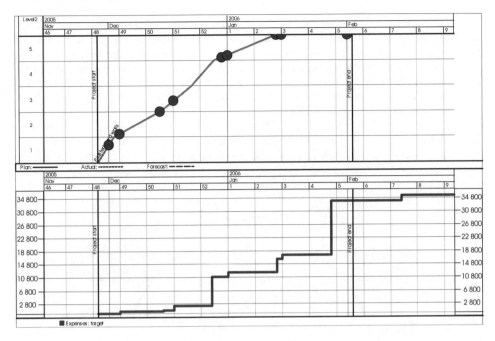

Figure 6.1 Cost loading the schedule for a task: the upper panel shows the task flowline, with payment stages, the lower panel is the cumulative cash outflow

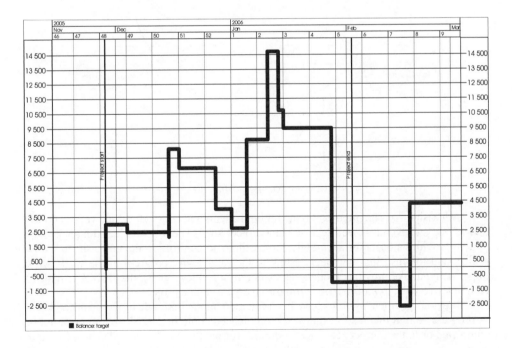

Figure 6.2 The net cash flow for a cost loaded schedule task

The payment calculations may be automated by assuming the same payment type for each major cost type. Because of the accuracy of the method and the large number of payments, the aggregated expense curves for all schedule tasks of the project will smooth out and resemble the familiar cash flow S-curve. Cash flow can be optimised for the general contractor by changing production rates and payment options for different subcontractors. Managing the contribution of a project net cash flow to a firm's working capital is a critical component of project success (Kenley, 2003).

The complex interaction of cash inflow and cash outflow provides a contribution to the working capital for the general contractor. Figure 6.3 (Kenley, 2003) illustrates the relationship between the net cash flow and the work-in-progress, which is dictated by the commitment made in the schedule.

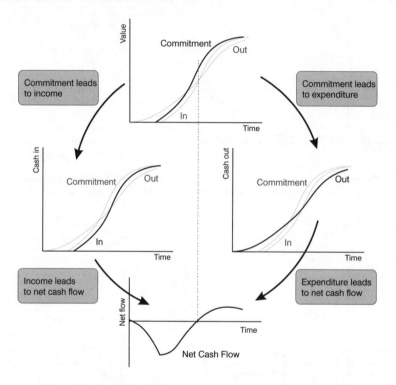

Figure 6.3 The origin of net cash flow (Kenley, 2003)

Net cash flow is an important issue for both clients and contractors, and yet it is poorly understood and even less well reported at a project level. The location-based management system provides a basis for accurately forecasting and controlling the net cash flow for a project and for ensuring the concrete link to the satisfactory progress of the work. Most importantly, unlike previous models (as explored in Kenley, 2003), the LBMS provides a firm link to *completion* of locations progressively through a project.

Previous payment systems have been founded on the principle of payment for work-in-progress, and in this respect the work commenced could be anywhere and need not be completed or in the correct sequence (required) for payment to be earned. Thus, for arguments sake, a project task could be 80% completed in all locations and the amount due would be 80% of the task total value, whereas under LBMS it is possible to ensure that 80%

of the task value is only payable when 80% of task locations have actually been 100% completed. General cash flow literature assumes income payments from the client to the contractor will be made monthly. This is an assumption based on widespread contracting practice.

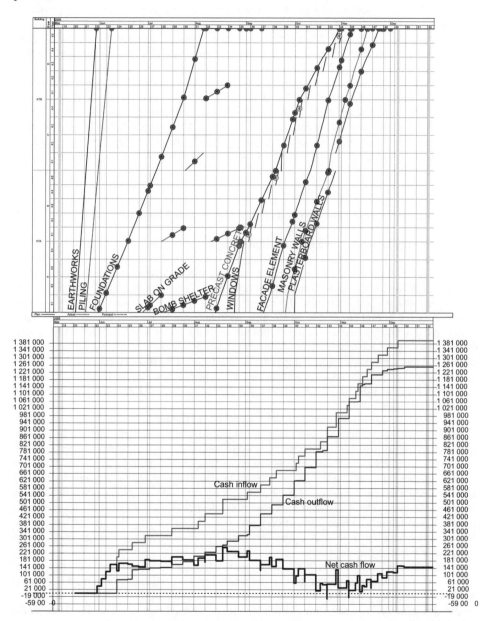

Figure 6.4 The payment schedule and cash flow charts for the early trades using data from two apartment buildings (with multiple risers and floors) from Finland

In Finland a different understanding of inward cash flow has developed. This has occurred because of the different planning philosophy which has applied there. Whereas generally the flow is 'lumpy', having been calculated from monthly lump sum payments, the location-based planning approach used in Finland allows payments to be tied to specific progress-related tasks and milestones. It is not unusual therefore to average ten payments each month. The linking of cash-in payments to the schedule in this way provides for a smoother S-curve on the inward cash flow and thus a smooth net cash flow. This has great significance for quantity surveyors and others who manage the certification of payments, and also for issues relating to the security of payments.

As an illustration of this, Figure 6.4 shows the cash-in and cash-out payments as well as the net cash flow for selected early trades from a real Finnish project. The circles on the flowline show the timing of the payments (although only the receipts linked to staged completion of the windows task are shown). The project has two buildings (A and B), each with two risers of four and five floors respectively. The net cash flow for these trades is clearly shown to become negative during the later stage of the work before returning to the planned margin.

Identifying production system cost using resource-based modelling

Cost loading the schedule using a cost estimate may enable optimisation of the net cash flow at the task level, but is silent on production efficiency. This method cannot allow for the effect of production on total cost (whether using direct labour or subcontractors) nor provide the basis for schedule optimisation to reducing production waste in the plan.

In traditional cost estimates, the cost is defined by using historical data to estimate resource use or lump sum prices. While this approach can be used to calculate cash flow, being linked to contractual terms, it does not provide any tools for measuring the production quality of a schedule. In contrast, production system cost models should have knowledge of the actual labour resources required for production (the real costs) as well as directly taking into account the waste factors, such as waiting time, relocation, double-handling, mobilisation and demobilisation, and it should calculate labour costs based on a composite model of resource consumption.

Ultimately, someone must pay the cost of an inefficient plan. Inefficient work will necessarily increase resource consumption compared to optimum production. In production system modelling, it is not necessary to know who will pay for the production cost, as it is a contractual problem to distribute the costs to different parties of the construction process. In particular, if savings can be achieved through optimisation then any distribution is once again a contractual issue. Nevertheless, long-term competitiveness is linked to the total production system cost and it is therefore desirable for contractors, and certainly clients, to sustainably reduce production system cost by improving the production system. On the other hand, it is at least in the subcontractor's interest to reduce production cost immediately and in the short-term.

The production system cost in a schedule is composed of direct labour cost and overhead cost. While material resources may be wasted and indeed such waste may be a direct product of the quality of the production system, only labour and time-related overheads can be directly modelled in a schedule as both labour cost and overhead cost are functions of the schedule.

Direct labour costs are mainly paid by the subcontractors and are a function of productivity. Currently, subcontractors may include an allowance of up to 50% for wasted time in the labour component of their bid. Essentially, they are making an allowance for being messed around by the general contractor. Location-based systems allow planners to see this waste and to fight it by planning better schedules. This will improve the profitability of subcontractors, improve relationships between the general contractor and the subcontractors and in the long run decrease their bid prices—therefore yielding direct cost reductions to the general contractor. In the short-term the general contractor will enjoy higher predictability of schedule, because all the subcontractors are motivated to reduce their wasted time.

Overhead costs are usually paid by the general contractor and are a function of the project, the construction phase durations or the duration of tasks using certain important resources (such as the tower crane and project management). Such costs are optimised by reducing the total duration of the project, and by reducing the duration of those key tasks.

Components of direct labour cost

Direct labour cost is composed of the following value-adding and non value-adding components:

- Working time
- Mobilisation, demobilisation and waiting
- Moving around on site
- Stockpiling, hauling, delivery and receiving materials.

Each of these components is described below.

Working time

Working time is the time while resources are engaged in the productive effort required to complete a task. This is effort devoted to value-adding activities in the lean production sense. It specifically excludes non value-adding effort (despite the fact that these may often seem necessary for production to occur) such as moving resources and materials, waiting, relocating, etc. In location-based management, the idea is to plan to maximise productive working time relative to other non value-adding activity time. Work is planned to enable resources to minimise the things which disrupt productive work, by forcing continuous use of crews in tasks through locations.

The total time spent working in order to produce a fixed output amount is minimised if the resources can maximise learning effects and work at their optimum speed. Learning will occur if there is repetition and continuity in the task, whereas learning will suffer if the task has breaks and, for example, the crew has to leave the site. For the purpose of the location-based management system we make the assumption that learning will fail if the crew runs out of work and leaves the site—the task will revert to the base production level as if no prior work had been completed. Direct labour cost can be minimised by allowing the crews to work on repetitive assignments uninterrupted for the duration of the task. Visually this is represented by a continuous line in the flowline diagram, with the learning effects shown by a change of slope as the task progresses through the locations. The complete learning model will be described later in this chapter (page 196).

Each crew is assumed to have an optimum pace which is reflected in the labour consumption rate of the quantity item (see page 131). It is possible to deviate from this optimum pace in the schedule, for example slowing down an overly fast task or by having more or less efficient resources. If the quantity of resources cannot be decreased—for example if there is just one crew of optimal composition—decreasing the production rate will increase the direct cost of labour because the workers are working slower than their potential.

Productivity can also be decreased by working longer days, during weekends or through insufficient rest periods. If longer hours are worked, the marginal productivity of each hour drops dramatically, while also incurring overtime payments.

Working time cost is calculated by the following formula:

$$C_P = \sum_{j=1,n}^{i=1,n} T_E \times S \times P_U \times P_O \qquad\qquad (6.1)$$

Where:

C_P = The sum of the working time cost for each location i, and for each resource j
T_E = The effective durations of locations
S = Shift length
P_U = Unit price of resource
P_O = The overtime multiplier (if non-standard working hours are used)

A simple example will be used to illustrate production system costs. Figure 6.5 shows a project with 2 tasks. Task 2 is faster than Task 1 and is 'as early as possible' without a 'continuity' requirement. It has one resource with unit price of €20 / hour. The effective duration of each location is shown by the planned line and is 0.6 days. Shift length is eight hours and overtime is not used. Working time cost in this example is:

(5 locations * 0.6 days / location) * 8 hours / day * €20 / hour = €480

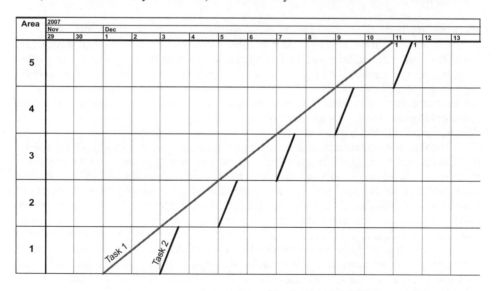

Figure 6.5 Production system costs arise from breaks in production

Note that if Task 2 was arbitrarily slowed down until it became continuous, it would require a duration of 2 days / location. In this case, the working time cost would increase to €1600 because resources cannot be taken away from a task which has only a single resource.

Mobilisation, demobilisation and waiting

Non value-adding time is spent every time a new worker or task crew comes to site or when they leave the site. First, new workers have to get to know the site and its peculiarities, they must move materials and equipment to the work site location, and familiarise themselves with the other workers and the location status: this is mobilisation. When crews leave the site, they clean up their work space and pack away their equipment: this is demobilisation. Mobilisation and demobilisation can easily consume more than four hours. Furthermore, there is further time spent travelling to another site. If the crew has to leave during the afternoon, an entire afternoon is often wasted.

Mobilisation time is generally unavoidable when mobilising resources for the first time and, similarly, demobilisation generally must occur at the conclusion of the work. Nevertheless, an activity-based planning system can result in many workflow breaks, with consequential mobilisation and demobilisation occurring as the work ceases and then renews. Without location-based planning for continuous workflow, there may be many such discontinuities resulting in significant amounts of wasted time.

Alternatively, it may be better for workers to be kept on site doing little (or more likely doing superfluous moving of tools and materials) and working out of sequence (most often when prerequisites, such as prior work, are not complete) during breaks in task flow. This usually has higher apparent direct cost due to double-handling, but avoids the risk of the resources not returning to site. It seems that, currently, the standard method is for subcontractors to be allowed to leave the site, with direct labour being used in secondary activities such as cleaning and hauling.

In the location-based management system, when there are forced disruptions in the workflow for a task, each crew may either leave the site or wait. For resources which leave the site, the wasted time cost is calculated by the following formula for each resource:

$$C_m = \sum_{j=1,n} \left[N_d \times \left(t_m + t_{dm} \right) \times P_U , T_d \geq 1 \right] \tag{6.2}$$

Where:

C_m = The sum cost of wasted mobilisation time for each resource j
N_d = The work breaks where the duration of the break $T_d \geq 1$ day
t_m = Mobilisation time
t_{dm} = Demobilisation time
P_U = Unit price of the resource

Continuing the simple example above, it is easy to read the number of starts and stops from the flowline. There are a total of five mobilisations and demobilisations and only one of each (the first mobilisation and the final demobilisation) are essential. Assuming two hours for each mobilisation and demobilisation, the wasted cost of mobilisation time is $5 \times (2 + 2)$ hours $\times$ €20 / hour = €400. In this example, the waste is around 45%. The waste could be reduced by increasing the production rate of Task 1 or by splitting the task to have only two segments, each with continuous work. The additional risk of having the workers leave the site will be examined in the section about production system risk (page 180).

In many cases, the same subcontractor works on multiple tasks. In large projects, the easiest way to illustrate waste related to mobilisations is to use the resource histogram. Every time there is an increase or decrease in the histogram for the same resource type, mobilisation or demobilisation time is triggered. Figure 6.6 shows the resource graph for the electrical subcontractor on a real project. This is quite a good resource profile because there are few unnecessary demobilisations. The gradual ramp-up does not cause waste if resources can be maintained in continuous production until they are released from the site. This example has 28 mobilisations/demobilisations of which seven are waste.

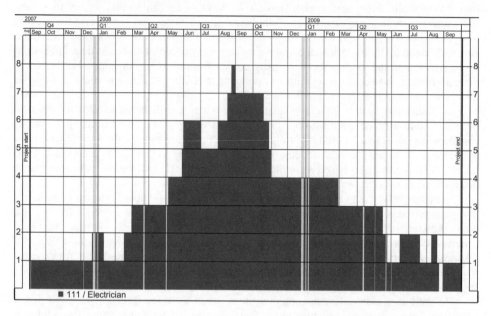

Figure 6.6 Electrical resources in a real project

The alternative to leaving the site is for resources to wait until productive work is released (as distinct from working out of sequence). For resources which wait on site, the total waiting cost is calculated by the following formula for each resource:

$$C_w = \sum_{\substack{j=1,n}}^{i=1,m} [T_w \times S \times P_U]$$ (6.3)

Where:

C_w = The sum cost of wasted waiting time for each resource j and break i
T_w = Waiting duration for each break
S = Shift length
P_U = Unit price of the resource

In our simple example, the waiting cost is calculated by adding up all the waiting periods. There are four waiting periods, each of them is 1.4 days. The waiting cost is (4×1.4) days $\times$ 8 hours / day $\times$ €20 / hour = €896. Note that this is a lot higher than the mobilisation or demobilisation cost for the same example (€400). However, the associated production risk is much lower, as will be discussed later (page 180).

Many managers might dispute the modelling of cost associated with leaving site or waiting time, believing they avoid such costs by ensuring that resources continue working in another location rather than leave or stop work to wait. This is a fallacy, as working out of sequence—similar to the 'making do' concept in the Lean literature (Koskela, 2004)—leads to inefficient production and rework. These are buried costs which are likely to exceed the modelled costs.

Moving around on site

When modelling production cost, it is necessary to recognise that some work patterns are more efficient than others. Poor planning can result in the *blow-fly effect*, where resources buzz from location to location, crisscrossing the site and generally wasting time in apparently important, but non value-adding, relocation time. When distances are great, such as in council road works or maintenance, such wasted time can actually exceed the time of production. It is necessary to calculate the cost of relocation to identify and resolve the blow-fly effect.

In normal location-based production, some time is necessarily wasted when a crew finishes work in one location and relocates to begin work in the next location. This can be greatly increased when work is performed out of sequence. In addition to just moving there, equipment and materials need to be moved. Crews often have a tendency to take a break, or even stop working early, when they have finished a location before commencing work anew. Generally speaking, the moving time depends on the distance of the move, availability of materials and equipment at the destination and the need to clean up in the location being vacated. More time is lost when locations are not similar because peculiarities of new locations need to be learnt prior to commencing work.

In the location-based management system, different waste times can be estimated for different hierarchy levels of the LBS. For example, moving from apartment to apartment on the same floor is cheaper than moving to different floors or to different buildings (a surprisingly common event). Another way to model this would be to actually model the time needed to move between each pair of locations, producing a wasted time matrix. However, this is cumbersome to set up without providing significant extra accuracy in most projects. Any modelling of relocation time is therefore necessarily an approximation.

The cost which is incurred by moving around on site is calculated by the following procedure for each resource:

1. Go through all the locations in location sequence.
2. Compare current and preceding location. Do they have the same 'parent' location?
   a. If yes, then use the waste cost factor for the task's accuracy level and move to the next location.
   b. If no, then go up through all the hierarchy levels until a common parent location is found and use the waste factor for that hierarchy level.
3. Sum the waste factors of all location pairs and multiply by the unit cost of the resource.

As an example of this calculation, let us consider a project of three location hierarchy levels: buildings, floors and apartments. It takes very little time to move from apartment to apartment. It may take an hour to move from floor to floor and four hours to move from building to building (remember that these assumptions include cleaning up, moving materials and equipment and learning the lay-off areas and peculiarities of a new location). The flowline schedule shown in Figure 6.7 has two tasks. Task 1 is flowing in a cost-effective way through locations while task 2 is exhibiting the blow-fly effect—moving in a random fashion through locations.

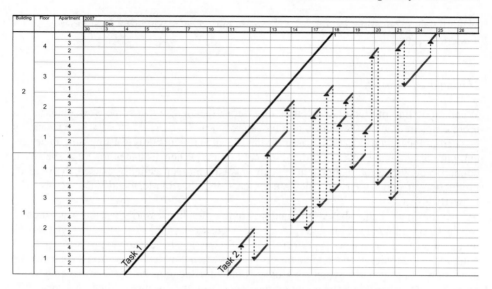

Figure 6.7 The blow-fly effect

Task 1 relocation time is 10 hours—1 movement from building to building and 6 movements from floor to floor. Task 2 relocation time is 54 hours—11 movements from building to building and 10 movements from floor to floor in the same building.

This procedure assumes that the locations are completely finished before moving on to the next location, an assumption that should hold for well-controlled location-based projects but it often does not hold for other projects. It also does not take into account movement required to rectify mistakes in earlier locations. Therefore, the method gives a lower boundary for the cost of moving between locations—the true costs might be expected to be a little higher. Location-based management offers a way to eliminate much of this waste by being explicit about locations and planning for crews to finish locations before moving on to the next location.

This waste factor can be very large on sites without location-based planning, as relocation occurs not just between locations (as Figure 6.7) but also part way through activities within locations—as work is often not completed before relocating due to incomplete prerequisites. For example, a Finnish superintendent once estimated that on a large job, where location-based methods were not used, over 20% of the production time was wasted on moving between locations looking for work. In this job, the locations were large and the control method was to send workers to available locations when they were found. Workers would work, on average, two hours in a location before running out of work and then going to ask someone where the next two hours work would be. This results also in increased coordination cost because the project's management must be looking for open locations all the time (to answer the question) instead of just knowing where crews should be moving next.

Stockpiling, hauling, delivery and receiving materials

Materials handling is a major source of wasted effort in construction. Stockpiling, hauling and receiving materials may not seem wasted time because it must be done to undertake production. However, it is non value-adding and in practice these operations are often not properly planned or the right resources are not used and waste occurs. The planning of deliveries and associated logistics will be described in a later section (page 188). However, some principles can be already stated here.

There is great significance in the planning decisions with regard to materials handling. Decisions about the location and size of material stockpiles, the timing of deliveries and the relationship between stockpile locations and moving around the site all have significant impact on project costs. Unlike other waste factors, which may be buried in the subcontract agreement, materials handling is often a direct and measured cost to the contractor.

It is a planning decision whether or not to use the same resources for receiving and hauling materials and for doing the actual work. These logistic operations cost time and money and often can be done cheaper by using somebody else such as direct labour. Often, this planning decision is not made explicitly but improvised on site (often the subcontractor's workers haul the materials). Optimally, these supporting activities could be used to either accelerate or decelerate the work to align the production rate depending on production system requirements. Cost is also an issue. How large an allowance has the subcontractor made for logistics in the bid? Is it cheaper to use direct labour, subcontractor labour, to hire labour or to buy logistics as a service? Should the deliveries come for every floor or every apartment? These cost differences should be compared to the production system production rate requirements to find the optimum solution for each task.

One further waste factor which follows this discussion is materials waste. Stockpiles lead to damaged materials. Moving materials over greater distances can also lead to increased damage. Damaged materials are either wasted, repaired (wasting more time) or may lead to rework due to quality concerns. Research is continuing to find out the magnitude of these effects and their relation to the schedule.

Identifying overhead costs

Overhead costs vary as a function of duration. They are usually proportional to the duration of the project, the duration of some particular tasks, or the utilisation of a specific resource. For example, a project engineer might be present for the duration of the whole project. A mobile crane is present only when needed as a resource in some schedule tasks, a fixed crane is required for the duration of a *hammock* of tasks (the total duration from the start of the first to the end of the last task in a set of multiple tasks relating to, for example, structure, roofing and façade tasks). For each overhead cost factor, overheads have a unit price, which is multiplied by a time variable calculated from either the overall duration of the project or from durations of selected schedule tasks or collections (hammocks) of tasks.

Overhead costs are minimised by compressing the schedule, or the relevant schedule tasks. Many schedule optimisation procedures optimise cost by cutting duration. This works in the ideal deterministic case when each task can be implemented just as planned. However, in the real world, there is a trade-off between overhead cost and other production system costs. Compressing the schedule leads to a higher risk of lost productivity. To enable optimisation of the schedule this risk has to be identified.

MODELLING PRODUCTION SYSTEM RISK

Production system risk in the planning system

Just as there are differing approaches required to model cost in the location-based management system, similarly there are differing methods of assessing the risk to production performance. In this context, risk is a stochastic assessment of the likely success or failure of a plan, with measurement of the associated consequences of real events. Just as it is important to model the costs which correspond with the design of the production system, it is necessary to model the risks which also flow from that design of the production system.

Events do not usually occur exactly as planned and deviations from the plan can occur for many reasons, even within a location-based management system. For example, subcontractors may not provide enough resources and may fail to achieve the desired production rates, deliveries might not arrive as planned, or subcontractors may leave the site and fail to return (an all too frequent experience in construction). There are hundreds of events and circumstances which must be tracked in a typical construction job and often some critical prerequisites are forgotten. Because of these uncertainties, just compressing the duration to minimise overhead costs and forcing planning tasks to be continuous is not enough. The planning system must also shield production from uncertainty. There is always a trade-off between duration compression and schedule risk.

This section describes a risk analysis model which can be applied to location-based plans. It uses probability distributions to model uncertainties, the production system cost model to evaluate any cost effects of risk, and planned logic to evaluate the effects of deviations in production on other tasks. Monte Carlo simulation[1] is used to calculate many scenario iterations, providing an aggregate outcome to estimate the risk inherent in the project schedule.

Simulation methods are not new to construction planning, having been well established in both CPM and PERT (see Chapter 2). The simulation model presented here is the first applied to specific location-based components—such as production rates and productivity—to achieve a risk model which more accurately approximates reality. The main difference when compared to earlier methods (where the developers of CPM used probability distributions and those for PERT used minimum, maximum and most likely values) is that activity durations are not manipulated directly by simulation. Instead the quantities, productivity and prerequisites are simulated, as well as certain production aspects such as commencement delay and return delay. This results in a more accurate simulation because these various factors are relatively independent and there are so many of them that the aggregate effect is difficult to capture in just one duration distribution. On the other hand, duration distributions of the same trade in different locations cannot be assumed to be independent, because the same resource gangs are generally undertaking the work through the multiple locations. The most critical difference is the explicit modelling of productivity impacts in the model, where the risk of damage to the production system arises from the discontinuous use of resources and consequent return delay distributions.

The various uncertainty types are described below with their potential effects on production. The risk model is used in Chapter 7, when defining the characteristics of an optimal schedule.

---

[1]   Monte Carlo simulation is a common stochastic modelling method for assessing the impact of uncertainty in a system, by iteratively evaluating a deterministic model (such as a cost or time model). It uses sets of random numbers as inputs for key variables (such as a price or time duration). The method is helpful for solving complex, non-linear models, or those with many uncertain parameters.

Uncertainty types in construction production

Uncertainty can be categorised into eight types with different characteristics. Each of these types is assumed to be independent.

- Uncertainty related to weather (environmental risks)
- Uncertainty related to the prerequisites of production
- Uncertainty related to adding resources
- Uncertainty related to productivity rates
- Uncertainty related to quantities
- Uncertainty related to resource availability
- Uncertainty related to locations
- Uncertainty related to quality.

With the exception of weather events, these factors are production system risks. Weather is an external environmental risk. The importance of this distinction is that production system risks form part of the production system and remain the contractor's responsibility. In contrast, external environmental risks such as weather, industrial disputes, acts of God, warfare, etc. form potential time extension claims and are external to the production system risk and generally are not the contractor's responsibility. Each of the uncertainty types is described below.

Uncertainty related to weather (environmental risks)

Weather causes a lot of deviations in construction projects, although some trades are more affected than others. Modelling weather effects can be done by a monthly probability for different kinds of bad or good weather. Each task can have a defined weather effect. The work can stop or productivity can decrease or increase by a given percentage for that day. If bad or good weather occurs, all tasks affected by the weather type in question are affected.

There is a relationship between weather effects and project contingency. Buffers between tasks belong to the production system, as a protection against production system risk, not to contingency. Weather effects, on the other hand, do have a relationship with project contingency.

Most environmental risks are ignored when modelling a location-based management system, however weather is unique in that it affects individual trades and specific locations differently (related to the effect of exposure). The amount of the effect may differ and it is possible to include data recording of weather conditions as part of the control monitoring system. For this reason it is a valuable component for risk modelling of location-based production. Managers must ensure that project contingency is adjusted accordingly.

Uncertainty related to prerequisites of production

This is a risk related to events that must be completed before a task can begin in a location. Prerequisites include events such as procurement, deliveries, the availability of resources, and the availability of relevant design information.

The prerequisite risks can be modelled by a probability distribution around the planned completion date for the prerequisite. A task can begin only after all its prerequisites are completed. This is a simplification from real production where tasks often begin before prerequisites are completed and are forced to leave unfinished work—resulting in the blow-fly effect of 'making-do' as described earlier in this chapter.

Uncertainty related to adding resources

Every time there is a mobilisation on site (because of new work starting, an increase in resource needs, or because the work was discontinuous), there is a risk of the resources not being available when needed. Many project engineers, who are responsible for project schedules, say that there is only a 50% probability that resources will return when required. The risk applies to the first mobilisation and subsequent mobilisations after resources have left the site when, for example, the return delay typically is one or two weeks. Subcontractors will often not believe there is enough work when they are first called and by the time there is enough work they will already be preoccupied on another site.

This uncertainty can be modelled by two distribution types. The first, delay on commencement, is more like a prerequisite and can be modelled in the same way for each point where resources are added to the task. The second, return delay distributions, have different properties and are only used when the resources run out of work and are forced to leave the site.

Uncertainty related to productivity rates

The appropriate productivity rates to use when planning production should correspond with good target productivity (and should exclude waste). But because of the huge individual differences in productivity rates (and even from the same worker over time), a probability distribution should be applied to simulate performance variation. For example, learning may have been assumed in schedules but does not necessarily occur to the same extent in practice, or the work might be more difficult or easier than originally assumed.

Each crew should have a productivity probability distribution of its own (rather than for the resources), because the productivity of crew members tends to be correlated. Crew productivity sets a trend production rate around which there may be independent location-based variation due to difficulty, quantities, weather or other location-based factors.

Uncertainty related to quantities

Planned quantities should hopefully be based on drawings current at the planning date. However, actual quantities can change as result of change orders or mistakes in the quantity take-off.

Each task should have a quantity distribution for the whole task and for location-specific variations. This is very important for work where there is a high level of uncertainty for the quantities, such as excavation—particularly on linear projects such as road or rail which have great uncertainty in material quantities and types.

Uncertainty related to resource availability

It is not always known whether or not the planned resources will be available. For example, the subcontractor may not have enough workers or might have other, more critical, projects which have higher penalties for delay, at the time when the task commences. Applying too few resources will lead to production rate deviation. Similarly, a subcontractor may not have enough work elsewhere and may seek to locate too many resources on the site. This will lead to discontinuous work due to the task pace being faster than the planned rate and its predecessors.

The resource availability can be simulated as the maximum number of crews available for each subcontractor between minimum, expected and maximum numbers. This may be a critical factor in planning work in special circumstances such as remote locations.

Uncertainty related to locations

All other miscellaneous uncertainty factors affecting durations can be modelled by using a distribution for each location. For example, this can represent uncertainty related to the difficulty of the location. It is assumed that these location distributions are independent of each other. Note that this is the only risk type in most CPM and PERT implementations: location duration is simulated independently of all other task durations.

Uncertainty related to quality

Quality errors can cause large schedule deviations. In the best case, they may only lead to rework affecting the workers doing the work (as long as the problems are identified early enough). In the worst case, the quality errors become hidden and require rework by multiple trades later in the project.

Quality is very difficult to capture in a simulation model. Construction companies in Finland report that, based on statistical information from hundreds of projects, following the original schedule has a very significant effect on quality. In the future, the simulation model could be expanded to include this result. However, more research on this relationship is needed. Therefore it is not yet recommended that quality risks be explicitly included in a simulation model.

Modelling control actions

For the risk model to be complete, it should be dynamic. Location-based production management is proactive and it is assumed that management can identify and react to deviations in production. These are control actions and may be simulated. The reaction time depends on how often information about actual production is gathered from site.

While control actions are described in Section Three, they are relevant to simulating the risk of a project plan because it is unreasonable to assume that a project's management will fail to respond to disturbances as they occur. Therefore it must be assumed that control actions will be implemented and that these will mitigate the impact of uncertainty. To model control actions, the following data is needed for each task.

Monitoring interval

Management cannot react before it knows that something is going wrong. The monitoring interval defines how frequently actual information is collected from site. However, management reaction is assumed to begin on the monitoring date, but only if the deviation is larger than deviation tolerance (see below).

In activity-based implementations, schedule updates are done monthly. In contrast, location-based control requires weekly or daily updates. This ensures that data will be available before the variation from uncertainty presents as a problem for production. Frequent monitoring is therefore assumed in simulating control actions.

Deviation tolerance

The deviation tolerance defines when a deviation from the plan is sufficiently large to warrant reaction. The tolerance can be defined either in terms of production rates or total delay in a location. The production rate tolerance is defined as a percentage of the planned production rate. Total delay is defined in days later than the planned date. Tolerance can be defined for both good and bad deviations. If one of these rules is triggered, control action can be assumed to follow.

Implemented control action

There are many possible control actions (see Chapter 8) and it is computationally expensive to optimise control actions during a risk simulation. However, a simulation model should at least examine two possible control actions: changing resources and working overtime. If resources are available (the simulated resource availability indicates that the subcontractor has available resources) and the control action is triggered by the production rate being too low, a crew can be added. The adding resources distribution is available for estimating the delay in implementing the control action. If the production rate is too high and there is more than one crew working, a crew can be demobilised without delay. If it is not possible to add resources, overtime may be applied to existing resources (with cost implications). The required overtime quantity is calculated and implemented immediately with increased cost. If overtime has been planned and the actual production rate is too high, then the overtime may be removed before decreasing resources.

The risk simulation model

While the detail of a simulation model is outside of the scope of this book, being a software implementation issue, the simulation process calculates iterations of plausible schedules by taking random samples from the probability distributions of the uncertainty factors described above, and applies them to the planned schedule. For example, each iteration will calculate different resource availabilities for subcontractors, different productivity rates and quantities, different weather conditions, and so on. The combination of these factors will result in a large number of possible outcomes, each iteration being a representation of one possible future reality. Computers can calculate thousands of iterations in a short period of time. When large numbers of these iterations are combined, a consistent picture of risky spots, probable finish dates and production system cost, will emerge. These results can be used to evaluate the reliability of a schedule and to optimise the schedule to find optimal trade-off between cost and time under conditions of uncertainty. In particular, they can be used to optimise the allocation of buffers between tasks.

Example of an iteration

To illustrate the idea of production system risk, the planned schedule for a small example of three tasks and five locations is shown in Figure 6.8. The small numbers in the schedule show the number of crews planned to be working in a location. The planned schedule assumes that all three tasks have two crews available.

Task 1 is planned to start in two locations with one crew in each, Task 2 starts with two crews which is reduced to one after the first location and Task 3 is planned to start with one crew and a second is added to the second location. The monitoring interval is five days for each of the tasks. In this example we ignore the weather and quality risks and concentrate on the resource issues. The first mobilisation is assumed to happen in the optimistic case five days earlier than planned, it is expected to happen on time, and mobilisation will be delayed ten days in the pessimistic case. Return delay is 0 in the optimistic case, expected to be five days and ten days is the pessimistic outcome. Task 3 is the only subcontractor which is large enough to confidently mobilise two crews. The maximum resource availability is assumed to be between one and three crews with Tasks 1 and 2 and two and four crews with Task 3. Task 1 is predictable by nature and the productivity can vary 10% in either direction. Task 2 has 40% variability and Task 3 has 30% variability. Each location for each task can have independent variability of −10% to +20%.

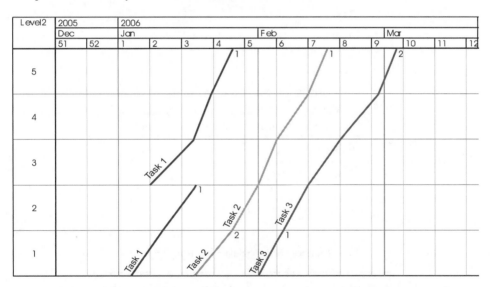

Figure 6.8 A planned schedule for risk simulation

In each simulation iteration, values are randomly sampled from the probability distributions between the values specified above and concentrating on expected values. For example, beta or normal distributions can be used. In an example iteration, samples from crew availability distributions indicate that Task 1 has resource availability of two crews, Task 2 has only one crew and Task 3 has three crews. The randomised mobilisation dates and productivity rates for each crew and associated location risks are shown in Tables 6.2 to 6.5. The project monitoring day is Friday.

Figure 6.9 shows the actual results of this iteration. Task 1 goes quite well in this iteration. The first part starts early, because the crew mobilises early, and finishes on time even though location-based productivity multipliers were below 1. The second crew for Task 1 mobilises late but is more efficient than the standard crew and so achieves a higher production rate.

Table 6.2 Productivity multiplier

	Crew 1	Crew 2	Crew 3
Task 1	1	1.05	
Task 2	0.7		
Task 3	1	0.8	0.7

Table 6.3 Mobilisation date

	Crew 1	Crew 2	Crew 3
Task 1	2.1.	8.1.	
Task 2	16.1.		
Task 3	28.1.	13.2.	

Table 6.4 Planned mobilisation date

	Crew 1	Crew 2	Crew 3
Task 1	4.1.	4.1.	
Task 2	18.1.		
Task 3	1.2.	7.2.	

Table 6.5 Location productivity multiplier

Location	1	2	3	4	5
Task 1	0.8	0.9	1.05	1.1	1.05
Task 2	1	0.88	0.95	1.05	1.1
Task 3	1	0.9	1.1	1	0.9

The problems appear with Task 2. Here, only one crew mobilises two days early. On the first Friday, the production rate is 35% of that planned, which is below the control action threshold of 70%. As additional crews are not available, the crew starts to work overtime, resulting in change of slope. On the following Friday, the production rate is still only 40% of that planned, so another hour of overtime is added. The next locations achieve almost the planned production rate with two hours overtime. Overtime cannot be increased further and additional resources are not available, so no more control actions are possible for Task 2.

The first crew of Task 3 tries to start work on 1 February. Because the preceding task is not yet finished, a return delay of four days is simulated. The crew comes on site four days after Task 2 leaves Location 1. A control action is triggered on Friday because the task is over five days late, but because another crew is already mobilising, no control actions are taken. The other crew, which was originally going to mobilise for the second location, starts work in the first location, resulting in a change of slope. The next control action point is on the following Friday 17 February. Task 3 has achieved 80% of its planned production rate but it is over five days late, so the third available crew is mobilised. Mobilisation delay is simulated as two days, so the crew becomes available on the following Tuesday—in time to begin the next location. On the next Friday, the task has exceeded its planned production rate by 30% but the task is still five days late, so overtime is utilised. The overtime is continued until the end of the project because the task remains behind schedule.

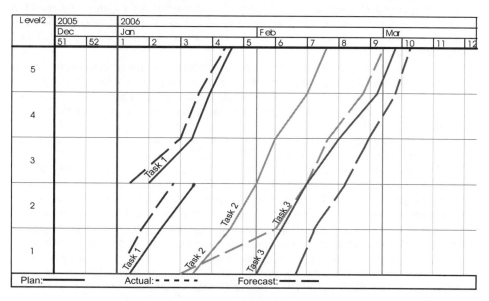

Figure 6.9 A typical iteration of the risk simulation with control actions

This example illustrates how risk analysis and simulation works. In real risk analysis, a thousand (or so) scenarios like this are created—randomly sampling the probability distributions defined by the planner—and the results are aggregated. The results can be used to optimise the plan—for example, by selecting a more dependable subcontractor for Task 2 or by adding buffers between Tasks 2 and 3.

Aggregated results of risk simulation

Aggregated results from risk analysis should also provide distributions for the production system cost, finish times of various milestones and the contract end date. It should show the critical parts of the project and points where control actions are most likely to be needed. Cumulative distributions are important because they show the probability of achieving certain dates, or cost targets, taking into account the uncertainty factors. Therefore, the schedule can be optimised to minimise the risk of schedule delays or cost overruns. Additionally, the critical parts of a project can be identified and more coordination effort provided to those activities. A complete example of optimisation of the schedule, using concepts of production system cost and production system risk, will be described in Chapter 7.

LINKING PROCUREMENT TO LOCATION-BASED PLANNING

Planning procurement using pull scheduling

Current activity-based scheduling practice, for linking procurement and design activities to the construction schedule, is to add the relevant events as predecessor tasks to the schedule. This method effectively pushes the start dates of production tasks, which is acceptable in CPM because activity logic drives production—CPM does not discriminate between different types of activities—thus the procurement activities are treated as being of equal importance to the production.

In contrast, procurement in location-based planning is calculated after scheduling the production tasks, and thus can use pull scheduling techniques. The production system, and in particular production continuity, is critical and the procurement process is treated as a prerequisite for starting production. Given that tasks are activities flowing through locations over time, procurement has the greatest effect at the beginning of the task and not so much later during the progress of the task. Quantities can be used to link a procurement task to one or more schedule tasks, because they are related to each schedule task.

Therefore, procurement and design are pulled by the master schedule when they are needed. This means that the procurement tasks are scheduled when required, rather than the current practice of planning an early start combined with activity float. The role for the procurement planner is to ensure that the procurement events are completed on time and have enough allowance for lead times of individual procurement events such as design, letting the work (bid documentation and evaluation), manufacture and delivery. Only in the rare case of a lead time exceeding the available time should the master schedule be adjusted.

In location-based planning the production flow is critical, therefore procurement activities should be pulled by production flow, rather than being allowed to push the production schedule.

Procurement tasks and events

A procurement task is composed of the organisation of material or subcontracted work packages that are able to be ordered from the same supplier or subcontractor. The quantities selected for a procurement task will form links to one or more location-based schedule tasks. Before the schedule task can begin, the procurement of materials and work needs to be completed.

The following typical procurement events, each with varying lead times, must be completed before deliveries can begin and work commence on the related tasks:

- Design finished
- Planning accurate task schedule
- Document and call for tenders
- Bid evaluation

- Contract
- Delivery order.

Each of these events may have a lead time after the previous event. The events are then scheduled in reverse time order working back from the commencement of the first-linked schedule task (and the demand for corresponding quantities) to establish the latest possible start date for the first procurement event.

The target cost for the procurement task is the sum of estimated costs of quantities. During implementation, the costs are controlled for each procurement task separately.

Example

Plasterboard sheets, insulation wool and frames are required in building plasterboard walls and suspended ceilings. Work is subcontracted and will be procured from two suppliers. Boards will be bought from a separate source to get the cost benefits which are available from the general contractor's annual supply agreement with the supplier. The insulation wool and frames will be included as part of the subcontracts. Thus there will be three procurement tasks associated with these two schedule tasks: one for the supply of plasterboard sheets and one for each of the scheduled tasks (work packages which include both work and materials). The target cost will be established for each procurement task based on estimated prices of work and materials.

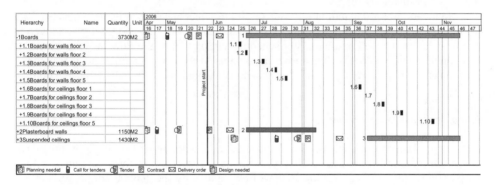

Figure 6.10 Three procurement sets of activities for two plasterboard tasks plus board

Figure 6.10 illustrates the procurement activities which occur in the procurement task for two plaster trades. The supply of the plasterboard has its own procurement task, and in this case the actual deliveries proposed for the plasterboard have been planned and displayed on the chart.

Logistics decisions

Delivery planning

Some logistics decisions should be made during the pre-planning stage. Important issues include:

- When to deliver the materials:
  For example, do you deliver in advance when the materials can be lifted with the crane while building the structure, or deliver just-in-time immediately prior to production in each or grouped locations?
- How many deliveries:
  For example, do you deliver for every apartment and/or office, for every floor or for every building?
- Time and resources needed for receiving and hauling (for each delivery)?
- Should the same resources be used for logistics as for production?
- What is the lead time before production can start?
- What will be the required storage time?
- What will be the cost of freight for delivery?

In the logistics plan, a target cost for logistics activities is established. This cost should be separated from other bid items because costs which are related to logistics may be minimised by making them transparent and controlling them as separate items. To reduce waste associated with hauling materials, it is necessary to decide beforehand who is responsible for hauling and how much time will be allowed for such activities. If the same resources are used for logistics as for production, the planned durations of locations with deliveries should be increased by the time spent in receiving and hauling. Being conscious about materials handling costs also enables the general contractor to decide whether to procure logistics separately or to use direct labour when it is cheaper than a subcontractor's inclusion for materials supply.

In location-based planning it is meaningful to select one of the hierarchy levels of the location breakdown structure to be the basis of deliveries. The planned timing of deliveries can be calculated based on the master schedule and plotted to the flowline (Figure 6.11). Plotting deliveries in the flowline allows the planner to see when materials are stored in the location and to plan storage so that it does not hinder other tasks within the same location.

Example of delivery planning

Boards for plasterboard walls have been planned to be delivered for every floor three days before the required time. Each delivery needs two worker hours for receiving and four worker hours for hauling. The logistics will be handled by the general contractor's direct labour because the subcontractor's bid for logistics was deemed too expensive and because the task is critical and the workers should be able to concentrate full time on production. Figure 6.11 shows two activities: concrete floor finishing work and plasterboard walls and deliveries of boards. Because of the buffer between the activities, stacks of boards delivered in grouped batches in time for the following work will not hinder the preceding task.

Materials storage and handling

Materials storage can be used to control the sequence of work and to minimise production disruption. If materials are delivered just-in-time and only for those locations where work is to follow, then crews will be unable to work out of sequence due to a shortage of materials. This is a powerful method to ensure the production sequence is maintained.

Stockpiles of materials are also a principal cause of damage and waste of materials and consequent rework. However planning to deliver materials precisely as required requires detailed knowledge of location-based quantities and associated timing of the works. This is a great strength of the location-based management system.

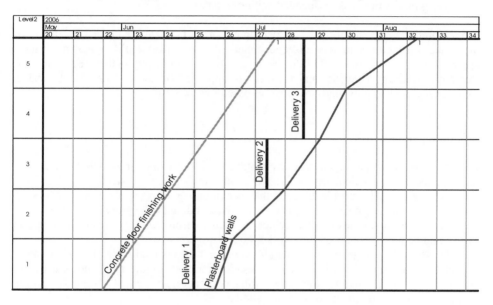

Figure 6.11 Deliveries for plasterboard walls grouped to minimise cost

SCHEDULING DESIGN IN LOCATION-BASED MANAGEMENT SYSTEM

In complex projects, getting the design on time can be critical to the success of the production system. Design sits between procurement planning and production planning because it shares some of the characteristics of both. Part of the design occurs as part of the procurement chain, before calling for tenders. On the other hand, detailed design is often not available at that point and will be designed in time for production to start in locations. Design also has a special requirement for gatekeeper functions: design is iteratively developed and often needs to be approved by multiple gatekeepers before it can be released to production.

Design tasks

In the location-based management system, design tasks are linked to one or more production or procurement tasks. Each design task has a supplier—for example, an architect or structural engineer company. Design has a process which includes one or more stages of design, gatekeeper functions and production and procurement tasks using information about the stage. Additionally, a design task can be linked to other design tasks. Each design stage can be location-based on any hierarchy level of the project's location breakdown structure.

As an example, the design task *ceiling drawing* has a link to the *suspended ceiling* procurement task, *painting* procurement task, *tiling* procurement task, *suspended ceiling* production task, *painting* production task and *MEP overhead* design tasks. Each task requires different information.

- *Suspended ceiling* procurement task: suspended ceiling types and quantities.

- *Painting* and *tiling* procurement tasks, *MEP overhead* design tasks: suspended ceiling height.
- *Painting* production task: suspended ceiling finishes.

From a production point of view, at least three stages of design are needed. Each one of these stages has one or more gatekeeper functions. Suspended ceiling types might need approval from the client. Height information needs to be approved by the MEP contractor (for constructability analysis) and finishes are approved by the end user of the space. Ceiling types and heights might be location-based for each building.

Scheduling design tasks

Each stage or gatekeeper function has a quantity, in worker hours, for each location and can use the same duration calculation rules as for production tasks. Stages and gatekeeper functions can be allocated to people to ensure that there is enough capacity for each process step. Worker hours for design tasks should be adjusted depending on the probability of failing to pass the gatekeeper. Some design activities have a higher probability of resubmissions and this should be taken into account in worker hour requirements.

Design tasks and gatekeeper functions can then be scheduled as described in Chapter 5, and flowline can be used as the visualisation method. Figure 6.12 shows a design schedule of five locations and three stages with gatekeeper functions in between. For clarity, production and procurement tasks are not shown.

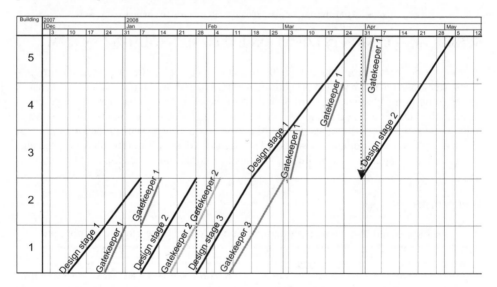

Figure 6.12 Design tasks and gatekeeper functions

Links to the procurement schedule

The first event in the procurement schedule is *design finished*. This event usually refers to information required in bid documents. Therefore one of the design stages (and passing the associated gatekeeper function) of the relevant design package can be linked to the first event of procurement. Because procurement has been pull scheduled, it provides the final deadline for related design tasks and stages. If it is discovered, during the process of design scheduling, that it is impossible to achieve this date, then the production schedule needs to be updated correspondingly.

Link to production schedule

Production schedule tasks cannot start productively before all required design information is available. The Production schedule establishes location-based milestones for the design schedule. Instead of enforcing hard scheduling logic, the schedules are softly linked—providing alarms when production deadlines are not met by the design schedule. This is because production requirements take precedence and thus automatic pushing of the production schedule by design activities should not be allowed. Only in the case where it is impossible to meet production deadlines, should production start dates be adjusted.

PLANNING FOR QUALITY

There is a direct relationship between the design of the production system and the quality of the work produced. In this sense, there is such a thing as planning for quality, which means a plan which is designed to ensure work can be completed to the intended quality levels. More importantly, time must be allowed for quality to be managed and there must be practical control mechanisms to ensure faulty work is not produced or retained.

Planning for quality means that enough time will be allowed for inspections and measurements to be completed and approved before the following trade comes to the location. In traditional planning systems, this time is required to be absorbed into the activity window. In location-based planning systems, the task is scheduled according to the calculated time required for production, so the buffer can be used to allow systematic time for quality inspections. As planning and control is based on locations, it is rational to control quality also by location. A key to this process is to pass 'ownership' of locations between the general contractor and the work crews.

Planning for quality has two aims: ensuring that following tasks have all the prerequisites completed prior to starting work, and checking that the quality of work already completed meets the required standards before the following crew commences work in the location. This might include special conditions such as satisfactory curing or setting of earlier trades. The location essentially belongs to the preceding trades until all these conditions are met.

In the pre-planning phase, the quality and prerequisite checks should be decided for each task and the locations where they will be required should be planned. There are some general quality events and prerequisites that apply to most of the tasks and checks which apply to individual tasks. Examples of general prerequisites for starting in a location are:

- Ensuring resource availability:
  If there are no breaks in flow, this applies only for the first location.
- Drawings completed, requests for Information completed and drawings delivered to workers:
  - For each location.
- Start-up meeting with the subcontractor:
  - Generally only for the first location, if there are long breaks in the workflow, then more may be needed.
- Accurate task plan prepared:

- Schedule, cost, quality, required for first location only.
- Handing over of each location from general contractor to the subcontractor:
  - This is important where progress is contractually controlled using incentives.

General quality events that apply to most tasks can include:

- Quality measurements and checks:
  - For example, concrete curing, drying, ensuring walls are plumb, floors level, etc.
- Handing over of location to general contractor:
  - Prerequisite for payment.

Handing over locations, first from the preceding subcontractor to the general contractor, and then from the general contractor to the next subcontractor, helps to prevent rework and ensures smooth workflow for the succeeding subcontractor. Measurements and checks help to prevent hidden quality problems which may result in increased maintenance costs.

PLANNING FOR SAFETY

Safety is similarly greatly enhanced by a quality schedule, as locations are handed over to crews exclusively and so there is a reduced chance of accident due to multiple trades working simultaneously within locations. The concept of ownership of locations enables a shift of responsibility for safety which can achieve general safety improvement on the site.

MODELLING PRODUCTIVITY WITH LEARNING

Learning theory

Learning curve theory is drawn primarily from manufacturing production, where there are high levels of repetition, and where competition is paramount. Also known as experience curves, learning curves relate to the reduction in time taken to do a task as an operator gains more experience. The theoretical relationship is that a work crew will benefit from repetition of a task, and that for each doubling of the number of repetitions, any crew performing an activity will achieve a constant rate of reduction in the time taken to complete that task. Thus the second is quicker than the first, the next two are quicker by the same increment as the second was over the first. The next four, then eight, then 16, etc. This relationship was first identified for the aircraft industry by Wright (1936) and the Boston Consulting Group found it also applied to other industries (Kerzner, 1984).

There is a simple mathematical relationship which is being described here. Typically the literature indicates that in manufacturing, learning achieves a cost and time saving of 10% to 30% each time a company's experience at producing a product doubles (empirical data from Finland suggest that the learning rate in construction is lower). For example, the time for a second unit might be 75% of the first unit. An example is given in the Table 6.6, for a process operating with a 75% learning curve.

Table 6.6 Table of production given a doubling rate of 75%

Cumulative production	Hours this unit	Cumulative total hours
1	812	812
2	609	1421
10	312	4538
12	289	5127
15	264	5943
20	234	7169
40	176	11142
60	148	14343
75	135	16459
100	120	19631
150	101	25116
200	90	29880
250	82	34170
300	76	38117
400	68	45267
500	62	51704
600	57	57622
700	54	63147
800	51	68349
840	50	70354

The unit time for production drops rapidly at first, but gradually more and more repetition is required to gain the same level of improvement. This is a logarithmic relationship. Figure 6.13 displays the experience curve and shows unit cost (time) plotted against experience (cumulative units of production).

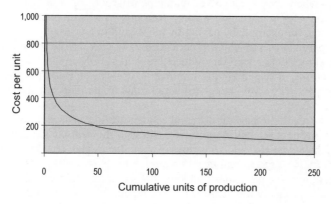

Figure 6.13 The experience curve in general production

Past research suggests that individuals performing repetitive tasks improve in performance as the task is repeated:

- The time required to perform a task decreases as the task is repeated.
- The amount of improvement decreases as more units are produced.
- The ratio of improvement has sufficient consistency for use as a prediction tool.

Learning predominantly affects the human component of the process, indicating improvements will be restricted where there is significant reliance on machine processing. Furthermore, not all the factors which lead to an improvement will apply to the repetition of work arising merely from moving through locations. Kerzner (1984) identified the following relevant general factors:

- Labour efficiency
- New production processes
- Equipment efficiency
- Resource mix
- Product Standardisation
- Product redesign.

Of these labour efficiency is the dominant factor in location-based modelling. Labour productivity improvement operates through the following mechanisms:

- Labour is quicker as it learns
- Less supervision is required for experienced labour
- Waste and lost time is reduced with experience.

Improvements in labour efficiency do not generally occur on their own. Management policy effects the potential gains, for example:

- Workforce stability—a changing workforce loses the gains already made and resets the experience curve. Construction has a very mobile workforce, with a resultant difficulty in gaining advantage from learning through experience. It is clearly desirable to maintain a work gang where they are involved with repetitive construction.
- Compensation—for construction workers to become more productive they may expect to be rewarded, for example by paying by piecework in construction.
- Reduced supervision—stable and experienced work crews require less supervision and therefore non-production staff can reduce their supervisory input.
- Work specialisation and methods improvements—specialisation increases workers' proficiency, primarily through increasing their rate of learning at a particular task.

This complexity results in several different models having been developed for different circumstances. Arditi et al. (2001) summarises these as the *straight-line power model* (log-linear model), the *Stanford-B model*, various exponential models (*basic exponential model*, *Dejong's model*, *Levy's model* and *Pegel's model*) and an *S-curve model*.

Construction poses particular problems for the application of learning theory. The most important is probably the identification of repetitive units of production. Unlike the assembly line production of widgets, repetition in a location-based construction process cannot be restricted to the completion of finished items or activities. This would be an unrealistically crude measure of repetition, as learning occurs throughout the production of large production activities.

Location-based learning

The location-based model assumes that learning is a function of the production worker hours spent by a crew in a repetitive activity within the one project. Repetitive activity is defined by quantity items contained in schedule tasks. Learning occurs separately for each quantity item, as they often represent different skills. Therefore, multi-skilling will decrease learning effectiveness for any given quantity item because many work types are being done by the same crew. Learning only applies within a single project because it is assumed that the learning does not improve the skill level of the crew in general—the workers are assumed to be professionals and so on top of the basic skills—but rather the knowledge of the special circumstances of the project task. Learning is assumed to be reset to base levels if a crew leaves the site because there is no guarantee that the same crew will return.

The *straight-line power model* of learning was proposed by Arditi et al. (2001) as being most suitable for location-based scheduling because Everett and Farghal (1994) found it to be the most reliable predictor of future performance of construction field operations. The authors found this approach to be further supported by empirical data from the Finnish productivity database, which provides average labour consumption data for different types of work. These were found to follow the straight-line power model almost perfectly.

The mathematical function of straight-line power model is:

$$Y = Kx^n, n = \log_S / \log_2 \qquad (6.4)$$

Where:

Y = The cumulative average worker hours required for the target work hour;
K = The duration (worker hours) required for the first target worker hour
x = The cumulative worker hour number
n = The *learning index* $(= \log_S / \log_2)$.

The learning index rate S is a ratio, with 1 signifying no learning and a number smaller than one meaning that learning occurs. In this application, K should always be more than 1 (meaning that the first target worker hour always takes more than one *learning-adjusted worker hour*). If there is a break in work and the crew leaves the site, the learning process restarts. Figure 6.14 illustrates Arditi's learning effect. This shows the rate of reduction in duration used in the example on page 198, where K is 1.48 and S is 0.925.

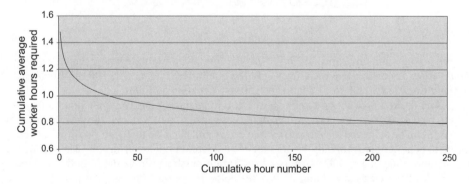

Figure 6.14 Arditi's learning effect applied to construction production

Standardised worker hours are used instead of actual quantities in order to standardise the quantities without having to define a learning function separately for each quantity item that has separate units. Standardised worker hours are calculated by multiplying the quantity by the labour consumption, as described in Chapter 5.

The learning rate can be established separately for each schedule task and the more time-consuming quantity items benefit from more learning.[2] There are many ways to show learning in the schedule. The first is by assuming that the number of crews stays constant. This will result in the schedule line bending left if there are no breaks. The second is by assuming that the amount of crews is adjusted to achieve the desired production rate. A crew can be removed from activity when the target production rate can be achieved with one less crew. This will result in the line bending left until the point where crew decrease becomes possible is achieved and bending right after that.

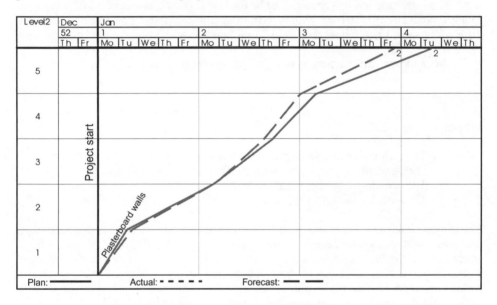

Figure 6.15 Task acceleration due to learning

Examples

Figure 6.15 graphically shows the effect of learning on an example where the task *plasterboard walls* is done in five locations (with the quantities: Location 1, 100 m^2; Location 2, 300 m^2; Location 3, 200 m^2; Location 4, 150 m^2; Location 5, 400 m^2).

The desired production rate is 70 m^2 per day, requiring the average labour consumption to be 0.46 worker hours per m^2. With the crew size of two workers, the desired production rate can be achieved on average by two crews and because there are two crews, each is going to benefit only from half the amount of total worker hours performed. The value of K is 1.48 and value of S is 0.925 (using values from Finnish empirical data). In the first location both of the crews are doing 23 worker hours (resource consumption 0.46 times the quantity 100 m^2 divided by crew size of 2). Based on the learning curve factors, they actually use 26.7 hours each to do 23 worker hours worth of work. After the second location, the learning adjusted production rate meets the average production rate and later the effects

---

[2] Arditi et. Al (2001) describe methods for using fuzzy logic to adjust learning rates. While outside the scope of this book, this is an interesting research discussion.

increase in every location. In this example, the total effect is two days saving of 17 days subcontract, 12% less than planned.

The same example with just one crew, and thus half the production rate, shows quite different results (Figure 6.16). The single crew is able to progress farther on its learning curve, so the total time savings are larger. The planned duration in this example is 33 days with the same quantities as in the last example. Total duration after taking learning into account is 27 days, a saving of five days or 19%. These simple examples show one reason why it is more expensive to cut duration by increasing resources.

There are further cost implications with increasing resources, which were explored earlier in this chapter.

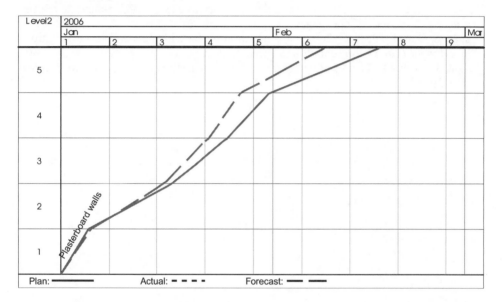

Figure 6.16 Task acceleration due to learning

In the authors' experience, learning curves often remain theoretical constructs in construction. In normal production there are start-up difficulties in the beginning when crews are learning the quality requirements of the site. These difficulties, combined with the fact that full crews do not usually become available at once, cause production to be slower in the beginning. After this initial and often dramatic improvement, very little learning seems to happen.

Instead of using elaborate learning functions, the authors suggest a simpler work-around is to use a production rate multiplier which is smaller than 1.0 in the early locations.

REFERENCES

Arditi, D., Tokdemir, O.B. and Suh, K. (2001). "Effect of learning on line-of-balance scheduling". *International Journal of Project Management*. **19**: 265–277.

Everett, J.G., Farghal, S. (1994). "Learning curve predictors for construction field operations". *Journal of Construction Engineering and Management*. ASCE. **120**(3): 603–616

Kenley, R. (2003). *Financing Construction: Cash flows and cash farming*. Spon Press, London.

Kerzner, H. (1984.). *Project management : a systems approach to planning, scheduling, and controlling*, 5th edition. Van Nostrand Reinhold, New York.

Koskela, L. (2004). "Making do—the eighth category of waste". *12th Annual Lean Construction Conference*. Copengagen.

Seppänen, O. & Kenley, R. (2005). "Using Location-Based techniques for cost control." *13th Annual Lean Construction Conference*. Sydney.

Wright, T.P. (1936). "Factors affecting the cost of airplanes". *Journal of Aeronautical Science*. **3**(4): 122–128.

Chapter 7

Using location-based planning methodologies

INTRODUCTION

It is one thing to know about the location-based theory and its associated methods, it is another to know how to use that knowledge to build effective schedules for a project and to both plan for production efficiency and provide confidence. As discovered very early on with activity-based CPM under conditions of resource management, there are many ways to plan any project, all of them effectively *correct*. The real challenge is to develop the optimum schedule, or better yet to take planning to another level and to use it as a strategic management tool which both optimises production and minimises risk.

It is possible to define methodologies to optimise a schedule in terms of cost and risk using the tools defined in Chapters 5 and 6. This chapter introduces the concept of a *good schedule* and presents ways to make and optimise a good schedule by using the tools and methods presented in previous chapters. While there are many approaches to scheduling, there are effectively two complementary methodologies for optimising production.

First, the *critical path methodology*[1] works without considering production system risk and concentrates on minimising project duration. This approach will therefore be familiar to existing activity-based CPM practitioners. This methodology approaches location-based scheduling in the traditional CPM manner—searching for earliest possible task logic but with continuous and aligned task production as the strategy. The differences from activity-based CPM scheduling include using quantities and production rates and using layered CPM logic to achieve continuity of the schedule. The CPM-based methodology achieves the shortest duration for schedules by strictly following the layered CPM logic, with tasks scheduled to follow their predecessors as soon as the logic allows, while minimising the cost of the planned schedule. Once production is aligned, the total duration is normally much shorter than in activity-based CPM schedules, because aligned production removes float from the production system and calculated durations are used instead of task windows. However, this methodology may lead to riskier schedules which may be difficult to implement in practice, as uncertainty and risk are not taken directly into account.

Second, the *risk management methodology* is an expansion of the critical path methodology but requiring that an optimal trade-off between time, risk and cost is sought by aligning the schedule, by planning tasks to be continuous and by buffering tasks against interference, variations and minor interruptions in production. Essentially the risk management methodology aims to be a more realistic model for project implementation: it is designed to provide greater confidence that the plan can be achieved. While the planned, deterministic cost of the schedule and duration is typically higher than in the critical path methodology, the risk of cost or time overruns is considerably smaller. Furthermore, there is evidence that the same duration as suggested by ordinary CPM schedule might be achieved with much lower risk and cost by using the risk management based methodology. This is achieved by planning for efficient production and then deliberately making allowance for the difficulties likely to be experienced on site.

---

[1] The critical path methodology of location-based scheduling is not the same as preparing an activity-based CPM schedule. This methodology assumes location-based tasks are formed from sequences of continuous activities organised to be as soon as possible, forming the semblance of a critical path.

It may be noted that traditional activity-based scheduling is not without its risk management strategies. First, a duration contingency may be provided which allows deviation in the project duration while remaining within the required target duration. Second, all activity windows may include an allowance for duration variation which is used to absorb variations in production. However, the former merely accepts problems and provides no confidence during production except as a safety buffer, while the latter becomes lost in normal practice and ceases to be an effective management tool. In contrast, the risk management methodology absorbs risk without increasing project cost or overall project duration.

It is strongly recommended that the critical path methodology is learned and implemented in location-based planning, to show the benefits over activity-based CPM methods and, with emphasis being placed on accurate work quantity estimation and consumption rates, to show the true production duration. However, before using the resulting schedule in actual production, those risk management methodology tools described in this book should be adopted to ensure successful project implementation.

What is a good schedule?

There is no such thing as a correct schedule in any form of scheduling. It is just a plan, and there can be good plans and bad plans. There are many ways that work can be planned and many solutions that will lead to a successful outcome. However, some schedules are clearly better than others. In particular, there are schedules which do more than merely achieve the shortest time for the project.

A good schedule maximises productivity, finds an optimal balance between risk and duration, and is feasible to implement. The location-based planning system includes tools for evaluating and optimising each of these three aspects.

Maximising productivity

Productivity is maximised by planning continuous resource use, with the plan being based on accurate scope and quantities, resources and productivity data. Each trade should use the optimum resources, organised using the most efficient work crews. Generally, the production rates of predecessors and successors should be aligned, and each location should be completely finished before moving on to the next location. Both critical path and risk management methodologies achieve schedules which maximise productivity but with varying degrees of risk.

Optimising risk and duration

The trade-off between risk and duration is difficult to solve because it involves deciding acceptable levels of risk for the project. In the deterministic case, assuming no variability in production (activity duration, late starts, etc.), it is always beneficial to minimise duration. However, in real production there is a high level of uncertainty involved, making a minimum time schedule risky. The trade-off between minimum time and reducing risk can be evaluated using production system cost and production system risk tools. Regardless of the chosen risk level, the selected solution should be efficient, meaning that a solution should be found which achieves a given duration without the ability to further reduce the risk.

The critical path methodology achieves schedules with minimum durations without taking risk into account. The risk management methodology is more complex because it tries to find a solution which achieves a minimum duration but one which does not exceed the selected risk level.

Ensuring feasibility

Both critical path and risk management methodologies aim at feasible schedules. A good schedule is a reliable model for production which can be practically implemented. This is ensured by planning with correct logic, taking into account resource constraints and by modelling subcontractor or work crew behaviour accurately.

LOCATION-BASED PLANNING PROCESS

Location-based planning is generally most effective when the following process is followed. Steps 1 to 4 use the basic location-based model described in Chapter 5. They are used by the critical path planning methodology. The risk management methodology is adopted by adding optional steps 5 and 6 to the basic process. Steps seven to ten involve the optional use of further tools on top of the planning process and can be used by either methodology.

1. Define the location breakdown structure.
2. Define location-based quantities.
3. Build tasks from quantities and define:
   a. Optimal crew
   b. Layered logic links to other tasks.
4. Align the schedule and optimise sequence and duration by:
   a. Changing production rates
   b. Changing sequence
   c. Breaking continuity
   d. Splitting.
5. Evaluate production system cost and risk (optional).
6. Optimise cost and risk (optional) by:
   a. Adding buffers
   b. Changing production rates
   c. Changing sequence
   d. Breaking continuity
   e. Splitting.
7. Cost load the schedule.
8. Optimise cash flow:
   a. Change payments
   b. Change production rates and start dates.
9. Approve the schedule.
10. Plan procurement and design schedule:
    a. Use pull scheduling techniques and soft constraints
    b. Do changes to the production schedule only if necessary.

The priority is to first plan all the tasks, each with one optimal crew, add all the logic links and only then begin to align and optimise the schedule. This is because the aligning process uses the flowline visualisation to find optimal production rates. If all the tasks are not there, it is difficult to see the big picture for optimisation. All steps of the process are described in detail below.

Optimal location breakdown structure

Defining the location breakdown structure (LBS) is one of the most critical planning decisions involved in location-based planning. This decision has far-reaching consequences, which include the following:

- The number of logic relationships required to model a project depends on the LBS
- The quantity take-off must correspond with the selected LBS
- Logistics and deliveries are planned based on the LBS
- Progress is controlled based on the LBS
- The clarity of the flowline visualisation depends on the LBS.

This section gives guidelines and hints about how to arrive at the best possible LBS. Hints which are specific to different project types are provided in the Chapter 13.

Guideline 1: Hierarchy levels should have a global meaning in the project

Many of the features of the location-based planning system, as described in Chapters 5 and 6, rely on the hierarchy levels of the LBS. For example, Layer 2 logic defines links on a level of accuracy. Cost loading calculates the timing of payments based on locations of the task on a given hierarchy level. Logistics and deliveries can be planned for a hierarchy level. Production system cost models assume that moving from any location to any other location on the same hierarchy level takes the same amount of time.

Thus each hierarchy level should have a global meaning in the project and no individual branch of the LBS should terminate at a higher level than the lowest level of detail for the project. However, all hierarchy levels may not have meaning for all sub-branches of the LBS. For example, in a residential building project consisting of a residential tower and multiple single-storey condominiums, the floor level does not appear relevant to the low buildings. In these cases, a location should be added to the LBS so that the overall logic can be maintained. In this example, a single floor level would be added to each of the condominiums. Similarly, detail may not always be required in some branches—for example a dining hall included in a student housing project may not require LBS detail to the rooms level—however the higher level can be replicated (for example as a single room) to provide the required hierarchy. A good way to check the validity of hierarchies is to try to come up with a name which describes all the various locations at the same hierarchy level.

Guideline 2: Locations must be physical and clearly defined

This guideline results from the fact that quantities need to be location-based and that locations must be usable for control. Often the person doing the quantity take-off is different from the person specifying the LBS. Therefore there should be no ambiguity about what the locations mean.

One example of how to apply this guideline is to define location boundaries clearly. For example, it is often very helpful to define the horizontal division between floors as the finished floor level (FFL). This ensures that most work associated with a level takes place within the appropriate space. Thus a suspended slab actually belongs to the floor below it, as it is constructed below the FFL (see Figure 7.1). This makes sense when it is considered that constructing the floor involves the scaffolding and formwork, and that this work predominantly restricts access to the floor below until it is removed. Of course, it restricts access to the floor above as well, but that is not a problem because Layer 3 logic requires the lower floors to be completed before commencing the same activities on the upper level. Once slabs are completed, work above FFL, such as finishes, will be undertaken in the upper location, while work below FFL, such as sub-floor plumbing or stripping formwork, will generally take place in the lower location.

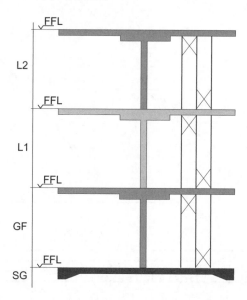

Figure 7.1 Suggested division of levels separated by FFL

This is a guide which particularly helps with understanding the significance of levels for horizontal elements such as floor slabs. Unfortunately, architectural conventions will often place the floor slab under a level into that level (in other words, the floor you are standing on will be regarded as part of the level in which you are in). This will require more care to schedule as the work to build the slab will interfere with work on the floor below. The key here is to choose the convention for the project and to draw a diagram such as Figure 7.1 to clarify the relationships.

Guideline 3: If possible, define vertical cuts for the project at the highest hierarchy level

Those locations which cut through a whole building vertically, or which are different buildings, are critical. Generally speaking, it is beneficial to have as many structurally independent vertical sections as possible in the project. This is best illustrated by a few examples.

The first example is a schematic having just three tasks: structure, roof and finishes. To reduce the risk of finishes suffering from rain, the roof must be waterproof before the interior work begins in the building. In Figure 7.2, there are two sections which are done in sequence. In Figure 7.3, one of the sections has been split into two structurally independent vertical sections (2:1 and 2:2) and they are then executed in sequence. The task durations of each of the new sections is half of the original. The resources used and the logic links are exactly the same. Finishes are able to commence and complete sooner while maintaining continuity.

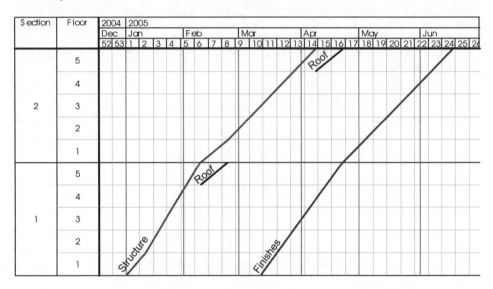

Figure 7.2 Two section sequence

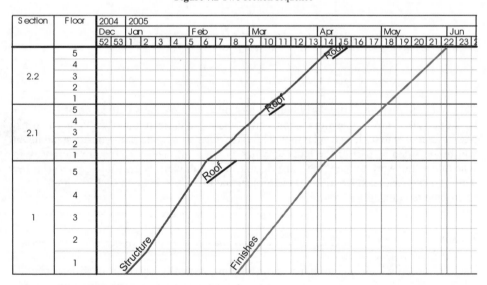

Figure 7.3 Two section sequence—one section split

In this simple example, it is possible to save three weeks (from 24 to 22 weeks, or 8% of total duration) without adding resources and by continuing to perform all the work continuously. In more complex examples, the effect is often much more dramatic.

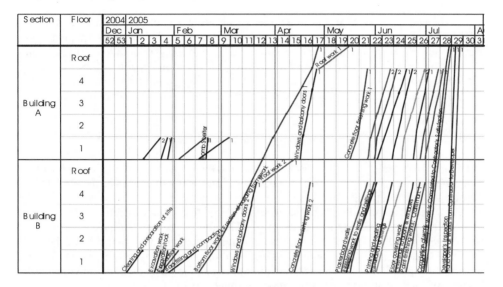

Figure 7.4 A two section sequence

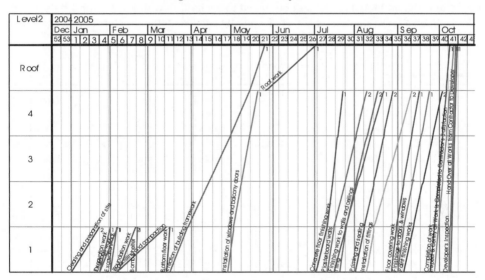

Figure 7.5 A project with similar sections combined

Figures 7.4 and 7.5 show two schedules for the same project but with different LBS. The schedules are otherwise identical but the two vertical sections have been combined in Figure 7.5. Combining has been done by combining the quantities of each building for each floor but using the same logic. This means that each floor takes longer to construct in the combined building. The duration of the combined solution is 41 weeks. The duration of the solution with two vertical sections is 29 weeks. This is a duration saving of 29% without adding resources or risk and using the same logic and quantities. Such is the saving that it can be well worthwhile to plan for temporary vertical waterproofing to the section interface (shuttering) while the second vertical section rises.

When planning vertical cuts, other trades such as mechanical, electrical and plumbing (MEP) should also be taken into account. The main mechanical plant room equipment will affect zones which must remain dust free during testing. If this [airconditioning] zone is not the same as, or within, the vertically cut LBS area, the duration benefits will be lost during the commissioning stages of the project. Some Finnish case studies have changed the design of MEP systems to align the affected zones with the LBS to achieve duration savings.

Guideline 4: The same LBS should apply to all or most trades.

If the same LBS can be used by most tasks, much of the logic can be modelled by location-based logic Layers 1–4. In this case the number of links that will have to be manipulated by the planner is dramatically decreased.

In some cases not all tasks fit the general LBS well, and trade-specific locations must be modelled by multiple activities in the original location. A better solution can be to create new branches in the LBS for these tasks. Generally speaking, new branches of the LBS should be added if many tasks follow the same location breakdown (such as having a branch for structure and another for finishes) and multiple tasks should be used for individual tasks having different locations (such as mechanical rough-in).

Typically, branching should happen below the floor level of the LBS hierarchy, so that tasks of different branches can still be linked to each other with Layer 2 or 4 links using floors. In many case studies, separate branches are created for the structure (by erection area or by pour), finishes (by space type) and façade (by grid line).

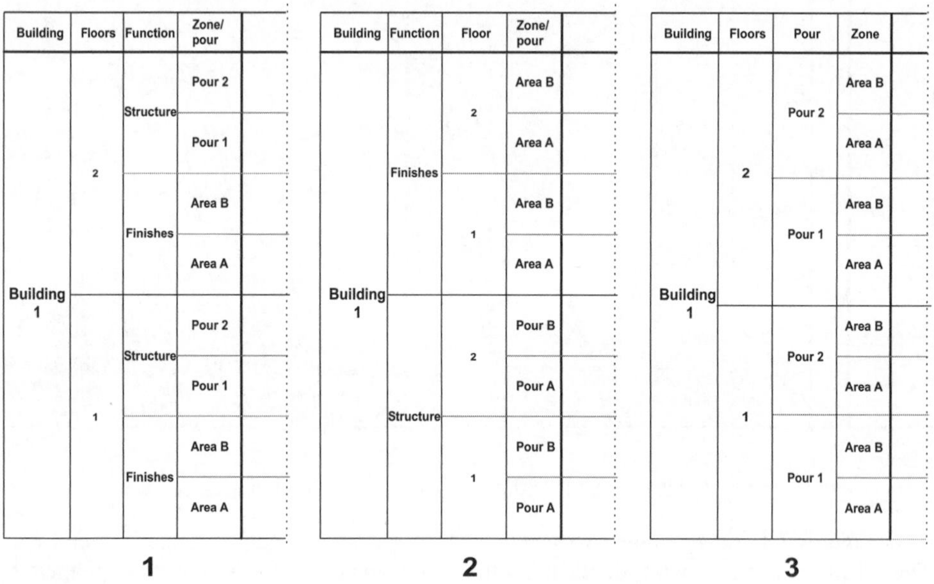

Table 1

Building	Floors	Function	Zone/pour
Building 1	2	Structure	Pour 2
			Pour 1
		Finishes	Area B
			Area A
	1	Structure	Pour 2
			Pour 1
		Finishes	Area B
			Area A

Table 2

Building	Function	Floor	Zone/pour
Building 1	Finishes	2	Area B
			Area A
		1	Area B
			Area A
	Structure	2	Pour B
			Pour A
		1	Pour B
			Pour A

Table 3

Building	Floors	Pour	Zone
Building 1	2	Pour 2	Area B
			Area A
		Pour 1	Area B
			Area A
	1	Pour 2	Area B
			Area A
		Pour 1	Area B
			Area A

Figure 7.6 Three alternative location breakdown structures

Example 1: In situ structure and finishes

The conflicting location requirements between in situ concrete structure and the following finishes trades are a common reason for requiring multiple branches in the LBS for structure and finishes. In building the structure, the work sequence per floor is usually divided into pours, with work sequence flowing through each pour. The pours are generally not relevant for the finishes trades, where the division into rooms or functional areas may lie across the structural pour divisions. However, both trade types need information about the higher levels of the LBS, that is the building and the floor, and they can be linked together on floor level of accuracy. Therefore the best way to model this is to have a separate functional breakdown below floor level and to have structural pour areas below the 'structure' branch and finishes rooms or functional areas below the 'finishes' branch (Figure 7.6 part 1). Finishes can be linked to structure using Layer 2 or 4 links within the floor level of accuracy. The structural schedule can be shown by hiding the finishes locations, and vice versa.

There are other, less efficient, solutions to this problem. The whole breakdown may be replicated with the structure and pours in the first case and finishes and functional areas or rooms in the second case (Figure 7.6 part 2). This then requires Layer 5 links from the completion of structure to the commencement of fit-out. Another solution is to squeeze the functional areas to fit logically within the zones formed by the construction pours (Figure 7.6 part 3). They may be treated as logical areas, but this method will have problems if there is interference between the pours and the fit-out areas. This method is best used when the fit-out can be logically divided within the pours.

If the fit-out is only to be managed at the floor level of accuracy, then it is sufficient to not display the pours when modelling the fit-out work.

Example 2: Façade

There is often a special case of trades which follow a different sequence or go around other locations, such as façade works. The façade is not usually built by floor but rather by elevation (such as north, south, east, west). There are links to the structure, roof work, site work and finishes.

Because the façade is important for building closure and often contains multiple activities, normally the best option is to create another branch of the location breakdown structure for façade under floors. In this way, the façade can be controlled by floor and by grid line, which will enable very powerful tools for controlling this critical part of the work. Because the façade shares floors with other systems, it can be linked to structure and finishes using Layer 2 or 4 logic.

If the façade does not have significant dependencies, it can be divided into multiple activities on the building accuracy level with finish-to-start links between them. These activities can then be linked to structure and finishes using Layer 2 or Layer 5 logic.

Guideline 5: Plan groups of similar spaces for finishes

Where possible, it is always beneficial to summarise information during the controlling phase by space type. For example, in an office building project it is useful to divide each floor into space groups such as offices, corridors, meeting rooms, auditoriums, restrooms and so on. In hospitals, it might make sense to have wards, clinical areas, offices, theatres and laboratories. Often, different space groups have different trades working in them so it makes sense to finish a space group before moving to the next space group. It is also easier to define logic separately for each space group. Moreover, the actual individual spaces will not even exist before walls have been built—this may happen very late in the process if the architect (or the client) wants to retain the option of moving the position of non load-bearing internal walls until the last moment.

Figure 7.7 presents a simple example of how grouping spaces may help in the planning phase. A school floor has been divided into classrooms, corridors and wet spaces. MEP work only affects the corridor whereas vinyl floor covering work is done in all locations. It is actually possible to begin vinyl floor covering work earlier than corridor MEP even though floor covering work must follow MEP in each location. Without accurate lowest levels of the LBS, these issues are not properly taken into consideration and planning time can be wasted by adding unnecessary start-to-start and finish-to-finish links.

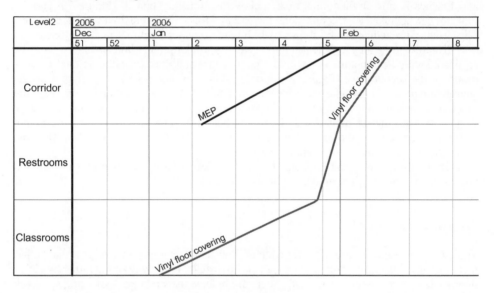

Figure 7.7 Special sequencing

Spaces can be grouped by location and then function. If the building's floor area is large, suitable location groups can be defined (for example north offices, south offices). The function should be placed higher in the hierarchy if most of the work of one function is done before moving to work on the next function. Similarly location should be placed higher in the hierarchy if work is expected to flow from location to location, finishing all of the space groups in the location before moving on to the next location.

Grouping spaces helps to manage change and uncertainty. If the actual spaces are not known in the preplanning phase, it is enough to estimate quantities on space group level and plan logic relationships on this level. The location breakdown structure can be made more accurate during implementation by using the location-based controlling model (Chapter 8).

Guideline 6: Handling large open spaces

It is often considered unnecessary to plan large open spaces using location-based methods because large open spaces are difficult due to many trades being able to work in the area apparently without needing to follow any sequence of physical space. There are also many ways to split up the large space into smaller logical locations. From experience, there are two approaches to handling large open spaces, each with its benefits and disadvantages.

The first approach is to plan and control the space using the higher level hierarchy (vertical cuts) to break it up. This location breakdown always makes sense because erecting the structure is a prerequisite for any work in the area. In this case, start-to-start and finish-to-finish links will predominate. Often, long lags are not needed because even if precedence exists when predecessor and successor are done in the same location, large locations often have areas where the tasks do not occur together. Trades usually have the possibility to work around each other. In a one location solution, a good tactic is to evaluate the total resource need for all tasks occupying the space and evaluate the space congestion based on that resource profile. If the space cannot accommodate the required number of people working at the same time, then tasks should be spaced further apart using buffers. With the single location solution, actual controlling of the work will be handled at the weekly planning level (see Chapter 9). Planned tasks will be used to establish the overall production rate requirement.

The second and, in theory, the best solution for modelling a large hall is to divide it into a grid of locations. Usually the best places to split are column lines if the hall is supported by columns. Locations should be large enough that adjacent locations do not generally interfere with each other and small enough that only one trade is able to productively work in the area. The grid can be defined in a hierarchical way if the direction of general workflow is known. However, it is often safest to have all of the locations on the same level to allow the visualisation and workflow to be reordered as necessary.

The benefit of splitting large halls is that it is difficult otherwise to evaluate the required dependencies, which easily results in logic errors or implicit buffers that are too large. Logic problems can arise if the hall is modelled as one large location, as quantities of work can vary greatly within a hall. For example, painting has a large quantity in the edge zones of the hall (wall painting) but a much smaller quantity in the middle (just column painting). The drawback of this approach is that different trades have different natural flows through a hall. Only some activities, such as painting of the ceiling and floor coverings, can be freely organised. Most of the tasks relate to mechanical, electrical and plumbing systems. This sequence of work is heavily constrained by design and there is no guarantee that all contractors prefer to work using the same locations. Without contractor buy-in, problems with working out the sequence will occur, with consequential loss of benefit.

Guideline 7: Lowest level locations should be small enough to remove implicit buffers and allow compression of the schedule

The forced time between the earliest practical start of a task and the logic-driven start of a task represents an implicit buffer. Implicit buffers arise when finish-to-start links are planned between tasks occurring in large locations, or locations at a higher level of the hierarchy (floors instead of rooms, for example) and they arise due to a lack of sensitivity or detail in the planning process.

Implicit buffers also occur when large amounts of sequential work, such as forming, reinforcing and pouring concrete, are scheduled within a single task, resulting in a forced wait for all work to be completed in one location, whereas in reality the first activity can commence in the next location while the succeeding activities are still being worked in the previous location (Figure 7.8).

In normal CPM practice, implicit buffers are avoided by having start-to-start and finish-to-finish dependencies. This practice could also be used for location-based planning however because location-based management does not require an increased number of links if locations are added, it is better to plan locations at a finer or more sensitive level of detail and use finish-to-start links with planned buffers. Also, because all the activities are created in a single task, it is very easy to schedule using more accurate task detail. Locations should be small enough that only one trade can effectively work in the location at the same time. Tasks should be small enough so that early trades are not forced to wait for later trades before moving into the next location, but not too small that it is meaningless to separate the work.

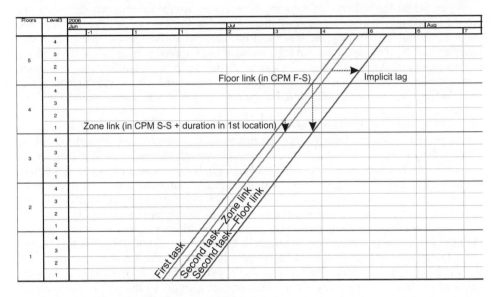

Figure 7.8 Implicit buffer arising from scheduling tasks at a higher LBS level

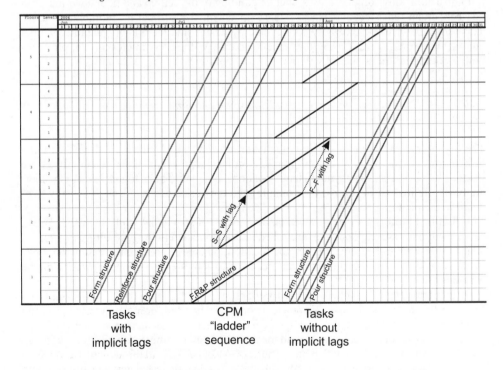

Figure 7.9 Three representations of the same work package to compare the relative effects

It can be seen that the common CPM practice of creating activity ladders based on activity windows can be avoided by appropriate location and task detail when planning the work. In this way, a much more accurate view of the flow of resources can be achieved. Figure 7.9 contrasts an activity ladder using a task window for framing, installing services and sheeting with plasterboard internal partitions, with a flowline of those three tasks through multiple locations—with and without implicit buffers.

Defining location-based quantities

The location-based management system functions best when it is driven by accurate quantities representing the work done at an appropriate level of location detail. Location-based quantities are required to make the schedule an accurate model of production. If the starting data is inadequate, there may be errors hidden in the plan which will emerge as problems during project execution. For example, the planned schedule might be implicitly assuming uneven resource use, even if the lines in the flowline appear continuous, because of actual variations in the quantity of work between locations.

The practice of drawing task lines with parallel slopes—a common error made by those who consider location-based scheduling to be merely a repetitive scheduling technique—is usually insufficient, because it involves an implicit assumption that either the quantities are the same in each location (the usual assumption) or that resources can be varied according to the demand in each location. Either way, this is a dangerous assumption because one of the primary objectives of location-based management is efficiency through planned and preferably continuous resource use. Unplanned fluctuations in either work progress or resource demands may damage the potential benefits of location-based production management.

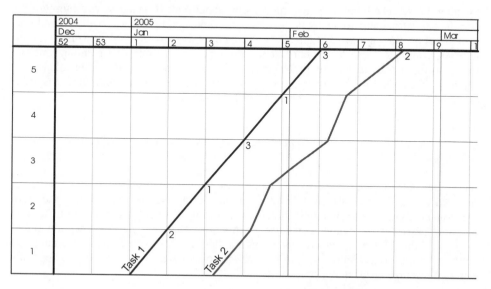

Figure 7.10 Unequal quantities reflected in uneven production

Figure 7.10 shows an example of a flowline schedule with two tasks, each having been planned with the same quantities that vary from location to location. In the case of Task 1, the planner has forced the line to have the same slope in each location by varying the number of work crews, reflecting what the site would have to do to maintain constant production, while Task 2 maintains constant resource use in each location, and thus the slope of the line varies. The small numbers at the top of each of the line segments of Task 1 show the resource needs for that location. While both tasks have the same total duration over the project, Task 1 will probably not be implemented as planned because of fluctuating resource needs. It is unlikely that the subcontractor will provide more personnel when required to achieve location-based production rate requirements. On the other hand, Task 2 can be implemented as planned if the subcontractor provides enough resources to consistently achieve the planned production rate. This results in far fewer adjustments and gives the subcontractor confidence that they can continue with the same resources until the end of their contract (Seppänen and Kenley, 2005).

The level of detail for quantities depends on available design information, the level of uncertainty and the methods that will be used in the project. If procurement control is desired, quantities should be defined so that the materials can be separated from the labour quantities. If quantities will just be used for scheduling purposes, less detailed quantities can be used. The simplest form of location-based bill of quantities (BOQ) just gives the driving quantity in each location for each task.

Usually the best approach is to make a location-based estimate and use that as a starting point. In this case also, the cost loading can automatically include costs from the estimate.

The following guidelines can be used in defining the items in the bill of quantities.

Guideline 1: It should be possible to assign a resource consumption estimate for each item

Location-based scheduling requires information about rates of resource consumption. In current estimates, there are many items which are estimated in batches (really just a lump sum price) or using approximations like building area. For example, MEP work might be lumped as one batch with quantity of 1 item. Another common example is painting which in Europe is often estimated based on the cubic volume of the building. Preferably, the quantity should be given in square metres as painters paint surfaces not space. In these cases the quantities required for production planning differ from estimating quantities.

Guideline 2: The complete work content of a quantity item should be producible by the same crew

It is not sensible to attempt to plan continuous work for work crews when different skills are required within the one task in different locations during the project, meaning the crew make-up needs to vary. If an item includes work done by different trades, it should be split into multiple items reflecting the quantity of work done by each trade. An exception to this might be if the different skills are always paired, thus effectively making a more complex work crew. Generally, however, it is best not to mix quantities of different trades with different resource needs into a single task.

Guideline 3: There should be enough detail to populate all tasks with quantities

Each task should have at least one quantity item in at least one location. If the quantity cannot be known, for example if design is not finished, at least the relative quantity in each space of the LBS should be evaluated. Then a rough estimate of duration can be split according to the relative quantity. In the total absence of quantities, as frequently occurs in the planning stage in many countries, it is common to schedule using time (hours or days) as the unit of quantity. While this method will build a schedule, it is little better than a CPM schedule, losing most of the location-based resource management capability and all the methods using location-based quantities. It also cannot be properly resourced. Another, better, solution is to get estimates of total hours in each location from the subcontractors, as this enables the allocation of appropriate resources and crews.

In practice, it is often difficult for a general contractor to know the quantities of work for the specialty trades, such as MEP contractors. In this case, the subcontractor may be asked to provide worker hour or worker day information. These practical problems related to the starting data will be discussed in the implementation chapter, Chapter 12.

Guideline 4: Quantities should be located according to workflow requirements

There are many quantity items which may be sensibly located in two or more spaces. For example, any item which divides two locations could belong to either one of the locations.

For scheduling purposes, the rules for locating these quantities should take into account the requirements of production. First, the work should be allocated to the space where there is the most interference. Second, locations produced first should take precedence. Typical examples with proposed solutions are described below.

Slabs

By convention, slabs are generally allocated to the lower level. This requires one less location than if they are allocated to the higher level. The general principle of allocating quantities to locations produced first holds here, but more importantly most of the work interference lies in the lower location. For example, the scaffolding and formwork remain in the lower location even after the completion of the slab. It is not sensible to have activities such as stripping and back-propping planned as level 4 if they take place on level 3.This issue is related to the guidelines for designing the LBS (see page 204).

Walls

Identifying the best allocation of wall quantities to the LBS requires knowledge of the construction sequence. The whole quantity of wall should be allocated to the location from where the production begins. This helps to ensure that the wall has actually been built before other tasks begin in either of the adjacent locations.

Where the location detail is very fine, such as at the room level, this guideline would apply only to the structure and the content (for example, services) in the wall. Wall components which follow the construction of the structure, such as finishes, could be allocated to the room. There is a trade off here between planning detailed locations and a greatly increased measurement requirement.

Defining location-based tasks

Location-based tasks are composed of location-based quantities. Selecting the correct quantities to be allocated to a task (remembering that work items in a BOQ can be assigned to individual tasks and that the planner has a choice about which items to join together into a task) is a critical planning decision because it affects the production rate and has implications for procurement. These are the main guidelines for determining which quantities can be included in the task.

Guideline 1: A task should require similar skills

The complete scope of the work should be able to be procured from the same subcontractor or by using direct labour. This is important because the dependencies between different quantity items are not modelled in the pre-planning stage. Therefore they cannot have buffers. Even if the schedule task is continuous, the work of different resources within the schedule task might be discontinuous. An optimal schedule task has a scope which can all be undertaken by the same resources.

There are exceptions to this guideline. If the sub-tasks of different subcontractors are completed in a tight sequence, it is possible to lump them into one schedule task. For example, formwork, reinforcement and concreting are almost simultaneous activities within a location. They can be allocated to the same location-based task without losing much accuracy. In this case, resources should reflect the average resource use on any given date. If this method is used, it becomes impossible to make conclusions on resource continuity just by examining the flowline diagram. However, the start and end dates and total production rates are reliable. This exception applies mostly to foundations and the structural phase when there is only one location where the crews can work at a time (for example, buildings with a small floor area such as towers).

Another exception is assisting work (for example, carrying materials, drilling holes etc.). This work can be lumped to the schedule task. Assisting work can require the use of fractions of resources—for example, having 0.5 labourers for each carpenter. In this case, the same assisting labourer could be assisting production of multiple tasks.

Guideline 2: Included scope should have similar external dependencies to other tasks

External dependencies are planned for the whole work package. Therefore all the quantity items should have the same external dependencies. For example, lumping together quantity items needed for framing the suspended ceilings and sheeting (or tiling) the ceilings should not be allowed in a single task if mechanical, electrical and plumbing (MEP) installations are happening in between. Another example of a common mistake which can cause serious problems is to lump all roofing works to the same schedule task. The critical part of the roof is actually making it waterproof. Other parts of the roof have different or non-existent dependencies to finishes. Therefore, they should be in different tasks or they will delay the project.

Sometimes a decision can be made that a certain dependency is not necessary to model, for example if a task in between is of very short duration. An example of this is electrical piping inside plasterboard walls which happens before the second board is installed. Instead of having three location-based tasks, it is possible to model this with one task without losing much accuracy. The basic rule of thumb is to think in terms of normal problems. If the electrical piping never causes problems in production, it does not need to be regarded as a separate schedule task.

Guideline 3: Level of detail should be based on available information

Often the schedule pre-planning must be done before the design is complete. Quantities or technical dependencies might not be known for all of the tasks. The lack of information mostly concerns MEP systems and finishes. In these cases, it is often best to make the lack of information explicit in the schedule. For example, if there is high degree of uncertainty about interior wall types, all walls might be lumped to a schedule task 'Interior walls'. If floor covering materials are not known, then a floor covering task might be created including all the various materials, with total covered floor area as the quantity. All the MEP horizontal ducts and pipes might be lumped to one task if their internal sequence or quantities are unknown. Work that occurs in the same time period, in these cases, can be lumped to the same task even if it will eventually be done by different subcontractors. There is no point in defining continuity very accurately because the quantities will certainly change for these tasks. However, the total production rate and start and finish dates for these activities should still be reliable because they are used as the basis of commitments for trades for which more information is available. In this case, larger buffers are needed both before and after the task because there must be room to plan more detailed tasks when more detailed information becomes available. See page 257 in Chapter 8 for information on current or detail tasks.

Guideline 4: Using workable backlog tasks

Items which do not typically cause interference to other tasks but require skilled labour can be scheduled as workable backlog tasks. To be scheduled as a workable backlog task, the work should have the following characteristics:

- It should be able to done flexibly, without mandatory technical successors
- It should require special skills.

Workable backlog tasks are used to level the resource use of skilled subcontractors to prevent return delays (arising from leaving the site when work is not available). Having workable backlog tasks in the schedule gives protection against risk. Each critical subcontractor should have some work available in workable backlog tasks.

Guideline 5: All quantities do not need to be scheduled in master schedule

Every project includes items of minor importance which do not require skilled workers to install. These items can be left unscheduled in the master schedule and used as workable backlog in case some workers have idle time during the project. To be left unscheduled, the work should have the following characteristics:

- It should be able to be done flexibly without mandatory technical successors
- It should not require special skills
- It should not have too large a work content (in worker hours).

The unscheduled work can be added to work content of any task during implementation to take advantage of too fast production rates or additional available resources. In any case, at least 80% of the project's worker hours should be scheduled accurately.

USING RESOURCES AND PRODUCTIVITY RATES

Defining productivity rates

The selected productivity rate should be the optimal rate for production of the work. This often differs from productivity rates used for estimating, where rates represent the average productivity rates including wasted time. Optimal productivity is the rate at which the workers can work when all the prerequisites of working (such as materials, design, space) are available, everyone understands what they are supposed to do and the workers can work without interruption by other trades—for long periods of time. The main objective of location-based management is to ensure these optimal conditions are achieved for the whole period of work for all major trade packages. Therefore, optimal productivity rates should be used in planning.

On the other hand, slowing down the production rate without decreasing resources will lead to inefficiency and may in turn lead to return delays and increased risk. Slowing down activities which have just one optimal crew should not be attempted unless other work is available for these resources. The best way to slow down the task in this case is to increase the scope of work by adding further quantity items. For example, the task 'installing soffits' may be too fast even when done with just one crew. To slow down the task, the scope of work could be increased by having the same crew also install the framing of drywall.

Defining optimum crews based on work content

Location-based tasks often contain work to be done using equipment, subcontractors and direct labour. In some cases, it may be tricky to decide which resources to include in a crew and which crew members define the duration. This depends on how consumption rates have been defined for the task's quantity items. This is best illustrated by a few examples.

Precast structure

Consumption rates for precast structure elements are often provided according to installation labour. For example, installing a precast hollow core slab takes (according to the Finnish productivity rate database) 0.52 worker hours per unit. If all the quantities of the task have similarly defined consumption rates (in worker hours instead of crane hours), then the crane should have a production factor of 0.0 and would not therefore be contributing towards the production hours. The optimal crew will have both the installation labour and the crane.

Excavation

Excavation and earthworks each generally have equipment as the driving resources. Consumption rate is often given in machine hours per unit assuming a certain size of machine. If a different machine is used, the consumption rate will change. Unless labour has been specifically included as part of the package, all labourers will have a production factor of 0.0 and will not contribute towards the task's hours. The optimal crew will have the machine and supporting labour.

Assisting labour

Assisting, or direct, labour is often directly employed by the general contractor. Because it is easy to get consumption rates for direct labour and more difficult to get them for subcontractors, it is easy to fall into the trap of defining durations based on assisting labour (especially when using automated importing from an estimating package and then not checking the consumption rates). Of course, assisting labour never drives the duration. Therefore, it is critical to use the actual installation resource consumption and the actual installation crew (which may include assisting labour). When using elemental quantities (see Chapter 5), it is possible to have only one consumption value for each quantity item. In this case, the installation consumption should be used and assisting labour should have a production factor of 0.0.

Using resource-based quantities

If resource-based estimating has been used, all resources related to production are explicitly described. In this case, hours for each resource type can be summed together and the optimal crew defined based on the ratio of hours to other resources. The resource with the most hours is defined as 'driving' and others will change in relation to that resource.

For example, the quantity of excavation might be 100,000 m³. The quantity item contains three resources: an excavator of 12 tons (subcontractor, including driver) 3,000 hours, assisting labour (direct) 2,500 hours and a measurement carpenter (direct) 1,000 hours. In this case, the excavator would be the driving resource with a quantity of 1. The crew requires 0.83 assisting labourers and 0.33 measurement carpenters. All these resources have production factors related to their actual productivity (normally 1.0). If elemental quantities have been used, only the driving resource would be defined and the other crew members reduce to a production factor of 0.0.

If the task contains multiple quantity items, hours are summed for each resource type for all items and the crew can be similarly determined.

USING LAYERED LOGIC

Layered logic should be used to minimise the number of required links to model a project. The power of layered logic is directly dependent on the project's LBS and how the quantities are allocated to both tasks and locations.

Layer 1

Layer 1 logic is the most powerful form of layered logic because it applies the same logic to all locations where the predecessor and successor occur together. Because many projects contain dozens or even hundreds of locations, use of this logic layer vastly reduces the number of relationships required. Most of the links in the finishes phase can be modelled exclusively using Layer 1 logic. Other typical examples for the use of this logic layer include links between structural tasks on the same floor and pour area.

When adding a relationship using this logic layer, care should be taken to make sure that the logic really applies in all locations. This is critical, especially in projects which are not repetitive and which have many different space types. Often the same relationship applies but the required lag is different. This occurs when a project includes both small and large locations. Small locations can have simple F–S relationships. Large locations might require a more complex relationship between the same tasks with laddered S–S and F–F relationships together with lags. In many cases, different space types have different logic. For example, the relationship between suspended ceilings and installation of lighting often depends on end user specifications, which can be very different in different space types.

Layer 2

Layer 2 logic can be used to link construction phases together. Typical examples of use include linking building dry-in to plasterboard walls in the building, or linking the installations of main mechanical room to tests and measurements of all mechanical systems in the influence area of machines.

The use of Layer 2 logic requires that the hierarchies of the LBS have been built consistently, so that they have a global meaning in all parts of the LBS. As with Layer 1 links, the resulting logic should be checked for correctness.

Layer 3

Layer 3 logic links locations together inside the location-based task. It can be used to force tasks to be continuous and to change the sequence of locations of tasks. Usually it is best to form all the tasks and their external links before making decisions about breaking continuity or location sequence. The initial assumption should be that all tasks will be done continuously. When optimising the schedule, this constraint can be released for selected overly fast tasks. Splitting tasks should be especially avoided before looking at the big picture.

Layer 4

Layer 4 logic is used when something affects adjacent locations (like locations above or locations below). This is usually needed for modelling the dependencies of structure between different levels or for modelling safety constraints for those tasks which create safety hazards for work being undertaken below.

The functioning of Layer 4 logic is heavily dependent on the location grouping that the logic operates on. Location grouping defines the related locations and how lags should be calculated. Unrelated locations should not form part of the same location group.

Layer 5

Layer 5 logic is activity to activity logic (linking specific task-locations to other specific task-locations) in the same way as ordinary CPM. It is rarely needed, because most of the dependencies in construction projects are location based. Layer 5 dependencies are used mostly when some activity does not directly fit the chosen location breakdown structure.

Normal examples include waterproofing certain floors in high-rise buildings in order to commence finishes earlier in the floors below. The structure of many wide buildings, or podiums connecting multiple towers, cannot be handled exclusively by Layer 4 logic, but require Layer 5 links to correctly link connection points of structure between different areas. Layer 5 logic can also be used to force connections between virtual locations, or connections which are not apparent from the LBS.

ALIGNING THE SCHEDULE: OPTIMISING THE SCHEDULE FOR DURATION AND CONTINUITY

The location-based schedule is first created with one optimal crew in each task and with all tasks continuous. This will generally result in some trades being slower than others. The continuity requirement pulls the start dates of first locations of fast tasks leaving empty space between tasks. Aligning the schedule means eliminating those empty spaces by changing the production rates so that the slope of preceding and succeeding tasks becomes similar. This allows tasks to proceed continuously without adverse effects on project duration. The methods described here apply mostly to activities where it is possible to freely increase resources, such as finishes or MEP. In more constrained environments, such as structure, a related technique of cycle planning may be used (see next section).

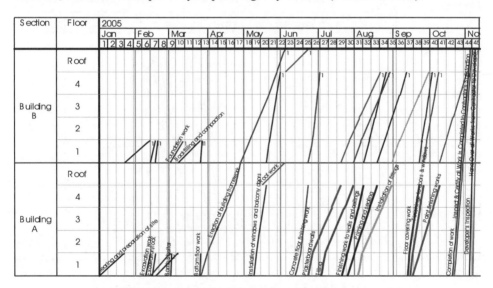

Figure 7.11 A simple example of an unaligned project

Figure 7.11 shows a typical, simple, schedule with one optimal crew in each activity and with all the tasks being continuous. The schedule is unaligned and the need for changes in production rates is apparent by looking at the patterns of empty space between activities. The schedule total duration is 45 weeks.

The schedule can be aligned by using the following tools (in order of desirability):

- Changing production rates by changing resources
- Changing production rates by changing scope
- Changing location sequence of tasks

- Changing soft logic links
- Splitting tasks
- Making tasks discontinuous.

Changing production rates by changing resources

The basic way to align a schedule is by adding crews. Crews can be added to the same location (doubling) or to different locations (splitting). It is helpful to begin aligning by assuming that all the crews work in the same location because then the trade can be visualised as a single line and the total production rate can be evaluated. This can be then changed into a more accurate model at any time by splitting the crews into their own locations.

If the locations are large enough, or the crews work in different locations, adding resources does not affect productivity. However, planning more resources increases risk because the subcontractor may not mobilise enough resources. Adding resources requires more supervision or productivity will be lost. The supporting activities of design and procurement need to be able to be similarly expedited. These factors should be taken into consideration even though it may seem lucrative to reduce duration as much as possible.

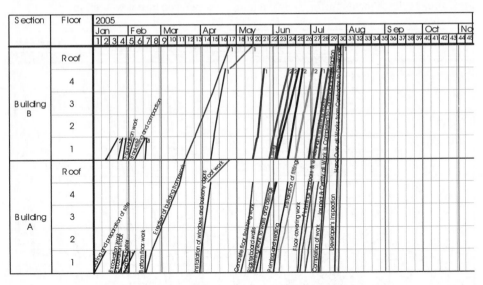

Figure 7.12 Adding a second crew to selected tasks to improve alignment

Crews should be increased until the task's slope is as near as possible to its predecessor's slope or until no more resources are available. In Figure 7.12, a second crew has been added to the following tasks:

- Clearing and preparation of site
- Foundation work
- Backfilling and compaction
- Plasterboard walls
- Tiling
- Priming and sealing

- Installation of fittings
- Paint finishing works.

As a result, the slope of all the earthworks and finishes tasks is approximately equal. Total duration of schedule is 30 weeks and the total time saving compared to the original schedule is 15 weeks. Concrete floor finishing work is a faster activity, so finishes could be further compressed by adding more resources. The bottleneck activity is erection of the building framework, which has the gentlest slope. However, adding another crane to this small project would be impractical for logistical reasons.

If the quantities vary in locations, it may be necessary to add crews just to certain locations. This can be done without additional risk if the resource requirements do not fluctuate but are steadily increasing and then decreasing. A solution where the resource load has multiple peaks is more risky. Adding resources gradually is standard practice in many complex projects. Once the first crew has learned the requirements of the project, others are mobilised, and these utilise the first crew's experience.

Figure 7.13 shows two tasks extracted from a large, real project with the number of crews balanced. Because the quantities vary, it is impossible to balance the production rates by using the same number of resources in each location. A second set of tasks shows otherwise identical tasks but one crew has been added to location D1 of the predecessor task. This change does not add too much risk because resources are first increasing and then decreasing, but it allows tasks to be executed much closer to each other, thereby preserving full continuity benefits.

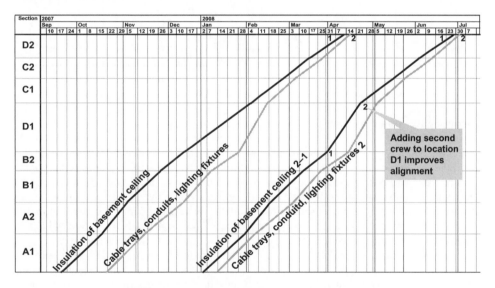

Figure 7.13 Using resources to balance variations in quantities

Changing production rates by changing scope

The production rate can also be increased or decreased by changing the scope of the task. This means that quantity items are added or removed from the task so that the worker hours required in each location change. This solution is very useful for decreasing the production rate of overly fast activities which have just one optimal crew working in them. However, it requires multi-skilled workers.

In the example, it is possible to improve the alignment by combining the bottom floor work (a too fast activity) with the erection of the building framework. This solution requires that the same resources are able to do both work types. If the work is done by a subcontractor, this will require changing the scope of the subcontract. Figure 7.14 shows the result of this. The total duration has been shortened by two more weeks. Erection of the building framework has slowed down and is now aligned with its predecessor, backfilling and compaction. The best way to recognise these opportunities is by reading the flowline chart horizontally from left to right.

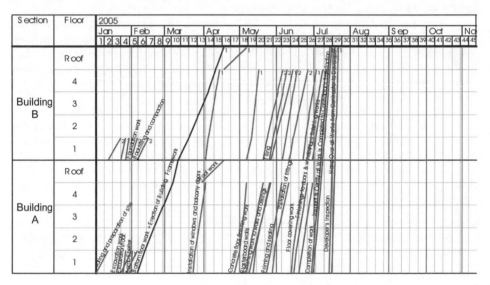

Figure 7.14 Adding sub-floor work to the erection of the building framework improves the schedule

Changing location sequence

In some cases, a schedule can be compressed by changing the sequence of locations. In the general case, the sequence of a successor should be the same as the sequence of its predecessor. However, sometimes a predecessor does not exist in all the locations of the successor. In such cases, it is often best to start the successor in those locations. In fact, if the sequence can be freely changed, the optimal sequence in terms of duration compression can be calculated by working in ascending sequence of location availability date, which is calculated from all the predecessors affecting that location of the considered task.

An even more powerful way to compressing duration is to change the overall project sequence. This means switching all Layer 3 logic links to a certain sequence, while preserving all the other links. This is usually possible only for structurally independent locations because the vertical sequence of the structure as it rises is fixed.

In our simple example introduced in Figure 7.11, this can yield only marginal benefit because the quantities are almost the same in both buildings. Figure 7.15 shows the schedule with Building B built first instead of Building A. In fact, the duration increases by two days in this example, because the quantities in the earthworks and foundations phase are larger in building B. In projects with different sized locations, the time saving can be significant.

An example that better illustrates the effect of changing project sequence to reduce project duration uses the same project as in Figure 7.3 (note that zones 2.1 and 2.2 become

Buildings 2 and 3 respectively) but the *foundations* task has been added before structure in each section. The duration of foundations is twenty days in Building 1, fifteen days in Building 2 and twelve days in Building 3. There are six possible sequences for building the project. Table 7.1 shows the duration for each of the sequences.

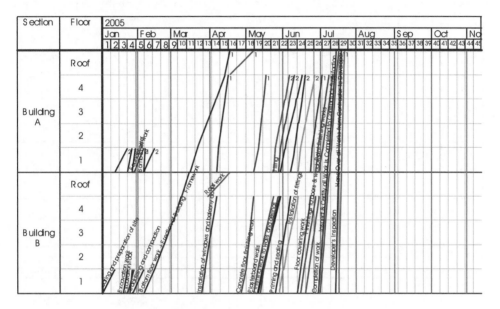

Figure 7.15 Changing the sequence of buildings does not bring benefits

Table 7.1 Table of possible location sequences

Sequence	Duration (weeks)
1-2-3	25.5
1-3-2	25.5
2-1-3	24.5
2-3-1	25
3-1-2	24
3-2-1	24.3

Even in this simple project, it is not trivial to work out the optimal sequence. The difference between the best and worst solution is 1.5 weeks. The optimal schedule is shown in Figure 7.16 and the worst in Figure 7.17.

A good heuristic—which works in this example but not in all complex real projects—is to begin work from the section with the least quantity of foundations, structure and roofing work and to finish in the section with the least finishes work. This is called Hoss' rule by its inventor Hoss (Kankainen and Sandvik, 1993). However, it is worthwhile to go through all the possible solutions to identify the shortest duration, because Hoss' rule does not work in all cases. This is feasible if there are four or less sections (four sections has 24 different alternative sequences). With more sections, the number of possibilities can be excessive, so it is necessary to use heuristics and test only the most likely sequences.

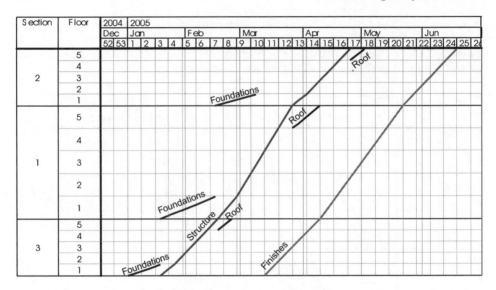

Figure 7.16 The optimal sequence

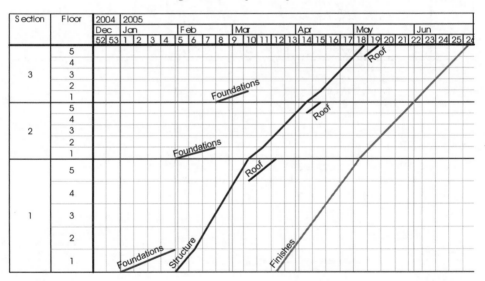

Figure 7.17 The worst sequence

Changing soft logic links

Links are often created in the plan for reasons of logistics or quality which are not essential. These links can be changed to better align the work, but there is often an associated cost or risk. Examples of common links which can be changed include the following soft logic:

- Undertaking concrete floor finishing work before the interior walls are installed to be able to use larger pour areas (breaking this results in the cost of doing smaller pours)

- Undertaking concrete floor finishing work after building the plasterboard walls so that boards which have been delivered earlier do not hinder the work (breaking this results in the cost of changing deliveries or storing the boards so that they do not hinder pours)
- Placing the floor tiling after any heavy MEP installations to reduce risk of tiles breaking (breaking this results in the cost of protecting floors, risk of tiles breaking and consequent rework)
- Installing floor drains before the finishing of the concrete floor to eliminate the cost of having to drill holes through the floor (breaking this results in the cost of drilling holes)
- Waterproofing the roof before hanging the plasterboard walls to eliminate the risk of boards getting wet (breaking this results in the risk of rework and quality defects)
- Installing floor covering work before the mandatory lag of concrete drying has elapsed (breaking this results in the risk of quality defects in floor finishes).

In addition there are dependencies which are really choices about which trade should go in first. In these cases there is no technical dependency but, because of space constraints, only one trade can work in the same location at the same time. These links can be switched without causing extra cost.

Another soft link type which can be changed is a resource link. The same resources can be working in multiple tasks and have been linked so that resource constraints can be met. This often happens for multi-skilled trades which work on multiple location-based tasks in the project. The sequence of their work can frequently be changed without increasing cost.

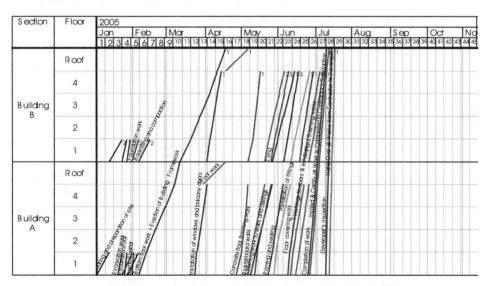

Figure 7.18 Changing links (relaxing the drying constraint) to achieve a more compressed schedule

Changing the links can result in better alignment of the schedule yielding associated duration savings. For links which cause extra cost, a trade-off analysis should be made to arrive at an optimal outcome. For resource links and chosen links, the changing of a link does not increase the cost, so they should always be changed if they result in better schedule alignment.

In the simple example, the following opportunities for compression of the schedule through changing the links are available:

- Concrete floor finishing work could start before the roof is waterproof (associated risk of concrete drying slower)
- Floor covering work could start earlier (currently a period of 6 weeks has been reserved as a mandatory lag for allowing the concrete to dry).

Changing these links would increase the risk of poor quality in the project. Figure 7.18 shows the effect of relaxing the concrete drying constraint to be 5 weeks instead of 6 weeks. One additional week of project duration has been saved.

Splitting tasks

Splitting tasks is the process of deliberately cutting a task into segments. Tasks need to be split in the following basic cases:

- Production rate is too fast, continuity is desired and it is not possible to increase the scope
- Crews cannot be added to the same location but other locations are free
- Some locations become available much later than others because of logic links and it is not possible to slow down earlier locations sufficiently to achieve continuity.

Almost every project has some tasks which are too fast when using the optimum crew. Having these tasks continuous will shift the task start date and cause wasted time in the schedule. These should be split, but split points should be optimised so that wasted time is minimised and there are as few split points as possible.

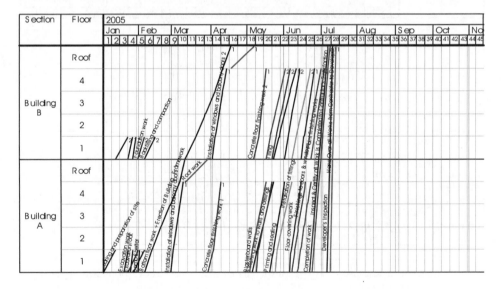

Figure 7.19 Splitting of roof work and concrete floor finishing work

Figure 7.19 shows the effect of splitting windows, roofing and concrete floor finishing work. The total effect on duration is not large (1–2 weeks). However, the concrete drying time is greater than before in Building A, reducing the risk of quality defects in floor covering work. For other finishes to be able to benefit from earlier availability of Building A, the production rate would have to be decreased.

Another common case which requires splitting is when locations are too small or crews require too much space to allow two crews to work in the same location at the same time. In this case splitting should be used to optimise the sequence of work through the locations for each of the crews. This type of splitting does not increase risk and cost compared to the alternative of having the crews affect the slope of a single line. However, the visual quality of flowline diagrams may suffer. Therefore it is suggested that the actual workflow of crews be planned in the implementation phase using the control tools, unless the workflow requires the crews to have very different sequences.

In complex projects it often happens that some locations become available much later than others. This may be caused by special logic links to non-repetitive elements or by predecessors having varying quantities in different locations. These factors often necessitate splitting the delayed part to its own sub-task. Otherwise, the start date of the available location would be delayed too much. However, other alternatives should also be explored because making this decision often results in all the succeeding tasks being forced to behave similarly.

Figure 7.20 shows part of a real hospital construction project where fireproofing of the third floor has roof work as its predecessor. The other floors do not have this constraint. Location availability dates for the third floor are shown by vertical lines. The fireproofing task and its successors must either be slowed down considerably or split into two parts. The decision will also affect all the other tasks that succeed fireproofing.

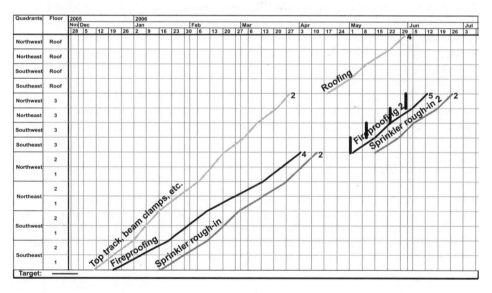

Figure 7.20 Tasks have to be split or slowed down if some locations become available later

Making tasks discontinuous

Tasks which are faster than their predecessors and successors with one optimal crew are problematic because forcing them to be continuous will pull the start date of the first location, causing large wasted space and extension of the project duration. If it is not possible to increase the scope and the total effect on the project duration is too high, often the only available alternative is to either to split the task or to make it discontinuous. Some tasks are, by their nature, possible to do discontinuously without any added cost or risk. Examples include tasks which do not require resources on site continuously (such as pouring concrete) or simple tasks which do not require professional labour (such us installing accessories). Other tasks should be split instead of being made discontinuous in order to minimise the amount of mobilisation and demobilisation activity (see splitting, pages 156 and 228).

Optimal rhythm of the schedule

Schedules have one or more optimal rhythms: that is, groups of tasks which have a common optimal slope for the group. This rhythm is often caused by bottleneck activities but may also be caused by the planner's decision to decrease risk by decreasing the resource requirements. There are some circumstances which cause a change in the optimal rhythm.

Optimal rhythm can change as a result of the following changes in the schedule:

• Most succeeding activities have faster production rates than their predecessor.
• Some locations become available much later than others necessitating splitting.

Most succeeding activities have faster production rates than their predecessor

This often occurs when shifting from a ddddhighly resource and space constrained activity to a series of more flexible activities, for example from structure to finishes. Structure can be accelerated only up to certain extent and often the optimal rhythm of finishes activities is faster. In this case either the whole group of succeeding activities needs to be split or the rhythm of the project is increased. If splitting is not done, changing the rhythm will result in some wasted time.

In the simple example project (Figure 7.19), there are two optimal rhythms. One is for the earthworks phase. The structure is aligned with the earthworks on the first floor. After structure, the finishes activities are much faster so they have another optimal rhythm.

Some locations become available much later than others necessitating splitting

If in some locations the majority of the tasks have been split to form new tasks, the optimal rhythm for the split locations can vary from the optimal rhythm of the remaining locations.

In the hospital example, fireproofing was split into two parts because of the roof constraint (Figure 7.20). The optimal rhythm of the third floor can be different from the optimal rhythm of the first and second floors.

Finding the optimal rhythm

It is difficult to define an algorithm for finding the optimal rhythm for the general case where quantities are allowed to vary in locations and splitting and non-repetitive logic is allowed. However, generally the optimal rhythm can be looked for by accelerating the slowest tasks until resource constraints do not allow more acceleration or until the risk level associated with getting the resources is deemed too high. When further acceleration of the currently slowest task is not possible, that task defines the optimal rhythm and is the bottleneck task. This task then dictates the duration of the project, which is a function of the slope of the slowest task and the duration of typical activities (see page 75). Optimally the slowest tasks can be accelerated to have the same slope as fast tasks which have just one crew.

CYCLE PLANNING

Cycle planning refers to the planning of highly constrained sequences of work involving multiple subcontractors. A typical example of cycle planning is in situ concrete structure. Here, formwork, rebar and pour follow each other in a tight sequence, while the previous floor must be finished before moving on to the next floor. Continuous flow can be achieved by planning multiple pours and cycling through the floor. Similar cycles occur in mining and tunnelling, or any project where it is mandatory to finish a location before anything can be done in the next location, and finishing each location requires the execution of multiple, dependent work stages, each done by different resources.

For projects with in situ structures, it is best to start with examining the cycle planning of the structure before scheduling other trades. The reason for this is that during cycle planning it may be necessary to change the location breakdown structure many times.

Location-based planning offers superior planning tools for cycle planning. Buildings, floors and pours form the locations, while the quantity of work defines the durations. The objectives of cycle planning include finding out the optimum number of pours and the optimum crew sizes to minimise the cost and duration and to maximise the flow of resources. The following things should be considered in cycle planning:

- Number and size of pours
- Bottleneck resources
- Crew continuity.

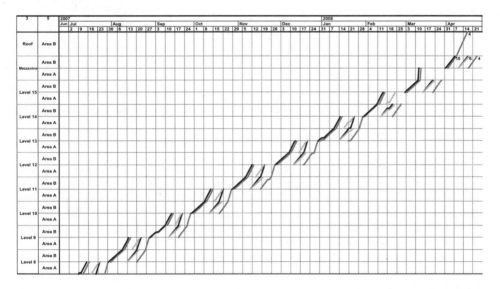

Figure 7.21 Two-pour cycle

Number and size of pours

The number and size of pour areas makes a huge difference to cycle planning. Smaller areas can have smaller crew sizes. However, the ability to create more space earlier during the construction of the next level may offset this difference. The cost of having more pours must also be considered. Figures 7.21 and 7.22 show a cycle planning exercise with two-pour and three-pour sequences respectively. Tasks on each area from left to right are:

- Concrete deck formwork (flyers) (black line)
- Concrete deck rebar (dark grey)
- Concrete columns—formwork, rebar and pour has been summarised to one summary task (light grey)
- Concrete core formwork (black)
- Concrete core rebar (dark grey).

To make the figure easier to read, the (almost vertical) pour activities after the corresponding rebar tasks have been left off the chart. The concrete deck formwork can continue on the next level once the corresponding area of columns below has been stripped. Additionally, concrete core stripping must be finished on the floor below.

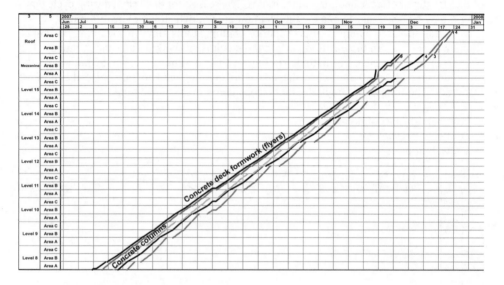

Figure 7.22 Three-pour cycle

In this example, the two-pour sequence must use larger crews than the three-pour sequence and even then it still takes 3.5 months longer to finish! This occurs because the quantities of Area B are much larger than those of Area A. This results in discontinuities for the crews. It is not possible to remove these discontinuities with just two pour areas of different sizes. Splitting and having multiple crews would help in total duration but would make the inefficient resource use even worse.

Cycle planning and resources

Bottleneck resources are often difficult to identify visually when cycle planning, especially when there are complex mixes of Layer 4 and Layer 5 logic. Finding the optimum resource balance for each trade requires some trial and error. Using computer software to implement location-based planning makes this a relatively easy task because all changes made to a task apply to all locations. Often, very small changes in the production rates will have huge consequences for the overall duration or the continuity of resource use. For example, Figure 7.23 shows the three-pour cycle example from above with one less formwork carpenter forming cores. This apparently modest change causes a delay of 2.5 months in the overall structure and makes most of the tasks discontinuous.

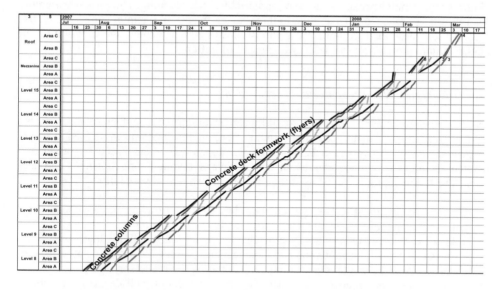

Figure 7.23 Changing the resources in the three-pour cycle

OPTIMISING THE COST, DURATION AND RISK TRADE-OFF

Cost is in many ways a function of schedule implementation. Subcontractor bids include allowances for wasted time. Material costs can escalate if rework is needed because of bad quality arising when work is hurried. Overhead costs—such as site trailers, the rents of equipment and machinery and salaries of project management—are functions of duration. Schedule overruns often result in penalties.

Schedule implementation is influenced by variability in the production system and by how the production system is buffered from variability. Risk can be reduced by decreasing the variability or increasing buffers. However, many of the risk reducing efforts have an associated cost. Decreasing the variability can lead to increased cost because of the increased coordination requirements. Buffers cost money because of increased duration. However, sometimes risk can be decreased without increasing expected cost. In these cases, the original solution was inefficient.

Optimising the cost, duration and risk trade-off means trying to find a minimum-cost solution which achieves the duration target and a selected risk level. Any solution which results in a higher cost with the same risk level is inefficient. Optimisation is not straightforward because either the acceptable risk level or the expected cost must be selected. Minimisation of risk always results in higher than expected costs and vice versa if the solution is efficient.

This section describes how to use the tools of production system cost and production system risk (see Chapter 6) to evaluate the cost-risk trade-off and to find efficient solutions. These tools relate to the risk-management methodology (see page 201). It is advised that methodologies from above this section are completely understood and implemented before implementing risk management approaches.

Planning tools to minimise variability

Many of the variability causes can be minimised by better planning. The various variability causes were introduced in Chapter 6. In this chapter, the use of planning tools to minimise variability is described. Note that it is almost always better to attack the variability itself rather than to protect from variability by the use of buffers.

Uncertainty related to prerequisites of production

Many of the prerequisite related uncertainties can be mitigated through pre-planning. If the quantity of work and material is known by location, the information can be used to ensure procurement and deliveries. Scheduling design and procurement based on production system requirements reduces the risk of start-up delays relating to prerequisites of production. The following tools can be used to minimise this source of variability:

• Planning procurement and design schedules based on the flowline master schedule
• Estimating quantities by location at a sufficiently accuracy level
• Taking into account the project-specific characteristics when defining logic
• Using small enough locations and F–S dependencies to force the preceding crew to completely leave and clean up the location before starting the successor
• Creating lists of prerequisites and scheduling them by location well ahead of production.

This type of uncertainty can cause start-up delays, work stopping and resources leaving the site, workers moving to the next location (while leaving unfinished work behind) and lower productivity.

Uncertainty related to mobilisations

This uncertainty can be decreased by planning continuous tasks with even resource needs and minimising the amount of new mobilisations. Subcontractors tend to show up on the contract start date. The usual behaviour is to start with a small number of resources and then to gradually ramp up. However, expecting additional resources during the period of contract is more risky. If multiple mobilisations are needed, the resource graph should increase gradually at first and then subsequently decrease—it should not be multi-peaked. The following tools can be used to minimise this source of variability:

• Planning the same resources to be used for the whole period of the contract and allowing for a ramp-up period
• Minimising the amount of mobilisations for any given subcontractor.

This type of uncertainty can cause production rate deviations. If the first mobilisation is delayed, a start-up delay occurs.

Uncertainty related to production rates

Production rate deviations are very common in practice. They can also be corrected easily during implementation by adjusting the number of resources or by working longer hours during production.

In the planning stage, the available tools to minimise this uncertainty are more limited:

- Adding detail to the quantities
- Using better productivity rates
- If the resources are known, adjusting the productivity estimate by their skill level
- Slowing down the rhythm of the project
- Selecting dependable subcontractors
- Specifying production rate targets (units per day) in the subcontract agreement.

Uncertainty related to quantities

Changes in quantities lead to production rate deviations if the consumption rates and resource amounts remain as planned. This risk can be mitigated by always basing the plan on the best current quantity information. The planning system must be updated if the quantities change.

Uncertainty related to resource availability

Small subcontractors or subcontractors with heavy workloads on other projects may not be able to mobilise enough workers to satisfy production system requirements. Any increase in production produced by additional resources adds risk to the project. Resource availability should be taken into consideration every time the production rate needs to be increased.

If enough resources are not available and the productivity is as planned, it is impossible to achieve the planned production rate. This is critical, because it can completely disrupt the location-based plan. The effects are largest when the difference between the actual resource availability and the planned resource need is a large percentage. For example, actually having one crew instead of two crews as planned is worse than actually having two crews instead of three planned crews.

Lack of information

Adequate information is often not available when the schedule needs to be planned and commitments made. Tasks without adequate information may lack quantities or may be very rough in detail, containing multiple sub-tasks for different subcontractors. In such cases the uncertainty level associated with the task is very high. Often these tasks have just a duration allowance and expected logic links. The effect of lack of information on production is unpredictable. The only way to mitigate this is to plan a detailed schedule when the information becomes available. Note that this implies that the current practice of activity-based scheduling when not based on quantities will have huge uncertainty factors.

Downstream effects of variability

When a risk actualises, there will always be an effect on the production system. The effect can be a start-up delay, a production rate deviation, interruption of the work or incomplete locations. Each of these effects can have downstream effects on succeeding tasks. Downstream effects can include lost productivity, start-up delays, and workers leaving the site or having to work out of sequence (including working around incomplete work).

A worker running out of work and leaving the site is the most prevalent of these down-stream effects and is the most critical for small subcontractors and small projects. Subcontractors will attempt to decrease their costs by allocating resources to projects where they can work most productively. The resources come back only when the subcontractor's management determine that they have sufficient work to return to site and be fully productive or when they are once again released from other sites. From any given project's viewpoint this is an unpredictable process. A crew that leaves the site and comes back after a delay often causes a chain of delays for succeeding trades. To break this chain, buffers are needed to protect the continuity of work from variability.

In large projects, subcontractors may not be aware that their workers are working with low productivity. Slowdowns typically take the place of leaving the site and having return delays. Slowdowns result in lower productivity and cause subcontractors to shift sequence and work in another location out of sequence, which will cause problems to other subcontractors. This cascading pattern of slowdowns may continue for an extended time period without the notice of management. Buffers can be used to protect the productivity of surrounding tasks and achieve better implementation of schedules, while providing an opportunity for management to fix the problem (see Section Three).

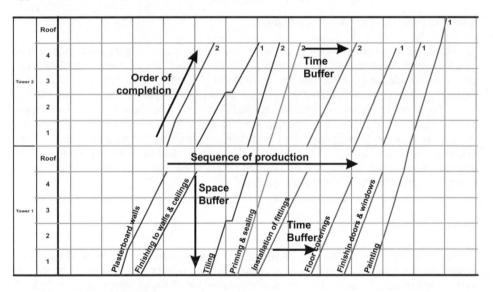

Figure 7.24 Time buffer and space buffer

Planning buffers to minimise the effects of variability

Downstream effects can be decreased or eliminated altogether by providing planned buffers in the schedule. As described in Chapter 5 (page 144), a buffer is an absorbable planner-defined time or space allowance between two tasks (Figure 7.24). In flowline diagrams, a buffer can be read by observing empty horizontal and vertical space between tasks.

- A **time buffer** is the period of time between two tasks in the same location while nothing is produced in the location. In a flowline diagram the time buffer can be read as the horizontal difference between tasks. Minor deviations, such as start-up delays, which are not expected to repeat in other locations are absorbed by the buffer and will only have down-stream effects if the delay is larger than the remaining buffer size. For repeating deviations, such as production rate deviations, a time buffer between activities allows time for control actions to restore the planned production rate (as long as control systems are in place, refer Chapter 8).

- A **space buffer** is the number of empty locations in production sequence between two tasks. In a flowline diagram the space buffer can be read as the vertical difference between tasks. Having a space buffer enables the following trade to start work somewhere else if it is not possible to start in the scheduled location because of some missing starting prerequisite. It also minimises the probability of workers leaving the site because there are many locations where the work can continue.

In routine production with repetitive locations, buffer planning is usually easy because if the production rates are aligned, the buffer size stays constant in each location. In non-repetitive special projects, buffer planning is more difficult because the locations can have widely varying quantities.

Assuming that variability cannot be further decreased, aligning the production and inserting buffers are the main tools for decreasing risk in the project. Buffers are always necessary in location-based planning if there is variability in the production system. Otherwise, even the smallest deviation will lead to a break of flow for the succeeding task and consequently a chain of return delays, increased cost and increased risk. The buffers decrease the risk level of the schedule.

However, buffers often have a direct cost effect because they involve a small increase in project duration. This cost can be calculated by calculating overhead costs without the use of buffers and then again using buffers. The critical question is to find the right buffer size for each activity, to adequately protect the production system from variability and to minimise the effect on overhead costs. Buffers are also a very efficient mechanism in location-based planning, as a buffer of two days in every location of a task still only adds a maximum of two days to the total project duration (depending on float), no matter how many locations there are. The impact of the buffer depends on the buffer size.

Deciding the buffer size for each task

As a general rule, the optimal buffer size for each task is a function of the variability of its predecessor, the dependability of the subcontractor and the location-based total float of the task. Some subcontractors have higher resource availability and thus can be more flexible in changing resources. Multi-skilled subcontractors and directly employed workers can often be productively employed in another task of the job. If locations are large and F–S dependencies have been used, there are implicit buffers in the system because the trades could physically work at the same time in the same large location. If locations are smaller, larger buffers are needed. Planning using more sensitive location detail will reduce schedule duration but increase risk, this should be mitigated with buffers.

The total float of a task determines the cost impact of adding buffers. If total float is zero, adding a buffer will increase the total duration of the project, thereby increasing cost. On the other hand, critical tasks are the most important to buffer because a chain of delays in critical tasks will cause the effects of variability not only to decrease productivity but to increase the project duration as well. Part of the reason for project schedules failing to hold is the lack of adequate buffers and continuity in the critical tasks of the project.

To sum up, buffers should be larger if:

- The predecessor has high variability
- The work is planned to be performed continuously

- The subcontractor is not known
- The subcontractor has a lot of work in other projects
- Locations are small
- The task has little or no total float.

Buffers can be smaller if:

- The predecessor has low variability
- The work has planned break points
- The subcontractor is known to be dependable or the project is especially important to the subcontractor
- The work is performed by direct labour who will be on site anyway
- The work is performed by multi-skilled labour and other tasks are available for them
- Locations are large
- The task has total float.

Simulation to find the optimal buffer sizes

A simulation model can be used as a tool to find optimal buffer sizes. Chapter 6 described the principles of the simulation model taking into account all aspects of variability. However, a simpler model can be used. The main requirement for a realistic simulation model is that it takes into account the adverse effects of discontinuity (that is, the return delay). The most important results for optimisation are aggregations from multiple simulation rounds. If a certain task causes problems in almost all of the iterations, a larger buffer size is required between the activities. The following procedure can be used to optimise buffer sizes.

1. Start with a schedule aligned and optimised with the critical path methodology.
2. Define the variability associated with each task.
3. Run the simulation.
4. Observe expected cost and risk level, if satisfactory stop.
5. Available buffer to be allocated to the critical tasks = Desired end date – Current end date.
6. Increase the buffer size for tasks with large probability of interference:
   a. Plan a greater buffer size for non-critical activities.
   b. The sum of buffer sizes for critical activities should be less than or equal to the available buffer to be allocated.
7. Go back to 3.

The cost risk trade-off question reduces to finding the optimum duration and risk level based on the preferences of the decision maker. After aligning the schedule, any decrease in duration will increase risk because achieving it requires either more resources or reduced buffers. Likewise an increase in duration reduces risk for the same reasons.

Other uses of simulation

Simulation can be used as a tool to analyse the cost, time and risk effects of various decisions. Examples include:

- Changing the design to minimise variability or to change logic
- Quantifying the risk associated with changing logic
- Selecting the optimum rhythm so that risk is minimised
- Evaluating different sequences of locations
- Selecting tasks which need more management time to reduce variability
- Supporting the decision to select a more dependable subcontractor for selected critical tasks even when more expensive.

The methodology for decision support is simple: simulation runs are made for both alternatives and the results are compared.

CHECKING THE FEASIBILITY OF THE SCHEDULE

Feasibility and predictability are important because otherwise the calculated potential savings in the production system cost will remain just that—potential savings, not achieved savings. A feasible schedule minimises the need for control actions on site; the production system is able to recover from minor deviations without management intervention. When major deviation arises, the management has more time available to react and localise the effect of the deviation. This section describes checklists which can be used to check the feasibility of a schedule.

Feasibility means removing unrealistic assumptions from the schedule. Typical unrealistic, often implicit, assumptions include:

- The ability of subcontractors to change production rates immediately (for example, by assuming the same durations even if quantities change)
- That resources can come to site immediately as and when required (discontinuous tasks)
- There is no variability in production rates (no buffers)
- Weather does not affect the schedule (structure raised with the same speed in winter and summer)
- Concrete has enough time to dry (floor covering work begins before enough time has elapsed after concreting)
- Work will be done during the holiday season (planning critical work for the holidays)
- The subcontractor will be able to provide sufficient resources (planning unrealistic resource demands).

By screening for these incorrect assumptions in a systematic way, it is possible to improve the schedule so that the risk in the production system is minimised.

Feasibility analysis

Feasibility analysis of the schedule includes at least checking the following things:

- **Quantities and task contents**: do the task contents accurately reflect how the project will actually be done? Are the quantities correctly distributed to locations?
- **Continuity**: are the tasks done continuously? Is there a possibility to change the schedule so that work is continuous?
- **Wasted time**: is there empty space anywhere which could be optimised by realigning the schedule?
- **Buffers**: are there enough buffers between tasks in each location?

- **Interference**: for each crossing of lines on the flowline, will the tasks interfere with each other when they are each undertaken in the same location?
- **Crews**: does every task have the optimal crew composition? Are enough resources available?
- **Resources**: what is the peak number of workers needed? Where is the peak situated?
- **Logic**: is there enough time for drying of concrete? Is the logic right?
- **Holidays and delay allowance**: have allowances been made for holidays and delays?

Quantities and task contents

Quantities are critical to the accuracy of a schedule. Each task in the schedule should be scanned (bulk check) to evaluate whether the quantities are approximately right. For the most critical tasks, it might be necessary to re-estimate the quantities, especially if they are uncertain.

Continuity

The basic assumption in the risk management methodology is that every planned break in the work continuity either costs money or adds more risk to the schedule. Therefore, each break in work should be evaluated to assess the cost or risk involved. The factors affecting cost may include:

- A potential increase in subcontractor bids: they need to employ crews elsewhere during the break
- It may become necessary to pay subcontractors for tasks external to the contract: for example, paying the subcontractor to keep the crew on site
- It may be necessary to find alternative work for direct labour during a break period: this often involves unnecessary hauling or movement around the site.

The factors affecting the schedule risk may include:

- Subcontractor crews leaving the site: it may be difficult to get them back when required and/or the same crew might not be available leading to start-up difficulties and loss of learning benefits.
- Subcontractor crew or direct labour slowing down their work: in order to avoid running out of space (for direct labour this will include additional cost).

The best way to mitigate these costs and risks is to plan a continuous schedule. However, it is often impossible to schedule a completely continuous project. In these cases the discontinuities should be limited to tasks which have the smallest cost and risk effect. It is cheaper and less risky to plan fewer break points. Many planned break points for the same crew will add to the risk unless the work is discontinuous by nature (for example, concrete pouring). The effects of discontinuity can be mitigated by the following means:

- For subcontractors:
  - Good contracts with the subcontractor
  - Explicit clauses about the discontinuities and when the workers should come back
  - Milestone before the break so that it is not possible to slow down the work
  - Choosing a subcontractor known to be dependable (partnering)

- Requires long-term cooperation and familiarity.

- For own work:
  - Availability of unscheduled work as a buffer (workable backlog)
  - Minor work that can be done at almost any time can be done during the breaks
  - Effective control of the workers.

Discontinuities are generally easier to handle with direct labour because unscheduled minor works can be utilised as a buffer. On the other hand, better control of workflow is needed because disturbances can lead to increased direct cost (which may be otherwise avoided by passing this onto subcontractors).

Buffers and wasted time

Buffers and wasted time should be examined from flowline charts which show those trades that make work ready for succeeding trades. Usually there are a few dozen trades which prevent other tasks from working in the space at the same time by having a mandatory technical dependency. Buffering these activities is critical because there needs to be sufficient time to react to take control actions if there are production rate problems.

When analysing the figures, uneven patterns of empty space between critical activities indicates tasks where adjustments can be made. It is often difficult, if not impossible, to attain a perfectly aligned schedule with all the trades working exactly at the same pace. The problem is usually caused by overly fast tasks which actually delay the project if they are forced to be continuous or are delayed by bottleneck tasks which are too slow. In complex projects, these problems are caused by quantities differing in locations—especially when predecessors and successors have different patterns of quantities. Such problems can be remedied by the following methods:

- For too slow activities:
  1. Examine the possibility of adding more crews:
     - Is there enough space?
     - Are enough resources available?
     - Is the subcontractor who will actually do the work known?
  2. Examine the possibility of splitting the work into two tasks:
     - Can the work start in other locations early enough to achieve benefits?
     - Are enough resources available?
     - Is it possible to split the work into two subcontracts or do part of the work with own resources?
- For overly fast activities:
  1. Decrease the amount of resources:
     - Is there more than one optimal crew in the current plan?
  2. Add more work to the task:
     - Items which can be done with the same resources and which have the same external logic to other tasks can be added to the work content of the task.
  3. Split the work and make it discontinuous:
     - To minimise risk, minimise the number of breaks to the flow.

Aligning the schedule better will enable cuts in project duration, while controlling the risk level by planning buffers between the critical tasks. The buffers should be planned so that their size is constant or increasing during the progress of the task. This is because the buffers are there to allow time for reaction should there be problems with the predecessor. If the buffers are diminishing, they do not help to prevent interference. On the other hand, the buffer at the start of a task defines when a subcontract should begin. It is often difficult for management to resist the temptation to start immediately as and when locations become available. Therefore, it is more realistic to have smaller buffers in the early locations.

Interference

Crossing lines in the flowline may indicate interference or missing logic. They should be checked to see if they can actually occur in the same location at the same time. This depends on the size of locations. If there is lot of space and there is no technical dependency, multiple tasks can work efficiently in the same space. However, the need for control may be greater and may result in extra cost. If there is a high likelihood of interference, the crew to go in first should be selected and a relationship should be added between the tasks.

This highlights one of the advantages of the flowline view—the ability to read the physical presence of multiple crews in a single location.

Crews

The crew of each task should be checked. Is the selected crew optimal for this work? Is there enough space in the location to do work efficiently with the planned crew size? If multiple crews cannot work in the same location due to space constraints, is it possible to split the task so that the crews are working in different locations?

Resources

If most of the resources are subcontracted, the total resource use does not need to be levelled. In this case, it is enough to ensure continuity of tasks which will automatically lead to even resource use for each subcontractor. However, the resource peak should be evaluated. If the resource peak is too near the end of project, it may signify a risk of finishing late. The optimum resource graph has the resource peak near the middle of the project.

There should not be any peaks in the use of direct labour or in resources of multi-skilled subcontractors who work in multiple location-based tasks. Resource use should be ideally increasing at first and then decreasing without multiple peaks. At the master schedule level, the resource use does not need to be totally level but any major peaks should be removed by levelling resource demands. This can be done by changing work contents of tasks or changing which tasks will be done by direct labour. Minor peaks can be levelled during implementation by using workers in logistics and to work on minor unscheduled work (workable backlog tasks).

Logic

Schedule logic should be checked to ensure feasibility. Common omissions include failing to incorporate concrete drying (in addition to curing) time into the schedule, leading to problems with dampness in finishings. Missing links can easily be visually identified in the flowline by reading each location from left to right and seeing that the lines do not intersect and are in right sequence.

Holidays and delay allowance

All known days off should be incorporated into the schedule. Allowances should be left for other days off. Separate delay allowances or staged buffers are not usually needed because their function is taken by the buffers of the production system. However, tasks belonging to a contractual milestone should be additionally buffered by planning them to finish earlier than indicated by the contractual finish date. Tasks between the interfaces in building systems (such as between the roof and finishes) should also be buffered more than other tasks.

EXAMPLE OF THE RISK MANAGEMENT METHODOLOGY

In the simple example presented earlier in this chapter, see Figures 7.11and 7.12, the duration achieved by aligning the schedule with the critical path methodology was 28 weeks. This duration was achieved without requiring any buffers. All of the tasks, except for roofing, installation of windows and balcony doors and concrete floor finishing, were continuous. Some of the tasks used two crews, while some needed just one crew to achieve the optimum production rate.

Let us assume that the overhead costs are €2,500 per week. The initial critical path methodology schedule thus has planned overhead costs of €70,000. Assuming 4 worker hours of wasted time for each additional mobilisation and demobilisation, the three discontinuous tasks cause additional €320 of wasted money. Money used in direct labour was €254,307 and the total production system cost was €323,199.

However, because the schedule does not have buffers, the total duration estimate may not be reliable. Any deviation in production rates will result in discontinuity for tasks with a likely cascading downstream effect. This effect can be evaluated by the use of a simulation.

For the risk analysis, assumptions about variability associated with each task are needed. To make it simple in this example, the same risk assumptions are used for all tasks. Risks are modelled only for start dates (optimistic 5 days earlier than planned, expected as planned and pessimistic 10 days later than planned), location durations (optimistic –20%, expected 0%, pessimistic +50%,), resource productivity (optimistic +20%, expected 0%, pessimistic –50%), and return delays in case a crew leaves the site (optimistic 0, expected 5 days, pessimistic 10 days). In the example, it is assumed that workers leave when they run out of work instead of the subcontractor being compensated for waiting time costs. In this case, there is a possible return delay. Simulated return delay probabilities for each trade are shown in Table 7.2.

Table 7.2 Simulated return delay probabilities for each trade

Task	Return delay probability
Excavation work	68%
Foundation work	67%
Excavation / rock	70%
Backfilling and compaction	90%
Installation of windows	62%
Plasterboard walls	95%
Finishing work to walls and ceilings	95%
Tiling	91%
Priming and sealing	93%
Installation of fittings	87%
Floor covering work	90%
Finishing to doors	88%
Paint finishing works	84%
Completion of work	87%
Inspect and certify work	95%
Developer's inspection	100%
Handover all work	100%

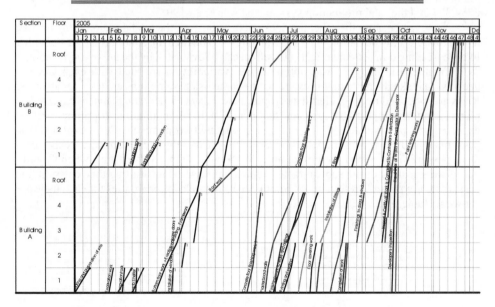

Figure 7.25 Risk simulation without buffers

Figure 7.25 shows how a schedule without buffers functions in a sample risk analysis iteration. Even small variations cause return delays which have cascading effects because there are no buffers. In this example iteration, the money used in actual work is €355,205, demobilisation costs are €7,120 and overhead costs are €116,429. The total production system cost is €478,753.

Instead of looking at individual scenarios when optimising a schedule, it is useful to perform a Monte Carlo simulation to aggregate the results of thousands of iterations. This provides information about most critical locations and tasks and about distribution of costs and end date. The ranges of finish dates and costs are shown in Table 7.3.

Table 7.3 Simulation results without buffers, resources leave site when delayed

	Minimum	Mean	Maximum
Duration (weeks)	39	46	53
Work cost	€321,979	€354,781	€388,869
Demobilisation cost	€3,280	€7,601	€11,720
Overhead cost	€97,500	€113,662	€131,071
Total production system cost	€432,894	€476,044	€520,409

Table 7.4 Simulation results without buffers, resources wait when delayed

	Minimum	Mean	Maximum
Duration (weeks)	30	33	38
Work cost	€297,713	€327,565	€363,725
Waiting cost	€94,054	€156,637	€242,814
Overhead cost	€72,500	€81,776	€95,357
Total production system cost	€497,091	€565,978	€683,443

A possible strategy to finish the project on time is to prevent workers from leaving and keep them on site instead. This approximates the behaviour of an aggressive project manager. This approach results in waiting costs instead of demobilisation costs, but removes the cascading return delay effect. However, the additional waiting costs are much higher than the savings in overhead costs caused by earlier finish dates. On site, these waiting costs would not necessarily manifest as waiting but rather by working unproductively, looking busy and 'making do' (Koskela, 2004). Table 7.4 shows the production system cost associated with this option.

Risk can be decreased by decreasing resources, doing work more continuously or by adding buffers. Figure 7.26 shows a solution where buffers have been included in the production system. The planned duration has increased to 48 weeks and the planned overhead costs to €118,000. A total of 20 weeks have been added as buffers. Buffers have been spread evenly between the tasks—this can be seen as empty horizontal space between tasks.

The aggregated simulation results show much less variability of both production system cost and total duration. In addition, the expected costs are €30,000 lower than in the initial solution. Thus finish date and production system cost can be achieved with a high probability because buffers eliminate the cascading return delay chain. Production system cost ranges for this example are shown in Table 7.5. Note that in this example, total duration is two weeks longer than expected duration in the first example without buffers. The worst case scenario with buffers is two weeks better than the worst case scenario without buffers. Effectively the buffers ensure that the planned duration can be achieved. Additional time contingency is not required and the probability of additional delays is smaller. To achieve this, buffers need to be spread between the tasks (like in the example in Figure 7.26) to prevent cascading delays.

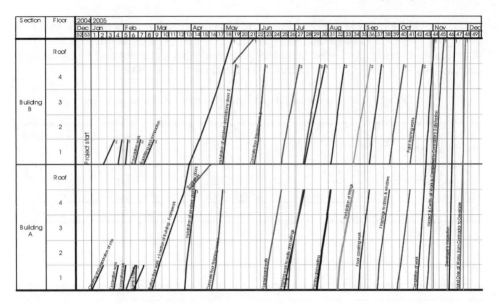

Figure 7.26 Risk simulation with buffers

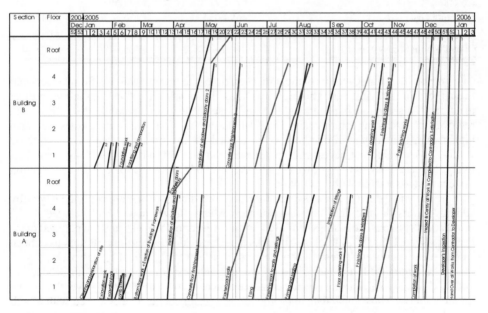

Figure 7.27 Risk simulation with buffers and slowing finishes

Risk can be decreased even further by changing the rhythm of finishes to be slower, thus requiring just one crew for most of the activities. The expected duration for this solution is 53 weeks (Figure 7.27). The distribution of finish dates has almost no variability. The production system cost for this solution is shown in Table 7.6. The expected result is €20,000 lower cost than in the faster solution with buffers.

Table 7.5 Simulation results with buffers

	Minimum	Mean	Maximum
Duration	47	48	51
Work cost	€291,542	€323,745	€355,090
Waiting cost	€520	€3,473	€8,560
Overhead cost	€113,571	€118,995	€127,500
Total production system cost	€410,879	€446,213	€485,178

Table 7.6 Simulation results with buffers and slower production rate

	Minimum	Mean	Maximum
Duration	52	53	56
Work cost	€271,937	€294,230	€321,864
Demobilisation cost	€800	€3,080	€6,120
Overhead cost	€126,071	€131,917	€142,143
Total production system cost	€405,914	€429,227	€461,977

COST LOADING THE SCHEDULE AND OPTIMISING NET CASH FLOW

If a schedule is cost loaded by adding a unit cost to each quantity item, it is possible to use this information to adjust the schedule and associated payments to optimise net cash flow. As explained in Chapter 6, payments to subcontractors and direct labour can be tied to either the completion of locations or to elapsing of time. The degree to which payments are tied to the quantity of completed work or to the actual use of resources can be designed to achieve favourable cash balance outcomes. Similarly, payments to the organisation doing the planning (the contractor) can be tied to either fixed times or to achieving milestones through finishing critical locations. In other words, combinations of different payment systems and altered production rates can change the inward cash flow as well as the outward cash flow profiles, and therefore the resultant net cash flow. The adjustment of production to achieve optimum cash flow is a legitimate business strategy.

Cash flow can be optimised by changing the payment logic or by changing production. Large incoming payments can be expedited by increasing production rates of selected activities. Expenditure can be delayed by changing milestones for when the payments are due.

Often there is a trade-off between optimum cash flow and efficiency of the production system. For example, expediting activities to get money earlier may increase risk in the production system. Changing milestones to be later for a subcontractor may reduce the incentive to achieve a steady production rate in earlier locations. Therefore, in location-based management, cash flow optimisation should be secondary to aligning the schedule and controlling the risk in the production system. However, sometimes cash flow can be improved without increasing risk. Cost loading is an important tool to detect these opportunities.

PLANNING PROCUREMENT AND DESIGN BASED ON THE MASTER SCHEDULE

In location based planning, by whatever methodology (critical path or risk management), procurement and design are scheduled last using pull scheduling methods. The aligned and optimised production schedule pulls procurement to the site when needed. Because the quantities are known by location and the schedule is feasible, it is now possible to know need times for all items within one week of accuracy. These need times are used to calculate the latest need times for all parts of the procurement process, such as design, call for tenders, evaluation of bids, contracts and delivery orders. When the need times are known, the actual procurement process schedules procurement-related activities so that each event happens before or at the need time date.

To be realistic, the procurement schedule should take into account the workload of design and procurement personnel. Procurement-related work should be evenly distributed so that all the need times are met. Procurement related to tasks with immediate need times, highest uncertainty and smallest location-based float should be prioritised. The design schedule should take into account the risk of failing to get acceptance from gatekeepers the first time and on each subsequent need for redesign.

In some cases it is impossible to achieve pull-scheduled deadlines in the procurement or design schedule. If this happens, the schedule tasks related to delayed procurement tasks should be updated with forced start dates. This often creates opportunities for decreasing the risk of preceding tasks because they can be slowed down or larger buffers can be added. If the delay threatens total duration, the rhythm of the first affected task and its successors can be increased. This creates a new, faster rhythm for the rest of the schedule.

REFERENCES

Koskela, Lauri (2004). "Making-do—The eighth category of waste". *12th Conference of the International Group for Lean Construction*, Helsingør, Denmark.

Seppänen, O. & Kenley, R. (2005). "Case studies of using flowline for production planning and control". *CIB–W70 Facilities Management and Asset Maintenance.* Helsinki, Finland, CIB.

Section Three

Location-based control

SECTION THREE—LOCATION-BASED CONTROL

The following chapters introduce and explore a new model for location-based planning and control systems for use during the construction phase of construction projects: location-based control. While the evolution of control systems was discussed in Chapter 4, the theory and associated methods and implementation described in Section Three is entirely new.

In this section, the term *planning* is used expanded to encompass the *re-planning* or *re-scheduling* of time-related models of construction work during the construction phase. This is *planning control*, which allows differentiation between the upfront pre-construction planning processes and the during construction reactive planning and control processes. The former were discussed in Section Two. In Section Three, the emphasis shifts to the theory of control and the tools and techniques for ensuring that the work goes according to the plan.

While the overall emphasis in location-based planning (Section Two) related to productivity, the emphasis in Section Three shifts to monitoring deviations from those plans and assessing the impact on productivity, in order to gain early warning of potential problems and enabling selection of corrective actions to restore efficient production.

The new model for location-based control is rich in detail, thus the discussion is also spread over three chapters. Just as in Section Two, it is helpful to separate the theoretical discussion (Chapter 8) from the analysis of detailed methods (Chapter 9) and the techniques for implementation of location-based control (Chapter 10).

Chapter 8 presents a new theory for location-based control of construction. The chapter describes a location-based control model designed to enable management to take better informed control action decisions and to be proactive in maintaining the original plan. The model includes four levels of planning: baseline schedule planning, detail schedule planning, control action planning and weekly planning. Progress data is compared to planned values and used to calculate forecasts. If forecasts deviate from plans and cause interference in the near future, a reactive control action planning process is triggered to prevent interference. Control action planning aims to restore the forecast start and end dates to planned values instead of changing the plan. Forecasts, adjusted with control actions, set targets for weekly planning which are used to actually guide the work. If the assignments selected for the weekly plan do not match the forecast production rate, the forecast is updated to show the long-term effects of the lower production rate.

Chapter 9 explores the methods required to use location-based control. It expands the theoretical discussion of location-based control by adding functional methods for improving production control, such as controlling cost, risk, procurement and quality. The chapter follows the structure of Chapter 6, because the methods introduced as planning methods within the location-based management system have been designed to be controlled at the same level of accuracy. In addition to those methods, it is also critical to be able to communicate the plans and their status effectively during implementation. Tools to visualise progress are described in this chapter. Methods to control and forecast costs are presented, based on both the conventional cost loading model and the production system cost model introduced in Chapter 6.

Chapter 10 discusses how to implement location-based control. It is one thing to know about location-based control theory and its associated methods, it is another to know how to use that knowledge to effectively manage schedules for a project and to control for both production efficiency and reliability. This chapter presents a process for project control, which utilises the tools described in the preceding two chapters. Location-based control processes include monitoring current status, accurate planning of implementation, forecasting progress, planning control actions, prioritising tasks, ensuring prerequisites of production, and executing the plan through good assignments and communication.

Chapter 8

A new theory for location-based control

INTRODUCTION

Control has always formed the end purpose for systems of planning and scheduling construction work. Kelley and Walker (1959) stated that the plan should form the basis for management by exception: management need only act when deviations from the plan occur. This acceptance of this purpose for planning and scheduling systems, whether it be CPM with its intention to provide a method for "those responsible for a job when to start worrying about a slippage and to report this fact to those responsible for the progress of the project" (Kelley and Walker, 1959), or PERT with its expressed purpose to review work programs to assess the likelihood of achieving targets.

There has been little change in the approach proposed by the first writers on activity-based control [or network-based control] (see Chapter 2). These systems, developed in the late 1950s, were a direct response to the lack of control provided by antecedent systems, which had little in the way of underlying logic and therefore had not been able to predict the consequences of delay or any other change on a project. The network-based systems were an enormous step forward. Nevertheless, they were, and remain, a blunt instrument for control. The process is to assess progress of activities against target acceptable date ranges and to assess the impact of time variation on the network through recalculation.

There is little sensitivity to the reality of construction in such methods of control. For example, there is usually no explanation for the cause of a time change, such as an increased amount of work, fewer resources than planned, altered complexity or altering the work method: all quite likely explanations for altered time performance.

The dominant extension of activity-based systems for control systems remains earned value analysis. This is a powerful macro tool for revealing overall trends. It is particularly useful on major projects and for company level review. However, such information is almost useless for the immediate needs of detailed day-to-day project management.

There is a major flaw in the application of activity-based control systems: they are a responsive mechanism. This means that they rely on a process of updating the schedule—based on time performance data and assessing the schedule for the consequences using the constraints of the logic network. This provides consequence information only, and it is provided long after the events which caused the problem have passed. The timing of the information is usually too late due to infrequent updates. When one takes a behavioural view of the construction management process, it is clear that decision makers do not want to admit that there are problems. The late provision of progress data and the failure to forecast future problems, enables them to push the problems into the future by deciding to solve them later rather than taking action now.

The traditional systems do not embed a concept of proactive control: making things happen according to plan. To make matters worse, there is no information embedded within the system that would reveal whether the cause of the problem may be continuing to influence the project. The possibility that the work may be being performed other than in the manner dictated by the logic network is effectively ignored under CPM, unless using techniques such as the progress override methods described by Murray (2007). Any experienced practitioner in construction understands progress problems are rarely isolated to a single instance or location, and there is a strong relationship between locations. Yet CPM also has no mechanism to manage the correlation between the duration of activities in different locations. In reality, all locations done by the same resources have heavily corre-lated production rates between locations. These should be taken into account by any control system. Overall, there is a problem with the timing and level of detail in the information available in traditional systems about the causes of delay.

Updating the schedule—a process where, typically, planned dates are replaced by actual dates and the durations of critical path activities are adjusted to show that project finishes on time—is a predominant feature of current control systems and supports the tendency to avoid confronting problems. The effects of this maladaptive process are often seen in the current outcomes of projects: everything seems to be all right until the finishes stage, when the schedule deviations and cost overruns suddenly become apparent in the last few months of the project—typically fixed by throwing lots of resources and money at the problem. A better method of control is long overdue.

Improving the control system requires embedding more information into the planning and scheduling system. The recognition of the difference between tasks and activities (as discussed in Chapter 5) along with the layered location-based logic provides the key to improved control. The information-rich environment which location-based planning and scheduling delivers allows a system to be constructed for controlling projects—and getting immediate feedback—as the detailed site planning occurs during construction. In this way, the delay consequences are understood immediately as detailed control plans are being made, rather than at some later date. Furthermore, failure to perform as planned can be forecast throughout the progress of a task, forcing early recognition of problems.

A major part of the power of location-based management comes from being able to make commitments to subcontractors about production continuity and predictability. The control system has to support that need, and continuous updating (replacement) of the schedule is inadequate as a control mechanism.

The location-based control system maintains four stages of information—*baseline, current, progress* and *forecast*. The baseline plan reflects commitment between the general contractor and the owner. It is updated only when the basis of this commitment changes—for example, because of change orders. The current plan reflects the way production will actually occur based on currently available information. It also reflects commitment between the general contractor and the subcontractors. Progress information describes how production has actually progressed. Forecast information uses information from all the other stages and describes the likely outcome should production continue in the same way. The forecast is adjusted by planning *control actions*. The adjusted forecast forms the loca-tion-based look-ahead schedule. All these stages can be compared to each other to find explanations for deviations and to forecast future problems.

While the new system of location-based planning and scheduling provides a powerful method for managing projects prior to construction, location-based control provides the essential tools to solve the currently intractable problems of responsive site management.

PRINCIPLES OF LOCATION-BASED CONTROL

The location-based control system uses locations to generate on-time response by manage-ment through visualisation of any problems before they happen. Forecasts are used to constantly remind management that a problem remains unsolved and that information is available to help take informed control actions. If the location-based control model is used correctly, management will be able to react to problems earlier and with better control actions. Instead of just recording deviations, the control system becomes a driver for action.

Production must be controlled so that it occurs according to plan in order to realise the benefits from location-based planning and scheduling. Even when there are changes in the project, the overall planned production rates and start and finish dates of trade packages should be reliable because they are used in the procurement process. The task of production management is therefore to find solutions to production deviations, to make things happen as planned, and to look for better solutions. To empower these steps, up-to-date status information needs to be maintained within the control system and it should be separated from the original baseline plans and the current up-to-date plans—in effect, it should not replace the planned values of the baseline or current plans. This progress information must include at least the same level of detail as that on which the production plan was formed.

A location-based model provides a lot more information in a plan than is available in an activity-based model, including:

- The flow of resources, which can be explicitly planned
- The quantities, which are known for each location
- The production logic, which is modelled more accurately using location-based logic
- The location-based model, which explicitly recognises that experience in one location is reflected in following locations—this is a form of learning.

The basis of control in activity-based systems is the control cycle (PDCA) as introduced in Chapter 4. The inability to model work repetition means there is little in the way of learning which can be applied in this cycle in activity-based systems.

The location-based planning model (Chapter 5) needs a compatible location-based control model in order to fully utilise and update its rich information base during the production phase. Location-based management systems require that there be earlier warning than provided in the traditional control cycle. This requires an extra loop, similar to Stacey's (1996) double loop learning model, which requires forecasting progress and then forming detail plans (including look-ahead plans). This is a double-loop control cycle (Figure 8.1).

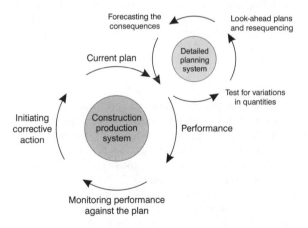

Figure 8.1 The double-loop control cycle

The plan should only move through to execution when forecasts are acceptable for production. The detailed planning system includes feedback to the production system of its performance data, including changes in quantities, work types, trade breakdowns, etc. Thus, the forecast is able to closely approximate the end result. When the forecast indicates a future problem, control actions must be planned to get the forecast back on track. Thus control actions affect the forecast information, not the original plans. The adjusted forecast, after control actions, forms the look-ahead plan and can be used to issue directives to the workforce.

In client and management reporting, both the unadjusted forecast (based on just the objective data) and the adjusted forecast together with planned control actions should be reported. This allows the client and management to get confidence that the project is taking constructive action to correct problems and, particularly, that it recognises that there are problems. The location-based control system promotes this kind of transparency. Problems are openly discussed and a strategy to address them is made explicit.

The location-based control model needs to provide accurate information sufficient to differentiate performance deviations (the traditional focus) from changes in circumstances. The sources of deviation may include:

- Quantity changes
- Start-up delays
- Production rate deviations
- Discontinuities
- Working out of sequence
- Production prerequisites.

Tracking this more accurate information, and having a system with sufficient flexibility to manage changes to implemented production plans, will lead to better management of the prerequisites for production, the availability of suitable resources and more detailed look-ahead planning during construction.

This chapter describes a new control model which utilises four stages of production information, each stage having its own schedule views, information and properties. The production management stages, in themselves, are not unfamiliar concepts in traditional activity-based control, however the ability to view the history of project performance on the flowline chart makes it possible to use the four stages consistently throughout the project—the normal process of maintaining historic schedules is relegated. The stages are baseline, current, progress and forecast. The features are summarised in Table 8.1.

Baseline

The *baseline stage* provides the founding set of project data, such as the committed plan for the project, against which all subsequent performance is compared. It functions in the same way as baselining a schedule in CPM control systems. The baseline plan cannot be changed, unless a new baseline is established, and it constrains the current plan. The location-based baseline model uses location-based quantities (*baseline quantities*) and tasks (*schedule tasks*) to plan the work as described in Chapter 5.

New baselines can be established when there are significant changes to the location breakdown structure or the quantities of the project, or if major changes result from owner-approved changes in the design. They should not be established if actual production rates deviate from planned production rates or for minor design changes which are not variations. Establishing a new baseline too easily loses much of the psychological impact of the location-based control system because updating the baseline always updates the performance metrics. Production should be held accountable to the original baseline plan as much as possible.

The baseline plan is used to plan procurement and to prepare the subcontract tender schedules and milestone information. To achieve these objectives, the start and finish dates and production rate requirements should be reliable.

Current

The *current stage* functions in a way similar to the baseline, however it specifically recognises the need for change in the project plan to take into account new information which was not available when the baseline plan was made; both changed project data and more detailed construction planning, including information from subcontractors. The current plan is changed whenever new information becomes available. This new information can include information about resource availability, prerequisites of production, quantity changes and changes in logic. However, even if there are changes, the original baseline places constraints on the finish dates in each location—necessary for the management of commitment to trade packages (commitments will be discussed in Chapter 10). This forces the process of updating the schedule to try to minimise any effects on other trades instead of allowing the dates to slip towards the end of the project, extending the critical path as they would under CPM. These constraints are *soft constraints*, resulting in *alarms* to signal if they are broken. The location-based control model establishes the mapping between these two planning stages, using a new set of location-based quantities (*current quantities*) and a set of current stage tasks (*detail tasks*) to manage the changes involved in current stage planning. Detail tasks also consist of *detail activities* in each location.

Two sets of quantities are maintained because production management needs to be aware of any quantity changes during the project. Baseline quantities contain the initial assumptions about quantities and productivity rates. When more information becomes available, the quantities may get more accurate, mistakes in quantity measurement may be revealed or there might be variations from the original design resulting in re-measures. All of these changes are updated to the current bill of quantities and affect the durations of current tasks. In contrast, the baseline bill of quantities is updated only when quantity changes result from client-approved variations.

Detail tasks drive the scheduling of work in the current stage. In its most basic implementation, a detail task is initially equivalent (apparently identical) to the baseline task, thus each schedule task has at least one equivalent detail task. As detailed planning of the work generally requires variation in the way the work is performed, detail tasks must be added to reflect the new plans for production. The detail tasks may be more accurate than the schedule tasks, but each detail task always belongs to a single schedule task on the baseline.

Implementation of the current phase might, for example, involve quantities broken down to a more accurate hierarchy level, a schedule task exploded into multiple detail tasks or changes of start dates or production rates based on subcontractor commitments. The logic of detail tasks is basically similar to the logic of schedule tasks described in Chapter 5. However, there are some important differences. Buffers are not so crucial on this level because it can be assumed that there is enough information for the production to be executed exactly as planned. The same resources may work in many different detail tasks which results in a need for an additional resource levelling mechanism.

Progress

The *progress stage* monitors the actual time performance of the project and therefore tracks data in the detail tasks. The progress of the production is measured by recording task start and finish times or completion rates in each detail activity location. Actual production rates for detail tasks can be calculated from this and, if actual resources are known, the actual resource consumption rates (man hours per unit) can be calculated.

Forecast

In the *forecast stage* the current plan and progress information can be used to calculate a *schedule forecast*. In the absence of control actions, the production must be assumed to continue with the actual production rate currently achieved, rather than that planned. Forecasting uses the planned logic network to evaluate the impact of deviations on following trades. This information can be used by the production managers to make informed decisions about suitable and immediate control actions that are required to restore planned production. This is done using alarms to alert management before interference has occurred. The model allows timely reaction instead of just recording the deviation and rescheduling, as required under CPM.

Table 8.1 Stage features

	Baseline	Current	Progress	Forecast
Quantities	Baseline quantities	Current quantities	Current quantities	Current quantities
Tasks	Schedule (baseline) tasks	Detail (current) tasks	Detail tasks	Detail tasks
Sub-task breakdown		Detail activities	Detail activities	
Flexibility	Fixed at start	Changeable		Uses baseline logic
Risk management	Alarms based on risk in the plan		Tracking progress	Alarms based on forecast

Location-based control

Together the components of the system form a comprehensive location-based double-loop control model based on location-based planning and scheduling. The baseline, which is based on assumptions (what *should* be done as well as the basis of commitment), is transformed into a realistic, current model, based on up-to-date information, detail planning and subcontractor commitments (what has been *promised*). Progress data and forecasts evaluate the implementation of both the baseline and current models. If forecasts deviate from the current model or assume more resources than available, it is likely that the model will not be achieved unless control actions are taken. The forecast adjusted with control actions and adjusted for resource availability forms the look-ahead plan (what *can* be done) which is directly relevant for implementation and can be used to issue directives for execution (what *will* be done). Directives include the actual construction elements that should be built (while look-ahead plans are concerned with production rates and finishing locations).

This chapter describes the mappings between the four stages of information and how they form a comprehensive, integrated model utilising the same information structure with different levels of uncertainty. The following chapters (9 and 10) build on that model to introduce additional tools and methods and methodologies about how the theory should be used in practice.

COMPONENTS OF LOCATION-BASED CONTROL

The theory of location-based control adopts all the components of location-based planning and scheduling and adds new components for control functionality. The essential components of location-based planning and scheduling are:

- Location breakdown structure
- Location-based bill of quantities
- Location-based productivity data
- Resource and crew data
- Schedule tasks
- The flowline schedule.

To these may be added the additional components required to establish the four stages of location-based control: baseline, current, actual and forecast. The new components are:

- Revised location breakdown structure
- Current bill of quantities
- Detail tasks
- Constraint dates
- Soft dependencies.

Soft dependencies are dependencies which may be overridden when work is actually executed out of sequence.

Revised location breakdown structure (LBS)

It is not possible to alter the hierarchy of the location breakdown structure without changing the baseline. Therefore, it should generally remain unchanged to allow the mapping of baseline and current plans. Changing the LBS would result in there being no common point of comparison for performance measurement. However, the location-based system allows more detail to be added to the LBS by dividing existing locations into smaller sub-locations.

This is useful in a situation where the most logical location breakdown structure cannot be known in advance for the most accurate hierarchy levels. For example, in many projects the end user of spaces will not be known when the project is pre-planned. In this case it is possible to pre-plan the schedule using only floors as the most accurate location hierarchy and then to develop the current plan using the actual spaces as the most accurate locations.

Another example is during the final stage of the project. At that time it is useful to control the finishing work and correct any errors for each individual room. In contrast, this level of detail would require too much work during production of the structure, yielding only a marginal value to the control process.

The effects of changes to the LBS during implementation are described below.

Adding new locations to an existing location hierarchy level

Adding a new location to an existing location hierarchy level is problematic because the baseline schedule tasks cannot have quantities in that location. Therefore, there can be no mapping between the baseline and current schedules in the new location. Mapping can be restored by updating the baseline schedule to also have quantities in the new location. If the new location breakdown structure enables better control, this can be a valid reason to change the baseline.

Adding new locations on a new hierarchy level

Adding a location on a new hierarchy level preserves the mapping between the baseline and the current schedule. The more accurate data of the current plan can be summarised to the baseline accuracy level, because all new locations on the new hierarchy levels must necessarily be hierarchically below existing locations. There is no loss of information and therefore the baseline schedule does not need to be changed. This is the most useful way of changing the LBS during production.

By assuming that the higher level location has begun when any of the new lower level locations has started, mapping the current to the baseline becomes possible. Similarly, the higher level location is finished only after all lower level locations have been finished.

Removing a location

Removing a location apparently necessitates changing the baseline schedule, but in practice no change is required. Removing a location is effectively the same as setting all the quantities in that location to zero (no work is done there). There is no need to remove locations from the LBS because the current quantities can be set to be zero in that location, achieving the same effect without changes to the baseline.

Current bill of quantities

The current stage uses a separate bill of current quantities (or current bill of quantities) as the basis for calculating detail task durations and logic relationships. The bill of current quantities would be identical to the bill of baseline quantities at the start, but with the bill of current quantities being updated as and when new information becomes available. Thus comparisons between the baseline and the current bills of quantities will show any quantity changes during the project. In addition to changes, quantity items may be defined on a more accurate hierarchy level (if new hierarchy levels have been defined in the LBS), they may be removed or new quantity items may be added.

The mapping between the baseline and current bill of quantities can be based on both a code and description of the quantity item. The quantity item is the same if the code or description *and* the unit match. All other task attributes, such as labour consumption, quantities in each location or location accuracy level, can be changed during the project.

The mapping is very important because it helps management to understand the source of deviations. For example, is the deviation due to a quantity change or alternatively an error in the original productivity estimate?

By assessing variation reports from earlier projects of the same type, the uncertainty related to each quantity item can be estimated, thus enabling better pre-planning in the future. It also provides much richer information to support time-related claims or disputes and is able to better show the effects of variations.

Quantity changes

If the quantities change in a location, or if resource consumption rates change, there will be an immediate effect on the current schedule. Quantity changes will effect duration only in the locations where there is change. In contrast, resource consumption changes will affect the task in all locations.

Figure 8.2 shows the visual effects of a quantity change. Generally flowline charts show the baseline plan (solid line) actual progress (short dash) and forecasts (long dash). In this figure, and also elsewhere in this book, whenever baseline and current schedules are presented in the same diagram, the current schedule is shown using the 'forecast' dashed line. In this example, the quantity of plasterboard walls has doubled but only on the second floor of each building. The quantity change effects the current durations of affected locations only. It also pushes other locations forward because of crew continuity, in this case causing a total delay of two weeks to the completion of the task—unless the production rate is increased to compensate.

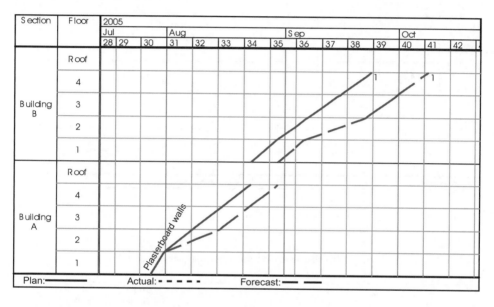

Figure 8.2 The visual effects of quantity change

Removing quantities

If any quantity items are removed from the current bill of quantities of a task, the current duration of the task will decrease. Taking the example of plasterboard walls (originally presented in Chapter 5, Figure 5.5), suppose a particular wall type is completely removed from the subcontract (due to a decision to change from a plasterboard to a masonry wall), resulting in a reduction in total quantities for the task. For the BOQ first shown in Table 5.2, the current BOQ for the revised task is shown in Table 8.2 with the removed quantity items shown as having been struck out. The corresponding current schedule for the plasterboard walls is shown in Figure 8.3. The duration has changed in each location, but the change is greatest in those locations where the removed quantities were greatest.

Adding quantities

Quantity items may need to be added to a task due to variations, changes in design or omissions during pre-planning. Quantity items are often added to provide further detail to the bill of quantities. For example, in the pre-planning stage, painting might have been roughly estimated in square metres. During implementation, management might want to control the quantity of wall, floor, ceiling and column painting separately.

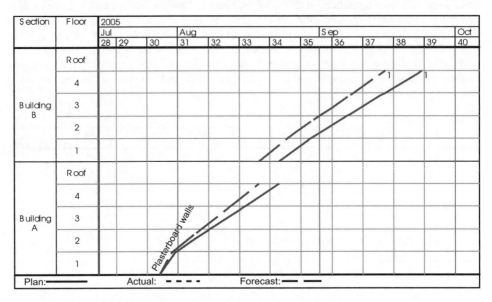

Figure 8.3 The visual effects of quantity deletion

Table 8.2 Deleting items (and quantities) from plasterboard walls

		Project:	Project
		Section:	Building A
		Floor:	1
Code	Item	Consumption man hours/unit	
456100	Erect plasterboard walls between apartments	0.65	1
456226	Mount 13mm special gyproc paneling on plasterboard wall	0.16	8.6
456216	Mount 13mm gyproc paneling on plasterboard wall	0.16	8.6
456146	Mount 79mm paneling on dwelling room wall adjacent to sauna	0.46	8.6
~~456136~~	~~Mount 92mm paneling on washroom wall~~	~~0.46~~	~~10.7~~
~~456126~~	~~Mount 79mm paneling on washroom wall adjacent to sauna~~	~~0.46~~	~~6.4~~
456116	Mount 92mm paneling on dwelling room wall	0.46	57.4

			Building B					Unit
2	3	4 Roof	1	2	3	4 Roof		M2
26	26	26	31	35.4	35.4	35.4	M2
26	26	26	25.5	30	30	30	M2
26	26	26	25.5	30	30	30	M2
~~22.5~~	~~22.5~~	~~22.5~~	~~14.1~~	~~19.6~~	~~19.6~~	~~19.6~~	M2
~~9.5~~	~~9.5~~	~~9.5~~	~~3.1~~	~~6.2~~	~~6.2~~	~~6.2~~	M2
118.6	118.6	118.6	109.8	134.3	134.3	134.3	M2

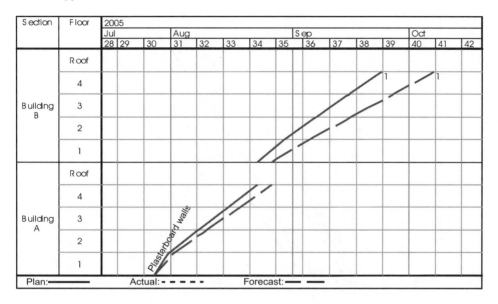

Figure 8.4 The visual effects of quantity deletion and addition

Table 8.3 Deleting and adding items (and quantities) to plasterboard walls

		Project:	
		Section:	Project Building A
		Floor:	1
Code	Item	Consumption	
456100	Erect plasterboard walls between apartments	0.65	1
456226	Mount 13mm special gyproc paneling on plasterboard wall	0.16	8.6
456216	Mount 13mm gyproc paneling on plasterboard wall	0.16	8.6
456146	Mount 79mm paneling on dwelling room wall adjacent to sauna	0.46	8.6
456136	Mount 92mm paneling on washroom wall	0.46	10.7
~~456126~~	~~Mount 79mm paneling on washroom wall adjacent to sauna~~	~~0.46~~	~~6.4~~
~~456116~~	~~Mount 92mm paneling on dwelling room wall~~	~~0.46~~	~~57.4~~
	New wall type	**0.80**	**30**

				Building B					
2	3	4	Roof	1	2	3	4 Roof		Unit
									M2
26	26	26		31	35.4	35.4	35.4		M2
26	26	26		25.5	30	30	30		M2
26	26	26		25.5	30	30	30		M2
~~22.5~~	~~22.5~~	~~22.5~~		~~14.1~~	~~19.6~~	~~19.6~~	~~19.6~~		~~M2~~
~~9.5~~	~~9.5~~	~~9.5~~		~~3.1~~	~~6.2~~	~~6.2~~	~~6.2~~		~~M2~~
118.6	118.6	118.6		109.8	134.3	134.3	134.3		M2
30	**30**	**30**		**50**	**50**	**50**	**50**		**M2**

To return to the plasterboard example, let us introduce a new wall type with a high labour consumption rate. There is 40 m^2 of this new wall type on each floor of the first building and 50 m^2 on each floor of the second building. The new bill of quantities is shown in Table 8.3. Figure 8.4 shows the result graphically. Even though the two quantity items were previously removed, the task will be late if resources are not added to compensate.

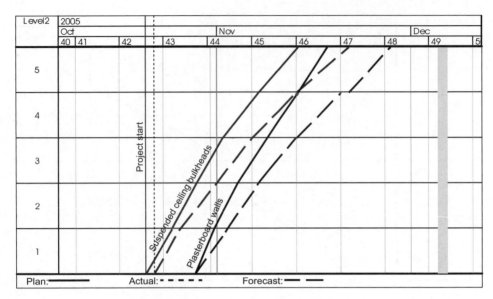

Figure 8.5 The visual effects of transferring and changing items

Table 8.4 Transferring and changing items (and quantities)

Target		Location:	1	2	3	4	5	
Code	Item	Consumption						Unit
Bulkheads								
	Bulkheads 1	0.5	25	12	15	25	22	M2
	Bulkheads 2	0.8	15	20	20	25	30	M2
Plasterboard Walls								
	Wall 1	0.46	25	30	35	35	12	M2
	Wall 2	0.5	12	15	20	24	42	M2

Current		Location:	1	2	3	4	5	
Code	Item	Consumption						Unit
Bulkheads								
	Bulkheads 2	1.6	15	20	20	25	30	M2
Plasterboard Walls								
	Wall 1	0.46	25	30	35	35	12	M2
	Wall 2	0.5	12	15	20	24	42	M2
	Bulkheads 1	0.5	25	12	15	25	22	M2

Changing task quantities

The optimal division of quantities to create trade packages (or indeed direct labour work) is often unknown beforehand. If multi-skilled workers are available to the project, it may be beneficial to transfer quantities to another crew during production. For example, it might be decided that the painting crew will also prepare taping and finishing on a floor before moving to the next floor. In the event that a crew works slower or faster than expected, these decisions are often made on site. The end result is that other activities may become slower (or faster) while the subject activity becomes faster (or slower).

Suspended ceiling bulkheads and plasterboard walls are presented as an example in Table 8.4 and Figure 8.5. The estimated productivity of one type of bulkhead was wrong and the current estimate is double the previous estimate (perhaps due to difficulty). In this case, it is also possible for the plasterboard crew to install the easier bulkhead type, so this quantity has been transferred to the plasterboard crew. The effect is shown in Figure 8.5. The current lines are now well aligned because of the transfer, whereas without the transfer the plasterboard walls would have been unable to work continuously, as the production rate of the predecessor was too slow.

Detail tasks

The baseline master schedule is often based on incomplete or inaccurate information and must be updated to make it relevant and suitable for controlling production.

Updating the current plan includes changing start dates and production rates of tasks to match the current information, adding more detail as well as planning flow across multiple sub-activities. The location-based control theory requires a new set of tasks—detail tasks—to handle the updating process, while maintaining the link to the baseline. Schedule tasks and baseline quantities together form the baseline schedule and its background information. Detail tasks and current quantities form the current schedule and are used to control actual production and to show the effects of deviations from the baseline. While baseline tasks define the commitment of the general contractor to the client, the detail tasks capture the mutual commitments of the general contractor and the subcontractors.

Most of the properties of detail tasks are identical to those of schedule tasks. The main differences are described below.

Calculating constraint dates from schedule tasks

Each detail task is associated with only one task on the baseline schedule. A schedule task can have multiple detail tasks associated with it. The mapping is used to establish constraints at the detail task level and to report progress compared to the baseline schedule.

In order to prevent schedule slippages, all detail tasks in any location should stay within the baseline schedule constraints for that location. The constraints set boundaries from the schedule (baseline) task for the scheduling of detail (current) tasks and are calculated for each location.

The start date constraint $S_i^{\,c}$ for Task i with n predecessors, is given by:

$$S_i^{\,c} = \max\left(S_i^{\,b}, \min(P_k, F_k^{\,b}) \right), 1 \le k \le n, \text{ where } k \text{ is relative to } i \qquad (8.1)$$

Where:

$S_i^{\,b}$ is the start date of the baseline task
P_k is the completion date of predecessor k in the location including the duration of the link's lags $D_k^{\,L}$ and buffers $D_k^{\,B}$
$F_k^{\,b}$ is the finish date of the predecessor baseline task.

The calculation of P_k depends on the link type. The lags and buffers are those between the task and its predecessor:

When F–S, P_k = *Finish date of predecessor + lag + buffer*

or $P_k = F_k + D_k{}^L + D_k{}^B$ (8.2)

When S–S, P_k = *Start date of predecessor + lag + buffer*

or $P_k = S_k + D_k{}^L + D_k{}^B$ (8.3)

When S–F, P_k = *Start date of predecessor + lag + buffer − baseline duration*

or $P_k = S_k + D_k{}^L + D_k{}^B - D_k{}^b$ (8.4)

When F–F, P_k = *Finish date of predecessor + lag + buffer − baseline duration*

or $P_k = F_k + D_k{}^L + D_k{}^B - D_k{}^b$ (8.5)

The finish date constraint $F_i{}^C$ for Task i with m successors, is given by:

$$F_i{}^C = \min\left(F_i{}^b, G_m\right), 1 \le i \le m, \text{ where } m \text{ is relative to } i$$ (8.6)

Where:

$F_i{}^b$ is the end date of the baseline task
G_m is the start date of successor m in the location, including the duration of the link's lags $D_m{}^L$ and buffers $D_m{}^B$.

The calculation of G_m depends on the link type. The lags and buffers are those between the task and its successor:

When F–S, G_m = *Finish date of baseline + lag + buffer*

or $G_m = F_i{}^b + D_m{}^L + D_m{}^B$ (8.7)

When S–S, Q_m = *Start date of baseline + lag + buffer*

or $G_m = S_i{}^b + D_m{}^L + D_m{}^B$ (8.8)

When S–F, Q_m = *Start date of baseline + lag + buffer − successor duration*

or $G_m = S_i{}^b + D_m{}^L + D_m{}^B - D_m$ (8.9)

When F–F, G_m = *Finish date of baseline + lag + buffer − successor duration*

or $G_m = F_i{}^b + D_m{}^L + D_m{}^B - D_m$ (8.10)

Equations 8.1 and 8.6 ensure that the schedule task buffers are made available to all the detail tasks within the schedule task, except in the final location where the buffer is released to the successor task.

In calculating constraint dates, the buffer is assumed to be owned by the predecessor activity. Thus the successor activity is not allowed into the *protected* space of the predecessor activity including its buffer. This gives more flexibility to controlling the predecessor and decreases the schedule linkages between different subcontractors and thus the risk of work stoppages. The finish date of the schedule task constrains the end date for detail tasks in the final location, because otherwise it is possible to plan detail tasks so that all the buffer is used in scheduling rather than as work buffers, which will leave no buffers to absorb any production deviations from plan.

If all detail tasks can be made to remain between these constraint dates, the original baseline schedule holds perfectly. The system provides constraint dates as a soft visual warning, instead of hard logic barriers, to show how much the schedule has slipped based on the current planning decisions. This forces planners to seek solutions to allow the baseline schedule to be implemented. However, if the original schedule is practically unachievable for a given a schedule task, the damage can be caught up by revising the detail tasks of the succeeding schedule tasks.

Location-based constraint dates can also be used as starting data for subcontractors so that they can develop their own schedules. If all subcontractors can work within their location-based constraint dates, there will be no interference from other subcontractors. In practice, this approach of visually reminding planners that they are behind has been seen to promote adaptive forward-looking behaviour aiming at catching up delays in a cost-effective way instead of merely crashing activities on the critical path, or allowing the delays to gradually accumulate and use up the project time contingency.

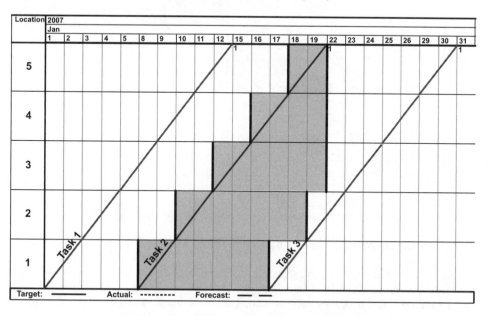

Figure 8.6 The constraint lines for detail tasks for baseline Task 2

When production is performing better than planned, detail tasks can be used to compress the schedule further by not using all the space between constraint dates. Then the detail tasks of the succeeding schedule task can start earlier than their own beginning date constraints. For the project to fully realise the benefits, all subsequent subcontractors must also revise their commitments.

Figure 8.6 shows a schedule of three tasks. Constraint dates for each location for Task 2 are shown by vertical lines. If all detail tasks and activities stay between these constraint dates in each location, the baseline schedule can be implemented as planned. Note that if the schedule tasks have been planned to be continuous in the baseline schedule, together with buffers, the constraint dates of the following locations will always overlap. If there are no buffers, all the detail tasks of a schedule task must be finished in a location before moving on to the next location. This is fine if all the detail tasks are done by the same resources. However, if different skills are required for different detail tasks, the absence of buffers will result in discontinuous work for some of the crews.

Figure 8.7 shows two baseline tasks and between them two detail tasks which belong to a third task. No buffers remain in the schedule. If Detail Tasks 1 and 2 require different skills and resources, the lack of a buffer results in discontinuous work for those resources. Planning buffers in the pre-planning phase adds flexibility in the implementation phase and reduces risk.

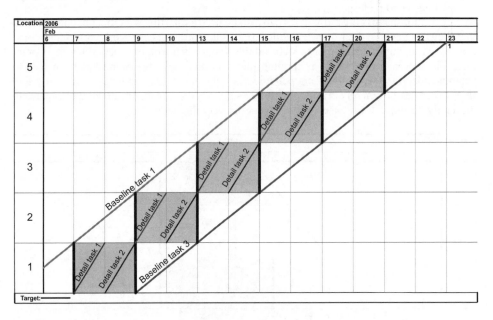

Figure 8.7 Two baseline tasks (1 and 3) and between them two detail tasks belonging to a third baseline task (2)

Reporting the current schedule compared to the baseline schedule

Mapping between detail tasks and schedule tasks can be used to review the current schedule against the baseline schedule. If there is no actual progress information available, and detail tasks have been updated, they represent the best available information about the current status and thus form the first baseline schedule forecast. In other words, without progress data, the current schedule is the baseline forecast schedule. The forecast schedule is calculated as follows:

- Forecast the start date in a location: the start date of the first detail task in the location (or any of the location's sub-locations)
- Forecast the finish date in a location: the finish date of the last detail task in the location (or any of the location's sub-locations).

This means that the schedule task is considered to have started as soon as the first detail task associated with it starts and is not considered to be finished until all detail tasks have been finished in a location. Therefore, when visualised in flowline, the forecast line may not be continuous. This forecast line is used to observe deviations from the original schedule and to assess adequacy of the buffers, not to observe whether the work was done continuously. Continuity of work is assessed by examining detail tasks directly.

Figure 8.8 shows an example of three schedule tasks. The bottom panel shows the Schedule Tasks 1 and 3 and between them the detail tasks now required for Schedule Task 2. The upper panel shows all three schedule tasks, but also displays the duration of the work being undertaken by all detail tasks in each location as forecast lines. If the current plan is carried out, the task will begin on time but the cycle time for each location is longer. However, there is no effect on the preceding or succeeding tasks. On the other hand, most of the buffer of the schedule task has been used to accommodate multiple detail tasks. In reality, to have continuity for each of the detail tasks, any buffer will be used, therefore a buffer must be planned where there will be multiple detail tasks which are each required to be continuous. This must be taken into account in planning schedule tasks.

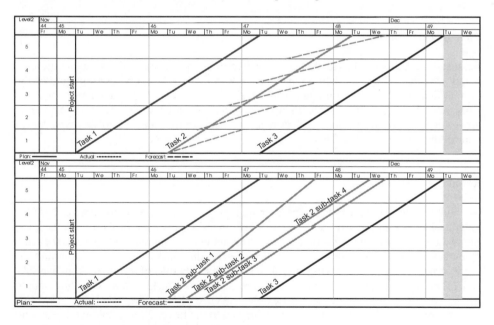

Figure 8.8 Four detail tasks for baseline Task 2

Quantities in detail tasks

Location-based quantities are utilised by detail tasks in the same way as with schedule tasks. However, current quantities are used instead of baseline quantities. This means that resource consumption rates or quantities can be different, resulting in different durations even when using the same resources. To preserve the mapping between baseline tasks and detail tasks, the quantities of a detail task should be on the same or a deeper hierarchy level as the schedule task's quantities. Otherwise it is not possible to calculate the forecast start and finish dates of the schedule task or the constraint dates for the detail tasks. There are no other constraints on quantities. Quantities can be removed, added or changed in any other way and the mapping will be unaffected.

Dependencies and buffers

Dependencies between detail tasks use the same five layers of location-based logic as their schedule tasks, however they exist only between detail tasks. Furthermore, schedule tasks do not restrict detail tasks in any way, so it is possible to change the logic of tasks when constructing the logic of detail tasks. Detail planning refers to a short-term planning method that uses detail tasks and hopefully utilises better information than would have been available when designing the baseline schedule tasks. In consequence, buffers do not play as great a role in detail planning. The role of buffers is also diminished by the fact that the detail tasks belonging to a specific baseline task are often completed by the one subcontract crew and thus are not so prone to disturbances. Buffers are still required between the detail tasks of different subcontractors, although they do not need to be as large as in the baseline schedule.

Float and criticality

Float and criticality can be calculated for the current model based on links and durations of detail activities. They can then be mapped back to the baseline, giving a measure of the current float and criticality for any baseline task. Baseline task float is the minimum of the float for each of its detail activities. The baseline task is critical when any of its detail tasks are critical. The use of float and criticality in location-based control is explored in controlling methodologies, Chapter 10.

Resources

Methods designed to derive durations from quantities, resources and productivity rates for detail tasks work in the same way as for schedule tasks. However, in the implementation phase there is usually more information available about the resources and, because the same resources can work in multiple detail tasks of a schedule task, the resource continuity is more complex than just observing unbroken flowlines. The flow of resources from one sub-task to another can be modelled by the use of resource links. Instead of making the resource links fixed, the current model uses a flexible sequence of tasks and locations to establish resource priorities. This is an important expansion to the Layer 3 logic link in the planning system (see page 135). Instead of choosing only a location sequence for a task, the planner can select combined location and detail task sequences for a resource or group of resources.

Figure 8.9 shows an example of four tasks: formwork, reinforcement, concreting and stripping of forms. The superintendent has decided to use just two sets of forms. The formwork crew should be able to work continuously through the building. The continuity of other trades should be optimised. The sequence for formwork resources is: Formwork 1 → Formwork 2 → Stripping 1 → Stripping 2 → Formwork 3 → Formwork 4 → Stripping 3 →

Stripping 4 → Formwork 5 → Stripping 5. The example has been modelled with this sequence and Layer 1 links between the activities in each location.

Resource links are needed, especially with multi-skilled trades such as most of the building services trades, where the same resources are participating in many tasks. Multi-skilling of the workforce significantly increases the complexity for a resource optimisation model. It is possible to use manual techniques to force resource levelling in such conditions, however this fine level of detail moves away from the simple principles of location-based control. It is suggested that resource levelling is used only to remove any peaks of resource use which exceed availability and any extended periods of low utilisation. Week-to-week modest variations are best handled in look-ahead schedules, using a weekly level of control and by using workable backlog as a levelling mechanism.

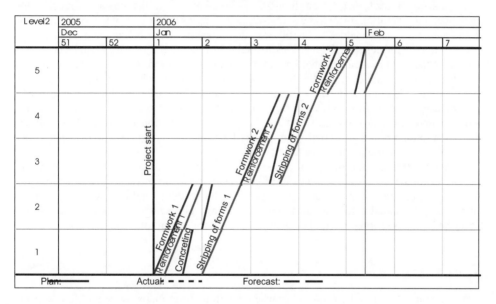

Figure 8.9 The visual effects of resource constraints at the detail level

Progress stage

During the actual production, or progress, phase the baseline and current information is gradually augmented by progress information. This information highlights any deviations from the plan, is used to calculate the forecast and is critical in the subsequent evaluation of the quality of the original plans. In the location-based system, status information should be tracked by location. More benefit is gained if progress is tracked for all components of the planning and control system. While additional optional components are introduced in Chapter 9, the basic components of the system (Seppänen and Kenley, 2005a) include:

- Actual quantities
- Actual resources
- Actual shift length and days off
- Actual start and finish dates.

This information can be used to calculate the following important values:

- Actual resource consumption
- Actual production rate.

These values should be monitored at the most accurate planning level. This is effectively the chosen location level of the detail task. The mapping between the detail and baseline tasks allows progress data to be compared with either of the two levels of planning. Each item is described in more detail below.

Actual quantities

The actual quantities for a location should be recorded as soon as the location is finished for each planned quantity item. If there are new quantity items that were not taken into account in the plan, they can be added to the relevant detail tasks with zero planned quantity.

Actual quantities reveal deviations from the current quantities which can be critical if they are expected to repeat in other locations. Quantity deviations can result from errors in measurement, an undocumented variation or an attempt by a subcontractor to charge for work not included in the contract. The quantities should be tracked by location in order to give management the time to react before it is too late. A subcontractor's invoice is usually too late to control deviations. Chapter 12, the implementation of the LBMS chapter, discusses some ways to get actual quantities by location, the most powerful of which is the use of 3D models in quantity measurement.

A deviation of current quantities should prompt immediate effort to identify the reason for the deviation and to prevent it from occurring again in subsequent locations for that task. In addition, it provides information for the schedule and cost forecasts and for calculating the actual resource consumption and production rates. These calculations are described in the section on actual consumption and production rates (page 274).

Actual resources

Actual resources should be tracked because deviations from plans can signal future problems or explain schedule delays. Statusing is often done weekly, so it can be difficult to get an accurate figure because the number of workers in a location might vary daily. However, checking the average crew size is often sufficiently accurate. Multi-skilled contractors, working in multiple tasks, further complicate the issue because their resources may be working on many tasks on the one day.

Deviations in resources are usually explained by one or more of the following reasons:

1. The planned production rate or resource consumption rate may have been wrong.
2. The resources may be more or less skilled than the average used in the plan.
3. The effective learning curve might be steeper or gentler than assumed.
4. The subcontractor may not have enough resources.
5. There may not be enough space for the planned number of resources.
6. A preceding trade may be proceeding too slowly, so fewer resources are needed to maintain steady production.
7. The contract does not have milestones that require the subcontractor to maintain a steady production rate, so the work will start with insufficient workers and then be rushed with additional resources toward the end.
8. The same workers are used to work in multiple locations or tasks at a time.

Actual resource consumption is calculated from actual used resources. It can also be used to warn management of any potential problems. As described in the risk model of Chapter 6 and the planning methodologies of Chapter 7, planning multiple crews increases the risk in a project. If the subcontractor comes to site with fewer crews than planned, the reason should be investigated immediately or large downstream effects may arise.

To get realistic information about productivity, the actual resources need to be distributed to detail tasks and locations. In some cases, it is possible to get this information from subcontractors. However, it is more common that only the total number of workers on site will be reported in subcontractor meetings. In this case, actual resources can be distributed to tasks and locations based on the value of actual work done. For example, if an electrical subcontractor had three electricians on site and, based on progress data, they did 60 worker hours of work in the *cabling* task and 30 worker hours worth of work in the *switchboards* task, it can be assumed that two electricians were doing cabling and one electrician was working on switchboards. However, if the subcontractor actually spent 120 hours on site to achieve 90 hours of work then total productivity would be smaller than planned.

Actual shift length and days off

Actual shift length and days off are important because resource consumption rates that emerge from simply calculating actual durations using the planned shift length will be incorrect in the event that the workers have been on holiday or have been working overtime or weekends. The actual work might be slower than planned and the subcontractor may have reacted by working overtime or on weekends. This should be recognised or there may be problems later should the workers refuse to continue overtime or further control actions were needed. Traditional control methods would focus on the fact that the activities were completed on time and ignore the actions taking place on site to achieve this result. This provides an incomplete picture and can mislead management and, indeed, may actually conceal cost. Furthermore, such productivity rates may often be erroneously used to plan future projects, making those projects in turn hard to realise.

Actual start and end dates and interruptions of the work

A daily level of accuracy is usually enough for the calculation of actual start and finish times for a location. For very short duration activities, this will distort the resource consumption rates but this will usually correct after enough locations. While a higher level of detail may be desirable, data collection requirements must also be realistic or the method may not actually be implemented on construction sites.

It should be known when individual locations were actually started and when they were finished, otherwise it is not possible to plot progress information to flowline diagrams or to calculate actual production rates or resource consumption rates.

Interruptions to work longer than a day should be recorded. Otherwise, the actual consumption and production rates will be overly pessimistic, providing misleading forecasts for this and future projects. The level of interruptions is also a good measure of success in location-based control, because the aim of location-based planning and control is to minimise interruptions. However, it should be noted that for multi-skilled contractors (such as mechanical, electrical and plumbing contractors) who work in multiple detail tasks, interruption in any single task is not so critical if the resources can productively continue work in another task or onworkable backlog tasks or locations. For those contractors, other metrics for measuring the success of the control system are proposed in Chapters 9 and 10.

Actual production rate and consumption

Using the information described in earlier sections, the actual resource consumption rate can be calculated using the following formulae (Seppänen and Kenley, 2005b). The total effective duration y^E is:

$$y^E = F^A - S^A - T^L - T^H \tag{8.11}$$

Where:

> F^A = actual finish date
> S^A = actual start date
> T^L = time lost through interruptions
> T^H = time lost through holidays and days off.

The actual production rate (*units / shift*) ϕ^A is:

$$\phi^A = \frac{Q^A}{y^E} \tag{8.12}$$

Where:

> Q^A = actual quantity (*units*)
> y^E = the total effective duration (*shifts*).

The actual worker hours for the task (*hours*) L^A is:

$$L^A = R^A \times y_s \tag{8.13}$$

Where:

> R_A = sum of actual number of resources (*number*)
> y_s = shift duration (*hours*).

The actual resource consumption rate (*actual hours / actual quantity*) χ^A is:

$$\chi^A = \frac{L^A}{Q^A} \tag{8.14}$$

Where:

> L^A = the actual worker hours
> Q^A = the actual quantity.

Actual production rates can be used to forecast progress for following locations of the task. Calculating the actual production rate does not require information about actual resources or shift lengths. If the actual quantity is unknown, the planned quantity can be used to arrive at a rough rate (which does not include the effects of quantity changes).

Actual resource consumption rates are much more powerful because they can be used to plan control actions (for example, how much overtime should be worked, or alternatively how many resources should be added, to catch up). They can also be used to show the effects of successful or failed project control because they are a direct measure of productivity. Additionally, they can be used to refine cost estimates and to aid in negotiations with subcontractors such as in evaluating subcontractor bids ("we can show that work is 20% more productive on sites using the new control mechanisms"). They expose the waste in the production process. Cutting the actual resource consumption rates will directly lower subcontractor costs. In fact, without gathering the data about actual resource consumption rates, it is very difficult to measure the effects of any process improvements. The location-based method for calculating the resource consumption rates does not require much extra work, yet yields great benefit—even if it is somewhat rough, relying on the knowledge that the errors tend to cancel out when there are enough locations.

Table 8.5 Transferring and changing items (and quantities)

Schedule task/ Micromanagement task	Location: Consumption	1	2	3	4	Unit
Target						
Plasterboard Walls						
Plasterboard walls - metal frame	0.46	100	200	200	100	M2
Current						
Framing and board one side						
Framing k300	0.18	100	200	200	100	M2
Board on one side	0.12	100	200	200	100	M2
Electrical piping						
Electrical piping	8	2	4	4	2	DAYS
Board one other side						
Insulation	0.08	100	200	200	100	M2
Board on other side	0.12	100	200	200	100	M2
Actual						
Framing and board one side						
Framing k300	0.18	100	250			M2
Board on one side	0.12	100	250			M2
Electrical piping						
Electrical piping	8	3	6			DAYS
Board one other side						
Insulation	0.08	100	250			M2
Board on other side	0.12	100	250			M2

Consumption rate example

This simple example illustrates the basic procedure. There is one schedule task, *plasterboard walls*, and this has been exploded into three detail tasks: *frames and board on one side*, *electrical piping* and *board on another side*. There are four locations in the example. Baseline, current and actual bills of quantities are shown in Table 8.5. The detail level schedule, with progress is indicated by dotted lines, is shown in the upper panel of Figure 8.10. The lower panel shows the corresponding baseline schedule. The progress lines represent start and finish dates in each location. At the schedule task level, the progress line extends into the percentage of hours completed over all of the detail tasks in the location. The flowline (Figure 8.10) shows that the *frames and board on one side* task has started early, proceeded slower and is currently late compared with the plan. Consequently, the *electrical piping* task has been interrupted in production with a break of one day between locations one and two. However, Figure 8.10 does not display the reasons for the delay.

To get more information about the underlying reasons, the actual resource consumption rates must be calculated. The data and calculations needed to estimate the actual resource consumption rates are the following:

- Planned for *frames and board on one side* (2 locations):
    - Planned resource consumption rate, $\chi^P = 0.3$ worker hours / m^2
- Actual for *Frames and board on one side* (2 locations):
    - Total effective duration, $y^E = 10$ days
    - Actual production rate, $\phi^A = 350$ m^2 / 10 days = 35 m^2 / day
    - Sum of the actual number of resources, $R^A = 2$ workers on each day
    - Actual shift length, $y_s{}^A = 9$ hours / day
    - Calculated worker hours, $L^A = 2$ workers × 9 hours per day × 10 days = 180 hours
    - Actual resource consumption rate, $\chi^A = 180$ hours / 350 m^2 = 0.51 hours / m^2

Therefore, in addition to a quantity overrun (actual quantities higher than the planned quantities in Table 8.5), the work has been less productive than planned: 0.51 hr/m^2 versus 0.3 hr/m^2. If the quantity overrun continues in the remaining locations and the same crew continues to perform with the same productivity rate, there will be serious problems in overall production.

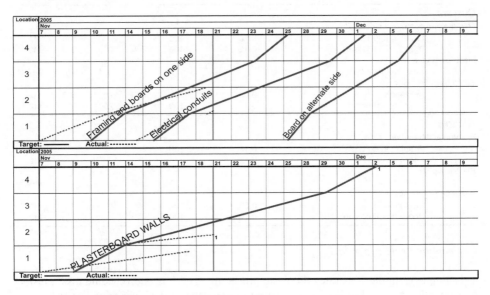

Figure 8.10 A comparison of task (lower section) and detail tasks (upper section)

The causes for the low productivity are not revealed by this data. The resources might be less productive than assumed, the work could be more difficult than standard or workers might engage in non value-adding activities, such as unnecessary hauling of materials, double-handling, rework, communicating with other people, etc. If the reason is not the skill level of the workers or an error in the estimate and there is no interference from other trades, productivity can often be rectified (increased) by devoting management time (supervision) to control the situation.

The next section describes how future production can be forecast using the calculated progress data and planned logic.

Schedule forecasts

The current and progress data can be used to calculate forecasts to predict the total effect of schedule deviations and variations, and therefore give early warning of problems. Forecasts should assume that production will continue with the achieved (rather than the planned) production rate, unless control actions are taken. It is not realistic to forecast based on planned production rates which are different from those achieved. Forecasts empower management to react to problems early enough to take effective action and provide the data required to support control action decisions. Forecasts can appear drastic when actual performance is well below planned, and such consequences would be hidden unless calculated performance data was used.

Forecasting is a process which utilises the best currently available information. In the early stages of the project, the original plan can be used. The forecast is then updated when new information about quantities or the schedule becomes available. During production, the actual production rates should be used as the basis for the forecast. The rich information content of location-based scheduling greatly increases the complexity of forecasting progress, and there are five types of forecast which arise:

1. Use the planned quantities, resource consumption and resources.
2. Use current information about schedule and quantities.
3. Use actual production rates.
4. Adjust for deviations from the plan.
5. Adjust for interference.

These calculations are described below in detail.

Forecast calculations

Each of the forecasting types uses the following calculations to take logic links into account and to forecast durations for locations.

- Forecast the start date in each location for either F–S and S–S logic sequence:

When F–S, $S_{i,j} = \max\left(S_{(i,j-1)}, F_{(i-1,j)}\right)$ (8.15)

When S–S, $S_{i,j} = \max\left(S_{(i,j-1)}, S_{(i-1,j)}\right)$ (8.16)

Where: $S_{i,j}$ = start date of the current task i in current location j.

- Forecast the resource consumption:
  Use the actual resource consumption of previous weeks. The length of the selected period affects the forecasting stability. If all historical production is used, the forecast is very stable and sudden improvements or problems in production rates do not affect the forecast unless they occur over a long period. If a very short interval is used, the forecast

will fluctuate weekly or even daily. In most projects, two or three weeks is a good interval for forecasting. Note that at the start of production, the forecast will usually look very pessimistic during first weeks of production due to common start-up difficulties. The effect is emphasised because information is only available from a short period of time. To make the forecast realistic, it is advisable to take these starting difficulties into account in the plan by using a lower production factor in the early locations.

• Forecast the duration $D_{i,j}$ of Task i in each location j:

$$D_{i,j} = \frac{Q_{i,j} \times RC_{i,j}}{R^P \times y_s^{\ P}} \tag{8.17}$$

Where:

Q = Quantity,
RC = Resource consumption,
R^P = planned resources, and
$y_s^{\ P}$ = planned shift length.

The numerator of the formula calculates the forecast worker hours required to finish the location. The denominator divides the hours by the available hours for production in a day. Note that planned values are used—the behaviour is assumed to revert back to the plan in the future. For example, if longer shift lengths have been used than those planned, the forecast will assume that planned shift lengths will be used from now on. Therefore, if it is decided to change the plan, it should be documented by changing the detail task or by planning a control action. Control actions are described later in this chapter (page 283).

• Forecast the finish date in each location for the F–F or S–F logic sequences:

$$\text{When F–F,} \; F_{i,j} = \max\left(\left(S_{(i,j)} + D_{i,j}\right), F_{(i-1,j)}\right) \tag{8.18}$$

$$\text{When S–F,} \; F_{i,j} = \max\left(\left(S_{(i,j)} + D_{i,j}\right), S_{(i-1,j)}\right) \tag{8.19}$$

Where: $F_{i,j}$ = the finish date of the current task i in the current location j.

Just as forecasting cannot use planned performance rates once production is underway, it is also not desirable (and unnecessary) to maintain production buffers around tasks that have commenced production and deviated from the plan. Thus the buffer can be absorbed by the deviation without delaying any following tasks until the buffer is consumed. This provides greater flexibility in the design of control actions, as forecasting only uses the mandatory lag and ignores the buffer when calculating the effects on following trades.

Forecasting should not force continuity on following trades, which must necessarily react to the delay of preceding trades by accommodation. Production continuity is a planned feature which cannot be maintained once progress varies from the plan in a way that impacts on the continuity of following trades—once workers are committed on site, continuity suffers when a task exceeds its buffers. Therefore, one of the main purposes for a forecast is to identify whether discontinuities will be predicted, thus highlighting disruption of the production system even though the project may not be delayed. Nevertheless, forcing continuity by altering the start dates remains a form of control action, as described in control actions section of this chapter (page 283).

The following methods are applied for each schedule forecast type:

1. Use the planned quantities, resource consumption and resources:
   The original planned production rates and quantities can be used if more accurate information is not available (that is, if detail tasks have not been planned and production has not yet started). However, the forecast start and finish dates should be adjusted if the predecessors' schedule forecasts are late. Start and finish dates are forecast using the calculations above. Durations and resource consumption rates are derived from the baseline.

2. Use current information about the schedule and quantities:
   There is more information available for forecasting once work crews have been selected for the job (either subcontract or direct labour) and quantities have been checked. A more accurate schedule is often planned together with the people responsible for the work. A more accurate schedule is modelled by using detail tasks. At this later stage, all the information required by the schedule forecast is available except for the progress data. The forecast of schedule tasks is based on detail tasks (page 265).

3. Use actual production rates:
   Progress data becomes available in the production phase. Actual start and finish dates can be used to calculate actual resource consumption rates. The production can be assumed to continue with the same rate unless control actions are taken. As described above, a duration forecast can be calculated by using the actual resource consumption rate from the last few weeks.
   Quantities can be forecast as follows:
   a. For each quantity item, calculate the quantity overrun ratio:
      total actual quantity / total planned quantity.
   b. Calculate the forecast quantity in each location which has not been finished:
      quantity overrun ratio × current estimate of quantity.
   Data for all of the calculations is available once a few locations or weeks of work have been completed.

4. Adjust for deviations from the plan:
   There are some deviations from the plan which must be considered in forecasting. They are:
   - Working out of a planned task logic sequence (Layers 1,2,4 and 5 logic)
   - Working out of a planned location sequence (Layer 3 logic)
   - Working in multiple locations at the same time with the same crew (that is, not finishing locations in sequence).
   A few guidelines about how to handle these situations are presented below:
   - Working out of sequence—external logic (Layers 1, 2, 4 and 5 logic):
     Working out of sequence in external logic means that the successor has begun before being allowed to do so by a logic link to another task. In this case, it is impossible to know if the link is still valid in other locations, therefore the forecast for other locations must follow the original logic. Alternatively, the detail tasks should be updated by the planner if the link is found no longer to be valid. In CPM applications, the term *progress override* is used when logic is ignored for specific activities. Similarly, in LBMS, only the currently ongoing location is allowed the override, locations not started are assumed to follow the original plan.
   - Working out of sequence—internal logic (Layer 3 logic):
     Working out of sequence in internal logic means that the location sequence for the

task varies from the plan. This is quite common in practice. The forecast can be calculated by assuming that those locations already commenced will be finished first and the production will then follow the planned sequence. If working out of sequence was caused by a plan change, the change should be documented by updating the detail tasks.

*   Working in multiple locations at the same time:
    Often crews do not finish their work in a location before moving on to the next location. This can result from contracts that are not location-based—the subcontractor will complete the easiest work first in order to get earlier payments. However, this makes it difficult to predict subsequent location sequences. The forecast can assume that any already started locations will be finished first and that the production will then follow the original sequence. The forecast production rate can be divided evenly between the started locations. To make the forecast more accurate, unfinished parts of work should be separated to new detail tasks or even new locations. If this is done, the finished parts can be marked as completed and a schedule of unfinished parts can be planned. This mechanism is similar to activity splitting in CPM.

Forecasting float

As stated on page 35, there are often multiple paths through a critical path schedule, but one or more of them will follow a line where the early and late start and finish dates will be the same (providing the project end date is the same as the earliest end date). This path is known as the critical path, as the jobs on the critical path cannot be delayed without affecting the end date of the project. Activities on the critical path are critical. Activities not on the critical path have float and Kelley and Walker (1959) defined four definitions of job float (Equations 2.8 to 2.11).

Float is a poorly understood concept. It carries great weight in critical path planning, particularly with regard to the impact of progress. Experienced planners frequently rile at the failure of inexperienced planners to understand the true significance of float. Perhaps it is fairer to say that, in reality, float is generally understood, but its application to delay analysis is poorly understood and schedules are poorly designed as a consequence. Thus, there is a fear that claims may fail due to poor application of CPM theory.

More importantly, despite their significance in claims assessment, criticality and float are blunt instruments which have little significance to the protection of production efficiency. Kelley and Walker's concept of float belongs in an activity-based methodology only. In location-based management, the continuity heuristics of Layer 3 logic change the meaning of float and shift its importance to the client and the contractor.

In the LBMS, *forecast float* is normal float as well as the float which arises in the forecast when progress is ahead or behind the schedule (it can be positive or negative). It can be used to evaluate the need for accelerating the schedule. Forecasting does not take into account continuity constraints, so standard CPM calculations can be used if there are no interference adjustments. Interference adjustments can be handled by assuming that they correspond to the earliest possible start constraint date for the location. Therefore, the already started locations before the point of discontinuity will often have forecast float. The logic links in the float calculation can be handled as follows:

*   Layer 3 logic links: the Layer 3 links between locations can be disregarded if the task is actually occurring in multiple locations instead of in sequence.

- Other links: if there is progress data which does not correspond with the planned logic, the dependencies in those locations can be disregarded. However, the links should be used to forecast float in other locations.

Forecast float is most useful when it is compared to a fixed end date, calculated from the baseline schedule. If the forecast total float is negative (forecast end date is later than planned), there is a need for acceleration or the project will be late. Mandatory lags only are used, instead of buffers, in forecasting total float. Thus, the project can show total float for most activities as long as there are no deviations from the plan and consequential forecast return delays.

Interference or float: identifying critical deviations

Protection of production efficiency is a key aim of the LBMS. From a production perspective, any forecast deviation is a critical deviation if it has the potential to affect the efficiency of work of a following subcontractor as well as the more familiar effect of delaying the activity or the project end date. Delay which impacts on production efficiency is termed *interference*.

Interference is an even more important short-term measure of criticality than float. In addition to using total float as a measure of long-term criticality, delays which cause interference may also lead to return delays or slowdowns. These are, by nature, unpredictable. In the short-term, the production management should react immediately whenever a deviation is going to cause interference to a task which is critical or near critical in terms of total float.

The following heuristics can be used to sort the production activities into criticality order with respect to the urgency of required control actions:

1. Temporal proximity to *interference point*:
   a. Calculate the number of days to the interference point:
      (the point where another subcontractor's work becomes discontinuous).
   b. Compare the *safe reaction time allowance*:
      (the planner's estimate of how long it will take, in the most pessimistic case, to implement a control action; for example, two weeks).
   c. If the interference point is closer than the safe reaction time allowance, go to step 2, otherwise go to step 3.
2. Calculate the seriousness of the interference:
   a. Find all the tasks and locations which will be impacted by the interference
   b. Find the minimum total float of locations impacted by interference.
3. Forecast the total float of the activity itself.

All ongoing tasks for which forecast interference is smaller than the safe reaction time allowance can be ranked based on the seriousness of the interference. Other tasks will be sorted after those interference-causing tasks in relation to their total float. The sorted list can be used to find tasks which have the greatest need for control actions and management attention. This is a similar method to the CPM method for smoothing in resource optimisation, where criticality provides the rank order (Gordon and Tulip, 1997). Here, however, the action is not to allocate resources, but to prioritise control actions and to protect production. In the LBMS, any task which causes interference in the short-term is more critical than other tasks, regardless of their float.

ALARMS

Alarms are early warnings of upcoming production problems. They are generated when the schedule forecast for a preceding task pushes the schedule forecast of a succeeding task—causing interference. Alarms should also be generated when work occurs out of sequence because this has the potential to cause interference in the future. For example, when a F–S relationship might have been planned but the succeeding task starts before the preceding task finishes. Even though work could start early, it is possible that the tasks will interfere with each other—causing slowdowns and other production problems.

Significant deviations in production rates tend to cause cascading chains of alarms, so only those imminent problems (in the near future) should produce alarms. To reduce the number of alarms, it is also best to only raise an alarm for interference which occurs between two different subcontractors (it can be assumed that each subcontractor can internally resolve their own production issues).

Alarms can be shown visually in flowline diagrams by (red) alarm dots. Figure 8.11 shows the plasterboard wall example, with two alarm dots caused by production being too slow for the *framing and board on one side* detail task.

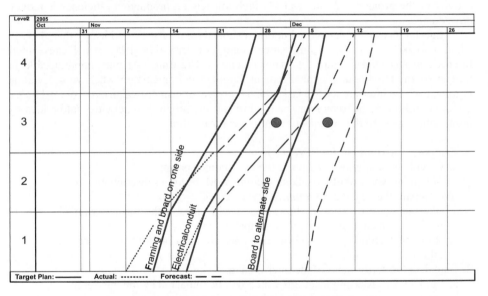

Figure 8.11 Alarms raised by slow tasks forecasting interference

For every alarm dot, the reasons for the problem should be investigated and a control action should be planned to prevent the alarm from turning into a real problem. For each alarm the following information should be tracked:

- The root cause of the alarm
- Control actions to prevent the alarm eventuating
- Claims related to the alarm
- Did the alarm actualise?

CONTROL ACTIONS

Control actions are the steps taken to recover from deviation in order to prevent further interference or to prevent project delay. Control action planning resembles rescheduling detail level tasks. However, there is a great difference between updating the plan and planning control actions at an implementation level. Control actions are needed when someone else's work will be interfered with, therefore there is a concrete goal for control action planning: finding a feasible solution to prevent interference. The list of available control actions is usually shorter than those available in planning both the schedule and detail tasks, because the control action must be implemented in the near future. Moreover, people close to the production should be included in the decision-making process. Carrying out control actions requires that everyone commits to the decisions. The following actions are available to remedy interference:

- Changing the number of resources (the same productivity will be assumed)
- Changing shift length or working overtime (on weekends or holidays)
- Changing the location sequence
- Splitting a task (which can be either to allow working in multiple locations at the same time or to allow a break with workers returning at an agreed date)
- Removing or switching technical dependency (this may cause interference in locations)
- Increasing productivity by reducing non value-adding activities (waiting, materials handling, rework, etc.)
- Shifting the start date of a successor task to make that task continuous.

With control actions, it is the forecast which is adjusted directly to correspond with planned control actions, rather than the plan. The plan is not changed because the fact that there was a deviation in the first place would then become hidden—possibly leading to a false sense of security. By updating the forecast instead of the plan, management accepts that there was a deviation but commits to action to remove alarms and to restore the original plan. The control planning process continues until all alarms have been resolved or a decision to do nothing has been documented. Because the forecast mixes control action information and information calculated from actual progress, it is desirable to maintain a log of the control actions taken.

Each control action changes one or more variables in calculating the forecast. The effects of the control action plan are calculated using the actual productivity calculated from progress data. New actual values will override the control action plan when they become available because they reveal whether the control plan was actually implemented with the planned effects. If no action has actually been taken, the period of feeling safe can last only for the sequence of a few completed locations before the new progress data again reveals the probable outcome.

All control actions, including the decision to do nothing, should be documented in case of a time-related dispute.

Example

To illustrate the use of control actions, the plasterboard wall example from Figure 8.10 can be continued. If the *framing and board on one side* detail task is delayed because of low productivity and a quantity overrun, it will threaten to make the following *electrical piping* task discontinuous. The resources are already working nine hours per day, so it will not be productive to add more overtime. An alternative control action must be sought. The following description is typical of the process to be followed.

After discussions and negotiations, the plasterboard subcontractor is ready to provide another crew and the electrical subcontractor (whose workers had already left the site to do other work) is willing to commit to a return date which allows their work to remain continuous for the remaining locations. However, the new plasterboard crew is not available for another week. To use the crew optimally, instead of accelerating *framing and board on one side*, the resource link between the *first board* and the *second board* is broken and the new crew mobilises to the first location for installing the second board. The electrical contractor can come back two days later and have continuous work. Meanwhile, management should devote time to improve the productivity of *framing and board on one side,* because even a single day's further delay will interfere with the electrical subcontractor's work. Note that if the productivity of *framing and board on one side* can be increased, the new crew will become unnecessary.

The control action plan would also need adjustment should *electrical piping* continue with a lower than planned production rate. Figure 8.12 illustrates the result of the control actions. The control action process will solve the problems if implemented, so the alarm has been removed.

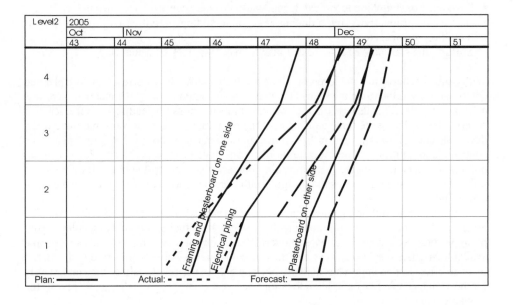

Figure 8.12 Control actions planned

LOOK-AHEAD PLAN

Location-based look-ahead plans are formed from forecasts once adjusted for control actions. The forecast should be adjusted for resource availability before using the look-ahead plan for creating weekly assignments. All forecast lines have resource assumptions, and the aggregated resource graphs for each subcontractor should be evaluated weekly to see whether either decreasing or increasing resources on site are required. The look-ahead plan needs to be modified if resource availability is larger than, or smaller than, assumed by the forecast. The final look-ahead plan, once adjusted for resource availability, can be used for planning weekly assignments.

WEEKLY PLANNING

Weekly planning is the lowest level of planning and it actually guides implementation. While higher levels of planning use production rates and locations, weekly planning uses well-defined concrete assignments. For example, the weekly production target for the masonry walls detail task could be to complete 80 m^2. The detail task could show that a location needs to be 50% complete by the end of the week. The weekly plan transforms this production rate target to one or more assignments by selecting which walls will actually be finished next week.

This selection process has a two-way link to the schedule forecast that has been adjusted by the effect of control actions. First, the forecast gives a production rate target to weekly planning by assuming that the same production rate will be maintained unless control actions have been planned. Second, the forecast of the next week will be adjusted if more or less work than that in the forecast is selected for the weekly plan. To enable these links, the location-based management system requires that each assignment has a detail task, quantity and location associated with it.

Production rate target for the next week

A forecast is based on the assumption that production continues with the same production rate unless control actions are taken. Because control actions should be planned together with the workers, the adjusted forecast gives a good indication of what *can* be done. The weekly planning process compares the quantity of selected assignments to the production rate target and gives an alarm if the production rate target is not achieved by the weekly plans.

Adjustment of forecast based on weekly plans

Because the final weekly plan should be a reliable plan of what will actually be done next week, it contains better information than the forecast which is just based on historical data. Therefore, if the final weekly plan contains less work than the schedule forecast, the schedule forecast of the next week should be updated to show the total effect. If less work is selected than forecast, the production rate often must be increased in the following weeks. By showing this fact to people responsible for the work early on, it is more likely that a production rate increase will become possible through successful control actions. Note that the weekly planning process may lead to new alarms, which would restart the control action planning process.

SUMMARY OF LOCATION-BASED CONTROL THEORY

The new location-based control theory states that using the described location-based model will enable management to take better informed control action decisions and motivate management to proactively seek out solutions which maintain the original plan. The model includes four levels of planning: baseline schedule planning, detail schedule planning, control action planning and weekly planning. Each level guides proactive control by setting constraint dates and targets to lower levels. These targets are made more accurate in micromanagement schedules utilising current information. Progress data is compared to planned values and used to calculate forecasts. If forecasts deviate from plans and cause interference in the near future, a reactive control action planning process is triggered to prevent interference. Control action planning aims to restore the forecast start and end dates to planned values instead of changing the plan. Forecasts, adjusted with control actions, form the look-ahead plan, which is then adjusted for resource availability. The adjusted look-ahead plan is used to set targets for weekly planning which are used to actually guide the work. If the assignments selected for the weekly plan do not match the forecast production rate, the forecast is updated to show the long-term effects of the lower production rate.

The following assumptions underlie the location-based control theory:

- Pre-planned baseline schedules cannot be used to control actual production because all the relevant information is not available before commencing the project.
- Location-based progress data can be used to realistically forecast future progress.
- By measuring quantity, production rate and productivity deviations, management is provided with better information to enable more informed control action decisions.
- Management should concentrate on preventing upcoming production problems instead of focusing on shortening the duration of the critical path.

Implementing the location-based control model should result in less production problems. The data gathering process requires more time than for an activity-based process but there is less need for firefighting actions or to find quick fixes to production problems. Instead, the production problems are prevented from continuing by the control action process. The control action process uses the available data to find out what needs to be done to prevent production problems. Instead of just artificially decreasing the durations of tasks on a critical path, the control action process defines whether new resources or overtime work is required. All decisions and claims related to an alarm are tracked systematically (*including* the decision to do nothing). This information has a critical role in time management and contractual disputes.

Location-based control provides an early warning mechanism and forces management to take actions immediately instead of later. When plans are being constantly updated, it is easy to cover up problems: any problems are pushed to the end of the project at which stage schedule updates become no longer possible. Location-based control allows problems to be solved as they become visible, which dramatically improves the typical end-of-project hurry and the associated cost overruns and quality defects. Projects can be reliably finished on time and within budget.

REFERENCES

Gordon, J. and Tulip, A. (1997). "Resource scheduling". *International Journal of Project Management* **15**(6): 359–370.

Kelley, J.E. and Walker, M.R. (1959). "Critical-Path Planning and Scheduling". *Proceedings of the Eastern Joint Computer Conference*.

Murray, W.B. (2007). *Faster construction projects with CPM scheduling*. McGraw-Hill, New York, NY.

Seppänen, O. and Kenley, R. (2005). "Performance Measurement Using Location-Based Status Data". *13th Annual Lean Construction Conference*, Sydney, IGLC.

Seppänen, O. and Kenley, R. (2005b). "Using Location-Based techniques for cost control". *13th Annual Lean Construction Conference*, Sydney, IGLC.

Stacey, R.D. (1996). *Strategic management of organisational dynamics*. London, Pitman Publishing.

Chapter 9

Location-based control methods

INTRODUCTION

Chapter 8 described the theory underlying the location-based management system's control sub-system and it's associated calculations. While the fundamentals of location-based control mostly concern the management of the project schedule, there are many further management functions which can be added on top of the basic system to enable the complete control of project production rather than merely controlling the schedule.

This chapter expands the discussion by adding functional methods for improving production control, such as controlling cost, risk, procurement and quality. The chapter follows the structure of Chapter 6, because the methods introduced as planning methods within the location-based management system have been designed to be controlled at the same level of accuracy. In addition to those methods, it is also critical to be able to communicate the plans and their status effectively during implementation. Tools to visualise progress are described in this chapter.

Methods to control and forecast costs are presented, based on both the conventional cost loading model and the production system cost model introduced in Chapter 6. The cost forecasting methods are forward-looking using location-based data and corresponding contracts with subcontractors, instead of being based on hindsight after receiving late accounting data. Rather than using actual accounting costs, they use location-based quantity and resource information and information about subcontract terms and conditions. This is necessary in order to establish real-time control, as accounting data necessarily lags project performance by a significant amount, possibly months, and is therefore too late to be used for effective control. Forecasting costs in real time also makes possible the comparison of subcontractor invoices with the forecast to identify costs which are not part of the original contract.

Once production is underway, more accurate information becomes available in terms of detail tasks, current bills of quantities and actual progress. This gradually decreases uncertainty for the project. The production system risk model can be updated to take into account information which is known (actuals) and which can be accurately forecast based on previous progress data (estimates and forecasts). This chapter describes how the location-based simulation model for production system risk is updated during the implementation phase.

Design and procurement-related activities are a significant contributor to actual production problems. Methods for controlling design and procurement include updating the design phase status and approvals by gatekeepers, and updating delivery times and amounts based on contracts and current information about demand timing. Methods to check the status of procurement events and to update the schedule forecasts based on actual status are described as part of the control of procurement in the location-based control system.

Both prerequisites and quality checks are important to the effective control of a project and must be managed. These should be tracked and all checks completed in a location before the location is accepted and the crew allowed to move on, otherwise there may be expensive long-term consequences.

It is not possible for the location-based control system to take learning into account in forecasting progress unless it is reflected in variation in actual productivity by location. The location-based learning model was introduced in Chapter 6 and the same model can be used to control detail tasks during implementation. However, in practice, it is very difficult to estimate actual learning from progress data as there can be many reasons for an actual production rate to differ from a planned rate.

This chapter concludes with the demonstration of four methods for representing project progress, from the familiar Gantt status line, through 'actual' lines on the flowline, to the innovative status control chart and production graphs. These provide powerful methods for managing projects in the implementation phase.

COST DURING IMPLEMENTATION

This section describes cost forecasting from the view point of a single actor in the construction process, the general contractor, and is adapted from Seppänen and Kenley (2005). However, before addressing the location-based cash flow model, it is useful to overview cash flow model theory.

There are several models for forecasting the cash position of a project for a head contractor, but "...models have been developed largely from assumptions and assembled rules, rather than from direct observation of project data" (Kenley, 2003). Models have been built using the following methods:

* Standard or ideal curves
* Balance sheet analysis
* Weighted mean delays
* The Kenley Logit net cash flow model.

While Kenley (2003) argued for the analysis of gross cash flow data for the construction of net cash flows, his was a model for smoothing project data *post-hoc* from accounting data. As such, it is unsuitable for real-time modelling as required for production control in a location-based management system. With such data, Kenley argued that models based on the construction schedule, with the terms and conditions of contracts and the project performance data are more suitable. This is what will be developed in this chapter.

Cash flow models

Practical cash flow models

The term *practical cash forecasting* has been used (Kenley, 2003: 40) to describe the method of modelling gross cash flows from schedules and cost data. Practical cash forecasting "is an idiographic approach requiring the preparation of a detailed, priced, work schedule for a project. The calculation of the cumulative costs according to the project work schedule provides the cash flow profile." Practical cash forecasting has been used by only a few authors to analyse project gross cash flow, such as Kennedy et al., 1970; Peterman, 1973; Reinschmidt and Frank, 1976; Ashley and Teicholz, 1977; Berdicevsky, 1978; Peer, 1982. Of these, only Ashley and Teicholz (1977) used the method for net cash flows. However, the method is commonly used for modelling the impact of the schedule on the profitability of the project for the contractor (for example, Elazouni and Metwally, 2005).

Net cash flow curves are actually the result of the various flows of cash in and out. Together these form the net cash flow. A representation of the construction of the net cash flow from the components is shown in Figure 9.1.

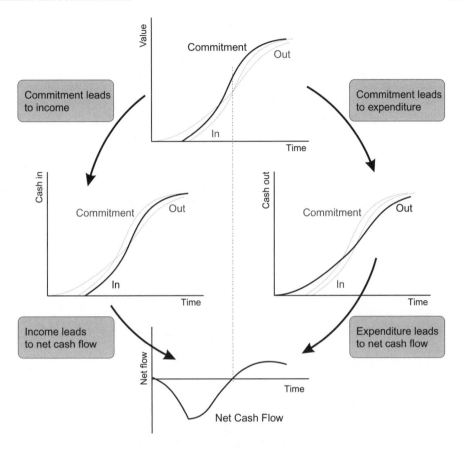

Figure 9.1 The source of net cash flow (Kenley, 2003)

Real time gross cash flow models

While generally recognised as the most accurate method for modelling gross cash flows, practical cash forecasting has, in the past, suffered from the problems of unreliable schedules and un-matched cost data.

An early work (Kennedy et al., 1970) developed a detailed project schedule with full costing based on the bill of quantities. Their model clearly showed the efficacy of developing cash flow profiles for projects from project schedules, and indeed the value of such methods for monitoring project performance during construction. However, they were very concerned about the cost of the method, which they calculated on their demonstration project cost approximately 0.96% of the total contract value; although they anticipated that better systems would reduce that cost to 0.47%. This is still a significant barrier to the method. Furthermore, they argued that a priced contractor's network was essential and 'should a realistic network not be forthcoming or not be planned in sufficient detail to be capable of providing a reasonably accurate cost analysis, the procedure cannot work' (Kennedy et al., 1970). A priced contractor's network is a

fully costed project schedule with all project costs allocated to the activities on
the schedule (Kenley, 2003).

The advent of the location-based management system, with accurate and detailed project
schedules combined with the ability to reflect accurate cost data by locations, now makes
practical cash forecasting of gross cash flows an economic proposition. This is also true for
practical net cash flow modelling.

Real time net cash flow models

The inward cash flow is stepped (Figure 9.2) according to the timing of payments. For the
head contractor, the inflow of cash is driven by progress payments from the client. Usually,
this involves monthly lump sums based on a calculation of the work-in-progress, but
increasingly alternative methods are used, particularly in Europe. For example, it is
common in Finland to derive progress payments from work stages and even location-based
milestones divided by trade. In a location-based management system, the latter offers
significant advantages for both client and contractor.

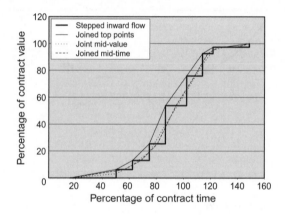

Figure 9.2 Representation of stepped inward cash flow (Kenley, 2003)

The outward cash flow is also stepped, but with many more payment events and thus a
smoother stepped curve results. Putting the two components together results in a
"sawtooth" effect as the cash flow position jumps up and slides down as cash comes in and
out respectively. The net cash flow which results is shown in Figure 9.3.
       The location-based cost control system is able to use data from baseline estimates,
project data and forecasts to calculate cash flow.

THE LOCATION-BASED COST CONTROL SYSTEM

The importance of the practical cash flow method is that it recognises that the only way to
get accuracy in a net cash flow system before the actual accounting data is available (in
other words, to forecast the cash flows) is to use a detailed and accurate schedule with
related reliable cost data. Location-based control requires real-time data and cannot wait for
accounting data, therefore it must be based on the location-based schedule and appropriate

cost forecasts. The cost control data must be available in sufficient time to forecast trends, to enable reaction before the forecast turns into reality. Therefore the cost estimate and forecast must be as real time as possible.

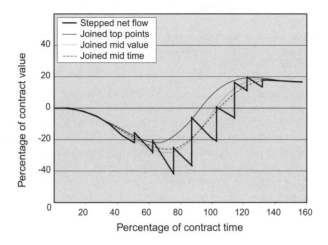

Figure 9.3 Representation of stepped net cash flow (Kenley, 2003)

The location-based cost control system therefore requires costs based on actual commitments from subcontractors and material suppliers, and the calculation of direct labour and equipment. This is a combination of estimates or targets, commitments, actuals and forecasts—closely resembling the scheduling phases of information: baseline, current, actual and forecast. The resulting cost forecast can be used to directly forecast the economic success or failure of the project.

As in all other parts of the control system, the cost forecast should be able to use data from the baseline, current, actual and forecast stages of information. Furthermore, the ability to use the location as the data container provides a level of accuracy not available in other cash flow systems.

Because the LBMS uses the bill of quantities to drive production control of location-based tasks, it is not necessary to monitor and forecast costs using exactly the same logic as the production schedule. For example, concrete is used in many tasks of the schedule. If it is procured from the same supplier, it makes sense to track and forecast costs of concrete separately from the foundations or pour deck tasks but using production data from those tasks. This requires a new concept of a cost tracking task, which is closely related but not necessarily identical to the baseline, detail, procurement and overhead tasks described earlier in Chapters 5, 6 and 8.

Cost tracking tasks

An emphasis on cost management is often considered detrimental to the production control aspects of schedules (for example Murray, 2007). To ensure this is not the case with the LBMS, a special task type should be used. Cost tracking tasks are additional special purpose tasks which use the same quantity information as procurement and schedule tasks but which are specifically designed to monitor and forecast costs. This new task type allows schedulers to do their work without worrying about other users such as cost managers.

Cost tracking tasks contain quantity items and resources, and are often identical in content to either baseline tasks or procurement tasks.

- A cost tracking task that is identical to a baseline task may be a high cost task which will be completely produced by one subcontractor.
- A cost tracking task that is identical to a procurement task may be used when the same material is being used in multiple schedule tasks and by multiple subcontractors or a mix of subcontractors and direct labour.

Sometimes costs need to be tracked in an entirely different way. A typical example is direct assisting labour—in this case, assisting labour could be grouped by construction phase, for example the assisting labour of the foundations phase could be tracked as a single entity.

Most often, cost tracking tasks are formed to track the costs of work and/or materials produced by a [single]supplier. However, in some project types it can make sense to form cost tracking tasks by production task or even by location. For work done completely by direct labour, material related to the task may be part of the same cost tracking task.

Cost tracking tasks are used to track and forecast the following data:

- Baseline costs
- Current costs
- Committed costs
- Actual costs
- Forecast costs.

The following sections outline the cost data for each of these stages.

Baseline costs

Baseline costs are the planned costs derived from the baseline schedule and the estimates of its resource commitment. It is calculated for each cost tracking task by summing the costs of all included baseline quantities. This provides the raw cash flow forecast and represents a best guess at the start of the project. The baseline cost serves as the budget for the cost tracking task.

The modelling of baseline costs based on quantities (cost loading) was described in Chapter 6 (starting page 167). In the pre-planning phase, accurate information about the quantities or production methods is often not available. In order to estimate costs accurately, it becomes critical to update the cost information in the planning system as new information becomes available.

The plasterboard wall example from Chapter 6 is continued in this chapter. The baseline schedule task assumed subcontracted work, supporting logistics to be sourced from direct resources, as well as materials sourced from suppliers according to the company's volume purchase agreements. In cost tracking, components of this task could be tracked in five cost tracking tasks:

- Plasterboard walls subcontract
- Boards
- Studs and frames
- Wool materials
- Finishes work assisting labour.

Each of these cost tracking tasks could also contain quantities from other schedule tasks—for example, the subcontractor cost tracking task could contain items from any other schedule task on which the same subcontractor was working. Similarly, finishes work assisting labour, boards and studs and frames may be used in multiple schedule tasks. For simplicity, the example presented in this section will only include items related to the production of one schedule task.

The estimated unit costs and quantities for both materials and labour subcontracts are shown in Table 9.1.

Table 9.1 Baseline costs

Cost tracking task	Item	Unit cost (€)	Quantity	Total cost (€)
Finishes work assisting labour (direct)	Supporting logistics	25.00	172.5 h	4,313
Wool material supplier	Wool	0.80	1150 m²	920
Boards material supplier	Boards	1.20	2300 m²	2,760
Frames and studs material supplier	Frames and Studs	0.50	4715 m	2,358
Plasterboard wall subcontractor	Installation of walls	22.00	1150 m²	25,300
			Total	35,650

Current costs

Current costs are equivalent to the baseline costs, except that the schedule is more accurate having been refined to suit the detail planning and broken down into detail work activities as necessary. Furthermore, the current costs are based on the actual commitments by the general contractor to the suppliers and subcontractors, and thus represent a better level of both detail (current tasks) and resource information (based on let contracts) as they use detail tasks, current bill of quantities and contract information. This remains an estimate as no actual cost information is available.

Current information is updated to the plan by using detail tasks and the current bill of quantities, as described in Chapter 8. Detail tasks contain more accurate information about the actual implementation. The cost model for baseline tasks (Chapter 6) can then be directly applied to calculate current costs in each location.

Committed costs

Committed costs are very similar to current costs. Committed costs are those costs which are bound by a contract, so they are in some sense more certain than cost estimates without contract information. The location-based management system handles committed costs by the combination of procurement tasks with contract information.

A procurement task, as described in Chapter 6, is composed of completed elements or the resources required to make elements. Procurement tasks can relate to multiple baseline or detail tasks but each always has a single supplier. Current information for the procurement task is based on current quantities until an agreement is let with a supplier, when the costs related to the procured items become committed. Committed costs are based on information about contracts let with the subcontractors or material suppliers. The information should always include:

- Type of contract: total price (fixed lump sum), unit price or resource-based
- Scope of work (quantity items)
- Agreed total price or unit price for each element or resource.

For better cost forecasting, the LBMS can optionally use the following information:

- For work items:
  - Location-based milestones for each schedule/detail task + penalty per day (or bonus for early completion)
  - Quantity-based milestones for each schedule/detail task + penalty per day (or bonus for early completion)
  - Compensation for waiting time (man hours)
  - Location-based bonus for the general contractor if the subcontractor can start on the planned date (location is clean, all prerequisites needed for that location have been completed by the general contractor)
  - Bonus (or penalty) to the general contractor for work continuity (discontinuity).
- For materials:
  - Initial delivery dates
  - Penalty for delivering late (per day)
  - Bonus to the general contractor if material can be delivered on the planned date
  - Agreed payment schedule.

The list above includes items that are specific to location-based projects. Their use in contracts will be explained in detail in Chapter 12.

   Committed costs use any available details of an agreement as the cost basis, once an agreement has been made. In the case of a unit price contract, the unit prices are locked for each quantity item. If the contract is a lump sum contract, the total price of the quantity items is locked. If the contract is resource-based, the unit price of resources is locked but the use of the resource is calculated from the schedule.

   Committed costs can vary from the baseline if either the quantities change (on unit price or resource-based contracts) or if the schedule changes sufficiently to trigger penalties or bonuses for the general contractor or subcontractor(s). Costs related to resource-based contracts, such as the use of the tower crane, change when the duration of the tasks requiring the crane changes.

   Committed costs for direct labour or mobile equipment are a special case. For example, the commitment for labour is an employment agreement. However, the committed costs for any single project are only the salaries related to the project which are based on actual hours which have not been paid. The same is true for equipment.

   In the plasterboard wall example, the agreement with the subcontractor could have the following information:

- The subcontractor will install boards, frames and work for the unit price of €26 per m². The subcontractor will also handle the deliveries of these materials to work areas. All additional work authorised by the general contractor will have a price of €30 per hour.
- For each floor which becomes available according to the schedule, and is clean for layout work to begin, the general contractor receives a discount of €250.
- For each floor which is available and clean for installing the second board, and for which insulation wool has already been delivered according to the schedule, the general contractor gets a discount of €250.
- Milestones for installing first board:
  - 2nd floor: 08-12-2005

- • 5th floor: 04-01-2006
- • Milestones for closing the wall:
  - • 2nd floor: 20-12-2005
  - • 5th floor: 17-01-2006
- • Each milestone has a penalty associated with not completing of €100 per day.
- • Payments for work completed will be due when milestones are realised. Half will be paid based on installing the first board and half based on installing the second board. If the start date of a location is delayed by factors outside the control of the subcontractor, the milestones will be adjusted based on the planned production rate.

Insulation wool will be bought according to the volume purchase agreement for the total price of €0.80 per m^2 and delivered to the work area by the general contractor before the scheduled start of work in each area. The agreed schedule (modelled using detail tasks) is shown in Figure 9.4. Plasterboard wall related tasks are shown in black. Immediate predecessor and successor tasks and electrical rough-ins between the sub-tasks are shown in grey.

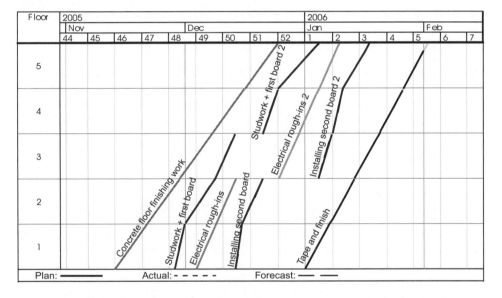

Figure 9.4 Schedule

Committed costs, based on the agreement and detail schedule, are shown in Table 9.2. and total €29,770, which is €5,880 less than budgeted (baseline). Note that as long as the detail tasks and their forecasts stay exactly as agreed, the general contractor is expected to gain the maximum discount available and the subcontractor is not expected to pay any penalties.

Actual costs

Actual costs are based on approved and paid invoices and paid salaries that are related to the project. Because invoices and salary payments necessarily lag production (with the exception of prepayments), it is important to know to which part of production the payments relate. This requires the payments to be allocated to cost tracking tasks and linked to completed quantities. Linking to completed quantities can be implemented in many ways depending on the information content of the invoice. The most common cases are described below.

Table 9.2 Current costs

Item	Type	Unit cost (€)	Quantity	Total cost (€)
Installation of walls + boards, frames and studs	Subcontract	26	1,150 m²	29,900
Subcontractor additional work	Subcontract	30	0 h	0
Wool	Material	0.8	1,150 m²	920
Supporting logistics	Direct labour	25	58 h	1,450
Location clean and available according to schedule	Discount to GC	250	10 locations	−2,500
Milestone penalties	Penalty	100	0 milestones	0
			Total	29,770

Prepayments

Prepayments do not, by definition, relate to actual production, but are paid before work or deliveries begin—usually to finance the subcontractor or supplier. They need to be allocated to cost tracking tasks but do not need to be linked to specific quantities or production. In forecasting, prepayments are added on top of the production-based invoices when calculating the total forecast for the task.

Material payments

Material payments are not necessarily related to actual production because deliveries can occur and be invoiced before the commencement of production. Material invoices normally refer to quantity items and unit prices. It is often unrealistic to expect that the information would correspond exactly to the bill of quantities used in scheduling the project. In this case, it is sufficient to say that a certain percentage of quantities related to the cost tracking task has been delivered and invoiced. In ideal circumstances, the invoiced and delivered items match with planned quantity items and the analysis can be done at the quantity item level. In forecasting, the delivered percentage of either the complete cost tracking task or the individual items can be compared to the physical completion rate based on the finished production. If deliveries are planned in the location-based management system, it is possible to link the invoices directly to deliveries.

Subcontract payments

Invoices that are related to subcontracts should be allocated to a cost tracking tasks and either the time period, the location or the quantity items should be used to establish a link to production.

Linking an invoice to a time period assumes that all actual production related to the cost tracking task and completed during the time period is included in the invoice.

If an invoice is linked to locations, for example a location-based payment schedule, then the costs can be linked to all production in the cost tracking task in selected locations.

In unit-price contracts, it is often possible to link the invoice to the completed quantity items directly. In this case, the payment should also be linked either to a time period or to locations to check if the actual quantities match with the invoiced quantities.

Equipment

Equipment rental can be internal to the company or external (based on invoices). Equipment is generally more complicated than subcontracts because it is often used by multiple cost tracking tasks. Equipment is most commonly paid based on time used, so the invoiced hours need to be allocated to the cost tracking tasks. The time period of invoiced hours need to be known in order to link the equipment usage to the correct period of production.

Salaries

Salaries are similar to equipment rentals. Paid hours need to be allocated to cost tracking tasks and to a time period.

Relationship between actual and committed costs

Committed costs can be decreased by the sum of the actual costs which have been allocated to a cost tracking task with its associated original commitment. Therefore, actual costs act to decrease the balance in commitment. Normally, actual costs decrease the committed costs by the same amount but in some cases, costs actualise which are outside the original commitment. For example, additional work could be done or a variation could result in invoicing outside the scope of original commitment.

Cost forecasts

Cost forecasts can be calculated based on both actual and committed costs as well as other available information. The cost forecast is calculated by assuming that production will continue in the same way, just as for other forecasts in the location-based management system. Cost forecasts are for all production which does not yet have linked actual costs, however costs for some completed work will also need to be forecast, because actual costs lag production.

The cost forecast extrapolates to future production based on actual costs up to this date. Thus it also takes the schedule forecast into account when forecasting consequential overheads and interference costs. This is the same as the forecast stage of information for the schedule. Effectively, this means that cost forecasting requires estimating costs and combining this with the timing from the schedule forecast—a sort of double forecasting effort.

Cost can be forecast for each cost tracking task. The project cost forecast can be then calculated by summing together all these various task forecasts. The forecasting procedure depends on the availability of information.

Cost forecast 1: tasks without commitment information

The cost tracking task's current quantities and unit costs can be used in forecasting as long as there is not more accurate information yet available. This is the case when subcontracts or material deliveries have not yet been agreed. The cost forecast can be adjusted if earlier tasks have consistently exceeded (or been consistently lower than) their budgeted cost and over 30% of the project has been completed (Pekanpalo, 2004). In this case, it is safe to assume that there is a common underlying cause to cost escalation (or savings) and the cause is likely to lead to further cost overruns (or savings) in the upcoming tasks.

Cost forecast 2: for tasks with commitment information, production not started

The second forecast is based on committed costs. The cost forecast is based on the current quantity and committed unit prices or committed total cost for the cost tracking task. Therefore, any quantity changes after commitment will change the cost forecast of the cost tracking task.

Note that if the contract is unit-based, the quantity items for the linked detail tasks should be updated to correspond with the quantity items in the contract.

Cost forecast 3: for tasks whose production has started

After production has started on a task, invoices and paid salaries combined with production status information can be used in forecasting, in the same way that earned value analysis uses percentage completion of activities. The forecasting process works in two phases: first, the quantities are forecast based on actual quantities in locations and second, the costs are forecast separately for each cost type: subcontract, material, equipment and labour.

In the location-based management system, actual quantities are measured for each location. If the quantity exceeds the planned quantity in a location, it can be assumed that the quantity overrun will repeat in the succeeding locations unless control actions are taken. After each completed location, the quantity overrun ratio can be calculated for each quantity item as follows:

$$R^Q = Q^A / Q^P \qquad (9.1)$$

Where:

R^Q = Quantity overrun ratio
Q^A = Total actual quantity
Q^P = Total planned quantity.

The same quantity overrun is expected to occur in each subsequent location. The quantity forecast for each location which has not finished is:

$$Q^F = R^Q \times Q^E \qquad (9.2)$$

Where:

Q^F = Quantity forecast
R^Q = Quantity overrun ratio
Q^E = Current estimate of quantity.

Subcontractor costs can be forecast based on forecast quantities in unfinished locations. First, the work already invoiced by the subcontractor is determined. This is calculated by using the time period and location data of the invoice allocation. Forecasting considers only quantities, time periods and locations which are not covered by approved invoices. After invoiced and remaining quantities have been identified, the actual unit costs for invoiced work can be calculated. Normally, actual unit costs should be the same as committed unit costs but can be different if additional work outside the scope of original commitment has been done. Actual unit costs can be used in forecasting if costs have been allocated to quantity items. If an invoice has not been allocated to quantity items, a cost overrun ratio is calculated instead:

$$R^C = C^A / C^{QC} \tag{9.3}$$

Where:

R^C = Cost overrun ratio
C^A = Total actual cost
C^{QC} = Total committed cost for included quantities.

This overrun ratio is applied to all remaining quantities to create the subcontractor forecast.

Material costs are more difficult to allocate to specific production because they often occur before production. However, they can be allocated to delivered quantities to identify actual unit costs. This allocation can occur either at the quantity item level or as an aggregate (for example, the percentage of materials delivered). Actual unit costs can differ at the building element level if the composition of the element has changed (for example, if a square metre of wall requires two boards on each side instead of one board, or if the material waste has been higher than planned). The quantity forecast and percentage of material delivered thus far can then be used to estimate future deliveries, applying the same unit prices. If delivery information is not available, the material cost forecast can be calculated by using forecast quantities and committed unit prices.

Labour and equipment hours are forecast by calculating the hours used per unit of production. This is calculated by allocating total salaries or equipment costs to the time period of production. The same pattern of hours with the same average unit rate can be used for forecasting future labour and equipment usage.

The total cost forecast for the cost tracking task is the sum of its subcontractor, material, labour and equipment forecasts.

Example

In the plasterboard walls example, the first location of studs and board on the first side has been completed and the second location is on the way as shown in the flowline in Figure 9.5, complete with actual and progress lines as well as alarms to highlight problem task-locations. Even though the payment schedule has the first payment only after the second location, the subcontractor has sent an invoice after the first location, because locations were not released to him on time (the general contractor also forfeits a discount of €250). The invoice involves work done between 1st December and 8th December and is time-based. Insulation wool has been delivered to the first location and €320 has been invoiced. Salaries have been paid for the time period of two days, 1st and 2nd December and a total of 30 hours have been allocated to the task. Actual costs related to plasterboard walls are summarised in Table 9.3.

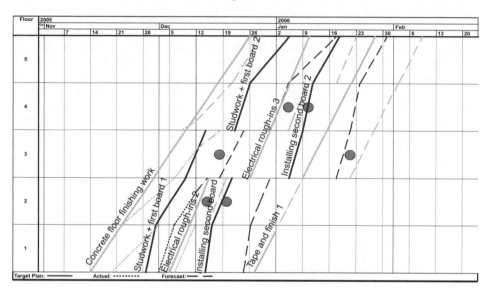

Figure 9.5 Actual progress and forecast (with alarms)

Table 9.3 Actual costs—Floor 1, installation of studs and first board

Item	Type	Unit cost (€)		Planned Quantity	Invoiced quantity	Planned cost (€)	Actual cost (€)
Studwork and first board	Subcontract	13	1–8 December	344 m²	320 m²	4,472	4,160
Insulation wool and second board	Subcontract	13		0 m²	0 m²	0	0
Subcontractor additional work	Subcontract	30	1–8 December	0 hr	30 hr	0	900
Wool	Material	0.8	Delivery 1	100 m²	400 m²	80	320
Supporting logistics	Direct labour	25	1–2 December	0 h	30 h	0	600
Location clean and available according to schedule	Discount to GC	250	Floor 1	1 location	0 locations	−250	0

On the surface, it appears that costs have badly overrun (actual cost of €5,980 compared to the planned cost of €4,302). However, by applying the forecasting rules above separately to the subcontract, material and direct labour, a different total picture emerges.

First, the quantities are forecast. The quantity forecast uses the measured quantity in each location and is not influenced by actual quantities produced during any time period of production. In this example, let us assume that the first location has been measured and actual quantity was 90 m^2 (planned 100 m^2). The quantity decrease was due to updated drawings, with one wall replaced by a masonry wall and changed locations for some other walls. The quantity forecast for the remaining locations is 90% of planned, assuming similar changes are to repeat in other locations.

The subcontractor cost forecast can then be calculated based on the quantity forecast and actual unit prices for quantities produced thus far. Because the subcontractor invoiced for work additional to that authorised by the general contractor (arising from a change in the wall location) the actual unit price is different from that committed. The actual unit price for studwork and first board is €15.81 per m^2 (invoice total €5,060 divided by quantity produced during invoice period 320 m^2). By applying this unit rate to the total quantity forecast, the cost forecast for studwork and first board is €16,367 and the cost forecast for insulation wool and second board is €13,455. Note that for insulation wool and second board, the original unit price must be used because there is no production information available at this point. The total subcontractor forecast is €29,821 compared to original plan of €29,900. The forecast assumes that unless control actions are taken, similar changes will occur in later locations.

Materials need to be forecast differently. Although much more was delivered than required for the first location, quantities cannot be allocated to production because no wool has been installed at the status date. Because the actual unit rate for wool is the same as the commitment, the forecasting method uses the total quantity forecast 1,035 m^2 and multiplies it by the actual unit rate of €0.8 per m^2 to arrive at the material forecast of €828 (planned €920).

The direct labour forecast allocates hours spent to actual production. Because salaries are paid fortnightly, lagging one week after production, hours are known for only two days of production. A total of 30 hours have been allocated to the task and the corresponding production is 62 m^2. This means that 0.48 hours of direct labour have been spent for every m^2 of wall. Actual salaries paid were €600, so the unit price has been €20 instead of the planned €25. However, if the present behaviour continues, 497 hours (0.48 hours per m^2 × quantity forecast 1035 m^2) will be spent during production of studwork and first board. The labour forecast is 497 hours × unit price of €20 per hour + 29 hours × unit price of €20 per hour for the second board (no production information yet). This equals €10,600, compared to plan of €1,450.

The total cost forecast for plasterboard walls is the sum of these forecasts. This equals €41,169, compared to the target of €35,650 and the previous forecast of €29,770, which was based on commitments. The use of direct labour in this task needs to be examined closely to find the reason for the cost drain. It may be that the subcontractor is using assisting workers to do part of their work, or logistics may have been organised ineffectively. It is critical to take action now instead of waiting for the situation to correct itself. The forecast based on commitments would not identify this problem.

Making adjustments for location-based penalties and bonuses

It is assumed that any deliberate workflow discontinuities that have been planned and shown to the subcontractor will already have been incorporated into the original unit costs or the total cost for the bid. However, if additional discontinuities arise because of the impact of late preceding tasks, or if prerequisites for construction are not ready, there will be extra costs incurred by the subcontractor. The flow-on effect on the general contractor's costs depends on the contract terms and milestones. Additionally, if resources leave the site, there is risk of a return delay, with associated effects on succeeding subcontractors and, ultimately, on overhead costs for the project and additional costs caused by rushing to complete on time at the end of the project.

Compensating for waiting time

If there is a clause in the contract compensating for idle time, the interference cost is the same as the added production system cost:

$$C^I{}_W = N^R \times C^L \times T^I \qquad\qquad (9.4)$$

Where:

$C^I{}_W$ = Interference cost of waiting
N^R = Number of resources interfered with
C^L = Hourly cost of resource
T^I = Duration of the interruption.

The interference cost can be directly added to the cost forecast.

Location-based milestones

Both the general contractor and the subcontractor can have location-based milestones with associated rewards or penalties. General contractor milestones ensure availability of locations at the promised time. If the location is not clean and available when promised to the subcontractor, the general contractor pays a penalty for each day or week of delay. This penalty can be calculated based on contractual milestones and forecast start and finish dates for each location. If the forecast start has shifted for some other reason than a variation in the production rate, the forecast cost should be adjusted to take the penalty into account.

The subcontractor commits to a location end date providing they are able to start that location as indicated in the contract. If the location is not finished by the subcontractor when required, they pay a penalty for each day or week of overrun. Again, it is straightforward to calculate the penalty based on contract information and forecast start and finish dates.

Modelling cost becomes more difficult if the subcontractor is unable to begin on time due to the general contractor delaying the handover and where multiple locations are released simultaneously. In this case, the planned production rates along with the same resources and assumed continuous work can be used to calculate new milestones.

The principle applied here is that the subcontractor should not be penalised for being late where they have not been provided access as planned. Milestones, the basis for calculation of penalties and bonuses, must be recalculated accordingly.

Previously, Figure 9.5 showed progress and forecast data after two weeks of production for the plasterboard walls example. The preceding task, concrete floor finishing work, has been delayed. This has caused plasterboard walls to start late in the first location. Also the start dates of many other locations are forecast to be late. In this case, the location-based milestones will be shifted forwards assuming the planned production rate and sequence. Figure 9.6 shows a new plan with the same logic, and it may be noted that the new plan raises fewer problem task-location alarms as more milestones can now be met. All the detail tasks have been moved forwards to their forecast start dates assuming the originally planned resource use. New milestone dates can be taken from the schedule:

- Milestones for installing first board:
  - 2nd floor: 12-12-2005
  - 5th floor: 12-01-2006
- Milestones for closing the wall:
  - 2nd floor: 29-12-2005
  - 5th floor: 26-01-2006

When new information becomes available about the actual finish dates of preceding tasks, the contractual milestones may be revised accordingly. Currently, the schedule forecast of first board of the second floor is delayed by one day, resulting in a forecast penalty of €100. The forecast for the fifth floor shows it delayed by three days, resulting in a forecast penalty of €300. Closing the wall on the fifth floor will be delayed by three days with an associated penalty of €300. This is effectively revising subcontractor commitments when control actions are unable to be taken to otherwise prevent problems.

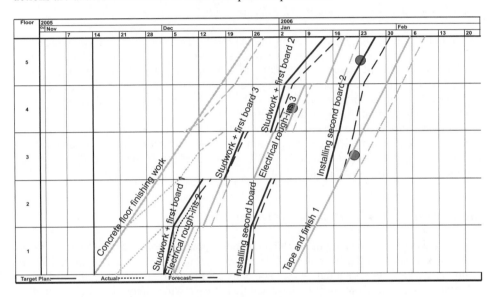

Figure 9.6 Adjusted forecast after two weeks of production

Production rate (quantity-based) milestones

Production rate milestones can be utilised instead of location-based milestones. In this case, the production quantity target for each day or week is defined. Achievement of milestones should be evaluated at agreed periods (for example weekly, biweekly or monthly). On evaluation, the forecast cumulative produced quantity should have reached at least:

$$Q = \sum \left(\frac{R^P}{T_i^P}, i = 1, n \right) \qquad (9.5)$$

Where:

Q_{P} = Cumulative produced quantity at period n

R_i^{P} = Production rate

T_i^{P} = Period interval

When production is constant.

$$Q = \frac{R^P}{T^P} \times n \qquad\qquad\qquad (9.6)$$

If the forecast quantity is less than the target, the penalty will be evaluated based on the number of additional days required to achieve the target quantity.

Bonuses for continuity and penalties for discontinuity

In addition to milestones, many location-based projects incorporate some form of incentive for the general contractor to release locations according to the schedule. If a location is available, clean and all prerequisites required for the location have been completed by the general contractor on the agreed date, the general contractor gets a discount. Alternatively, if the next location is not ready to be worked when required by the schedule, the general contractor may commit to paying a penalty to the subcontractor.

Figure 9.5 shows the forecast of the preceding task in the example. Because the start dates of all locations are delayed, the general contractor is forecast to lose all bonuses for installation of first board, a total of €1,250, and bonuses from second to fifth floor for installation of second board, a total of €1,000.

Non-location based contracts and come-back (return) delay

If the contract does not say anything about compensation and does not have location-based milestones, it is likely that workers will leave immediately once work is unavailable and return only when there is enough work to be done continuously. In this case, the cost effect is indirect and results from shifting the schedule forecast, which may cause delays to other tasks and finally an increase in the forecast overhead costs. The effect on the schedule forecast depends on how aggressive we want to make the assumptions about come-back delay. Some possible examples:

- a fixed come-back delay (such as five days from interruption).
- a location-based come-back delay (such as when the schedule forecast can go continuously through three locations).
- a time-based come-back delay (such as when the schedule forecast can go continuously for three weeks).

Cost forecasting of overheads

After all the schedule forecast modifications have been made, the durations of the overhead tasks may be recalculated based on the schedule forecast. The cost forecast is then based on the committed unit prices of overhead items.

As an example, the committed monthly price for rental of the tower crane (including operator) is €20,000 per month. The original schedule has a total duration of 12 months for those tasks requiring a tower crane (structure, façade, roofing). The production of structure starts slowly and the forecast roofing end date is consequently delayed by three months. The forecast cost for the tower crane increases from €240,000 to €300,000.

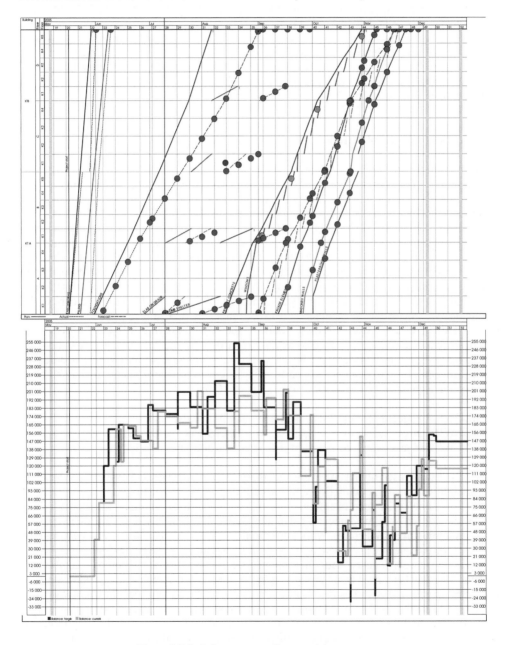

Figure 9.7 Cash flow corresponding to activity progress

Forecasting cash flow

Finally cash flow can be modelled by using the forecast start and finish dates of locations and the forecast production rates. The forecast amount of expense payments can change if quantities have changed, if overhead task durations change or if the subcontractor is compensated for idle time. Income payments usually stay the same but their timing may change based on schedule task forecasts.

Figure 9.7 shows the cash flow example from Chapter 6 (see Figure 6.4) with actuals and forecasts. There have been some deviations, for example foundations are going slower than expected due to an error in the quantity take-off. This causes higher costs because of higher quantities, and also delayed expense and income payments. The effect on the net cash flow balance is shown in the bottom part of the figure.

COMPARISON OF TASK-BASED FORECASTING TO EARNED VALUE

Chapter 4 introduced the development of schedule-related cost control methods. In particular, the very common method used for balancing the conflict between time progress and cost progress was discussed in the section commencing page 104. It can be argued that the earned value method is an approximation required to control production in the absence of sufficient progress data. This approximation is no longer required when using the location-based management system—task-based forecasting provides much more production-related progress data than is available to activity-based earned value forecasts, although there remains a place for the calculation of earned value as an indicator. Whereas earned value is more concerned with the overall production cost performance, task-based forecasting gives early warning of cost leaks, thereby allowing rapid reaction to correct the situation well before it would otherwise be seen in the earned value charts.

Earned value measures the cost performance by comparing estimated costs of the project schedule (budgeted costs) to actual expenses until the current date (actual costs). Schedule performance is measured by comparing the planned schedule (work scheduled) to the work completed in reality (work performed). These four measurements are combined to create three curves (see Figures 4.6 and 4.7, page 106):

- BCWS (budgeted cost of work scheduled—the baseline)
- BCWP (budgeted cost of work performed—schedule performance)
- ACWP (actual cost of work performed—cost performance).

In the location-based management system, these curves can be generated using the baseline stage information and cumulative actual accounting costs up to status date.

It can be easily seen from the above measures that they are much rougher (generic) than the context-specific forecasts calculated by the location-based management system. For example, the actual cost of work performed (ACWP) only considers total cost, regardless of the commitments on which the costs were based. Because some costs occur before production and some costs occur after production, it is very difficult to compare actual costs to the schedule because it is not possible to know which costs match which production. The schedule performance indicator of earned value does not take into account where work has been performed or if the correct work was performed—it standardises everything based on money units and the worth of work performed, without considering criticality or production flow. It is also silent about production efficiency, because actual costs are based on commitments and contracts which may assume wasted productivity. Earned value only evaluates performance against budgeted performance, but that budget may have contained a lot of waste. Earned value makes forecasts solely based on historical progress. It does not consider future commitments or forecasts—for example, the effects of longer task durations on overhead costs are not considered at all.

To see the difference more clearly, let us consider again the plasterboard walls example. Table 9.4 shows calculations of BCWS, BCWP and ACWP based on the baseline. Figure 9.8 shows the EVA diagram. Three variations can be calculated:

- Schedule variance   BCWP – BCWS =   –€3,946
- Cost variance   BCWP – ACWP =   –€220
- Spend variance   BCWS – ACWP =   €3,626

From these variances it seems that less money has been spent than scheduled, the schedule is behind the baseline and production is costing a little bit more than expected.

Table 9.4 Earned value data (plasterboard example)

	BCWS (€)	BCWP (€)	ACWP (€)
2 December	4,313	1,240	0
9 December	9,706	5,760	5,980
16 December	15,099		
23 December	20,492		
30 December	25,885		
6 January	31,279		
13 January	35,659		

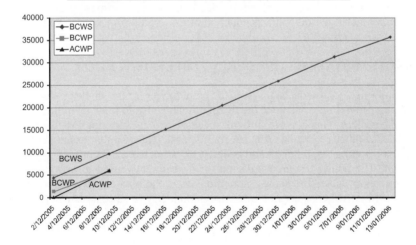

Figure 9.8 Earned value comparison for plasterboard

Instead of deviations, most applications of earned value use performance indicators in forecasting. The most common measures are schedule performance index (SPI) and cost performance index (CPI):

- Schedule performance index   BCWP/BCWS =   0.59

- Cost performance index   BCWP/ACWP =   0.96

Both of these indices are below 1, so both schedule and cost performance are lower than expected. These indices can be used to calculate the schedule and cost forecasts. A commonly used method is to divide the original budget by the cost performance index. In this example, the end result forecast would be original baseline of €35,659 divided by CPI 0.96 which gives €37,145. This can be compared to the location-based cost forecast of €41,249 which is 11% higher.

The example shows that location-based cost forecasting reacts much faster to deviations, by considering more information in creating the forecast. For schedule forecasting, the original duration divided by SPI has been occasionally used. In this example, this would be 33 days divided by the SPI of 0.59 which equals 55.9 days. In comparison, the location-based forecast calculation is based on the detail tasks that have actually been following the planned production rate. The location-based forecast was 36 days. In this example, the earned value cost forecast is less than the location-based cost forecast, and the schedule forecast using earned value is more pessimistic than the location-based forecast.

It is clear that earned value, while a good guide, does not consider precedence or any external factors affecting the schedule, so it cannot be used for accurate schedule forecasting. The location-based schedule forecast uses the same principles of comparing actual production rate to plan but it also considers precedence and resources. The schedule performance index itself is not a very useful measure, even in aggregate, because it gives the same weight to all activities without considering criticality or downstream effects. The cost performance index can be useful for pointing out problems with the overall cost performance of the project. However, it should be used with caution because it assumes that future production behaves similarly as past production without considering task or construction phase differences or actual commitments. There is also the difficulty of some costs occurring before, and some lagging behind, production. However, it may be argued that these effects cancel each other out when production of the overall project is considered.

PRODUCTION SYSTEM COST DURING IMPLEMENTATION

To make informed control decisions from the point of view of the whole production system, it is necessary to not only calculate the direct costs but also to track the production system cost for the project. This is more difficult, because it is usually not possible to know the actual costs of other companies (the subcontractors). Thus, similar assumptions as those used in the planning stage have to be adopted to estimate the labour and equipment cost component. The actual resource utilisation can be directly used to estimate the actual production system cost. The same assumptions as in the plan can be used in forecasting, combined with actual consumption rate information. Overhead costs can be forecast in the same manner as above, based on the schedule forecast.

Current production system cost

The current production system cost can use exactly the same rules as the baseline production system cost (see Chapter 6). However, detail tasks are used instead of baseline tasks for the calculations. Comparing the difference between the baseline and current production system cost, for each subcontractor, shows how well the baseline productivity targets can be achieved with current plans.

Actual production system cost

The actual production system cost can be calculated by multiplying the used labour and equipment hours from each subcontractor by the assumed unit cost of labour or equipment. For subcontractors, this is often an approximation because it may not be feasible to track crew sizes daily. Also, this measure does not distinguish between working time and the various waste factors. Actual production system cost can be compared to baseline and current production system cost to see how well productivity targets were actualised. Actual production system costs include waste (time spent doing non-productive activities such as relocation [location demobilisation and mobilisation], double-handling, moving between locations, etc.) which may be affected by planning quality as well as implementation.

Forecast production system cost

The production system cost forecast can be used to reveal future loss of productivity, assuming no control actions are taken. This is the most important method during the implementation phase because it can evaluate the cost effects of any planned control actions. The calculations are the same as the baseline calculations described in Chapter 6, but use the detail task forecast start and end dates and any control action resources in the calculations.

MODELLING THE COST EFFECTS OF CONTROL ACTIONS

While the schedule forecasts are focused on forecasting the consequences of current performance into the future, with a particular emphasis on achieving milestones, cost forecasts can be used to optimise for cost as well as time.

Schedule and cost forecasts are not useful unless they result in corrective action. Control actions may be planned to minimise both the time and cost impacts of deviation. Each control action incurs a cost to one or more actors in the production system. This cost may be caused by the need to work overtime or to add more resources, or through one of the waste factors related to inefficient production. To evaluate all these effects, the control actions should be evaluated both by using the production system cost and by measuring the actual cost effects. Production system costs can differ from actual costs for any actor, depending on the contractual relationships. For example, a subcontractor working overtime results in a direct cost to that subcontractor. The general contractor may not have to compensate for this additional cost, depending on the contract and the particular circumstances. However, it is useful to understand how the costs for other actors are affected by the control action plan. It is always easier to implement a control action plan when it is optimal for everyone. If the savings are larger than the costs, and all actors are rational, a way to distribute the cost will usually be found.

Alternative control actions, with their associated impacts on the schedule and actual cost and production system cost, should be evaluated separately for each affected detail task. Production system cost and schedule effects can be modelled by recalculating the production system cost forecasts with different resource quantities, sequences and shift lengths depending on the possible control actions. The actual costs, and who bears the true production system cost, are a matter of negotiation and depend on the contracts made with subcontractors or workers. The subcontractor will certainly be motivated to implement a control action if the forecast cost of the control action is less than the forecast penalty if the control action is not taken. Cost and schedule forecasts are a great way of communicating the need for action. Note that if the subcontracts do not include location-based milestones, the means of control are greatly diminished. The cost effects of various control action types are described below.

Changing the number of resources (assuming the same productivity)

Productivity of individual resources can be assumed to stay constant, thus control actions which vary resources with constant productivity will not directly affect costs. However, from the subcontractor's point of view, there remains an opportunity cost to be considered—the resources might be working in other, more critical jobs. Additional resources may not be available without cost when the resources are committed elsewhere. An alternative for implementing this control action is for the general contractor to supply direct labour to assist the subcontractor, or by contracting an additional subcontractor. This is effectively revising the subcontractor commitments to prevent problems when control actions are unable to be executed successfully.

Changing shift length or working on weekends or holidays (overtime)

Increasing the number of hours which may be worked, as a control action, usually causes additional costs and reduced productivity. The productivity of overtime is significantly less than can be assumed for a regular (8 hour) workday. Furthermore, workers may need to be compensated for overtime work, leading to an increase in cost. However, this control action may still be worthwhile as it may lead to a reduced production system cost, if it reduces disturbance of the workflow.

Changing the location sequence

Altering the location sequence (for example, moving to the third floor rather than the second floor after finishing the first floor) is a control action which may increase the production system cost if the resulting sequence increases the distance between locations. It impacts on both relocation costs and the time for moving between locations.

Splitting a task

Tasks may be split as a control action to allow work in multiple locations at the same time, or to allow a break in the continuity of the task, with workers returning at an agreed later date. Working in multiple locations at the same time by splitting a task into multiple tasks is a valid control action which will not incur additional production system cost as long as it is implemented by mobilising additional crews. However, if the original resources are able to spread over multiple locations simultaneously, production costs will increase because of lost productivity. Additionally, forecasting progress becomes more difficult. Continuity breaks cause both cost and lost time because of mobilisation and demobilisation.

Removing or switching technical dependencies

Sometimes dependency relationships are not followed on site or new ones emerge. However, changing dependency relationships may decrease productivity, increasing the time spent doing the work. For example, switching the order so that painting is commenced after the carpet is installed will decrease productivity because of the need to protect the carpet (and will likely cause quality problems). This cost can be modelled by changing the productivity of the affected detail tasks—the impact is on labour—assuming quality problems are prevented with adequate care. In some cases, relationships can be changed without any loss of productivity, such as work which may occur in any sequence, but not at the same time, in the same location.

Increasing productivity by decreasing non value-adding activities

Optimally, productivity can be increased without incurring additional costs by removing non value-adding activities such as materials handling, moving around the site, etc. However, additional management time is usually needed. Removing such activities only works as a control action if the task is being done inefficiently by including such work.

Shifting the start date of a successor task to make that task continuous

Where a task is forecast to become discontinuous, shifting the start date may present a possible control action to prevent discontinuity arising. This control action may decrease the waiting and mobilisation costs for either this or successor tasks. However, it can require renegotiation of contractual milestones. It can also adversely affect overhead costs if there are no buffers in the production system to absorb the delay.

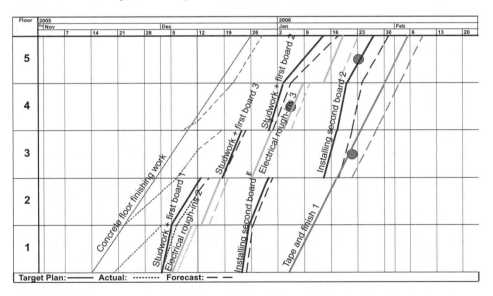

Figure 9.9 Project forecast before control actions

Control action example

The plasterboard walls example can be used to illustrate current, actual and forecast production system costs. The initial baseline task of plasterboard walls had a duration of 33 days with a plan requiring two resources with completely continuous work. Assuming an hourly cost of €25 per hour and an 8 hour day this results in a baseline production cost of €13,200 for the task. In practice, the subcontractor works with just one resource and the detail plan shown in Figure 9.9. The detail plan has been adjusted from the original because the predecessor task (concrete floor finishing) was delayed and a new plan was approved following negotiation. Current production system costs are €7,200 for productive work time (just one person working for 36 days instead of the two assumed in the baseline) added with waiting cost of 3 days (€600) totalling €7,800. The actual production system cost thus far is €1,400 because one worker has been working for 7 days. The forecast production system cost is calculated based on the actual and forecast resource use. Because the installation of studs and first board has been slower than planned, the forecast has shifted. Total work time will be 39 days assuming the current production rate. This results in a working cost of €7,800. On the other hand, waiting time has decreased from 3 to 2 days, so the total waiting cost is €400 and the total forecast production system cost is €8,200.

Table 9.5 Initial cost forecast calculations before control actions

	Baseline (€)	Current (€)	Actual + Forecast (€)
Plasterboard walls			
Work	13,200	7,200	7,800
Waiting	0	600	400
Electrical			
Work	6,600	3,700	3,700
Waiting	0	2,900	3,300
Mob and demob	0		
Tape and finish			
Work	4,600	4,600	4,600
Waiting	0	0	600
General contractor			
Finishes superintendent	9,800	13,953	14,535
Haulage equipment	3,300	3,900	4,200
Total	**37,500**	**36,853**	**39,135**

Table 9.5 summarises the initial calculations for the plasterboard walls contractor, electrical contractor, tape and finish contractor and the general contractor. Both electrical and tape and finish are done by one person with a €25 per hour cost. Overhead costs include the finishes superintendent with a monthly cost of €5,000 and haulage equipment for the plasterboard walls task with a daily rent of €100 per day. Because incorrect productivity assumptions were made in the baseline, the baseline working costs are much higher for plasterboard walls than in the current plan which is based on actual commitments. Duration of the complete work package has increased quite a bit, so the current and forecast overhead costs are much higher than in the baseline.

Control action processes should be employed to try to minimise the total forecast production system cost. One possible control action plan intended to the minimise overall production system costs could include the following changes:

• Change of sequence for plasterboard walls

- Overtime for the plasterboard walls subcontractor to achieve planned production rate in locations 3, 4 and 5
- One planned longer break for the electrical subcontractor.

Figure 9.10 shows the flowline figure with these control actions modelled. Factors affecting the task sequence are incorporated into detail tasks because they will also affect contractual milestones, assuming the initial production rate. Control actions affecting production rate are modelled by changing the slope of the forecast. Table 9.6 shows the total production system cost of this control action plan. The baseline values are unaffected by control planning. The current values reflect current commitments assuming that everyone achieves the planned production rate. The total committed production system cost has decreased for the electrical contractor, because this work was made more continuous and their resource will leave the site for two weeks instead of waiting on site for work. The forecast work cost of plasterboard walls has increased because of overtime work (values assume 50% additional pay for overtime hours). However, changes in sequence have removed all forecast waiting time from the plasterboard walls contractor.

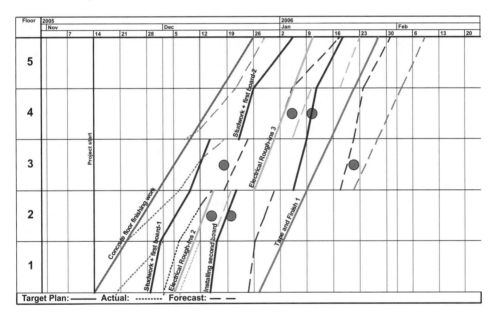

Figure 9.10 Project forecast after control actions

The tape and finish subcontractor had forecast waiting time due to the plasterboard wall subcontractor being delayed. With this control action plan, the forecast waiting cost reduces to nil. Also, the general contractor benefits because the forecast total duration of the work package is shorter, which reduces overhead costs related to the superintendent salaries and haulage equipment. The resulting production system cost is less for all contractors involved, so it should be easy to get to an agreement on the plan. This is facilitated by the contract between the general contractor and the plasterboard walls subcontractor, which includes location-based milestones with penalties. The subcontractor can avoid those penalties by working overtime according to the control plan. The electrical subcontractor should also be happy because the general contractor can make a commitment that work will be continuous after the electrician's come back from their break of two weeks.

Table 9.6 Cost forecast calculations for the control action plan

	Baseline (€)	Current (€)	Actual + Forecast (€)
Plasterboard walls			
Work	13,200	7,200	8,175
Waiting	0	600	0
Electrical			
Work	6,600	3,700	3,700
Waiting	0	400	400
Mob and demob	0	100	100
Tape and finish			
Work	4,600	4,600	4,600
Waiting	0	0	0
General Contractor			
Finishes Superintendent	9,800	13,953	13,953
Haulage equipment	3,300	3,900	3,900
Total	**37,500**	**34,453**	**34,828**

PRODUCTION SYSTEM RISK DURING THE CONTROL PHASE

The production system risk model must be continuously updated when more information becomes available in order to remain a useful tool during production. In the location-based control system, new information is modelled by using detail tasks, actual time and cost data, as well as estimates and forecasts. This information should be available for risk analysis in the control phase as well.

The implementation stage risk analysis need not simulate information which is known or which can be reliably forecast based on current information. Issues that continue to have uncertainty arise throughout a project and should be updated weekly in order to have the most reliable simulation results. The effects of new information are described below for each uncertainty type (also refer Chapter 7).

Uncertainty related to weather

Forecasting future weather is subject to great uncertainty throughout a project, and generally maintains the uncertainty level adopted when pre-planning the project. Until weather forecasts get more accurate, the risk analysis model can continue to sample the original weather distributions for the remaining days of the project. Weather does not need to be simulated for days prior to the simulation date because there is no variability associated with past weather.

Uncertainty related to prerequisites of production

Uncertainty related to the prerequisites for a task can be removed for each prerequisite once work associated with the prerequisite has actually been finished. For example, when design is complete and/or a delivery has been completed for a given location, there is no risk related to those prerequisites any more. For unfinished items, the risk analysis should be updated with the best currently available information about their dates. As a general rule, risk should reduce as the date approaches the committed date.

Uncertainty related to adding resources

When additional resources are required, there is uncertainty that they will eventuate. The uncertainty related to adding resources can be removed once the site data shows that the resources have actually arrived and are working on site. Additional information about resources may include promises by subcontractors to bring more resources (or come back on site) on a given date. Variability associated with these items is dependent on how much the subcontractor can be trusted.

Uncertainty related to productivity

After a crew has been working for a few weeks, the estimated productivity can be replaced with actual productivity. All other crews that have not yet mobilised should have their risks adjusted based on the productivity of crews currently working. For example, if a subcontractor provides a crew which achieves 80% of the desired productivity, it is safe to assume that the next crew will not achieve the required productivity either. Therefore, the expected productivity of future resources can be adjusted based on the actual productivity achieved.

Uncertainty related to quantities

Quantities can be simulated for unfinished locations. If earlier locations have had deviations from planned quantities, the remaining locations should have their expected quantities adjusted accordingly.

Uncertainty related to resource availability

There is always uncertainty that resources will arrive on site when required. When the resources actually come on site, the resource availability risk related to them can be removed. However, for future mobilisations, there remains risk although it may be reduced as more information becomes available through discussions with subcontractors.

Uncertainty related to locations

Uncertainty distributions should generally be independent of location (the same risk may be assumed in every location). Thus, only the general estimated risk values usually need to be updated. There are, however, circumstances in which the risk profile may become different in one location and would need to be updated. For example, it may become known that there is more difficult work in one location, such as increased services, when compared to other locations and this had not been taken into account in the original plan.

Simulation model in implementation phase

The simulation model does not need to simulate events which have already occurred or where the information is known. Such risk items should be replaced by actual data in the simulation. The simulation should only be performed for future production. By doing this, the simulation becomes a constantly updating management tool, providing information about future productivity risks and the probable end date if no corrective action is taken. During implementation, the results of the risk analysis augment the information from the deterministic schedule forecast.

SUPPLY CHAIN: DESIGN, PROCUREMENT, DELIVERIES AND LOGISTICS

While maintaining the efficiency of production is important to a successful project, the off-site activities can be equally important. Site production is vulnerable to delays caused by the supply chain functions of design, procurement, delivery and the associated logistics.

Controlling design

Delays in design are often the cause of large deviations in production. Location-based management links each design activity to one or more procurement events or production tasks as a prerequisite. In addition, the gatekeeper functions which are an intrinsic part of the design process are taken into account explicitly (see Chapter 6). As in other parts of the management system, the current, actual and forecast data should be available at the same level of detail as when planned.

The current design schedule reflects the current information about the project and is constrained by the baseline design schedule just as for production tasks. New information related to the design schedule is often related to changes in user requirements or variations. These, along with their gatekeeper functions, can be updated to detail design tasks.

The actual design schedule also has great similarities with the actual production schedule. The only difference is that the gatekeeper functions can cause the design tasks to jump back to the preceding design phase. If the design phase is not accepted by the gate-keeper, the actual percentage completed of the design task needs to be updated to reflect the work remaining in revising the design. The gatekeeper function should not be allowed to be marked complete until the gatekeeper has given final approval for that design phase.

The forecast can be calculated in the same way as for production tasks. Design tasks have an estimated work quantity and actual production rate. These can be used to calculate the forecast production rate. However, the design schedule should only use the actual production rate of the same design stage in different locations. The production rate cannot be generalised for different design stages because they may contain different types of work. If design fails to pass a gatekeeper, the forecast will be automatically adjusted because failure will cause the design phase to lose part of its actual completion rate to reflect work remaining before completion.

Controlling procurement

Controlling procurement is a critical part of controlling production. An important part of controlling procurement is to control the on-time completion of procurement events such as the completion of design, documentation, calling for tenders, bid evaluation, contract documentation and delivery.

The lead times for these components (events) in a procurement package will have been planned in the pre-planning phase (see Chapter 6). During implementation, this information needs to be continuously updated and made available for production management. If an event is delayed from its latest possible finish time, the schedule forecasts of related detail tasks should be shifted by an equivalent time. The total effect should be calculated based on the latest information about the event, although if a succeeding event is completed on time, the effect of the previous event delay on the schedule forecast disappears.

However, if an event is completed early—which often happens because procurement engineers and designers need to balance their workload—the forecast should not be updated to start early. This is because the production requirements of continuous work, and agreements with subcontractors, take priority. In accordance with lean principles, the procurement events are scheduled using pull scheduling—they are driven (demanded or pulled) by the site task, not the completion of prior procurement events. However, the consequential possibility of starting production earlier may be modelled by deliberately changing the timing of the associated detail tasks.

Using location-based data for delivery planning

Deliveries should be flexibly pulled to the site as required. Depending on the delivery method, the actual delivery date should usually be fixed one day to two weeks before delivery. This is a critical real-time decision because deliveries made too early may lead to waste of material and unnecessary hauling, or they may disturb other trades, while deliveries made too late will inevitably delay production. Some materials, such as concrete, are not able to be stored and there may be penalties if a planned delivery is not called up. Management may play it safe and only order concrete when the delivery is certain, which may lead to the concrete being delivered too late for production, leading to production delays. Of course, ordering concrete too early and then cancelling the delivery will also lead to delays.

Location-based control data can be utilised to improve these decisions. Up-to-date quantities in a location give indications about how much should be ordered and detail task start date—current and forecast—give the current best information about probable need times. Using the location-based control system tends to make the production system more reliable, so material buffers or inventories are rarely needed. On the other hand, location-based plans often allow for a space buffer which means that if the production continues as planned, there are free locations where materials can be stored.

The methods for planning deliveries are identical to those used in pre-planning (Chapter 6). However, current and forecast information is used instead of baseline information.

To continue the plasterboard example from Chapter 6, the concrete floor finishing work has started late. The quantity of plasterboard walls has increased on the second and fifth floors and there is more actual production rate information. Figure 9.11 shows the current delivery schedule (dashed line) for plasterboard compared to the baseline plan (solid line). The planned deliveries for the first and second floors have been threatened by a delay to the concrete floor finishing work, so they have been shifted forwards by three days. Because the production rate is slower in the detail plan, it has become possible to aggregate deliveries for all the other floors together, and the exact timing has been planned based on forecasts. This delivery plan should be continuously updated if there are deviations, such as slower progress of a predecessor or faster progress of a successor.

Delivery dates should be updated when more information becomes available and they should affect the forecast start dates of detail tasks needing the materials. For example, if a delivery for a certain location is delayed, the forecast for that location should be shifted. This may lead to control action planning such as stopping work earlier or slowing production down so that materials are not exhausted.

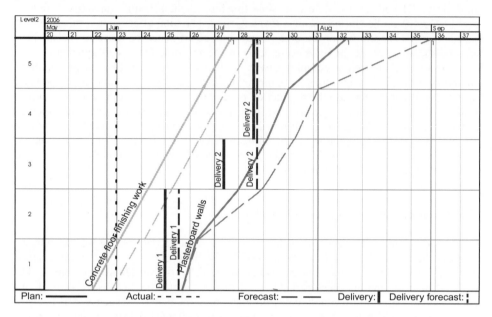

Figure 9.11 Current delivery compared to baseline

CONTROLLING PREREQUISITES AND THE MAKE-READY PROCESS

To facilitate location-based control, the prerequisites for starting work in a location should be planned and controlled. The other tools of this chapter make many of those prerequisites explicit, such as design, procurement, deliveries and preceding activities. However, there are many more things which must be completed before production can commence or continue. Examples of other prerequisites are submittals, requests for information, equipment and the availability of resources.

The location-based management system shares the fundamental philosophy of the Last Planner System (Ballard, 2000). To avoid the risk of incomplete work, workers should not be allowed to begin work in a location unless all the prerequisites have been made ready. The principal difference between the Last Planner System and location-based planning in the handling of prerequisites, is that location-based control takes explicit care of many of the prerequisites. The LBMS also recognises that it is easier to achieve the prerequisites of continuing work than beginning work. For example, because the same work will already have been done in previous locations, resources and equipment are generally already available, procurement activities will have been completed and material is more likely to be available for the subsequent locations—unless there are work stoppages.

Those prerequisites which are not explicitly taken into account, by other parts of the system, should be controlled by location-based checklists. Each prerequisite should have an associated person responsible and a planned date for completion. When all prerequisites for a location have been completed, the location becomes part of the workable backlog for that trade, together with any specific workable backlog tasks. Workable backlog is described in Chapter 10 (page 145) and its use is described in Chapter 10 (page 353).

CONTROLLING QUALITY

To minimise the cost of rework or moving around (including mobilisation and demobilisation), locations should be completely finished before moving on to the next location. It is important that 'completion' includes all the necessary inspections and quality checks. Usually, there should be enough time for inspections and quality checks before the next trade wants to enter a location because of the buffers between schedule tasks.

Locations which have incomplete or failed quality checks should never be considered finished in the location-based management system. If location-based payments are used, the location should only qualify for payment if and when all the quality checks have been passed. It is usually better to accept a delay of few days in order to complete required quality checks, than it is to leave unfinished or poor quality work behind in any or all locations. The cost effects of unfinished work will become apparent in the final stages, when all the visible quality damage must be fixed. Furthermore, during the maintenance or warranty period, the invisible quality defects will manifest.

Schedule forecasts will become distorted and unreliable if they are calculated based on production rates such that poor quality work results. Unfortunately, there will be an additional round of corrective work required, which often is executed with the same resources that are causing delays in production in other locations. This can become a vicious circle.

LEARNING DURING IMPLEMENTATION

Unfortunately it is almost impossible to measure the contribution which learning-based performance improvement actually makes to progress and it is therefore very risky to assume such in any forecast. Therefore, the location-based control system does not make assumptions about learning when calculating forecasts.

If learning has been assumed in the plan, it can be updated at the detail task level by using the same model. If learning actually affects production, then productivity will improve as production proceeds. This will be reflected in the forecast. If the learning rate is higher than assumed, the work will be forecast to finish early, enabling the utilisation of this improvement in productivity by either completing early or by decreasing resources where the preceding tasks do not allow early completion.

COMMUNICATING SCHEDULE AND PROCUREMENT STATUS

This section describes the various ways to visualise and communicate the current schedule status. Each tool is shown, together with an example, to illustrate the respective benefits and disadvantages. The tools are:

- Gantt chart status line
- Actual lines in the flowline
- Project control charts
- Production graphs.

Gantt chart status line

The traditional Gantt chart can be used to show whether a task is complete, or not, and its current status. The status is visualised by drawing a *status* or *control* line. This line runs vertically following the current date, but curves left or right, depending on the status of the task bar, as it passes through tasks. The status line segments each task bar by an amount corresponding to the extent of completion of the task. Typically, the status line zig-zags around the current date to accommodate the progress of tasks. The basic heuristics for determining the line position are simple:

- If the task is in progress, the status line cuts the bar depending on the completion rate of the task.
- If the task has not begun, the status line is drawn on the left side of the bar (if the task should have begun, the line will bend left, otherwise it continues straight at the current date).
- If the task has been finished, the status line is drawn on the right side of the bar (if the task should not have been finished yet, the line will bend right, otherwise it goes straight at the current date).

The method is easy to understand. A status line which passes through a task bar to the left of the current date (earlier) at some position indicates that something is late (that point should have been reached earlier). A status line which passes through a task bar to the right of the current date (later) indicates that something is going better than expected (that point should not have been reached until a later date). Figure 9.12 shows a small project. Tasks 1 and 2 have been completed (the status line is on the right side of them), Task 3 is a bit late (the status line bends left) and Task 6 is ahead of schedule (the status line bends right).

The method looks simple, and so it is for those tasks which are either finished or not yet commenced. However, for tasks which are in progress, the accuracy of the visualisation depends on how the completion rates have been calculated. If the Gantt chart is hierarchical (some bars are summaries of other bars), the completion rate of a summary task needs to be carefully evaluated.

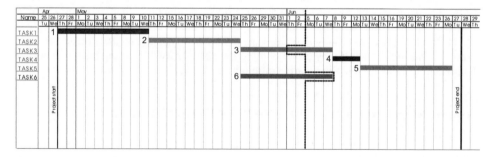

Figure 9.12 Progress viewed in a Gantt chart using a *status* line

Completion rate of a task

It is often very difficult to say how much of the work has been completed. The estimates are usually guesses, at best, unless the actual quantities are tracked accurately. The quantities are usually known when the subcontractor invoices and the work is measured. However this is normally too late to provide benefits in understanding the current status of the project.

Completion rates are easier to evaluate if the work has been split into smaller entities, such as locations. It is easier to say that a location of a task has been finished and to calculate the completion rate of the task from this information, rather than to just evaluate the completion rate directly. If location-based principles have been used in the planning stage, an estimate of quantity in each location is also known. If the location size is small enough, the completion rate can be evaluated by just recording which locations are finished and which are under way. Finished locations contribute all their quantity to the calculation of the completion rate for the whole task, and the commenced locations which are not yet finished contribute half.

Completion rate of a summary task

The calculation of the completion rate for a summary task is not trivial. The completion rates for schedule tasks in lower hierarchies need to be weighted to give an accurate picture of the status of a summary task (typically a construction phase or contract). This is because schedule tasks are not typically all measured using the same units. For example, foundations in the construction phase might include formwork in square metres, reinforcement bars in kilograms and concreting in cubic metres.

One way to circumvent this problem is to only draw the completion line according to the latest sub-task. While this method is often used, it provides an overly pessimistic view of production because there might be one or two locations, or a non-critical sub-task, where the work has not been started and this results in a representation where the overall construction phase appears to be delayed.

If location-based planning has been used and locations, quantities and consumption rates are available, more accurate methods may be utilised. The summary task completion rate can be calculated on the basis of completed quantities. The quantities should be weighted by the work content (man hours) of the tasks. In this way, tasks requiring lots of resources, and locations with more work, are weighted more heavily and a more accurate picture of summary task status emerges.

Benefits

- The Gantt presentation format will be familiar for most users
- The information is easy to understand
- It is quick to see at one glance which tasks are delayed and which are ahead of schedule.

Disadvantages

- It is an uneconomic way of presenting information—a Gantt chart of a project may have thousands of rows
- The information generally needs to be summarised. If quantities and worker hours are not used, the summary level information of status will not be accurate
- It does not show how deviations affect the production overall
- It cannot differentiate between start-up delays, interruptions and slowdowns
- It is not possible to view progress one month and compare with another month without showing multiple status lines. Even then, it is very difficult to visualise the production rate by just comparing two status lines.

Actual lines in the flowline

Location-based scheduling has location information available, which enables more information to be represented on a single chart. The flowline view already introduced can
include *actual* progress lines for actual work done, most commonly using dotted lines,
plotted along with the planned lines. This is similar to forecast lines, which are dashed. This
form of representation is a very powerful tool for project control because it shows the
schedule deviations graphically. The basic heuristics of drawing the actual lines are simple:

- The actual line starts in a location on the date when the work started in the location
- If completion rates have been estimated for locations, the actual line is drawn into the location according to the calculated percentage completion on the status date
- If no completion rates have been calculated, the line will be drawn halfway through the
  location (indicating progress)
- If work is interrupted in the location, the line is drawn horizontally through the days of
  interruption
- The actual line ends in the location on the date when the work actually ends in that
  location.

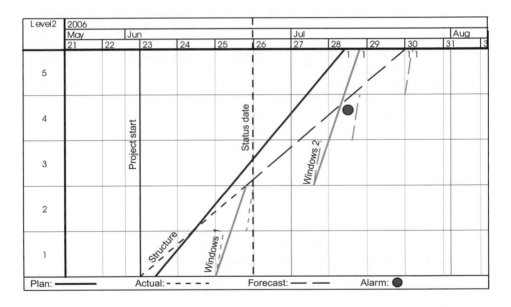

Figure 9.13 Progress viewed in a flowline using *actual* lines and *forecasts*

This method of control requires more accurate status information. In addition to knowing
whether the task is finished or not, it must be known when the work started and finished in
the location and whether there were any interruptions during the work.

It is possible to 'read' the actual lines to interpret project progress. In addition to
serving as a status report, the method gives information to support control action decisions.
It shows the reasons for deviations and how they affect the following trades. Figure 9.13
shows an example progress flowline schedule with actual lines drawn. It is possible to read
the following information from the diagram:

- The structure has started ahead of the schedule but is going slower than expected
- Window installation started the second location late but has been faster than expected, however it has been interrupted for a while because the structure was not finished (and will continue to be interrupted).

The actual lines allow the management to take control actions to get production back on track. In this example, site management could accelerate the structure task or slow down the window installation (to optimise the time when workers come back).

The method is capable of revealing many other common problems, including overlapping of work in sections and failure to complete locations. This type of information is easily revealed by the actual lines and corrective action can be taken.

Benefits

- The presentation is familiar to location-based planners
- It is a graphical and intuitive way of monitoring progress
- The presentation is efficient
- It facilitates control actions
- It allows management to visualise the actual production, compare to plans, and to learn from previous mistakes
- It separates start-up delays from interruptions and production rate deviations
- The chart can be viewed 'historically' as the progress at previous dates is easily viewed in a single chart. No archives need be referenced
- The historic nature of the view makes it a powerful tool for claims assessment.

Disadvantages

- The presentation is not familiar to CPM-based planners
- If there are lots of lines in the flowline diagram, drawing actual lines may make it hard to read and understand (a 'jungle' of lines). Selected task views may be necessary[1]
- It is difficult to show the status of the whole project in one chart
- If there are lots of deviations in the project, it is difficult to see which actual line relates to which planned line
- It needs more status information and work to keep it up to date.

Production control charts

A production control chart is a location-based tool for showing the status of the whole project on one print-out. It is a matrix of tasks and locations. Tasks are on the horizontal axis and the location breakdown structure is on the vertical axis. Each cell of the matrix shows the status of a task's location. This representation is often attempted by CPM-based planners using a spreadsheet and manually colouring in progress on site. Excel charts are sometimes used, with data being refreshed following inspection. The LBMS provides a mechanism for automating this according to the project's location breakdown structure.

---

[1] While filtering the views is a good way to solve this problem, another solution is sometimes to reproduce the plans at a large scale, for example on A0 sized paper. An amazing amount of information can be displayed this way.

Status can be shown with colour-codes and numbers in the cells, for example using the following colour coding:

- Green indicating that the location has been finished
- Blue indicating that work is on way and on schedule
- Yellow indicating that work is underway but the location is late
- Red indicating that the work should have started but is delayed
- Partial colours can be used to signify interrupted locations
- Numbers can be used to show planned and actual start and end dates or costs.

An example of a control chart, with only two tasks shown, is shown in Figure 9.14. The dates shown in the top part of each cell show the planned start and end dates. The dates shown in the lower part show the actual start and finish dates. It can be read from this chart that the structure task is late. It is in progress, but running late on the second floor and has a late start on the third floor. As a consequence, the windows task has a late start on the second floor.

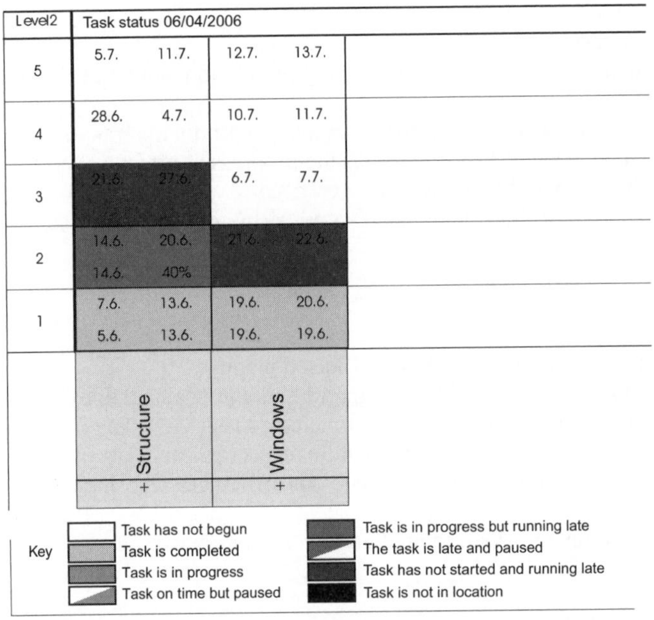

Figure 9.14 A simple control chart

To make the control chart work effectively, customised control charts should be provided to the various superintendents and subcontractors for the site. The client may want to see the status of the most important tasks at the summary level. Because of the variety of information needs, information must be able to be either summarised or reported in detail.

When working with summary tasks, similar issues need to be resolved as for Gantt chart status reporting. For example, how do you calculate the completion rate of a summary task in a location? How do you define the colours of summary task squares? The logic could be as follows:

- Define the colour according to the worst sub-task
- The start date of the summary task in the location is the start date of first of the sub-tasks in that location
- The end date of the summary task in the location is the end date of the last of the sub-tasks in the location
- The completion rate will be calculated from the completion rates of sub-tasks weighted by using the total hours of work in each of the sub-tasks.

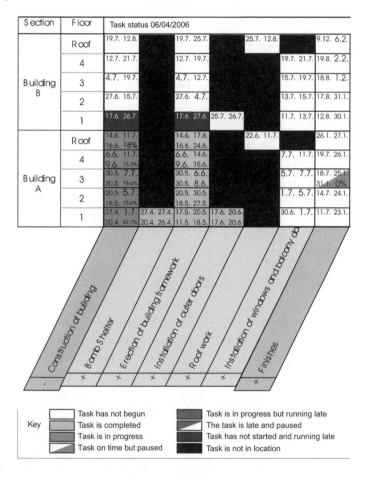

Figure 9.15 A hierarchical control chart

A hierarchical control chart is shown in figure 9.15. Construction of the building is a summary task formed of the tasks bomb shelter, erection of building framework, installation of outer doors, roof work and installation of windows. Finishes is another summary task, which has been closed (not displaying sub-tasks) in preparing this report and therefore only shows the summary. Work has started but is currently interrupted on the third level.

Control charts and stages of information

Actual data can be compared with either the baseline or the current plan formed by detail tasks. Each approach has merit, depending on what needs to be communicated. The client is generally interested in seeing the progress against the originally approved baseline. The same information should be used by the superintendent to keep the project on track. However, subcontractors who commit to detailed schedules should have a control chart which compares their progress against the detailed schedule.

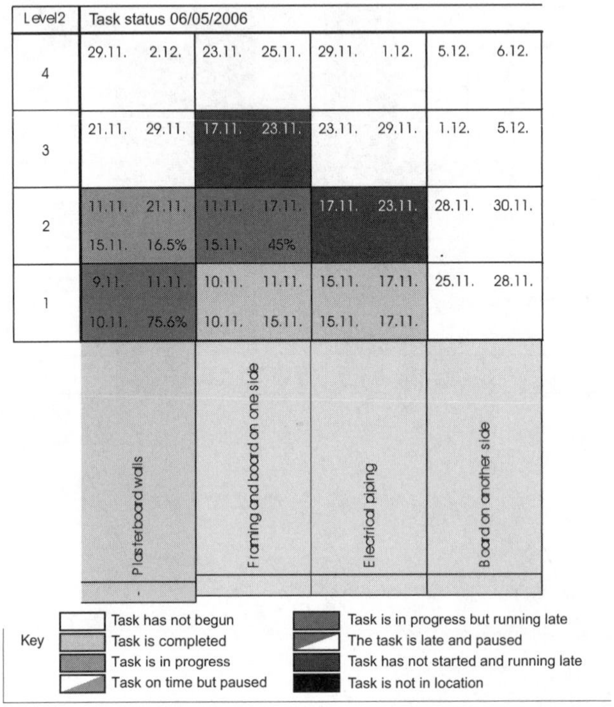

Figure 9.16 Baseline and detail tasks

Sometimes it is useful to show both levels in the chart. Baseline tasks can contain detail tasks as a sub-hierarchy. The detail task cells' colour is calculated from the detail task timing and the baseline task colours are calculated from the baseline task timing. Therefore, detail tasks can show that the schedule has been achieved, while the baseline task shows that the task is late in the same location. An example illustrating both baseline tasks and detail tasks in the same diagram is shown in Figure 9.16. There are three detail tasks under the plasterboard walls baseline task. The baseline task has been started but is delayed on the first floor, even though all its detail tasks have been completed on time. The second floor is not delayed at the baseline phase, even though two detail tasks are delayed on that level. The baseline task phase is for management and client reporting and to set boundaries for detail planning. The detail task phase is useful for superintendent weekly planning, as well as tracking subcontractor performance compared with commitments.

Controlling prerequisites with a control chart

The status of prerequisites in a location can be visualised by using a control chart. If all prerequisites are clear, the cell can be 'freed' or shown in white. If some prerequisites are lacking, it remains 'constrained' and should be shown in grey. This allows the superintendents and subcontractors to immediately see which locations are available for work.

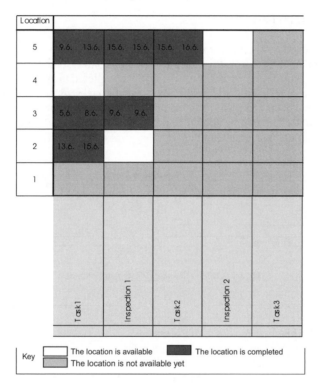

Figure 9.17 Chaos control chart

Chaos control chart

A *Chaos control chart* can be used to save a project which has plunged into chaos. It shows all the activities in a location at a very accurate level of detail. All the inspections and quality measurements that have to be done should be shown as separate columns in the matrix. Locations that are free should be shown in a different colour. A free location means that the location is available for work and that all the prerequisites and earlier tasks have been cleared. The task of production management is to have at least one free location for every trade at all times and to ensure that locations are finished before moving to the next location.

Using this very accurate control chart is amazingly effective in practice. However, it demands lots of time to update and a lot of management effort to make it work, so it probably does not provide sufficient extra benefit for practical use on non-chaotic sites. It makes a good tool to be taken out of the bag if the project has got into trouble—particularly in the finishes phase.

Figure 9.17 shows an example of chaos control chart. Available locations are shown in white. Locations which are not available (incomplete prerequisites) have been greyed out.

Green (or in this chart the darker grey) means that the location is finished. Other colours are not used because in chaotic projects the baseline schedule is no longer a useful tool. Instead of controlling the schedule, the finishing of locations and freeing of locations to the following trades is the only thing that matters. Only actual start and end dates are shown in the cells of the completed locations.

Procurement Gantt chart

A procurement schedule is often displayed as a Gantt chart that shows the delivery time as a bar derived from the master schedule and the events as symbols before the delivery bar. The symbols show the latest time when the event may be completed in order to meet the required delivery time. Thus reflecting the demand-pull of the tasks on the procurement events. Events which have been completed can be displayed as boxes which are greyed out, while the events which are late can be shown as boxes with red borders. Figure 9.18 shows the procurement schedule in such a Gantt chart.

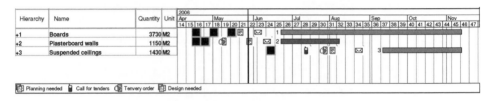

Figure 9.18 Procurement schedule in Gantt view

Benefits

- Easy to see the planned sequence of events.
- The time of delivery can be seen in the same chart.
- Late events can be illustrated.
- The chart can also show included materials/work and their need times.

Disadvantages

- Becomes difficult to read if there are lots of events and procurement packages.
- Symbols used to denote events may not be familiar to the user.

Procurement control chart

It is not possible to display procurement on a flowline, however a procurement control chart can be extremely effective. The procurement control chart (Figure 9.19) shows the status of procurement events in a matrix similar to the schedule control chart. The procurement packages are shown on the horizontal axis and the procurement events are shown on the vertical axis. Late events are shown in red, events that should be done in the near future (during the next two to three weeks) are shown in yellow and completed events are shown in green. The colours direct the procurement engineers to do their work in the correct sequence. Numbers in the cell can show dates or week numbers for the latest (last responsible moment) by which procurement must occur.

Benefits

- Easy to read and understand
- Colour coded visual warnings direct the work
- One sheet of paper can usually show all the information required.

Disadvantages

- Does not show time of individual deliveries or quantity of materials included in them.

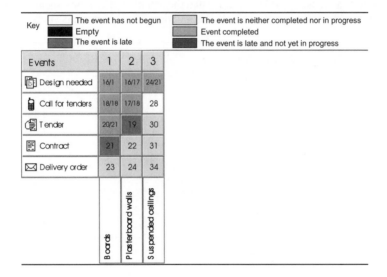

Figure 9.19 Procurement control chart

Production graphs

Production graphs are an extremely powerful way to show production rate information. They differ from the flowline by considering all similar production, regardless of its location. On the vertical axis is the unit of production, it can be a physical unit (such as m^2 or m^3), a labour related unit (for example man hours), a completion rate related unit (% completed) or cost unit ($ or €). On the horizontal axis is time, in the same way as any schedule.

Information in production graphs is shown by cumulative graph lines, or bars, to indicate the quantity produced for any given day, week or month. The same graph can contain baseline, current and actual/forecast lines and bars. Any production with the same unit can be combined to a single line. For example using labour, percentage completed or cost units allows the combining of any production to a single graph.

The most powerful use of production graphs is to show the momentum (a term used by Murray, 2007) of a subcontractor for the construction phase or for the complete project. This can be done by summarising all production undertaken by that subcontractor and using the worker hours or percentage completed as the comparison unit. Because the graph is based on all production, it will give an accurate indicator of how the performance of that subcontractor has varied as a function of time. Any dips in performance can then be investigated. This is different from the flowline because a flowline considers production of separate tasks in locations and allows visualisation of interdependencies of tasks. A flowline shows progress through locations, the production graph shows progress through quantities.

The production graph shows overall performance which can then be optimally divided to tasks and locations by using information in the flowline. It is a very powerful tool for controlling the production of multi-skilled subcontractors working in multiple tasks, and in cases where work is undertaken simultaneously in multiple locations. Figure 9.20 shows a production graph for the plasterboard walls example, using worker hours of the plasterboard walls subcontractor as a production measure. It can be seen that at the status date, production is delayed but, as result of planned control actions, the committed production (current) will be achieved during the next week of production. Note that the only difference to earned value using worker hours is that the actual worker hours spent are not shown in the diagram—instead only the value of production (hours earned) is shown.

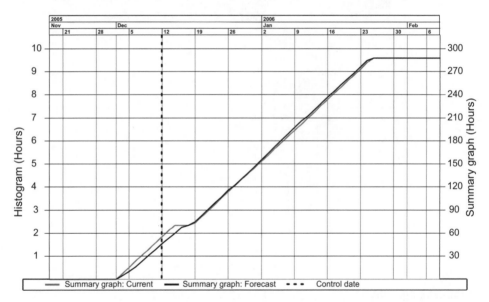

Figure 9.20 Production graph for plasterboard

Production graphs using physical units (for example, m² of formwork) can be used to optimise future deliveries and to check any invoices against the actual physical data. They can also be used to check the feasibility of the plan or forecast, based on the material availability information. In the plasterboard walls example, insulation wool was the responsibility of the general contractor. Figure 9.20 illustrates the demand for insulation wool (m²) as function of time. The current and forecast lines become horizontal when the subcontractor is completing work which does not require wool.

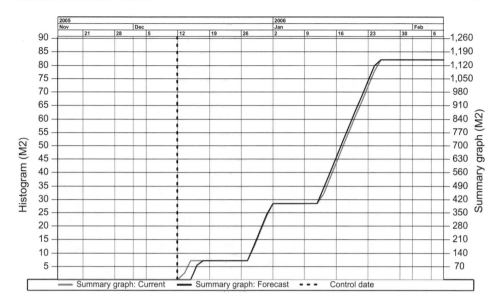

Figure 9.21 Production graph for wool

Benefits

- Easier to read and understand for subcontractors and clients than flowline diagrams.
- A powerful way of combining production information for different tasks and subcontractors.
- Information can be used for planning deliveries and checking invoices.

Disadvantages

- High information requirement—quantities, worker hours and their completion dates are required.
- Does not show downstream effects of lower production rates.

REFERENCES

Ashley, D.B. and Teicholz, P.M. (1977). "Pre-estimate cash flow analysis". *Journal of the Construction Division, ASCE*, Proc. Paper 13213, **103** (CO3): 369–379. See also discussion by Gates, M. and Scarpa, A. (1978), **104** (CO1): 111–113, and closure by Ashley *et al.* (1978), **104** (CO4): 554.

Ballard, G. (2000). *The Last Planner system of production control.* PhD Thesis. School of Civil Engineering, The University of Birmingham, Birmingham.

Berdicevsky, S. (1978). *Erection cost flow analysis.* Unpublished MSc Thesis, Technion, Israel Institute of Technology, Haifa, Israel.

Elazouni, A.M., and Metwally, G. (2005). "Finance-based scheduling: Tool to maximize project profit using improved genetic algorithms". *Journal of Construction Engineering and Management, ASCE*, **131**(4), 400–412.

Kenley, R. (2003). *"Financing construction: Cash flows and cash farming".* E&FN Spon, London.

Kennedy, W.B., Anson, M., Myers, K.A. and Clears, M. (1970). "Client time and cost control with network analysis". *The Building Economist* **9**: 82–92.

Murray, W.B. (2007). *Faster construction projects with CPM scheduling.* McGraw-Hill, New York, NY.

Peer, S. (1982). "Application of cost-flow forecasting models". *Journal of the Construction Division*, ASCE, Proc. Paper 17128, **108** (CO2): 226–32.

Peterman, G.G. (1973). "A way to forecast cash flow". *World Construction*, October: 17–22.

Reinschmidt, K.F. and Frank, W.E. (1976). "Construction cash flow management system". *Journal of the Construction Division*, ASCE, Proc. Paper 12610, **102**(CO4): 615–627.

Pekanpalo, H. (2004). "Rakennustyömaan kustannusvalvonta ja kustannusten ennustaminen (Cost control and cost forecasting of a construction project)". Master of Engineering thesis, Construction Economics and Management, Department of Civil and Environmental Engineering, Helsinki University of Technology. Espoo, Finland.

Seppänen, O. and Kenley, R. (2005). "Using Location-Based techniques for cost control". *13th Annual Lean Construction Conference*, Sydney, IGLC.

Chapter 10

Using location-based control methodologies

INTRODUCTION

It is not sufficient to design a control system for location-based management. Those involved with the planning and control of construction, by whatever planning method, will recognise that it is difficult to get project staff to record status data, let alone follow a specific plan. However, a successful project is much more likely when a location-based plan is actually followed. Management, having worked hard to design a feasible schedule, must be proactive in ensuring that the plan is followed.

Lean construction accepts the common definition for *control* as *making things happen* rather than *monitoring results*, and this interpretation is equally applicable in location-based management. Instead of just recording deviations and 'managing by exception', it is desirable to actually follow the plan to its level of detail. This chapter presents a systematic process for project control, which utilises the tools described in the preceding two chapters. Location-based control processes include:

- Monitoring current status
- Accurate planning of implementation
- Forecasting progress
- Planning control actions
- Prioritising tasks
- Ensuring prerequisites of production
- Executing the plan through good assignments and communication.

Monitoring current status provides location-based management with real-time information about production in order to facilitate control decisions. Monitoring can be organised in many different ways, from self-reporting by subcontractors to totally centralised monitoring. In time-critical jobs, with little or no buffers, daily monitoring may be necessary, while for typical jobs, weekly monitoring is usually enough. This chapter discusses the information that is needed by the location-based control system and suggests possible ways of gathering this information.

Accurate planning of implementation is achieved by using detail tasks based on the best information available at the planning stage. Baseline schedules are often not directly relevant for production because they were planned using incomplete information. The level of accuracy becomes critical in detail planning—too much detail will result in tedious status reporting and updating, while too little detail may cause important dependencies to be overlooked or result in a loss of control. Detail tasks should only be updated within the constraints of the contractual relationship and commitments between the general contractor and the subcontractor, and should reflect the current consensus about how the work will actually be carried out on site. If the subcontractor is delayed, the detail tasks should not be updated but, instead, control actions should be planned to get the project back on track. If a delay occurs that is outside the influence of the subcontractor, the detail tasks can be updated to reflect the current understanding between the general contractor and the subcontractor. A methodology to help decide when to update plans will be introduced.

Schedule and cost forecasts utilise both the status information and the current set of detail tasks. In the location-based control methodology, they are used to force reactive action as early as possible and to communicate the effects of any delay to stakeholders of the project. If forecasts differ from the plans, appropriate control actions need to be taken to restore production.

Control actions can be optimised by using information from contracts and the calculated production system cost. These are short-term control actions designed to restore production to correspond with the planned detail tasks. The main priority is to minimise any interference between various subcontractors. Most control actions either increase or decrease the production rates of tasks. However, in some cases, changes of sequence, delaying the start dates or making tasks discontinuous may all be valid options.

In addition to reactive control measures, such as control actions, proactive control is a critical part of the location-based controlling methodology.

The aim of proactive control is to anticipate and prevent problems before they occur. This involves finding out what needs to be done to ensure that production can continue with the planned production rates and that tasks can start on time, then giving the responsibility for any required action to someone who can control the on-time implementation of the selected actions. Proactive control is the most critical part of controlling because if proactive control is effective, there should be no variability in production rates caused by issues which could have been prevented by timely action, and the need for buffers to absorb variation would be diminished in the production system. In other words, establishing proactive control acts to reduce risk in the production system.

The proactive part of the location based control system consists of tracking prerequisites of production for both the starting and continuing locations—by using checklists— controlling procurement and deliveries, updating current quantity information and planning more detailed task schedules. Use of a proactive controlling system should always result in taking action to make things happen. For example, detailed task schedules should be communicated to get commitments from the various actors to do the work as planned. The tools should be used to get better understanding of the production process and to enable management to make more informed decisions. The tools should not be used to just model decisions which have already been made based on intuition and experience. Plans should drive the work, not vice versa.

Production management generally competes for limited resources, therefore proactive control should be prioritised around focus points. There is an infinite number of ways to reduce schedule risk through better planning as well as by spending more time in ensuring prerequisites are in place, but the resources available on site define how many of these actions can be effectively implemented. There are some things which are mandatory to prevent problems resulting in the near future, such as restoring production of a delayed task and ensuring free locations for ongoing tasks. Any remaining time can be used to ensure on-time commencement of upcoming tasks. Each of these three groups can be ranked in priority order by using a combination of schedule forecasts and float measures in prioritisation.

LOCATION-BASED SYSTEMATIC CONTROLLING PROCESS

To support the implementation of the location-based plan, a systematic control process should be implemented. The systematic control process can be implemented daily or weekly, depending on the requirements of the project. The suggested control process has ten steps:

1. Monitor the current status:
   a. Actual start and finish dates

b. Actual resources
c. Actual quantities.
2. Compare forecasts to plans to detect deviations.
3. Control action planning.
4. Evaluation of resource needs.
5. Creating reports for site meeting.
6. Site meeting.
7. Management and client reporting.
8. Detailed planning:
   a. Update baseline tasks if necessary
   b. Check and update the location breakdown structure
   c. Check and update quantities
   d. Plan new detail tasks
   e. Update existing detail tasks if necessary.
9. Monitor and prioritise prerequisites of production.
10. Weekly planning and communicating assignments.

Each step is described in detail below.

MONITORING CURRENT STATUS

The first step includes updating the status of each task and location to the control chart and plotting the progress to the flowline diagram. The process of monitoring includes collecting the necessary information by location and making sense of the current status of the project.

Collecting location-based status data

There are many ways of collecting the status data. They can be divided into two main categories: centralised and distributed.

Centralised information collecting

In centralised strategies, the person responsible for monitoring status should tour every location to observe and record the status of work. This gives an accurate and objective snapshot of the progress as of the status date. However, it does not give information about what has happened within the monitoring period. For example, if status information is collected weekly, the controlling system will not have accurate information about the exact date or duration of any location within that period. These would have to be estimated, which may cause problems with accuracy of information.

Distributed information collecting

In distributed strategies, multiple persons or organisations supply information related to actual progress. For example, subcontractors may self-report their work status including accurate start and finish dates for each location. Alternatively, superintendents can report the status of their work. The information from distributed information gathering strategies may be more accurate but there is a higher risk that information is not correct. For example, a subcontractor may report that a location is completed but there may be some minor part of the work which is not yet completed.

Combination of centralised and distributed strategies

Normally it is best to have a combination of centralised and distributed strategies. For example, subcontractors might report their progress weekly and the general contractor can verify the information on the status date.

Progress reporting principles

Consistent principles for recording progress should be followed and should be agreed before the start of the project. Principles relate to recording the start dates, finish dates, completion rates, work interruptions, actual resources and shift lengths. It is always possible to collect more accurate information. However, at some point the cost of additional accuracy is greater than the benefit. Until completely automated ways of getting progress data become widely available, it is usually necessary to resort to rough approximations.

Actual start dates

Actual start dates should depend on the scope of work used to calculate durations. If the labour consumption rates that were used while developing the plan included hours devoted to hauling materials and other preparatory work, the location must have started when these assisting activities started. However, if logistics activities are not part of the allocated activity duration, then a location has started only when the actual production has started.

If actual start dates of locations are unknown, they can be approximated backwards from the current status. For example, if three locations of the same size have been completed during the week, their start and finish dates can be set two days apart, assuming the planned sequence, if better information is not available.

Completion rates and quantities

The need to measure completion rates in a location depends on the size of the location and the total duration in the location. If locations are generally small, it is enough to record their status as not started, in progress, interrupted or finished without using completion rates. However, if a location's quantities are large, then a completion rate measure may be useful. The accuracy level desired in completion rates should be decided on a project-by-project basis. If 3D models of the building are available, it is easy to accurately calculate actual quantities completed. Otherwise, it is normally enough to use rough measures such as started, 25%, 50%, 75%, finished. Anything more accurate than that is often a guess and should be used only if the actual installed quantity has been measured or is available from a 3D model. Results provided by a quantity surveyor as part of an assessment of work-in-progress should not be used unless they occur on the control date. Otherwise, the progress information comes too late to be beneficial for control action planning.

It often happens that a small part of the work in a location cannot be completed because of missing design details, insufficient materials, delayed inspections or because a previous subcontractor did not complete a location. Instead of leaving the tasks in these locations open at 95% complete, it is advisable to mark the location complete and create a new detail task for the remaining part of the work with the correct dependency relationships to explain why work was not completed. Otherwise, the flowline chart will have long and somewhat misleading 'tails' for actual lines until the last 5% of the work can be finished. On the other hand, the remaining scope has to be documented in a new activity because otherwise these small remainder activities will accumulate in the last months of a project, leading to delays and quality defects. Of course, completing the last 5% of a task is better done with the rest of the task, so resolving the problem remains a better solution.

Work interruptions

Work should be marked as interrupted in a location when there is no progress or there is too little progress compared to the last status date. Work interruptions are important to record because they potentially have large effects on forecast calculations. For example, if work started on Monday and finished on Thursday but no work happened on Tuesday and Wednesday, the effective duration of the work is actually only two days, not four. If the duration of four days was used to compare the actual production rate to the planned production rate, the forecast would look much worse than the reality.

Actual resources and shift lengths

It is easy to get the average on-site workforce for each subcontractor because that is usually available in subcontractor weekly reports. More accurate information can be received if workers are required to register at the gate when arriving and leaving (for example, by swiping a card at the gate). It is much more difficult to track for each subcontractor what the resources were actually doing and where they were working. However, as explained in Chapter 8, it is possible to compare the actual work done by the subcontractor to the actual average workforce to arrive at reasonable rough estimates of actual task and location resource levels. The actual shift length is often easier to obtain because usually subcontractors who are working longer than the standard hours will be known to the project team because of their need for special arrangements for locking the site and notifying security.

COMPARING ACTUAL PRODUCTION TO PLANNED: DETECTING DEVIATIONS

To be able to make informed control decisions, the current status of the project needs to be visualised and compared to both the detail and baseline plans. The main tools for visualisation are flowline diagrams with actual lines, production graphs and control charts (see Chapter 9). Flowline diagrams can easily become crowded by too many lines if both actual and planned lines are shown. Therefore it is a good practice to define filtered views of the project to highlight a particular need. The following filters are applicable for most projects:

• Baseline view showing critical tasks
• View of detail tasks currently in progress
• Location-based views, for the main areas of the project
• Time-based views, for example the current date ± 4 weeks

- Subcontractor views, showing all detail tasks of a subcontractor and immediate predecessors and successors
- Superintendent views, showing all detail tasks of a superintendent and immediate predecessors and successors.

The use of each of these views is shown by an actual case study of a 15,000 m² office building in Finland (see also Chapter 15). All of the examples below show different flowline views of the status in the middle of April 2005. Ordinarily, the figures would be printed to A4 or larger sheets of paper. This is not possible in this book, but we have tried to make the figures as legible as possible without removing too much detail.

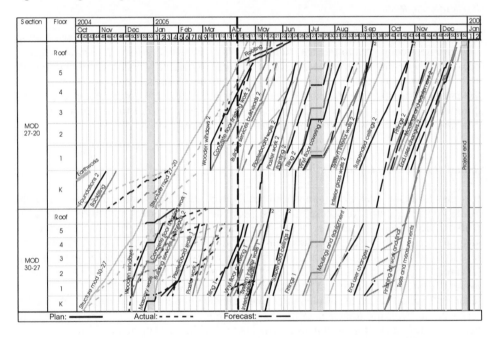

Figure 10.1 Selected schedule view for controlling the Opus project, Finland

Baseline view showing critical tasks

The baseline view when filtered to display critical tasks shows the effect of current deviations on the project end date, the buffers and how production has occurred compared to the original baseline. This view is useful for client and management communication.

In the schedule shown in Figure 10.1, with only a selection of tasks shown in the view due to the available page size, it can be seen that there have been some deviations on structure over the first phase (MOD 30–27) but that those have mostly been caught up by absorbing the planned buffers. Other forecast delays for the first phase arise because of changes in detail tasks. In contrast to the original plan, the work starts on the second floor and the first floor is completed last in sequence. In the second phase, the delays of the structure have been mostly caught up, but the forecasts differ from the original plans—most likely because of changes in the detailed schedule. The last month of the project does not have much buffer left, so it is risky. However, the current forecast does not extend the finish date of the last activity.

View of detail tasks currently in progress

A view of the current detail tasks can be used in site meetings in communicating the need for control actions to subcontractors.

In the example schedule shown in Figure 10.2, all the ongoing detail tasks and their actual lines and forecasts are shown. At the detail task level, the deviations are quite small, which means that detail task controlling has worked quite well on this project. The structure task for phase MOD 27–20 is ahead of the detail task schedule. Window installation is going faster than planned and the crew has been forced to leave the site (note the alarm symbol in the flowline). The first concrete pour in section MOD 27–20, Floor 2, has finished according to plan but the preparation of the second pour is delayed, causing forecasts for all future pours to shift by one day. In section MOD 30–27 there are deviations with building services bulkheads which, in turn, seems to be suspended. This has caused the mechanical duct work to stop and a slowdown in the installation of horizontal heating pipes and radiators. Tiling has a slower than planned production rate. Vinyl floor covering has not started according to its detail plan.

By using simple figures like this, it is easy to explain what is currently happening on site and to make comparisons to the agreed production plan.

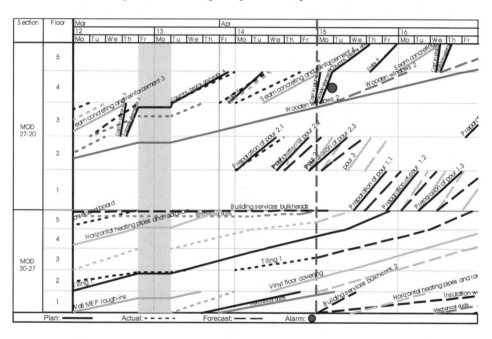

Figure 10.2 Detail view: schedule—Opus, Finland

Location-based views

Critical locations can be examined in more detail in separate views of the schedule. This makes it possible to show deeper location breakdowns. If the room level is shown, it is possible to give these schedules to crews to show the sequence of their work.

A roofing detail schedule is shown in Figure 10.3. The roof of the office building was on four different levels and the roof structure varied in each area. In the example schedule, tasks and sequences were changed on different zones. For example, zones 4 to 7 have a mechanical room and thus have more work than zones 1 to 3. The actual lines follow the detail task plans almost perfectly, which again shows the results which can be achieved by good schedule control. Although many tasks do not seem to be continuous, the same subcontractors have other work inside the building. Their work has been optimised using subcontractor views and resource levelling. Overlaps of one day in the actual lines result from using daily accuracy in monitoring and control (the start time has been drawn to be always at the beginning of the day and the finish time to be always at the end of the day, however the actual change would be somewhere in the middle of the day).

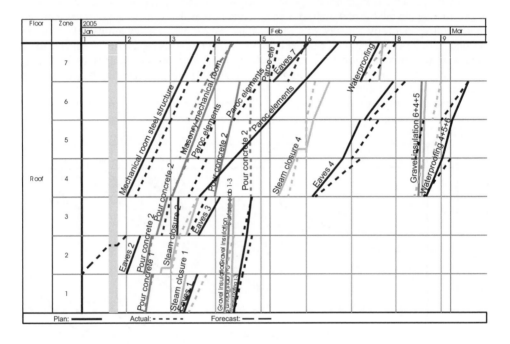

Figure 10.3 Location view: detail of roof construction—Opus, Finland

Time-based views

Time-based views can be used to ensure that the starting prerequisites for forthcoming tasks are ready and detailed task schedules have been planned. For example, a rolling six week view for the next six weeks can be used to ensure that all tasks can start on time in the look-ahead window.

Subcontractor views

Subcontractor resource continuity can be examined in subcontractor views. These views should show all the detail tasks for a subcontractor and all immediately preceding and succeeding detail tasks.

Building services bulkheads and plasterboard walls were done by the same subcontractor in the example project. This subcontract had a lot of problems. Figure 10.4 shows that the preceding task, concrete floor finishing work, did not prevent building services bulkheads from starting on time. The same workers are being used to do both building services bulkheads and framing and first board detail tasks. The actual lines show that while one sub-task is going on in a location, the other sub-tasks have stopped. Every time production management required that work be finished, all the resources moved to that location. When the second section (MOD 27–20) is examined together with the first section (MOD 30–27), it turns out that the subcontractor is trying to level resource use. He is slowing down so that the same resources can continue straight into the first detail task of the second section.

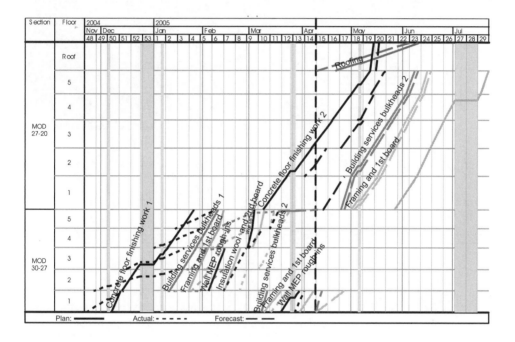

Figure 10.4 Subcontractor view: trade detail—Opus, Finland

Superintendent views

A control chart can be filtered to just show relevant tasks, and this is often the best visualisation for a superintendent. Red and yellows must be remedied (finished) and grey squares need to be made ready for production.

A control chart for a superintendent is shown in Figure 10.5. The crossed out locations have been completed. Mechanical ducts and building services bulkheads on the fifth floor have been interrupted. The darker colour in mechanical ducts means that something external to the subcontractor is preventing the task from continuing. Tasks with one line from left to right have started but are delayed and dark (normally red) tasks have delayed start dates. The diagram immediately shows the current status and what needs to be done.

Production graphs

Production graphs usefully supplement the information from flowline charts and control charts. While flowline charts highlight any interference in locations or task production rates, and control charts are powerful in illustrating locations which are delayed when compared to either the baseline or detail schedules, production graphs can be used to evaluate the overall performance of any given subcontractor. They can also be used to evaluate changes in the overall momentum of a project. The best way to standardise information for the purpose of constructing production graphs is to use the value of worker hours produced as the vertical axis unit.

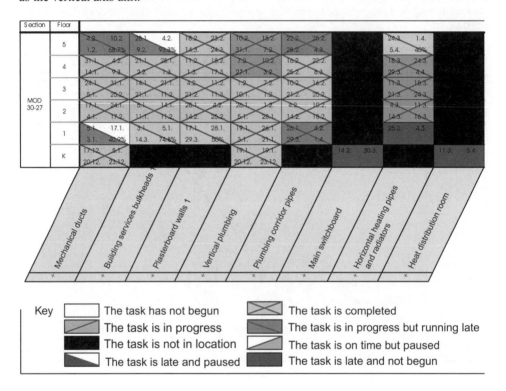

Figure 10.5 Control views: superintendent mechanical—Opus, Finland

Figure 10.6 shows the production graph for the suspended ceiling bulkheads and plasterboard walls subcontractor (compare to the corresponding flowline in Figure 10.4). The target (baseline) graph shows that the subcontractor was originally supposed to work in two clearly separate work segments, with a break in between (the flowline shows a break between sections). Each one of the continuous work segments has a slower start, then a ramp-up period and then a slower finish. This happens because additional resources are assumed when two baseline tasks (bulkheads and plasterboard walls) are worked simultaneously. The current graph shows the current commitments with the subcontractor. The start date is 1.5 weeks later than the baseline, there are more work breaks of shorter duration and the production rate is roughly half the baseline. The actual line has fallen short of commitment but production is expected to catch up by the end of April (the actual line becomes a forecast line after the current date). The current delay, compared to commitment, is 2.5 weeks (the horizontal difference between the actual and current lines). The delay is about 100 worker hours (the vertical difference between the current and actual/forecast lines on the production graph).

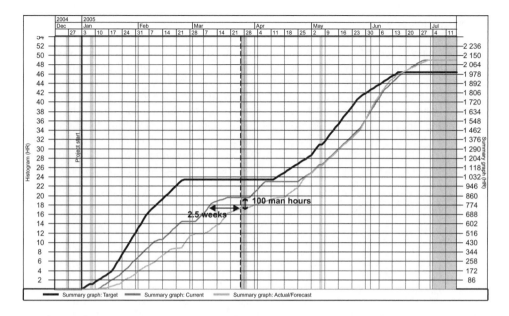

Figure 10.6 The production graph for suspended ceiling bulkheads and plasterboard—Opus, Finland

CONTROL ACTION PLANNING

In the location-based controlling methodology, forecasts are used to provide early warning of problems. If a forecast is different from the plan, it should always trigger an analysis of the underlying reasons with corresponding corrective action proposed. This should occur even if the total forecast for the project duration stays within the contract limits. This is because delays use up the schedule buffers, making controlling gradually more difficult and expensive towards the end of the project (the project becomes more rushed). Additionally, excessive schedule pressure at the end of the project causes large cost overruns and quality errors, so it is preferable to catch up gradually and to commence immediately once deviations are noticed.

When deviations occur, the following list of questions should be addressed:

- What happened?
- Why did the deviation occur?
- What is the effect of the deviation?
- Which control actions can be used?
- What is the optimal control action plan?

What happened?

The deviations can be broadly classified into five groups (Seppänen and Kankainen, 2004):

- Start-up delay or starting too early
- Production rate deviation
- Unplanned splitting of the crew into multiple locations
- Change of work sequence

• Interruption of work.

These are easy to identify from reading the flowline diagrams and control charts based on the location-based status data. Start-up delays are shown by a horizontal deviation between the planned and the actual lines in the flowline or by a red colour in the control chart. Production rate deviations are revealed by the actual line having a gentler slope than planned. Unplanned splitting of the crew (that is, not finishing locations) is shown with overlapping lines in multiple locations on the flowline, or by multiple locations being incomplete and in progress on the control chart. Sequence changes can be read from the flowline as differences of patterns in the actual and planned lines. Work interruptions are seen as horizontal actual lines. Figure 10.7 shows these basic deviation types as a flowline.

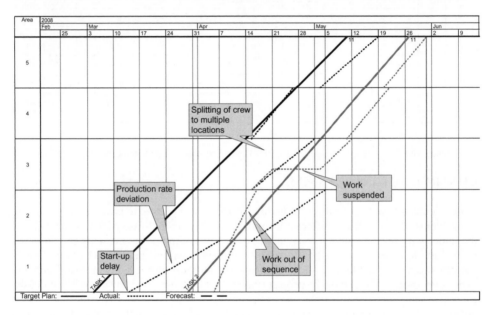

Figure 10.7 Basic deviation types

Why did the deviation occur?

It is critical to accurately assess why a problem occurred. By using quantities, planned production rates and actual resources, more information about the deviation can be assessed. For example, the reason for the production rate deviation might turn out to be working shorter days or with fewer resources than planned. These methods have been described in Chapter 8.

Root-cause analysis often requires going deeper than just using the progress data from the location-based control system. For example, a start-up delay can be caused by resource problems, bad weather, missing design details, variations or materials. The explanations affect the potential ways to react to the problem.

Some typical reasons for deviations are described below for each deviation type.

Start-up delay

Start-up delays usually relate to the prerequisites of starting (delay of procurement, deliveries, design not ready), preceding tasks being delayed or resources not being available.

Production rate deviation

The reasons for a production rate deviation can relate to having too few resources on site, increased quantities, lower productivity or more difficult work than expected. However, these are not yet root reasons but require further investigation. For example, a subcontractor might decide to start with a smaller crew because he thought that there was not enough work ready for him to produce at the planned production rate. Alternatively, resources may not have been available. By starting this discussion immediately, the problem will already be half solved.

Increased quantities also requires further investigation. Is this a one-off increase or is it going to occur also in future locations? Was the quantity increase caused by mistake, rework, design change or error in quantity take-off? The answers to these questions affect the total effect of the deviation and possible corrective actions.

Lower productivity can be caused by multiple factors. Workers may be slowing down because there is not enough work available. Alternatively, there can be problems in material supply or other trades can be interfering in the same location. Sometimes there are problems which relate to work methods—productivity may have been overestimated. Again, finding out the actual root cause helps to solve the problem.

Splitting of work to multiple locations

A single crew working in multiple locations at the same time usually results from one of a few causes. First, the crew can be doing part of their work in many locations before coming back to finish the work. For example, the plasterboard wall crew can be doing bottom and top plates on multiple floors before coming back to install studs and boards. If the plasterboard walls task has been planned as a single detail task, the progress data show the crews working in multiple locations at the same time. In this case, either detail tasks should be updated to reflect the actual flow of work, or the work methods should change to conform to the plan—the work method may interfere with following trades.

Second, there might be some factor preventing the finishing of the locations. For example, quality problems in preceding work might make it impossible to finish work in a location. Missing details in the design can sometimes prevent finishing the work. Alternatively, the subcontractor may simply be doing the easy work first to achieve an initial high degree of completion—that results in incomplete locations (and problems finishing).

Third, splitting sometimes occurs because the subcontractor has increased resources but they are working in multiple locations rather than the same location.

Fourth, in large open spaces, the location boundaries may not be optimal for actual workflow and workers may not perceive a benefit in following the boundaries. This can be partially resolved by physically marking location boundaries on the floor. However, especially with MEP rough-in work, some overlapping of locations will often be inevitable because of the requirement to completely finish a duct or pipe which may span multiple locations before starting work on other ducts or pipes.

Change of sequence

There are many potential reasons for crews working out of sequence. Sequence changes can result from poor communication of the plan. The workers may not know where they are supposed to work. A sequence sometimes changes due to some locations being easier than others and these then allow the subcontractor to finish more work if he starts in the wrong location. Prerequisites can often cause changes of sequence if they are not ready in time.

Interrupted work

Work can be interrupted for many reasons. One reason can be that the preceding trade is proceeding too slowly. Interruptions can result from bad weather or the subcontractor needing the crew on some other site. Sometimes it is impossible to finish the location.

Learning from mistakes

The reason all schedule deviations should be documented is to enable learning from the past to be applied to similar projects in the future. This is especially important if the deviation could have been prevented by controlling the prerequisites of production more carefully. Common causes for deviation include that something has been overlooked in the design and a request for information has not been sent in time, or materials were incorrectly ordered. These could be prevented quite easily by including accurate checking of the design through a prerequisite checklist of all tasks. Virtual construction using 3D models (visisalisation of the schedule in 3D) provides another effective method for resolving many design problems.

The documented reasons should be used in planning the prerequisites of other tasks in the same project and when doing the pre-planning phase risk analysis in future projects.

What is the total effect of deviation?

The effect on a project of a start-up delay depends on the time buffer planned between the preceding and succeeding tasks. If there is no buffer, the start of the succeeding task will be delayed too. If there is a buffer, it can absorb a delay while the delay remains smaller than the buffer.

Figure 10.8 shows this scenario graphically with just two tasks. The same analysis can be used to forecast interference if the succeeding schedule task begins too early. Start-up delays can be prevented by good control of procurement and planning and controlling of starting prerequisites. If a start-up delay is smaller than the buffer between tasks, it does not cause problems by itself. However, because it uses up the buffer between tasks, it increases the risk in the schedule.

Production rate deviations can be caused by a preceding task going too slowly or a succeeding task going too quickly. These deviations cause the succeeding task to lose its prerequisite of continuity in the future. This type of deviation can be prevented by accurate quantity take-off by location, and by accurate productivity information. The actual amount of resources on site should also be tightly controlled. Figure 10.9 shows the effects of production rate variation. If the problem is not immediately corrected, it will cause problems with production in the future.

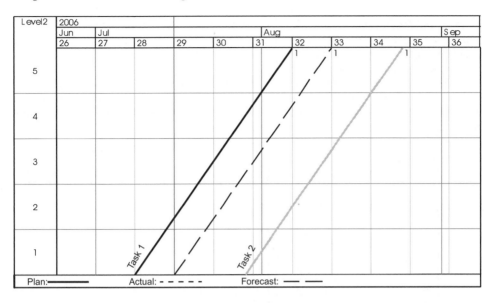

Figure 10.8 Effect of a start-up delay without a buffer

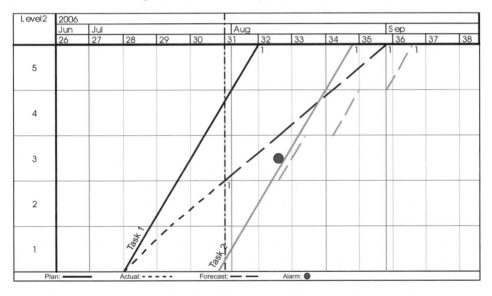

Figure 10.9 Production rate deviation

Splitting work to multiple locations at the same time is very common behaviour for subcontractors. Instead of finishing a location and then moving on to the next location, the subcontractors tend to begin work in multiple locations at the same time. If the schedule indicates that the subcontractor should have two crews working to achieve the desired production rate, it is likely that the crews will work in two locations unless they are actively controlled. This behaviour is often caused by the level of detail in the schedule being too rough. Instead of finishing all sub-tasks in the location, the subcontractor may do part of the work in multiple locations, which leads to not finishing locations on time. This deviation can be prevented by planning more accurate task schedules, taking into account the actual production of the subcontractor. The task schedules should be planned together with the subcontractor giving his input.

This deviation results in locations being unfinished and disrupts the work of the next trade. Figure 10.10 shows the effects of splitting. The total production rate is almost the same as planned (forecast in Location 5). However, the changed pattern of work prevents the succeeding task from working according to plan.

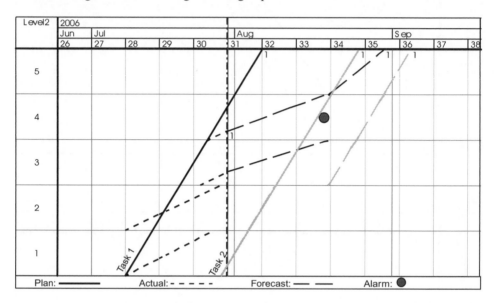

Figure 10.10 Effect of splitting

The fourth deviation type, working out of sequence, is also very common. Location-based planning explicitly plans the sequence in which each trade goes through the building. If a trade breaks the sequence, it will force the following trades to also break sequence. If there are buffers, they can usually absorb a few breaks in sequence without adversely affecting other trades. Without buffers the sequence of the following trade has to change or the flow will be interrupted. If the sequence is not tightly controlled, the project becomes more difficult to control. Working out of sequence also results in cascading delays because it usually leads to multiple crews working in the same location with consequent slowdowns and discontinuities of work.

Working out of sequence can be prevented by prioritising the make-ready process. For example, if materials are not delivered to a location, it is impossible or costly for the subcontractor to start working there out of sequence. Figure 10.11 shows the effects of working out of sequence. If Task 2 does not respond by changing sequence, it will have problems. If it changes its sequence, all following trades in turn will need to change their sequence as well.

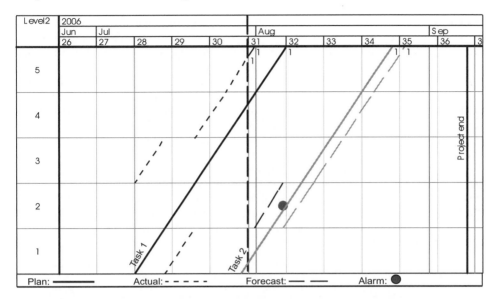

Figure 10.11 Effect of splitting

Which control actions can be used?

After a deviation has occurred and its total effects are known, all possible control actions should be listed. Which control actions can be used and on which tasks? This depends, for example, on the contracts with subcontractors, resource availability and the willingness of workers to work longer days. Control actions should be evaluated for currently ongoing baseline tasks.

Possible control actions may be broadly classified into five types (Seppänen and Kankainen, 2004):

- Changing the production rate
- Changing the work content
- Breaking the flow of work
- Changing the location sequence
- Overlapping production in multiple locations.

Additionally it is possible to accept the delay and take no action.

Changing the production rate

Changing the production rate of a delayed task is usually the best type of control action because it maintains work continuity. The production rate can be increased by adding more resources, by working longer days or on weekends or by increasing the productivity of the crew by reducing non value-adding activities.

The option to add resources is constrained by resource availability and by contracts. Indeed, all of these control actions may require support from the contract to enforce, or persuasion will become necessary. If the desired production rate has not been explicitly mentioned in the contract (usually by reference to location-based completion dates or quantity milestones), the subcontractor is rarely contractually bound to add more resources or provide another immediate solution. If there are no location-specific milestones, the contractor does not need to catch up immediately. In these cases, contracts limit the available control actions. However, if the subcontractor can be shown the benefits of adding resources now (for example, achieving an even workload, preventing the forecast delay of a milestone or forecast disturbances in the future, even promising more work from the general contractor in the future) it is often possible to persuade them to do so. Location-based forecasts are an effective communication tools in these discussions. If the work is being done by direct labour, these control actions are easier to implement as long as more resources are available.

For example, the subcontractor could be shown the flowline in Figure 10.12, highlighting just their task and the contractual milestones. If they continue with the current production rate, both milestones will be delayed and they will have to pay penalties—3 weeks for the first milestone and 6 weeks for the second milestone. The general contractor and the subcontractor can figure out together the best way to avoid the penalties and to restore the production rate.

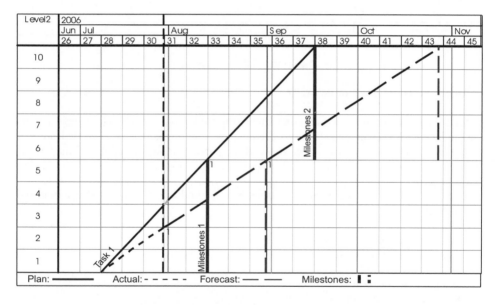

Figure 10.12 Effect of low production rate on future milestones

Changing the work content

Changing the work content is another possible way to change the production rate. It means that the specific trade crew will do more (or less) productive work in each location. For example, the crew installing plasterboard walls could also install the frames for the suspended ceilings in the same location before moving on to the next location. Adding work will slow down an overly fast trade. This is easy to do in practice with direct labour but contracts with subcontractors might make this more difficult to achieve once construction is underway. In normal cases, the subcontractor will be willing to do other work at an hourly rate, but this may lead to an increase in the overall cost. Changing the work content also requires multi-skilled workers unless the work is of a very similar nature (for example, the frames for plasterboard walls and the frames for suspended ceilings).

Breaking the flow of work

Breaking the workflow means making the succeeding trade discontinuous. With subcontractors, this usually occurs automatically as soon as the crew notices that they do not have enough work made ready by the earlier crew. If the crew leaves without explicit decision by the project manager, this phenomenon is called a disturbance instead of a control action. However, when taking this as a control action, the crew will be instructed to come back on a given date when there will be continuous work available. Instead of having multiple breaks of work, they can plan to have one break of longer duration. Nevertheless, this control action may cause more problems later because the crew may not come back when it is needed, or other workers may come in their place—losing the benefits of learning.

Direct labour or multi-skilled workers (such as MEP) make this control action easier to implement where there is other work available, such as a workable backlog. Production management should also make this decision with direct labour because the workers will have a tendency to slow down when they are running into a preceding crew—causing a loss of productivity while appearing to remain busy.

Changing the location sequence

Changing the sequence of locations is always easy to implement. However, it often causes more problems than it solves. Shifting the sequence too often will cause the subcontractors and workers to lose sight of the overall pattern. This may destroy the overall sequence and workflow is very hard to restore. Often, production prerequisites are made ready in the planned sequence which may also increase the risk of work stoppages for future trades. Changing the sequence also requires sufficient buffer to remain such that the new sequence will not move into unready locations.

Overlapping production in multiple locations

Overlapping production in many locations is the traditional way of catching up. Resources are added to all available locations. This causes problems in production control because it is difficult to know who should be working where. Crews become more difficult to track, and it becomes harder to control production rates and to plan control actions in the future. The control action will almost certainly have the undesired consequence of reduced productivity (workers end up walking around looking for work), while other trades may follow suit. If well controlled, this is a valid control action in those cases where resources cannot be added to a single location (for example, due to space constraints). However, the subcontractor should be notified that the workers are allowed to work only in the locations designated by project management and a different crew should work in each location. Otherwise, the subcontractor may simply split the crew to multiple locations and each will be completed slower. This control action can also be used in the case where the general contractor agrees to supplement the resources of a subcontractor whose task production rate is too slow, by hiring an additional new subcontractor or by using direct labour. In this case, it makes sense to divide responsibilities of the different subcontractors by location.

Optimal control action plan

An optimal control action plan is that plan which will deliver the optimum outcome for the project overall. After all available control actions have been listed, an optimal combination of control actions (an optimal control plan) should be selected. The forecast production system cost and risk are the best criteria for optimisation because they can be used to calculate the total effects of control actions. By knowing the impact on the costs of different actors, it is easier to find a solution which is fair for everyone.

Control actions costs, from the standpoint of a general contractor, depend on whether the crews are directly employed or subcontracted. Direct labour is easier to control but any break of workflow, and associated loss of productivity, will directly result in increased costs. Subcontractors are more difficult to control but the cost effects of control actions depend on the contract and the negotiating power of the general contractor. There is often no direct cost associated with breaks of the subcontractor's workflow—however because of the possibility of a return delay, the production system risk increases. Direct cost effects for the general contractor are summarised in Table 10.1.

Table 10.1 Direct cost effects for the general contractor arising from control actions

Control action type	Cost: direct labour	Cost: subcontract
Adding resources	• Additional mobilisation decreases productivity and results in higher direct labour costs.	• Costs depend on negotiating power and contracts. • No cost if milestone or production rate target set in contract is endangered or if the subcontractor has resources available. • Otherwise normally payment by hour; subject to agreement. • Sometimes a new subcontractor can be added with costs subject to agreement
Working overtime	• Overtime compensation.	• GC may have to compensate for overtime. • No cost if milestone or production rate target set in contract is endangered.
Changing work content	• No additional cost effect if the workers have the necessary skills and can achieve the planned productivity.	• Contract needs to be revised to add / remove work or hourly rate of the subcontractor can be used.
Breaking flow	• Workers should have other work available (either in this project or somewhere else) — otherwise a direct cost effect.	• Depends on the contract. Usually the subcontractor doesn't need to be compensated for breaking of flow (he has already factored this inefficiency into his bid).
Changing sequence	• Usually no direct cost effect	• Usually no cost effect. If the payment schedule has been tied to finishing locations, the subcontractor may need to be compensated.
Increasing productivity	• Decreases direct work cost because productivity increases. May cause increased coordination costs depending on actual control action.	• No cost effect. May cause increased coordination costs depending on actual control action.
Overlapping production	• Usually no direct cost effect.	• No cost if milestone or production rate target set in contract is endangered or if the subcontractor has resources available. Otherwise normally payment by hour; subject to agreement.

Production system cost and risk tools are more useful to optimise control actions, because they help to optimise the production system, resulting in 'win-win' situations. This optimisation considers the constraints related to control actions and the cost and risk effects in the complete production system, instead of just the direct immediate cost effects to the general contractor and subcontractors. Table 10.2 lists the typical production system cost and risk effects and constraints related to typical control actions.

Table 10.2 Production system impact of control actions

Control action type	Effect on production system cost	Effect on production system risk	Constraints
Adding resources	Assumes the same productivity → no additional production system cost effect.	Risk of not getting the resources, additional production rate risk, waiting time or demobilisation in future affects more resources.	Resource availability, contracts, location size may prevent adding new crews to same locations.
Working overtime / on weekends	Lower productivity for overtime, overtime payments → increased production system cost.	Small effect on risk because the same workers are used.	Labour unions, willingness of workers to work long hours, contracts.
Changing scope of work	Less learning → small increase in production system cost. May have lower productivity for additional work— increased production system cost depending on skills.	Production rate variability for additional scope may be higher.	Requires suitable work, needs multi-skilling.
Removing non value-adding activities	Increases productivity → saving in production system cost.	No effect on risk.	Current process must be inefficient.
Breaking flow	Additional mobilisation and demobilisation → increased production system cost.	Increased production system risk because of return delay risk.	
Changing sequence	Relocation costs.	Additional risk for future trades who also need to change sequence.	There must be a workable backlog location available.
Overlapping production	Assumes the same productivity → no additional production system cost effect.		There must be a workable backlog location available.

The schedule risk effect of a control action plan generally depends on two factors: first, how likely is it that the control action can be carried out as planned and second, will the control action cause problems to other trades.

The first factor should be evaluated separately for different tasks. For tasks done with directly employed resources, the reliability is often higher. For subcontracted tasks it all depends on the willingness and motivation of the subcontractor to implement the control action. Often money is the best motivator: if a payment to a subcontractor is endangered or a penalty threatens, it is usually easy to persuade the subcontractor. This is simple to show by using flowline diagrams and comparing the planned, actual and forecast production rates. Otherwise, good leadership and negotiating skills are needed. If a control action is not possible because of an unsuitable contract (for example, no milestones for locations or payments are based on completed quantities instead of completed locations), it is still possible to save the situation through good communication.

Alternative control action plans should be quantified, in terms of both production system cost and risk, to arrive at an optimal solution. Often the solution including risk is different from the solution that would result if only optimised on the general contractor's direct costs. For example, requiring additional resources may have a high risk factor associated if the subcontractor has lots of work on other sites. Instead it might be better, from the point of view of the entire production system, to get another subcontractor in or to allow direct labour to help and to accept a short-term cost increase to prevent future problems.

Example of control action optimisation—Opus

Earlier in this chapter, an example was presented that analysed two tasks—building services bulkheads and plasterboard walls for the Opus project. The subcontractor for both tasks was an Estonian company working in Finland. It had many large projects and few possibilities for adding more workers in a foreign country. The subcontractor had two milestones, one for each section.

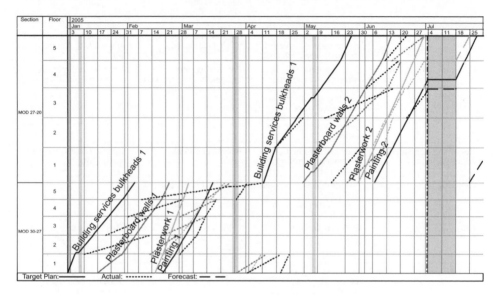

Figure 10.13 Control action optimisation using detail tasks—Opus, Finland

The subcontractor was delayed in the first section such that they could continue with the same resources into the second section. However, because the same resources were used to cover both of the baseline tasks, they achieved about half the production rate. The production management team did not believe the subcontractor had the ability to provide further resources to resolve their production problems, so another control action was selected. The succeeding tasks, plasterwork and first painting, were to be done by the painting subcontractor. The detail tasks for that subcontractor were changed so that, on the first round, they just plastered the exterior walls. Taping and finishing walls was done by the painting crew, which started after the plasterwork crew. This allowed the first baseline task (plasterwork) to break the dependency to plasterboard walls and thus most of the delay could be caught up without an increase in production system cost or risk. This is shown in Figure 10.13. In section MOD 30–27, plasterboard walls finished four weeks later than the baseline. However, the successor tasks finished just two weeks late because the dependency of plasterwork to walls was broken by the scope change. In section MOD 27–20, there was a delay of three weeks. However, the control action meant that successor tasks only had small delays.

Updating forecasts with control actions

After the control action plan has been selected with the cooperation of subcontractors, the adjusted forecast becomes the plan to be followed. The adjusted forecast will be used in subsequent steps in the weekly control process to communicate resource needs, to plan upcoming commitments, to prioritise make-ready processes and as a basis for weekly assignments for plan implementation.

EVALUATING RESOURCE NEEDS

To get the most benefit from the location-based controlling system, the resource needs of the most important subcontractors in the project should be evaluated and communicated weekly. To reduce the risk of additional mobilisations, subcontractors should be notified well in advance of additional resource requirements. This is especially relevant for multi-skilled contractors (such as, MEP contractors) because it is very difficult for them to know when new resources should be mobilised. MEP contractors often create accurate schedules for their own work but, because they do not have good enough information regarding other trades, their resource forecasts have a tendency to be inaccurate. The general contractor can do a lot to improve the resource forecast to both parties' benefit.

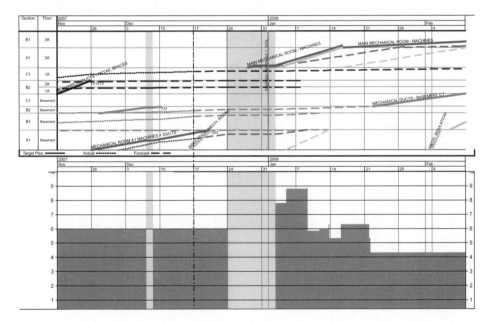

Figure 10.14 Resource needs forecast: top—detail tasks; bottom—resource chart

The resource forecast is calculated based on the detail task forecasts and control actions. Forecasts assume that actual behaviour will follow the detail plans in the future unless control actions change the plan. For example, if a detail task assumes four workers yet three workers have actually been working on the task, the forecast will assume that one worker will be added unless a control action has been planned to make the change.

Figure 10.14 shows a typical example of a resource need forecast for forthcoming weeks. The top part of the figure shows the detail tasks and the actual production of the mechanical subcontractor working in a large project. The bottom part shows the actual resource graph and resource forecast for the forthcoming six weeks required to achieve the schedule forecast. Because most of the locations are unfinished, and many tasks are going on in various locations and the original plan has not been followed, it would be very difficult to control what should happen without the resource forecast. The forecast shows that most of the unfinished locations should be finished after the Christmas break by adding three more workers for two weeks. In the long run, the resource needs decrease to four. This resource forecast is updated weekly based on actual production and forecasts.

CREATING REPORTS FOR A SITE MEETING

A big part of production control occurs in weekly site meetings, where the subcontractors report on their progress, deviations are analysed, the status of design is reviewed and control actions are decided. This is the best forum to present location-based production information and to get commitments to control actions because all the main parties related to production are usually present in site meetings. However, location-based plans and control information tend to be very complex unless restricted to views of information relevant for decision making. Therefore, creating distilled reports for a site meeting is a critical part of the weekly controlling process.

Typically, a combination of control charts, flowline figures, resource diagrams and production graphs can be used. This section explains the most effective reports for site meeting communication. Reports should be sent to participants well before a site meeting so that participants have time to understand their content.

Control charts

Control charts are very easy to understand and can convey a lot of information on one sheet of paper. The most important things to consider when creating control charts for communicating at site meetings are:

- At least one control chart should show all the important tasks
- Provide charts with only views of specific tasks and their related tasks
- The layout should remain consistent from one site meeting to the next
- Site meeting control charts should compare progress to detail task level commitments.

All important tasks in one control chart

The status of all important tasks should be reviewed at every meeting. Importance can be defined by criticality and float but also by considering the economic importance of the task and its effects on other production. Only tasks which are very small in terms of cost, have a lot of float and have no effects on other production should be left out.

Views of related tasks in locations

Control chart print-outs are most effective if they can show cause and effect. This can be accomplished by selecting dependent tasks in chronological order along the horizontal axis and related locations on the vertical axis. This means that each control chart print-out has the work of multiple subcontractors in it. Cause and effect can easily be read horizontally. If a predecessor subcontractor has a yellow square meaning a delay in the location, then the succeeding subcontractor most probably will be red, as it will also be delayed. A similar understanding cannot be achieved by simply looking at control charts for each subcontractor in some other sequence.

Related tasks are easy to choose by construction phase and location type. Good examples include:

- Earthworks, foundations and structure control chart
- Structural and enclosure control chart
- Corridor MEP and finishes control chart
- Office room MEP and finishes control chart
- Suites MEP and finishes control chart
- Stairwell room MEP and finishes control chart
- Restroom finishes control chart.

The same task can be shown in multiple control charts, depending on its logic. For example, structure could be the last task group in the earthworks, foundations and structure control chart, yet all finishes control charts and the enclosure control chart should include this information to make clear when a floor is ready for the following tasks. Figure 10.15 shows a control chart for drywall and related tasks. The top row of numbers in each cell indicate the planned start and finish dates, the bottom numbers are actual start and finish dates, or the completion rate in those locations where the location has started but not yet finished. This chart immediately shows areas that are not completely finished, which has resulted in the next trade being unable to completely finish the same location. Also, setting out is going well ahead of other production and work is not being released by the drywall subcontractor responsible for the first three tasks to the electrical and waterproofing subcontractors. The sequence of tasks displayed has been selected so that logic flows from left to right. Theoretically, the completion status of a cell on the right should never be better than the status of the cell on left.

The layout should remain consistent

Participants will gradually begin to understand the information presented in the new format as they attend multiple site meetings. To gain the maximum benefit from this learning, the format of print-outs should be decided as early as possible and remain consistent during the project. The same print-outs with the same locations and the same tasks will then be presented in all subsequent site meetings, until all tasks in that control chart have been completed. The only things that should change are the colours and dates in cells.

This solves the problem with the current practice of printing out a new set of Gantt charts every week with different information content. Getting different reports every week can make life very difficult for the subcontractors.

Location-Based Management for Construction

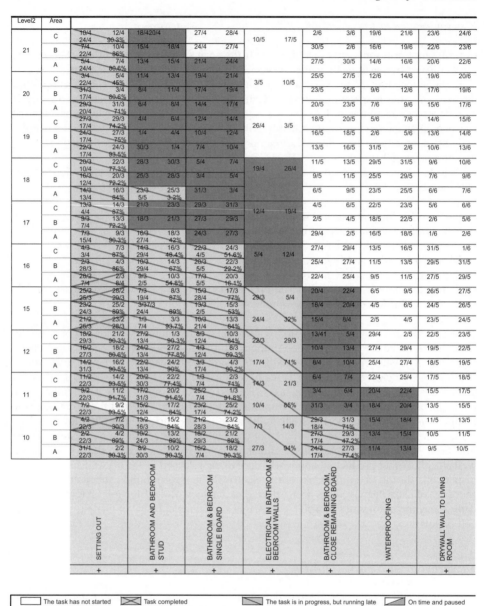

Figure 10.15 Control chart for drywall

Progress should be compared to detail task commitments

Site meeting control charts should compare progress to the detail task commitments. The baseline may not have much to do with current reality and its role is more important in reporting progress to management and the client. To make this work, guidelines for updating detail tasks and commitments are needed. Control charts will not work if the plan is always updated weekly—merely to correspond with the actual progress.

Flowline figures

In large projects, flowline figures tend to become quite complex without filtering. Subcontractors who are not used to reading flowline figures will heavily resist if they find that the diagrams are difficult to understand. Good practical guidelines for flowline figures to be used for site meetings are:

- Concentrate on showing problems and control actions—show problem task, predecessor and successor tasks.
- Avoid figures with too many lines and figures where lines are crossing.

Show problems and control actions

Flowline figures are best used to show current and forthcoming problems caused by deviations in start dates, production rates or the suspension of tasks. To illustrate the problem and its immediate effect, it is usually sufficient to show the problematic task and the immediate predecessor and successor tasks. Two versions of the figure should be presented—one assuming that no control actions will be taken and another with the planned control action. By showing the problem and the agreed solution to the problem, the belief of the participants in the production control system will be strengthened.

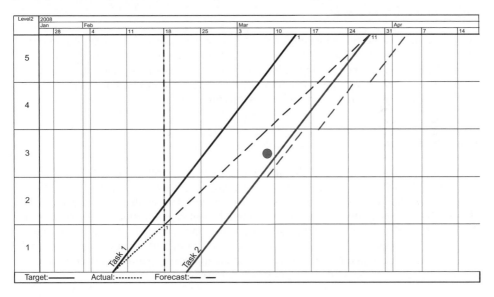

Figure 10.16 A production problem illustrated in a flowline

Figure 10.16 shows a flowline figure for a production problem. Figure 10.17 shows the same situation but with an adjusted forecast based on a documented control action to add another crew for locations 3, 4 and 5. Peer pressure from other subcontractors, to undertake the control action as agreed, will follow presentation of the agreed control action.

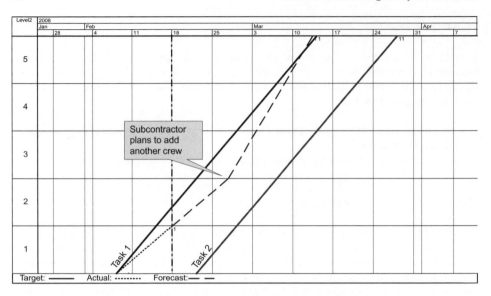

Figure 10.17 A production problem with adjusted forecast for planned control action

Make flowline figures simple

Flowline figures work in site meetings only if they are very clear and simple. A good basic rule is that each single A4-sized figure should not contain more than 20 locations and ten tasks. There should not be crossing lines or very complex work patterns because these will not be easily understood and will result in resistance. Ideal figures have locations sorted so that workflow proceeds from bottom left towards top right.

Production graphs

Production graphs are useful if the actual workflow has been very different from the planned workflow. This will result in the corresponding flowline figures looking messy. Figure 10.18 shows the actual progress flowline of the steel contractor compared to the planned progress. The steel contractor is erecting half of each location to roof level before doing another half. Because the schedule has not been updated to correspond with the actual workflow, the floors in these locations do not seem to be finished. The best approach would be to update the schedule to correspond with the actual logic, but the production rate can also be verified from the production graph, shown in Figure 10.19.

Gantt charts

In projects implementing the location-based management system, Gantt charts should only be used for visualising weekly plans. Assignments and weekly plans will be discussed later in this chapter.

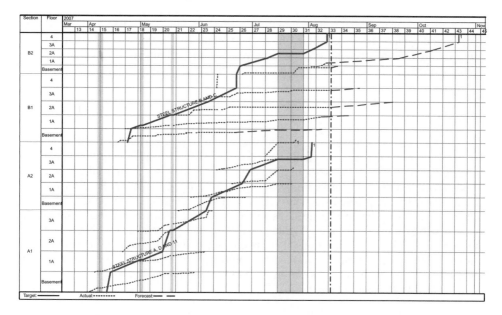

Figure 10.18 Flowline of actual progress of the steel contractor compared to the planned progress

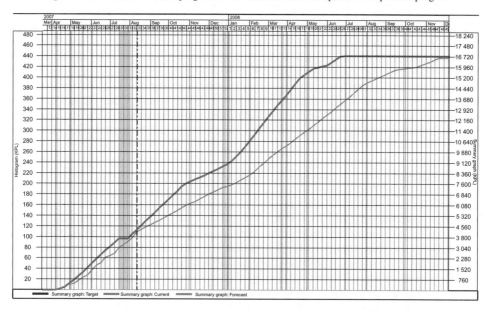

Figure 10.19 Production graph of the steel contractor compared to the planned progress

SITE MEETINGS

The parts of the site meeting which discuss the schedule and production need to change when implementing the location-based management system. It should not be necessary to waste time explaining what has been done during the previous week, because control charts will have been sent to participants before the meeting. It is more important to discuss deviations from the plans, any emerging problems and proposed solutions. This is best done by going through the reports and documenting reasons and actions to the diagrams. These modified documents may then be appended to the meeting minutes.

Control charts

Control charts can be reviewed by area. All yellow (started but delayed), red (not started and delayed) and suspended squares should be addressed to discover the reasons for either delay or being unable to finish the work in a location. Often, factors outside of the production schedule will be preventing tasks from starting or finishing in a location. By discussing all locations which have problems, a comprehensive list of reasons may be generated. In the authors' experience of many site meetings, many issues that were not previously understood by the general contractor were highlighted by this process. Control charts should never be used to plan or inform control actions, they are only meaningful to visualise deviations from commitments. Control actions informed by control charts typically fail because control actions need information about location availability and production rate—which cannot easily be read from a control chart.

Flowline figures

While control charts are great for managing by exception, flowline figures can prompt reaction before a problem occurs. Similarly to control charts, identified problem spots should be discussed. Because the control action process will have been carried out prior to the site meeting, the purpose of discussing agreed control actions is to make them part of the official minutes and to make sure that everyone commits to the control action plan. Because control actions often prevent problems in another subcontractor's production, discussing and committing to the control action in the site meeting increases trust in the management system. Production graphs can be used in the same way to discuss production issues.

Prerequisites of future production

Time should be devoted to discussing production during the upcoming three to six week period to make sure that all issues which are outside of the production schedule are dealt with in each location before production starts. It is very common that production rates follow plans, nevertheless a missing design detail or material delivery may prevent work from starting in a location. The only effective way to deal with these issues is to create a location-based checklist for each task, listing all known issues affecting the start of a task in a location and the completion of a task in the location. This list will be used later in the control process in prioritising management actions to resolve any issues.

Site meeting minutes

Annotated control charts and flowline figures should be appended to site meeting minutes. Meeting minutes should concentrate on forthcoming production, and any agreed control actions to prevent problems and ensure smooth production in the future. In current practice, site meeting minutes all too often tend to focus on the past (this was done last week) or the present, with very little focus on the only part of production which can be changed—the future. In location-based management, discussing the past is only relevant if it helps to prevent similar problems from occurring in the future. Most of a meeting (and the official meeting minutes) should be used to discuss how to make the next few weeks as productive as possible and to prevent problems before they occur.

CLIENT AND MANAGEMENT REPORTING

Client and line management require project progress reports focusing on schedule, cost, safety and quality. Often, management and each particular client have a standard form in which they require information. This discussion assumes that reporting can be changed to utilise location-based management concepts.

All client and management reporting is done by comparing the current, actual and forecast information to the baseline information. The best reports for client and management reporting are:

- Completion rates
- Production graphs by construction phase and earned value
- Summary control charts
- End result forecasts
- Variation reports.

Completion rate reports

Completion rate reports compare the physical degree of completion, at status date, to the baseline degree of completion at status date. This information can be reported at the project, construction phase, subcontractor or task level. Completion rates are calculated based on work complete, at status date, by weighting each completed task and location with the number of worker hours.

The benefits of using completion rates include that they give a good summary of schedule progress compared to the plans. However, completion rates only provide a static view of production and do not indicate whether progress is accelerating or decelerating. Additionally, they are silent about the total effects of delays. Therefore, this information needs to be augmented with other, more dynamic data. When reporting completion rates, it is good practice to also show target and actual completion rates for the previous report and to give an indication of whether the situation is getting worse or better.

Production graphs by construction phase and earned value

Production graphs give similar information as completion rates but in a dynamic format. The most useful production graphs for client and management reporting have the value of production on the vertical axis, measured in worker hours, and time on the horizontal axis. Such a graph shows the actual speed of production compared to the plan. By analysing the slope, management (or the client) can immediately see if production is getting back on track or if it is getting further behind. The horizontal difference between the baseline and the actual line shows the current schedule delay. If the actual worker hours consumed are added to the figure, the graph changes to the earned value and thus also allows analysis of the productivity in the project (how many hours have been used compared to finished production).

Care must be taken when interpreting production charts. The schedule delay shown by the production graph is not totally accurate because it does not take into account out-of-sequence work or the downstream effects of tasks. For example, if non-critical work has been done faster than in the original plan and critical tasks are delayed, the production graph can still show that production is on time, even though the schedule is not. However, it gives a good indication of the overall production and most importantly the current overall production rate. If the overall production rate falls, it signals big difficulties for the project as lost momentum is not easily regained.

Summary control charts

Control charts can be used for management and client reporting purposes by comparing baseline dates to actual dates. Information can be presented at the summary (for example, the construction phase) or task level. The key is to compress the information enough so that it is readable on one sheet of paper.

Control charts can graphically present the status of both tasks and locations. However, similar to completion rates, control charts are static reports and cannot show the total effect of deviations.

End result forecasts

A report of the forecast finish dates for a project, its construction phases, milestones and its individual tasks will answer the deficiencies of other reports. Such reports take into account the production rates and the schedule logic. However, forecasts can be artificially manipulated to make a false impression by changing detail tasks and by adding control actions which are not based on reality. If reporting is to be transparent, the forecast should be reported before the effect of any control actions and after any control actions and should include documentation regarding all agreed control actions. These figures should also be reported compared to the same forecasts for the previous week. Note that showing forecasts before and after control actions requires that some training be provided to management and the client, otherwise the difference between adjusted and unadjusted forecasts tend to look very scary, especially on larger projects. The idea is to show that project management is aware of a problem, they have a plan to fix it and can show the total effects after the fix.

Variation reports

To show the effects of variations (following a variation or change order), baseline tasks need to be updated. Variations can change the quantities of tasks, add new tasks, remove tasks or change logic. The schedule effect of a variation can be shown by updating the baseline tasks without changing any other logic. This will keep the buffers, resource requirement and flow intact and shows the effect of variation without any loss of productivity. In cases where commitments have already been made to detail tasks with subcontractors, it becomes more difficult to show total effects.

DETAILED PLANNING

Detailed planning includes updating existing tasks to correspond with current commitments and information, and planning new detail tasks. This section presents guidelines about when plans should be changed in preference to planning control actions to restore production to an existing plan.

Updating baseline tasks

Baseline tasks describe the commitment of production management towards the client and/or company management. Thus they should not be updated unless agreed to by the client or management. A baseline update is allowed in the following situations:

- Approved variation
- Uncontested delay caused by client
- Logic error in the baseline
- Irrecoverable delay of one month or more
- Approving the baseline in stages.

Approved variation

If a client has approved a variation, it should be updated to the baseline schedule either as a new task or a change in the quantities of an existing task. If the variation naturally belongs to the scope of an existing baseline task it is better to change the quantities of the related task. This will show the effect on the schedule while preserving the continuous resource use, assuming the same resources and not consuming any planned buffers. The project can then be re-planned to achieve the original duration, or alternatively the duration can be extended. This depends on negotiations between the client and the general contractor. Usually, any increase in resources or decrease in buffers will cost money or increase risk for the general contractor. This can easily be shown using production system cost and risk tools. Location-based management allows the effects to be effectively communicated—but only if location-based planning was used to plan the original baseline.

Uncontested delay caused by the client

Delays caused by the client, or by force majeure which the general contractor could not anticipate, can be added to baseline (this depends on contracts and national issues—for example in Finland, general contractors are required to assume at least 12 days of severe weather during winter). This requires a mutual understanding that the delay was caused by the client or acceptable force majeure. The delay can be added as a new task or as a work break on the days affected. As above, if the client requires the project to catch up, the cost effects of this can be shown using the original assumptions.

Logic error in the baseline

The baseline can be updated if there is a clear error in the logic. For example, the baseline schedule can be updated if the selected location breakdown structure does not work optimally for controlling the project. Issues which are commonly overlooked when deciding a location breakdown structure include the operation areas of mechanical equipment which need to be finished and cleaned all at once. If the location breakdown structure does not take this into account it needs to be changed, because this is revising the basic structure not just adding more detail to the LBS.

Updating the baseline due to planning mistakes should not result in an extension of project duration. Additional resources may need to be added, as necessary, to keep the original planned duration for each contractual milestone—unless the client agrees to additional time.

Unrecoverable delay of a month or more

For psychological reasons the baseline can generally be updated if there are delays which cannot be caught up. If the baseline is not updated, the control charts and flowline diagrams based on it will continue to raise alarms about the same issues—which cannot be solved. In a Finnish case study, this decreased the motivation of the team because control charts were mostly red all the time. The solution was to update the baseline and then control the remainder of the project so well that nothing was subsequently allowed to turn red. This helped to restore the project spirit and was thus beneficial for overall project success.

However, the baseline should never be updated for minor delays. Another Finnish case study was delayed by one month because of unexpected additional work in the foundation phase but the project team was able to catch up by controlling the project well and by careful use of planned buffers so that finishes could begin according to the baseline.

Approving the baseline in stages

In some large projects, the entire schedule cannot be approved at once. Schedules can be approved by construction stages (for example, substructure, superstructure and finishes) or by locations for example, Building A and then Building B). More detail can be added to unapproved parts of the baseline while controlling the approved parts using detail tasks.

Revising location breakdown structure

The location breakdown structure usually needs additional detail during the implementation phase. It might be sufficient in the baseline schedule to schedule at the floor level of detail, however in schedules guiding implementation, a room or space group level of detail is usually required. In hotels, the suite will be required; for residential construction, the appropriate level of detail is each apartment; in office buildings, space groups such as north offices, south offices and corridors may be used. If the appropriate level of detail is not in the baseline plan, the location breakdown structure needs to be changed.

Sometimes the location structure changes dramatically so that the baseline cannot simply be exploded into smaller sub-locations. In this case, the baseline needs to change as well, to maintain a link between the baseline and the detail tasks. This decision usually requires permission from management and the client.

Optimal level of detail for locations

The controlling process is more powerful when there are more locations. Ideally, the lowest level locations should be small enough that no more than one crew can work in the location simultaneously without interfering with another crew. However, locations must also be logical so that a crew can completely finish a location before moving to the next location. If crews finish locations partially, it will lead to a messy display of actual lines and problems in calculating schedule forecasts. Because the same location breakdown structure is applied to all crews sharing the space, this constraint should apply to most of the crews.

Sometimes, it is better to create new detail tasks instead of creating new locations. This applies for tasks such as pouring finishes concrete. It is not beneficial to create a location for every pour area because these locations are not likely to be needed by other tasks (other tasks use rooms instead, which are defined by walls which will not have been built yet). Instead, a detail task can be added for each pour—on a hierarchy level immediately above the room level. Alternatively, pours can use the room level of hierarchy which requires estimating which rooms are affected by which pour.

Updating quantities

Quantities for forthcoming detail tasks should be checked weekly. The reasons for quantity changes usually fall into the following categories:

- An error in quantity
- A variation
- More detailed quantities were needed for detail tasks.

Error in quantity

If there is an error in quantity, the correct quantity should be updated to the current quantities, not to baseline quantities. Baseline quantities can only be changed if the client has given the information, is liable for it and accepts that it is wrong.

Variation

Variations (or change orders) can be updated to current quantities until they are approved by the client. Then the corresponding change should be made also to baseline quantities. It is good practice to flag all unapproved variations to be able to report the monetary and time value of all unapproved variations.

More detailed quantities needed for detail tasks

The most usual reason for a quantity update is that the level of detail of baseline quantities is not accurate enough for detail tasks needed to model actual production. For example, in the baseline schedule there might be one quantity item for plasterboard walls. The required level of detail during production could be to have separate items for layout, frames, insulation wool and boards. Once again, this change should be made to current quantities rather than the baseline quantities.

Planning new detail tasks

Task schedules are formed by detail tasks which relate to a baseline task. They define how the target set in the baseline task will be achieved. The initial version of a new detail task should be planned before any call for tenders so that the information can be utilised at that stage. Optimally, before making a call for tenders, planning a task schedule should be incorporated into the list of required procurement events for all procurement packages. If work is going to be performed using direct labour, the task schedule initial draft should be planned before resources come to the site.

Creating a good task schedule requires the following information:

- Design
- Quantities
- Knowledge about alternative construction methods
- Links to other tasks
- Resource availability.

Task planning is an iterative process. Usually it is best for the project engineer to prepare a draft task schedule and then to refine the draft in subcontractor meetings and with the responsible supervisor. The task schedule is locked when a subcontractor (or workers) commit to it. After that, the detail tasks are not updated except in special circumstances (for example, if delays occur or better working methods emerge). The task scheduling process has six stages:

1. Estimate the quantities of the task by location on a more accurate level:
   a. more accurate locations
   b. more accurate quantities
   c. current information about unit costs.
2. Explode the baseline task to smaller detail tasks.
3. Explicitly plan the resources used in detail tasks (who is going to do what, naming the resources if known; if not, numbering them).
4. Define the logic of the detail tasks.
5. Iterate until a feasible plan satisfying the baseline constraints has been achieved *or* until it is certain that baseline cannot be achieved. Minimise interference to other tasks.
6. Monitoring, controlling and updating task plans.

Each of the stages is described in more detail in the following pages.

Calculating quantities

Quantities used in the master scheduling phase are rarely accurate or at a sufficiently detailed level to support task scheduling. As the first stage of task scheduling, quantities are recalculated to reflect the current situation and how the work will actually be done.

Example 1. Concrete floor finishing work

In the original bill of quantities, the quantities can include just the total cubic metres of concrete needed and the total kilograms of reinforcement (rebar) needed on each floor. Task scheduling requires planning how big to make the pours and separating the rebar for each pour. Sometimes, the existing location breakdown structure needs to be augmented, by subdividing floors to spaces and estimating quantities using the more accurate LBS. Often, the type of concrete assumed in the baseline plan can be changed, which will change the relative rebar quantities (or remove the rebar altogether).

Example 2. Interior walls

The original bill of quantities may make assumptions about the types of interior walls of the building. The exact division into masonry and plasterboard walls and other wall types should be checked when doing the detailed task schedules. The wall types need to be checked to estimate the quantities of material needed.

Example 3. Roof work

The roof is often divided into work zones which are not known when planning the baseline, because they are a matter of preference for the subcontractors. For example, subcontractors may protect their work against weather by working under a tent and zones are dependent on the size of the tent and how easy it is to move them. The quantities should be recalculated correspondingly.

Exploding baseline tasks to detail tasks

New, accurate quantities are used to plan detail tasks for the baseline task. Before the actual planning can begin, the list of detail tasks should be decided. The detail tasks should accurately describe how the work will actually be done.

Example 1. Precast concrete structure

The precast concrete structure baseline task can be exploded to the following detail tasks:

- Columns, walls and beams
- Slabs
- Reinforcing and concreting slab joints
- In situ concrete areas
- Exterior wall elements.

Example 2. Concrete floor finishing work

The concrete floor baseline task can be exploded into the following detail tasks:

- Concrete floor finishing
- Stoppers / temporary walls
- Heating equipment (if needed)
- Reinforcement
- Pouring concrete.

<u>Example 3. Roofing work</u>

The roofing baseline task can be exploded into the following detail tasks:

- Smoke vents
- Pouring concrete
- Eaves
- Waterproofing
- Thermal insulation.

Planning resources

Resources for each of the detail tasks should be decided. Instead of using just the type of resource, all the resources can be assigned names or numbers so that it is possible to track any individual resource through the task schedule and see if the worker or equipment has waiting hours. This also makes it easier to develop personal assignments for crews based on the schedule.

Defining logic

The logic is defined between the detail tasks (sub-tasks) in the same way as with the baseline tasks. At the detail task level of accuracy, splitting a task does not necessarily result in increased production system cost because the same resources can be used in multiple sub-tasks. In normal task schedules some trades work continuously on a single sub-task across multiple locations while some resources work on multiple sub-tasks in a single location before moving to the next one.

Optimising task schedules

The flow of resources and the use of space should be optimised so that the baseline schedule objectives can be achieved. There are numerous variables that can be changed in order to optimise the task schedule:

1. Resources:
   a. Changing the number of resources
   b. Changing resources used in a sub-task
   c. Changing the flow of resource (the sequence in which the resource is utilised in the various sub-tasks and locations).
2. Detail tasks and their sequence:
   a. Changing detail tasks and their logic
   b. Making exceptions to the logic if necessary.
3. Splitting and continuity:
   a. As soon as possible
   b. Continuous work
   c. Splitting detail tasks
   d. Planning resource continuity across multiple sub-tasks.
4. Sequence of locations:
   a. Changing the sequence in which locations are built.

5. Workday lengths and days off:
   a. Planning overtime or weekend work for critical activities
   b. Taking agreed days off into an account.

There is less uncertainty associated with task schedules because they are done with good information regarding the work to be done. Thus large buffers are not generally needed. Instead the baseline buffers that are still unused at the time of planning can be used to provide more flexibility for task schedule planning.

   An optimised task schedule considers the following issues:

- Current schedule status of preceding tasks.
- Contracts and other commitments.
- Updated design information.
- Knowledge about the subcontractor and his resources.
- Knowledge about special requirements and risks associated with the work.
- Continuous resource use for all the resources.
- No resource overlaps—resources are used in only one detail task and one location at a time.
- Constraints set by the baseline schedule in each location.
- Costs less or equal to the budgeted cost for the baseline task both in terms of actual cost and production system cost.

Task schedule planning is illustrated with two examples from the case project (for case details, see Chapter 15). The case project is a business park of 15,000 m². It has been divided into two sections and has seven floors in each section. The sections are built in sequence. The task schedules of structure and concrete floor finishing work are presented.

Example of a task schedule—structure (Opus)

The first version of the task schedule for structure of the first section was planned three weeks before the start of the task. The quantities had only been estimates in the baseline stage. Quantities were taken-off accurately and productivity information from the subcontractor was used in task scheduling. Figure 10.20 shows the original baseline line and the vertical planning area boundaries signifying constraints from the baseline schedule. If these lines are exceeded by any detail task, the future trades may not be able to perform their work as planned in the baseline. The vertical lines do not follow the baseline exactly because of the buffers planned in the baseline schedule.

   The baseline task was exploded into five detail tasks on each floor: columns and walls, slabs, joint reinforcement and pour, cast in situ concrete and façade elements. The critical resource was the crane which was used in all the sub-tasks except for joint reinforcement and concreting. The use of the crane was planned to be continuous. It was not possible to plan continuity for concreting or joint rebar work because only one section was built at the time and elements could not be installed fast enough to provide continuous work. In November, one day each week was reserved for cold weather and during December and January, two days each week (In Finland workers do not have to work if the temperature falls below −15°C.) The resulting initial task schedule is shown in Figure 10.21. It can be seen from the figure that the original baseline schedule task was not feasible. Floors three, four and five cross the planning area boundary line, which means that the detail tasks of succeeding baseline tasks will have stricter constraints in those locations. There are empty spaces with no work happening because detail tasks have been updated due to delay outside

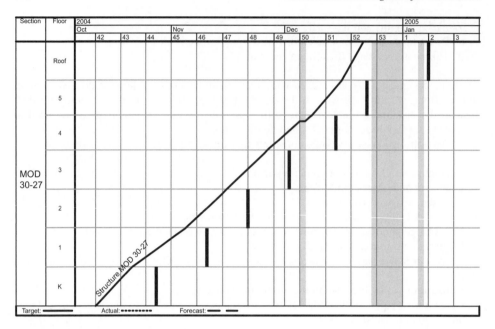

Figure 10.20 Baseline task line and associated planning area boundaries for Structure—Opus, Finland

the control of the subcontractor (missing connection design detail between the basement and the first floor).

The task schedule for the second phase was not planned at the same time because the project team wanted to use the actual production rates of the first phase once obtained.

Example of catching up previous delays by detail task scheduling—concrete floor finishing work (Opus)

Initial versions of detail task schedules are planned before any call for tenders, so they enable catching up on the previous delays from the baseline schedule. If the detail tasks of preceding baseline tasks exceed their baseline boundaries, it means that succeeding baseline tasks need to be produced faster. Production management can select the tasks which are cheapest and least risky to catch up. Catching up can be done incrementally (across multiple tasks), for example a delay of one month can catch up one week in each of four succeeding baseline tasks. When preparing detailed schedules for later baseline tasks, it is important to take into account the actual status of preceding baseline tasks and catch up if possible.

In the Opus example, concrete floor finishing work has structure and wooden windows as predecessors on floors one to five and steel structure of mechanical room on the roof level. The task schedule also has resource dependencies to the slab-on-ground task schedule, which utilises the same resources, and to the roof work task schedule because the same concreting crew pours concrete on the roof, preferably on the same day as they pour inside. Successors include all the other interior finishing work, such as suspended ceiling bulkheads. The mechanical room needs to be poured on the roof level before mechanical installation starts.

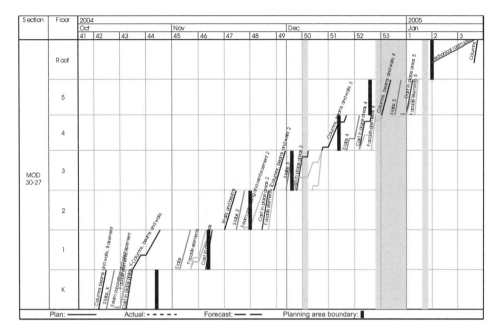

Figure 10.21 Detail tasks for Structure which fail to remain within baseline boundaries—Opus, Finland

When the task scheduling started, there had been start-up problems with the structure. This forced the concrete floor finishing work to complete in a shorter time period on each floor than in the baseline schedule. Pour areas were determined for each level and their total floor area calculated. Floors one to three had three pours, four and five were planned to be poured in two zones and the mechanical room was done with one pour. Each pour had the pour preparation task on the floor level and the actual pour day defined. Continuous flow was planned for the pour preparation crew.

Figure 10.22 shows the task schedule for concrete floor finishing baseline schedule lines and those of its immediate predecessors and successors. Planning area boundaries are shown with black vertical lines. The beginning of Pour preparation has been delayed by almost a month because of delays in predecessors and a change of sequence to finish floor two before floor one caused by missing design details for the first floor. Floors one and three go beyond the planning area boundaries but all the other floors can be finished according to the baseline schedule. The succeeding task, suspended ceiling bulkheads, will have a delay of less than one week. Delay caused by the missing design details and delay of structure can be caught up almost completely by accelerating the pour schedule.

UPDATING DETAIL TASKS

The detail tasks of any subcontract can be updated until they represent mutually agreed plans for implementation by both the subcontractor and general contractor. After this, they should only be updated if there is a change in work methods, a variation or a delay outside the control of the subcontractor. Basically, detail tasks function in the subcontractor-general contractor relationship in a similar way as baseline tasks function in the general contractor-client relationship. It is important to track commitment to detail tasks. The general

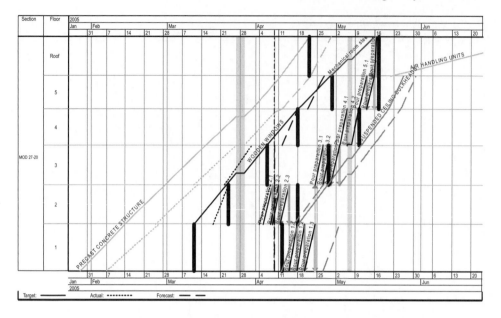

Figure 10.22 Task schedule of concrete floor finishing and baseline schedule lines
of immediate predecessors and successors—Opus, Finland

contractor commits to make the locations available for work according to the schedule and
the subcontractor commits to finishing locations on time. If there are deviations from
committed plans, control actions should be planned instead of updating the schedule.
Updating can occur in the following cases:

- Subcontractor negotiations
- Start-up meeting
- Delay caused by external factors
- Change of scope / variations
- Change in work methods.

Subcontractor negotiations

The first versions of detail tasks are often planned by the general contractor to be used in the
call for tenders. When the subcontractor is selected, the detail task schedule is updated
based on the subcontractor input. For example, work methods, resources and production
rates can change. However, the changes should not be allowed to affect other, already
committed detail tasks or cause additional delays compared to the baseline.

Start-up meeting

The start-up meeting is a good time to do a final revision of the task plan. After this, the
detail tasks related to the subcontract are locked by commitment and controlling begins.

Delay caused by external factors

If a subcontractor will be delayed by an issue outside their control, the detail tasks can be updated correspondingly. The updating process depends on the contracts between the subcontractor and the general contractor. In current practice, the general contractor may have more negotiating power and the subcontractor may be forced to catch up the delays of others without additional compensation. However, location-based management advocates a fair approach which assumes similar resource continuity and resource use. Delay is either accepted or the subcontractor is paid more money to catch up.

Change of scope or variation

Sometimes the scope of the subcontractor's work is changed to include other tasks. In this case, all the detail tasks of the subcontractor can be updated to achieve the best possible resource flow. A variation (change order) also triggers the updating process. An additional quantity is updated to detail tasks and detail tasks are then re-planned to take into account the new commitments. The finish date of the subcontract may or may not change, depending on the agreement.

Change in work methods

Finding a better way to do the work is a valid reason for updating the plan. New ideas often emerge after start-up difficulties are over. These may affect productivity, sequence of detail tasks or even logic links to detail tasks of other subcontractors. Detail task schedules should be used as a tool to see if the change is beneficial for the project before it is approved.

A work method change does not automatically necessitate changing the detail schedule. For example, if the change is expected to affect productivity, wait for actual information to show the actual effect, rather than changing the detail tasks. Task schedules should be updated if there are changes in the logic or resource flow through the sub-tasks.

ENSURING PREREQUISITES OF PRODUCTION

Each task and location has prerequisites associated with it before production can productively begin: preceding tasks must be finished and to the required quality resources need to be available, all issues related to design need to be solved. In the lean construction literature, ensuring these prerequisites by screening and pulling is called making work ready (Ballard, 2000). In the location-based management methodology, make-ready assignments are planned after all plans have been updated, the actual status is known and tasks have been prioritised based on their performance and float.

In the LBMS, prerequisites can be categorised into five basic groups:

1. Availability of resources and equipment.
2. Design.
3. Preceding tasks.
4. Procurement and deliveries.
5. Potential problems.

Availability of resources and equipment

In the ideal case, the work will have been planned to be implemented continuously with the same resources from the beginning to the end of the task. In this case, the resource prerequisite needs to be checked only for the first location or if there are breaks in the workflow. If resources are added during the task, their availability should be ensured in time. In practice, it is very common to plan increased resource needs for a complex location but then to never actually ensure that the extra resources will be on site when needed.

This prerequisite should be checked for every new planned or forecast mobilisation of resources, by calling the subcontractor and the workers and ensuring that they will be on site on time. If it turns out that the resources are not actually available, the forecast should be updated accordingly.

This prerequisite does not affect the workable backlog because, as long as there are some resources on site, work can be continued albeit slower than planned. However, the fact that there are fewer resources than planned should be updated to forecasts to examine the effects of slower production.

Design

The design and details should be completed and all the requests for information (RFIs) resolved before any work can be allowed to proceed into a location. Otherwise, there is a high probability of being unable to finish the location. The design should be pulled by the production such that a workable backlog of at least a few locations is always available.

Preceding activities

Preceding activities can be seen explicitly in flowline diagrams. If they are proceeding on the planned production rate, they will automatically create workable backlog equal to the space buffer in the schedule. Production problems and their effects on the workable backlog can be seen from the schedule forecast of the preceding activity. If the forecast indicates that the work will interfere with another trade, control actions should be planned to prevent a clash.

Procurement and deliveries

Depending on the chosen delivery method, deliveries might come separately for every location or every floor. The materials needed to finish a location should be available before starting work. If the same resources that are used in production are used in logistics, they should be available so that materials can be hauled to the work site. Otherwise, the materials should already be stored at the work site before the workers arrive.

Materials can be used to create a workable backlog by delivering in advance of production. However, in most cases they should not be delivered to the location before the preceding task has finished working there. Storing materials of the next trade often hinders production of the preceding trade, as well as risking damage to materials. The result is often that the preceding trade will be unable to finish their work completely in the location. The schedule forecasts can be utilised to optimise the actual delivery date (see Chapter 9).

Potential problems

Risks associated with a task should be identified before the start of production in a location. They should be communicated to stakeholders and preventive actions taken. Otherwise, the risk can actualise during production—causing work stoppages, increased cost or damage to workmanship and quality.

WEEKLY PLANS AND ASSIGNMENTS

Production is executed on site based on assignments. Assignments can relate to either work of production management (making work ready) or production-related tasks. To maintain productivity, production-related tasks need to go on continuously so that each crew needs to have sufficient available assignments to keep them occupied for the week. The planning of make-ready operations is based on prioritisation of tasks.

Defining production assignments

The target for production assignments can be taken directly from the schedule forecast, which has been adjusted by agreed control actions. The adjusted schedule forecast gives reliable information about how much work the crew can complete in a week based on historical production rates. If locations are small, assignments can be given in terms of locations to be finished in a week (for example, finish carpet installation in rooms 1, 2 and 3). However, sometimes work occurs in large locations and only part of a location can be finished during the week (or day, if controlling is done daily). In these cases, it is known how much should be produced but not exactly what: the assignment is not well defined. Drawings or 3D models can be used to highlight the production that should take place on any given day using the forecast production rate to see how much should be allocated for each day. The quantity of selected assignments can then be compared to the schedule forecast. If fewer assignments have been planned than the schedule forecast, the forecast can be updated correspondingly to show how much more needs to be selected for upcoming weeks. Planning well-defined assignments helps also with trade coordination if multiple crews are working in the same location.

Defining production management assignments

Tasks need to be prioritised so that management attention is focused on the correct things. The task of production management is to ensure that the prerequisites of production are available for all trades, procurement goes according to plan and information, and tools and resources are available before starting work. Additionally, management support and attention is needed for tasks which are not meeting their production objectives and which threaten production of other tasks.

Traditional measures like criticality are not enough in the location-based environment because production system cost and production system risk are affected by many other factors. Instead, tasks can be divided into three classes with each class having its own measure of criticality:

1. Tasks forecast to cause disturbances in the next few weeks.
2. Ensuring prerequisites for continuing tasks in progress.

3. Ensuring prerequisites for starting new baseline tasks.

This does not mean that group one is more important than group two, or it is more important than group three. All groups need actions weekly—only the priority measure within each group varies.

Tasks forecast to cause disturbances in the next few weeks

Disturbances increase cost and cause loss of productivity. They may also lead to cascading return delays. The location-based management methodology aims at minimising interference. Even if a task is not critical, it is important to get it back on track because return delays can cause succeeding tasks to become critical in the future.

In addition to planning control actions, management attention should focus on implementing those actions. Instead of controlling progress weekly, the problem tasks should be monitored daily. Otherwise, there is a risk that control actions will not be implemented as planned (for example, new resources do not show up, workers do not follow the agreed overtime hours, etc.). Following work methods closely also makes it possible for control actions to be aimed at increasing productivity. Prerequisites related to problem tasks should be made a priority to increase productivity.

Those tasks currently ongoing which are suffering from disturbances should be prioritised based on the forecast time until the next disturbance—the time until the next trade will be affected. Optimally, all delayed tasks receive daily management attention to ensure that control actions are implemented and the production can be restored to plan.

Ensuring prerequisites for continuing tasks in progress

For ongoing tasks which are not suffering from production problems, seamless continuation of production should be ensured. Production management needs to create enough ready locations that the work can continue as planned. This includes confirming that additional resources come on site if additional resources have been planned, making sure that materials come to the correct locations as needed and that design is complete and all requests for information have been resolved.

For ongoing tasks, the priority order is defined by the remaining buffer to succeeding tasks—the location-based free float. This is because low free float means that any deviation will cause earlier problems to other subcontractors.

Optimally, it should be possible to ensure at least two weeks of available work for each crew on site. If there is additional time, more work can be made ready based on location-based free float.

Ensuring prerequisites for starting new baseline tasks

The prerequisites for starting forthcoming tasks need to be assured in advance. These relate mostly to completing procurement-related tasks on time—completing design, planning detail tasks, calling for tenders, completing the tendering process, entering a contract and conducting a start-up meeting. The make-ready process for forthcoming tasks should be prioritised by using the location-based total float of the task. This is because, for tasks not yet started, detail tasks have not been locked. Most probably, succeeding detail tasks have not been locked either. Therefore, disturbances can be avoided in the detail task planning process, for example by delaying the start of non-critical tasks. Criticality, in terms of total project implementation, is the correct prioritisation basis for these tasks.

The planned procurement schedule should be implemented in all cases. If resources are available, procurement-related tasks can be started early based on the total float of the related baseline tasks.

COMMUNICATING AND IMPLEMENTING THE PLAN

Assignments should be clearly communicated to subcontractors, workers and production management to get their feedback and commitment. If it turns out that assignments cannot be implemented as planned, the forecast for the next week should be adjusted to take production realities into account. This may necessitate another round of detail task updating and control action planning, targeting coming weeks. In the end, all participants should agree and commit to the set of assignments for the next week (or day).

Some useful tools for communication include showing assignments for the next week in floor plans or 3D models. It is also good practice to show which actual elements belong to each location so that everyone has the same understanding of locations. In complex projects, location boundaries may not be clear and it is beneficial to clearly illustrate each location by using floor plans.

PUTTING IT ALL TOGETHER

This chapter has described how the controlling tools can be used to improve production control of the project. However, it is one thing to know about the tools and another thing to actually use them. To implement the system, all the parts need to be incorporated into a weekly routine. A sample routine is presented below. This discussion assumes a medium-size $100 million project with a project engineer having the responsibility to gather data and produce reports and a site manager and superintendent participating in decision making with subcontractors. In larger projects, more staffing may be required and roles might be different.

Monday am:
- Subcontractor meeting (site manager, superintendents, project engineer supporting):
  - Discussion of control actions and resource requirements
  - Final buy-in to weekly plan.
Monday pm:
- Project engineer updates look-ahead and forecast based on meeting results
- Project engineer sends updated look-ahead plan to participants.
Tuesday:
- Planning new detail tasks with subcontractors (project engineer, superintendent, site manager, subcontractors).
Wednesday:
- Getting commitments to non-committed detail task plans (project engineer, superintendent / site manager, subcontractors)
- Notifying subcontractors whose commitment need to be updated (project engineer).
Thursday:
- Other production system related activities (project engineer, superintendent, site manager):
  - Additional detail planning

- Prerequisites of starting and continuing upcoming locations
- Updating procurement and design status.

Friday am:
- Subcontractors send their production status by 8 am (subcontractors)
- Updating production information based on subcontractor reports (project engineer)
- Walking the site to confirm production status (project engineer, superintendent, site manager)
- Project engineer creates reports for subcontractor meeting:
  - Control charts
  - Flowline diagrams of production problems.

Friday pm:
- Identified production problems discussed with superintendents and site manager
- If resource requirements for the following week change from actual resource use this week, subcontractors are contacted to see if they are going to add / decrease resources
- Weekly plans are initialised from forecast adjusted for resource availability
- Superintendents work on weekly plans and get commitment
- Project engineer sends subcontractor meeting reports to meeting participants.

REFERENCES

Ballard, H.G. (2000). *The Last Planner system of production control*. Unpublished PhD thesis, School of Civil Engineering. The University of Birmingham, Birmingham.

Seppänen, O. and Kankainen, J. (2004). "An empirical research on deviations in production and current state of project control". *The 12th International Conference of Lean Construction*. Elsinore, Denmark.

Section Four

The location-based management system

SECTION FOUR—THE LOCATION BASED MANAGEMENT SYSTEM

The following chapters of Section Four re-introduce the location-based management system and draw together the components in a practical way to enable successful implementation. The first three sections of this book have been dominated by location-based planning and control—key components. This section moves beyond to describe the entire LBMS, the challenges in implementing the system and the differences in application which may be encountered on different project types.

The section builds a concise guide to the LBMS, a gathering of the essential components and how to use them. This is necessary as while the system is rich and complex—providing power—it is also elegantly simple in its essentials—providing ease of use and comprehension. It has been found that there are many tricks and techniques that can be applied in its use, and these are addressed in the later chapters of this section.

There are two broad facets to the LBMS which each need to be addressed in order to build an effective system. These may be termed the hard and the soft components, as they differentiate between the technical and functional capacity of the system and the social and team building components that are required for success. Each are as important as the other. In this section, the hard and soft component parts of the LBMS are discussed, addressing the associated high-level business advantages which flow.

- Chapter 11 describes the hard components and effectively provides a summary of the LBMS. This will provide a clear picture of the functions which need to be implemented.
- Chapter 12 discusses the resistance to change which will be confronted by management and the methods to use to overcome this resistence and successfully implement LBMS. These are the soft components of the LBMS. This makes use of the practical experience of companies that have adopted LBMS systems to various degrees.
- Chapter 13 discusses some of the various project types which may be managed using LBMS and considers the relative advantages. Different strategies relate specifically to different types of projects, such as residential, office, retail and health projects. The chapter also discusses special project types such as civil engineering projects and maintenance work and shows how they may be managed using LBMS.
- Chapter 14 looks at the special case of linear projects. These are projects for which location is intrinsic and linear, such as road, rail, tunnelling and pipe projects. The particular case of mass haul optimisation is considered in this context.

In this section, it is not possible to discuss the LBMS without reference to the technical solutions which have developed in partnership. The discussion will refer to the suite of software developed by Vico Software, as it is the only developer to specifically provide an integrated location-based planning, control and management solution for commercial construction. In 2009, Vico Software released its Vico Office suite—an integrated location-based software suite involving the BIM, estimating, scheduling and visualisation. This discussion also makes reference to the 2008 products, specifically Constructor, Estimator, Control, and 5D-presentor. In particular, Control is the location-based planning, scheduling and control tool developed by Vico Software. More information on Vico Software may be found at:
http://www.vicosoftware.com

Chapter 14 while considering the specific case of linear scheduling, makes use of DynaRoad. This software is dedicated to the production of T–D Charts for the construction of linear projects involving mass haul, such as road and rail projects. The 2009 release of DynaRoad 5 has greatly enhanced power for the production of efficient project schedules. More information on DynaRoad may be found at:
http://www.dynaroad.fi/pages/index.php?lang=en

Chapter 11

Location-based management system

INTRODUCTION

The book has introduced many concepts new to construction managers or which they will only rarely have had an opportunity to apply. Up to now, the book has concentrated on the essential location-based planning and control components of the system, and there has not yet been an opportunity to present the entire LBMS, or to deal with practical issues of implementation and the flow-on implications of changing management roles and strategies.

This chapter describes and presents the components of a fully functioning LBMS which is much more than merely planning and control. These are the components which an organisation should establish in order to manage projects using the location-based management system. This includes the technical systems required to support LBMS in construction, and the interface with other management systems such as design and estimating. Chapter 12 will then discuss the soft implementation issues.

There is very little new in this chapter, rather it is an overview and gathering of previous material—however it makes an excellent summary or introduction for new readers mainly concerned with implementation rather than theory.

What is LBMS?

The location-based management system (LBMS) is an integrated network of management system components potentially involving all stages of construction, from design through to completion. The system components are unified through their knowledge of location. Location allows the integration of many data components into a knowledge-base for a project. This makes the LBMS rich in integrated data. It is not a building information model (BIM) but rather a methodology for interacting with a BIM, placing demands on the BIM for both properties and characterisation (breakdown).

Using the location-based management system will lower costs due to design errors, improve constructability, reduce risk in project delivery and deliver reduced production costs when applied consistently across successive projects.

The LBMS is not a competitor to existing documentation and management systems but rather an extension or overlay. Much existing technology can be used, albeit in a different way. It must become location-aware and sensitive to production efficiency—and lean principles must be adopted to remove waste from the production system.

The LBMS provides finer control and improved reuse of data through using location as the unit of analysis. It provides more powerful project control through using tasks as the method of control, which in turn provides a production system upon which to build lean production project strategies. Lean production principles of workflow provide the underlying production philosophy.

It is the integration of many components, including organisational and project systems, that makes the LBMS a management system.

Location—unit of analysis

Location as the unit of analysis is at the heart of the LBMS. Location provides the container for all project data, and is used as the primary work division through a location breakdown structure (LBS)—rather than the more familiar work breakdown structure (WBS).

Location is the container for data which relates to the quantum of the project. The LBS is hierarchical so that a higher level location logically includes all the lower level locations. Each of the location hierarchies has a different purpose. The highest level is used to optimise construction sequence, because the structures of such sections are independent of each other and therefore it is possible to start them in any sequence or to build them simultaneously. The middle levels are used to plan production flow of structure (and often reflect physical constraints). The lowest levels are used for planning detail and finishes. This allows data to be collected at different levels within the hierarchy. The location contains the following types of data:

- **Building objects or components such as elements and sub-systems**
  Traditionally, building objects or components have been only available in drawings. Optimally, these should be documented in a 3D object oriented construction model. This may also include IFC (industry foundation classes) data for documentation data reuse[1].
- **Planned and actual building component quantities**
  Quantities should preferably be measured directly from a 3D model. Measurement by a quantity surveyor or estimator is also possible, although manual measurement should be redundant with 3D model-based technology such as Vico Software's Vico Office. Variations in quantity which occur during construction should be able to be tracked.
- **Building system production assemblies**
  In a construction system, the assembly of components is important, including the selected method of construction. Thus the assembly should include support components such as scaffolding and plant requirements such as cranage. These should be measured together with the measurement of materials quantities.
- **Planned and actual material costs**
  Costs associated with planned and actual material quantities.
- **Building system costs**
  Costs of all the components should be aggregated for each location and hierarchy level within the LBS. This process must be iterative, because labour costs should always be calculated based on the actual resource use reflected in the location-based plan.

Task—method of control

The task is the method of control and is the container for data which relates to the production of the project, in particular labour resources, time and cost. A task is the aggregation of all activities of the same type that repeat in multiple locations. Tasks have common resource requirements but quantities, crews and productivity will vary between locations. The task contains one or more quantity items, from all locations in the project, for work which can be aggregated into logical work packages (for a discussion of location-based quantities refer page 393). The decision to aggregate quantities comes down to deciding how the work will be carried out. If work is managed as one package all the quantities that relate, rather than work which is clearly to be managed separately or at a different time, should be aggregated.

---

[1] Increasing interoperability between the specialist BIM systems such as Revit, Tekla and ArchiCAD which all now export directly to Vico Office indicates that IFCs are reducing in importance.

Quantities define the scope of the task and locations for the task (work may not exist in all locations). Some tasks necessarily become single-location activities.

Each task is defined at (and belongs to) a hierarchy level of the location breakdown structure. For example, the structure is raised one floor at a time, so the logical hierarchy level is floor. Finishes are done one apartment at the time, so the logical hierarchy level is apartment. The task contains the following types of data:

- **Standardised production data**
  Standard consumption rates for ideal crews and resources. These provide a default for the resource planning process.
- **Planned and actual resource requirements**
  Planned resource demands and actual resource consumption for the task.
- **Work crews**
  The make-up and number of work crews for the planned and actual performance of the work. Work crews aggregate the resources for a given task and may themselves have properties.
- **Logical constraints**
  The tasks have a logical relationship and sequence through the project. The layered CPM network logic belongs to tasks.
- **Prerequisites for production**
  Any prerequisites for production such as procurement, precedents, materials supply, etc. belong to the task as the method of control.
- **Performance and forecasts**
  Past performance for the task is recorded and used to estimate future performance.

Underlying production philosophy—lean principles of workflow

The pre-construction phases—design, documentation and planning—should be undertaken to maximise the production efficiency of the construction phase and to reduce waste in production. The construction should be planned to achieve flow of work through locations and certainty in timing and logistics. In the construction phase, progress should be monitored and control actions taken to ensure that the planned production efficiency is delivered.

Waste is minimised by planning for workflow through resource continuity. Management by location allows the following:

- Continuous workflow wherever possible within project constraints
- Planned breaks or multiple crewing to achieve project objectives
- Alignment of production rates to achieve rhythmic production
- Space and time buffers between trades
- Reduction of interference or disturbance between trades
- Preventing cascading delays of the schedule
- Confidence in schedules, particularly for subcontractors
- Flexibility and variation in location requirements (repetition can be variable).

Flow is considered the continuous flow of resources through locations with all prerequisites completed for each location in sequence. Sequencing within a trade is according to the task internal logic, as location completion releases work to the next location.

LBMS simulates an assembly line for production in construction. Here however, rather than the unit of production being moved past the workers and machines (which assemble items sequentially), the workers and machines are moved through the locations of the item of production treating each location in the same way an assembly line would treat the next unit of production. The following principles of lean production (page 107) are included in the LBMS:

- **Value**—defined in terms of the end user, value means that the good or service must meet the needs of the end user at a specified price and time. The LBMS is designed to achieve client objectives with lower risk.
- **Value stream**—an holistic concept of design (problem-solving), information management and production (physical transformation) including all steps and actions required to deliver the product. The LBMS achieves this through better design and the provision of improved planning for logistics where supply is planned by location.
- **Flow**—the organisation of work such that the work flows rather than using a series of high-speed batch processes. This is driven by the sequential completion of locations and the prevention of batching fast (or well paid) activities.
- **Pull**—the end user pulls the production such that it is only produced to suit their requirements. In LBMS, each location may be different and resources are pulled by the specific needs of each location (quantities are variable), thus allowing flexibility. Resources are not added to congested locations. Empty locations pull resources.
- **Perfection**—the previous principles interact in a virtual circle to improve towards perfection. When all tasks are in harmony, site production and final results will approach perfection.

Integration into a system

The formation of a management system requires components that work together. The location-based management system is designed to enable components to emerge to support the system. Currently, commercial systems are available for:

- Design development and modelling (Revit, Tekla, ArchiCAD etc.)
- Measurement of quantities (Vico Office)
- Assembly formation and estimating (links to production data, Vico Office)
- Scheduling for lean production (Control)
- Forecasting (Control)
- Control (Control)
- Reporting through 5D visualisation (3D + time + cost) (Vico Office).

The LBMS hinges on the capacity to reuse data *for production* through all these phases. In current industry systems, the focus of data reuse has been on documentation exchange—the ability to use design data in multiple design phases (such as architectural, services engineering, etc.) and through production to as-built documentation. While these are important, other than for LBMS there has been little interest in data modelling to support production efficiency.

Consider the technology commonly in use today to carry out the above functions:

- Documentation systems have been vector-based, generally 2D. These contain no information of value to the production system. Recently, there has been movement toward integration, with the development of object oriented models and initiatives such as the industry foundation classes (IFCs). Object oriented modelling and data interoperability have been promoted by the International Alliance for Interoperability (IAI). With all the 3D documentation systems, the emphasis has been on information flow, data exchange between design professions and data reuse. There has been little emphasis placed on the production system itself.

- Cost modelling systems are based on priced unit rates, elements, or bills of quantities. There has been professional resistance to the automation of measurement and a stubborn reliance on manual take-off. While demonstrating a disconnection from the movements toward data interoperability, this emphasis reflects a distrust of automated systems, a distrust which is entirely justified given the failure of the documentation system to create the type of objects that cost systems need to measure. Driven by an auditing imperative, cost management ignores the components of cost which relate to production efficiency. There has never been any interest in identifying and managing the cost of production waste in the production system. There has also been a reluctance to specifically include location as a parameter for measurement, presumably because of the risk of error and the increased likelihood of error discovery with the increased sensitivity which location provides.

- Estimating systems are based on unit rate priced quantities or, more commonly, prices provided by subcontractors. These are built on historic data and necessarily encapsulate previous poor performance. Thus waste is locked into historic data forming a vicious circle which reinforces wasteful production behaviour.

- The dominant production planning tools are activity-based. Their emphasis is on the execution of a series of discrete activities with logical connections between. The reliance on critical path methods forces a focus on activity windows—the scheduling of sufficient time for completion of an activity, with activity overlaps to reflect the movement of resources between activities. There has been no linking to either the documentation or the cost management systems, except as priced activities, and a structural inability to plan and manage for efficient production. Indeed, there has been no concept of production efficiency to rely on, except for those developed in lean construction. The result has been that planning appears to have been relegated to the function of defending time-related disputes, rather than the more noble aim of improving production performance.

- Forecasting and control are limited to activity-level forecasts and statusing. Forecasts assume remaining planned durations still apply, despite current poor performance, and therefore fail to adequately forecast the production system. This does not provide adequate information for predicting future performance.

- Visualisation is limited to the linking of activity-based schedules, such as CPM, to a 3D object oriented model, with manual definition of the relationship between the objects and the activities. This is both demanding and inflexible. This visualisation only offers a snapshot of data and when the 3D model or the schedule changes, everything needs to be relinked.

The common thread through the above discussion is that the entire production system uses information without reference to the impact on the production efficiency of the construction process. The failure to grapple with production efficiency means that the construction industry has been unable to make the productivity improvements found in other industries. It has not adopted lean production methods as there is no underlying production mechanism, or production line, upon which to resolve lean methods. Without this key ingredient, supply chain methods struggle, just-in-time methods misfire and productive effort continues wasting up to 50% in non value-adding activities.

The location-based management system has components which specifically provide for data reuse to improve production efficiency:

- Documentation systems are object oriented and aimed at production information. Rather than being focused on being able to properly represent information in 2D when preparing documents for construction, the emphasis is on creating objects which properly represent that which will actually be built. In the process, design errors can be identified and objects are resolved which are suitable for measurement as a builder requires the information. Most importantly, the design model is location-aware—all objects are allocated to a hierarchy of locations (in a location breakdown structure) for data integration and reuse.

- Measurement systems are based on the measurement of objects within locations as an intrinsic component of the method of measurement. Objects include properties which relate to production, such as methods of construction, consumption rates, plant and equipment, labour resources and gangs or crews.
- Estimating systems are based on the application of the object properties to the location quantities, thus allowing easy pricing for tender or other purposes. Historic rates recognise the difference between optimal and normal production. Planning can therefore be for optimal production wherever possible and the impact of disturbance can be modelled.
- Production planning consists of location-aware scheduling methods such as flowline scheduling. Quantities are available for each location as is the required production information such as construction methods, consumption rates, crew sizes, plant requirements, etc.
- Forecasting and control are empowered by location-based schedules and task-level forecasts and statusing. This provides suitable information for predicting future performance of tasks and therefore project performance.
- Visualisation is provided by the linking of the location-based schedule with the originating location-based 3D model. Awareness of both time and cost (or resource) data provides the capacity for 5D modelling and visualisation.

The important things to remember about the LBMS is that the location is the unit of analysis and the task is the unit of control. This enables a management system which is designed with lean production principles for production efficiency.

LBMS COMPONENTS

There is a lot of detail to the components of a location-based management system. In the following sections these components are summarised to help build a comprehensive picture of the LBMS. This will assist with implementation, described in the following chapter.

The discussion covers the basic components: location breakdown structure, location-based quantities, location-based estimating, location-based planning and scheduling, location-based control, location based reporting, location-based quality management and location-based financial control. While not strictly a location-based method, visualisation will be mentioned as this plays a significant role in bringing it all together and, significantly, plays a major role in communication of results.

Location breakdown structure (LBS)

Locations in a project are defined by a location breakdown structure (LBS). It is possible for the project to be broken down in many different ways, however, locations must be hierarchical so that a higher level location logically includes all the lower level locations.

Each of the location hierarchies has a different purpose. The highest level is used to optimise construction sequence, because the structures of such sections are independent of each other, therefore it is possible to start them in any sequence or to build them simultaneously. The middle levels are used to plan production flow of structure (and often reflect physical constraints). The lowest levels are used for planning detail and finishes. The guidelines for commercial projects are:

- The highest level location hierarchies should consist of locations where it is possible to build the structure independently of other sections (for example individual buildings or parts of large buildings).
- Middle levels should be defined so that the flow can be planned across middle level locations (for example, riser floors in a residential construction project, where a floor is usually finished before moving to the next floor).
- The lowest level locations should generally be small, such that only one trade can effectively work in the area (for example apartments, individual retail spaces, corridors). The lowest level location should be able to be accurately monitored (that is, the foreman must be able to assess whether or not the work is completed in that location).

Each task is defined at (and belongs to) a hierarchy level. For example, the structure is raised one floor at a time, so the logical hierarchy level is floor. Finishes are done one apartment at the time, so the logical hierarchy level is apartment.

Location-based quantities

Quantities are an integral part of the logic of location-based scheduling, and in particular the internal logic of a task. Quantities drive the production process and determine the amount of work to be done in each location. The bill of quantities (measure) of a task defines explicitly all the work that must be completed before a location is finished and the crew may continue to the next location. Many different items of work may be undertaken in a single work package, or a sequence of work packages in a summary. Quantity items are aggregated according to the following guidelines:

- The work can be done with the same crew
- The work has the same dependency logic outside the package
- The work can be completely finished in one location before moving to the next location.

While it is possible to work flexibly using manual methods for deriving quantities, it is far better to have a system for automated quantity calculation from a 3D model such as Vico Office.

Manual sources

Quantity-based estimating can be approximated if actual quantities are not known. For example, when importing an activity-based schedule from a CPM schedule, it is usual to have access only to duration data. In its crudest form such quantities reduce to measures of days, with consumption rates of the shift length (usually 8 hours). This is of low value because the assumed crew sizes are not known and therefore production cannot be optimised or controlled. Resource-loaded durations are more valuable because, by using worker hours or worker days, all calculations of the location-based management system can be used. However, without quantities it is difficult to accurately estimate worker hours or worker days and such data can be misleading.

More commonly, a manual quantity take-off can be prepared. This can be prepared by an estimator and would normally concentrate only on the major activities to be included.

Elemental methods with appropriate rates are useful in such situations. This allows simple quantities, such as floor area or number of columns, to be used to approximate quantities.

While approximate methods are never as accurate as a proper measure, they are rapid and useful for indicative scheduling—such as early estimates—and frequently they are the only option available. It is helpful to remember that almost all CPM schedules are currently derived predominantly from guesses of total lapsed time and only use quantities and resources for critical activities. if at all.

Where quantity surveyors are engaged on a project, they may manually prepare a detailed measure of project quantities according to some form of method of measurement (these vary from jurisdiction to jurisdiction). While not as accurate as an automatic take-off, these form a very reliable basis for scheduling. However, unfortunately, such bills of quantities rarely include the location breakdown, despite quantities normally being measured in locations and annotated as such. Aggregation is rarely provided with location detail. In this case the quantities may still be used by dividing the quantities according to elemental quantities such as floor area. This can only be an approximation as quantities rarely strictly follow elemental quantities, nevertheless it is a very quick way to generate a schedule from a BOQ. The biggest challenge is to organise the BOQ items into a form which can be used in scheduling.

Automated sources

Modern technology has the potential to dramatically impact on the quantity surveying profession, not least through automating the production of accurate measures of quantities for building projects. Many systems can produce quantity measures of varying accuracy, software such as Vico Office now has the capacity to produce a complete and accurate measure of all quantities for a project and to include location information for the quantities, thus providing quantities according to a LBS.

The methodology is to shift the effort in measuring from manual take-off from drawings of a project, to building object oriented models of the project using 3D modelling software such as Revit, Tekla or ArchiCAD and then to create location-based measures automatically in minutes using a tool such as Vico Office.

In this new world, the quantity surveyor will be forced to shift emphasis from measurement to using 3D models and modelling task data—a significant challenge.

Location-based estimating

One of the most important drivers of production feedback is cost control. A necessary front-end to this is the estimation of costs for the project. In a location-based project, it is desirable that these costs are estimated according to the LBS.

Combining the power of location-based quantities with task assembly information (recipes) provides the ability to achieve cost estimates rapidly. It is also much easier to identify pricing or bid package errors when using location-based cost models.

Currently, standard methods of measurement do not support location-based quantity measurement and so estimates are generated in aggregate, broken down by trade or work package. The location-based management system enables an additional layer, the LBS, to be overlaid onto the estimating process to provide much richer information. This information is also much more useful for cost control of projects. Tools such as Vico Estimator enable full assembly-priced estimates to be calculated from the location-based quantities from Vico Office.

Location-based estimates can be used in estimating, tendering, bid packaging and in the analysis and correction of all of these. They are a powerful additional tool for LBMS.

Location-based planning and scheduling

Location-based planning assumes that productive work consist of a set of tasks which flow through locations as continuously as possible. The intent is to plan projects for production efficiency, create a schedule that is workable and sustainable in accordance with lean principles, and finally to provide the basis for location-based control.

Location-based scheduling allocates quantities to tasks and applies task properties such as CPM logic, resources and crews. Location-based planning includes the planning for production efficiency, such as aligning production rates, determining crews or gangs, splitting tasks and cycle planning. There is richness of detail in determining a location-based schedule, including learning, quantity variation, cost loading, minimising production system cost and planning for procurement, logistics and quality.

Tasks, flowlines and logic

Tasks are the aggregated activities which represent an item of production in a series of locations. Whereas activity-based planning would treat activities as discrete units of work, in location-based planning they are treated as a single entity—task—which can be managed by location. Thus tasks rather than activities contain the properties of the work.

In location-based scheduling the tasks are represented in a flowline. This is a single line representing the task as it flows through locations on a graph which consists of the LBS on the Y-axis and time on the X-axis. This form of representation allows a single line to represent the work in many locations—which would otherwise take many lines on a Gantt chart.

Tasks derive their duration from multiplying the quantity by the consumption rate. Duration for each location is provided by the quantity in that location. Total duration is calculated by the accumulation of all locations, or by using the total quantity multiplied by the consumption rate.

$$Total\ hours = \sum_{i=1}^{i=n} (Quantity_i \times Consumption_i) \tag{11.1}$$

This calculation can be influenced by other factors, such as difficulty in a given location.

The location-based schedule requires that tasks can be linked using CPM logic. In fact, location-based logic requires that five layers of CPM logic be applied at the activity level. These are:

1. External logical relationships between activities within locations.
2. External higher-level logical relationships between activities driven by different levels of accuracy.
3. Internal logic between activities within tasks.
4. Phased hybrid logic between tasks in related locations (location lags).
5. Standard CPM links between any tasks and different locations.

Layered logic automatically generates CPM logic between two tasks by using locations—for example, a task can be planned to happen before a successor in every location where they occur together. Layered logic is not hierarchical, instead all logics apply equally and the resultant forward and backward pass calculations are necessarily iterative and conditional in their function depending on the layer logic which applies in each situation. Standard CPM calculations can be used with the addition of heuristics which enable continuous work by pulling the start dates of early locations.

In addition to the use of CPM logic lags and leads, location-based scheduling includes buffers. Buffers may be in either duration or location.

- Lags and leads: the required fixed duration of a logical connection between two activities or tasks.
- Buffers: The absorbable allowance for disturbance between adjacent tasks or locations is a component of the logical connection between two tasks but which may absorb delay.

Buffers appear very similar to lags, except they are there to protect the schedule and are intended to absorb minor variations in production.

Detailed planning

The detailed management of activities is important during production. Scheduling at this level introduces a great deal of detail and may sometimes even be considered over-detailed. Nevertheless, where control of individual crews is required, detailed planning is necessary.

Detailed planning can be achieved in two ways depending on the specific needs. First, you may plan for individual crews which work independently following the location-based logic. This method results in many more flowlines, although these may be aggregated into summary tasks. This is the preferred method where multi-skilling is not an option.

Second, a group of activities may be planned as a single task and then broken into detail tasks during the control phase as current tasks. This method is ideal where multi-skilling is required or where a single trade contractor is in charge of a group of activities, such as services. However, tasks of multiple subcontractors should never be lumped together.

Cycle planning

There are usually a few critical task sequences which must be performed in a cycle, particular with vertical structure. The establishment of a cycle for tasks which have circular logic (they eventually depend on a later task at a lower or previous location in the cycle) is important to establish the maximum rate of production.

Contractors are usually excellent at establishing structural cycles using manual techniques. Location-based scheduling allows the rapid creation and modelling of construction cycles using location lags. The planner is able to rapidly optimise labour requirements or crew sizes to align production for different components of the cycle.

Templates

There is a myth that every project is different. In reality, projects share a great deal more than they are different. While materials, methods and quantities may vary, task components and logical relationships are generally the same from project to project. This allows the use of templates to generate rapid schedules.

The use of the task as the unit of control, rather than individual activities, results not only in greatly reduced numbers of control items to plan but also in general relationships which hold through the project. These form patterns which may be applied as the starting point in the generation of a later project schedule. A past project (or an ideal project) may be used as a starting point to determine which quantities go in which tasks, and the logical relationships between tasks. The result may then be adjusted for project circumstances.

Templates are only possible in location-based scheduling and present a great opportunity for rapid planning.

Aligning production: assigning resources

Whether or not templates are used to plan a project, the initial result derived using one crew only will generally not be buildable within time constraints. The planning phase in location-based scheduling becomes a process of aligning production to achieve maximum continuity consistent with timely completion.

The process is to speed up or slow down tasks to achieve as close as possible the natural project cycle—usually dictated by the optimum construction cycle. In some cases the natural cycle itself may need to be faster to achieve project targets. Where tasks cannot be accelerated or slowed without undue impact on efficiency, tasks may be split (when too slow) or broken into stages (when too fast). Such deliberate planning allows for decisions about production efficiency to be made as a trade-off against project constraints.

Decisions made in the planning phase to align production should be tested with the contractors responsible for the work at the earliest possibility.

Learning and production variation

The effects of learning are easily identified in location-based planning due to the apparent repetitive nature of tasks flowing through locations. Learning may be automated by using a formula approach to apply a learning factor to tasks to reduce the consumption rate progressively, or can simply be modelled using a difficulty factor applied to early locations.

The most common example of the use of learning is in cyclic activities involving structure. It is usual for the early typical floors to take longer than later floors. In reality optimal production, as contained in the production factors used in planning, should be adjusted for the difficulty of doing it the first time, and in each situation where there is substantial variation from a pattern. Typically this is achieved by applying a difficulty factor greater than 1.0 for early locations in a typical cycle.

Similarly, variations in difficulty in specific locations can be handled by applying a factor to slow the work in that location.

Cost loading

Location-based quantities make it possible to cost load the schedule directly, using the cost estimate for a project. Every quantity element has an estimated unit cost and, if desired, the resources can have costs applied directly rather than relying on built-up rates including labour. By summing the material costs and the cost of resources it is possible to calculate the cost for the entire schedule task or for any activity location of the schedule task.

A cost loaded schedule (which includes production rates, start and finish dates, and quantities for each location) can also be used to calculate the timing of payments, progressive cash flow and earned value.

Production system cost

In traditional cost estimates, the cost is defined by using historical data to estimate resource use. While this approach can be used to calculate cash flow, it does not provide any tools for measuring the production quality of a schedule. In contrast, production system cost models should have knowledge about the actual labour resources required in production (the real cost) as well as directly taking into account the waste factors, such as waiting time, relocation, double-handling, mobilisation and demobilisation, and it should calculate labour costs based on a composite model of resource consumption.

Ultimately, someone must pay the cost of an inefficient plan. Inefficient work will necessarily increase resource consumption compared to optimum production. Direct labour cost is composed of the following value-adding and non value-adding components:

• Working time
• Mobilisation, demobilisation and waiting
• Moving around on site
• Stockpiling, hauling and receiving materials.

Each of these components should be included in a production system cost calculation. The deliberate planning of location-based planning and scheduling allows these factors to be included in a cost estimate—even if only to determine the lowest production cost and therefore the optimum solution balancing time-related costs such as overhead and production cost, where overhead costs vary as a function of duration and are minimised by compressing the schedule, or the relevant schedule tasks.

Production system risk

Production system risk is a stochastic assessment of the likely success or failure of a plan, with measurement of the associated consequences of real events. The following risk factors apply to location-based schedules:

• Uncertainty related to weather
• Uncertainty related to the prerequisites of production
• Uncertainty related to adding resources
• Uncertainty related to productivity rates
• Uncertainty related to quantities
• Uncertainty related to resource availability
• Uncertainty related to locations
• Uncertainty related to quality.

With the exception of weather events (an external environmental risk), these factors are production system risks.

Location-based production management is able to use the planning system to identify and react to deviations in production, and such actions may be simulated. Thus it is possible to identify a schedule which does not provide sufficient production buffers to mitigate against production risk factors.

Procurement

Procurement is the first step in managing the supply chain and timely procurement is critical for achieving lean production as the procurement process is treated as a prerequisite for starting production. There are two methodologies which may be applied to procurement planning and location-based scheduling allows either method.

The normal method used for CPM scheduling is to schedule procurement activities as activities, and therefore to push the schedule of dependent tasks. In location-based scheduling this takes on a special meaning as the pushing procurement activity only pushes the first location in the sequence.

The alternative methodology is pull scheduling of procurement. Location-based scheduling also allows the production tasks to pull the scheduled activities for procurement, based on duration, on a just-in-time basis. Thus the procurement tasks will be scheduled to start as required to meet the needs of the production schedule.

Procurement is generally required for the commencement of the first location, but some activities may also be required for each location. This may be simply managed with a location-based approach. However, the management of delivery, or logistics, is more likely to be an issue at each location.

Logistics, materials storage and handling

Logistics controls the actually delivery of materials to site, the extent of materials handling—such as storage on site—and such production waste as double-handling and damage. In location-based management, it is possible plan for logistics at a very detailed level. This is because a properly completed location-based plan includes quantity information which can be used to determine the logistics. The logistics planning can address:

- When to deliver the materials?
- How many deliveries are required?
- Time and resources needed for receiving and hauling (for each delivery)?
- Should the same resources be used for logistics as for production?
- What is the lead time before production can start?
- What will be the required storage time?
- What will be the cost of freight for delivery?

In location-based planning, the hierarchy levels of the location breakdown structure are used as the basis for deliveries. Plotting deliveries in the flowline allows the planner to see when materials are stored in a location and to plan storage so that it does not hinder other tasks within the same location.

Materials handling can be used to control the crews to the planned location sequence. If materials are only delivered just-in-time and only for those locations where work is to follow, then crews will be unable to work out of sequence due to a shortage of materials. This approach also reduces stockpiles of materials, which are a principal cause of damage, waste of materials and consequent rework.

Planning to deliver materials precisely as required requires detailed knowledge of location-based quantities and associated timing of the works.

Planning for quality and prerequisites

Location-based management allows deliberate planning and control of quality. Planning for quality means that enough time will be allowed for inspections and measurements to be completed and approved before the following trade comes to the location. In location-based planning systems, it is possible to explicitly plan for quality by using the buffer to allow systematic time for quality inspections. On the other hand, quality is enhanced during production by planning to have only one contractor in the same location at the same time and minimised material storage on site.

Controlling quality has two aims: ensuring that following tasks have all the prerequisites completed prior to starting work, and checking that the quality of work already completed meets the required standards before the following crew commences work in the location. A key to this process is to pass 'ownership' of locations between the general contractor and the work crews, and vice versa.

The process of passing ownership can also be used as a signal for approval of payments. For example, in Finland it is common to tie all payments to satisfactory completion of locations. While this results in a great many payments, it provides tremendous mechanisms for quality control.

Prerequisite checks should be decided for each task, and the locations where they will be required should be planned in the pre-planning phase. Prerequisites are those conditions which must be met before work can commence and include such things as prior task completion, prior location completion, resource availability, documentation, task planning and quality checks.

Planning for change: buffers

Client change has great impact on a project schedule and may be handled in two ways in location-based management.

- A buffer can be used to allow for uncertainty where this is expected. This shields the production from design variability. Buffers can provide room to re-sequence the work, for changes in quantities or for control mechanisms.
- In conjunction with the buffers, the data available in location-based management allows for detailed documentation of the cause and effect of variation, particularly when compared with the baseline plan. It is very easy to demonstrate the effect of client changes on the production efficiency of the project—and thus to establish the cost of lost efficiency if required.

Adopting a risk management approach

While most planners will be familiar with the detailed activity planning of CPM and the apparent accuracy this entails, there is an entirely different approach to project planning which has been run successfully in places like Finland for many years. This is the risk management strategy—the project plan delivers the shortest duration while minimising the risk of overrun.

This method requires that tasks be planned with sufficient planned buffers to minimise the risk—risk profiles and a risk simulation may be used to identify tasks and locations where the project is at risk of disturbance. Using a risk management approach suggests that a slightly slower project (slower due to increased buffers, not reduced rate of production) will reduce the risk and therefore lead to a more reliable project outcome. It similar to the case of the fabled hare and the tortoise—slow and steady wins the race.

Using location-based planning and control

Ultimately the key to location-based planning is to recognise that many correct schedules are possible and indeed many of these will be good ones. A good schedule maximises productivity, finds an optimal balance between risk and duration, and is feasible to implement.

Productivity is maximised by planning continuous resource use with the plan being based on accurate scope and quantities, resources and productivity data. Each trade should use the optimum resources organised using the most efficient work crews. Generally, the production rates of predecessors and successors should be aligned, and each location should be completely finished before moving on to the next location.

The trade-off between minimum time and reducing risk can be evaluated using production system cost and production system risk tools. Regardless of the chosen risk level, the selected solution should be efficient, meaning that a solution should be found which achieves a given duration without the ability to further reduce the risk.

Summary of the location-based planning process

The following summarises the steps involved in location-based planning:

1. Define location breakdown structure.
2. Define location based quantities.
3. Build tasks from quantities and define:
   a. Optimal crew
   b. Layered logic links to other tasks.
4. Align the schedule and optimise sequence and duration:
   a. Changing production rates
   b. Changing sequence
   c. Breaking continuity
   d. Splitting.
5. Evaluate production system cost and risk (optional).
6. Optimise cost and risk (optional):
   a. Adding buffers
   b. Changing production rates
   c. Changing sequence
   d. Breaking continuity
   e. Splitting.
7. Cost load the schedule.
8. Optimise cash flow:
   a. Change payments
   b. Change production rates and start dates.
9. Approve the schedule.

10. Plan procurement and design schedule:
    a. Use pull scheduling techniques and soft constraints
    b. Do changes to the production schedule only if necessary.

Location-based control

Location-based control assumes that planning has maximised productivity, found an optimal balance between risk and duration and is feasible to implement. It remains therefore to ensure that the work is undertaken such as to achieve this plan. Deviation from the plan will, necessarily, result in a less than optimum solution. Control has always been the end purpose for systems of planning and scheduling construction work.

Kelley and Walker (1959) stated that the plan should form the basis for management by exception: management need only act when deviations from the plan occur. There has been little change in the approach to control as proposed by the first writers on activity-based control, however these control methods provide little in the way of detailed information about progress and what they do provide is generally too late. Location-based control provides tools otherwise not available, and provides early warning of emerging problems.

A major part of the power of location-based management comes from being able to make commitments to subcontractors about production continuity and predictability, so the control system has to support that need.

The location-based control system uses location to generate on-time response by management through visualisation of any problems before they occur. Forecasts are used to constantly remind management that a problem remains unsolved and that information is available to help them take informed control actions. If the location-based control model is used, then management will react to problems earlier and with better control actions. Instead of just recording deviations, the control system becomes a driver for action. The task of production management is therefore to find solutions to production deviations, to make things happen as planned, and to look for better solutions.

There is a lot more information in a plan using the location-based model than is available in an activity-based model, including:

- The flow of resources, which is explicitly planned
- The quantities, which are known in each location
- The production logic, which is modelled much more accurately using five layers of CPM logic.

The location-based control model needs to provide accurate information sufficient to differentiate performance deviations (the traditional focus) from changes in circumstances. The sources of deviation may include:

- Quantity changes
- Start-up delays
- Production rate deviations
- Discontinuities and working out of sequence
- Production prerequisites.

Tracking this more accurate information and having a system with sufficient flexibility to manage changes in implemented production plans will lead to better management of the prerequisites of production, the availability of suitable resources, and more detailed look-ahead planning during construction.

The location-based control model utilises four stages of production information. The stages are: baseline, current, progress and forecast. The location-based management system tracks data in these four stages using two sets of tasks: *schedule tasks* and *detail tasks*. Schedule tasks are the tasks and activities which make up the baseline schedule and are sometimes called *baseline tasks*. Detail tasks are the exploded current plans for the actual production of the baseline task. Detail tasks always belong to a single schedule task and make up the current view of the baseline task, and may be called the *current tasks*. Schedule tasks and detail tasks are used to construct the four stages of the control schedule.

Baseline schedule

The *baseline stage* provides the founding set of project data, such as the committed plan for the project, against which all subsequent performance is compared. It functions in the same way as a baseline in a schedule in CPM control systems. The baseline plan cannot be changed, unless a new baseline is established, and it constrains the current plan. The location-based baseline model uses location-based quantities (*baseline quantities*) and controlling tasks (*schedule tasks*) to plan the work.

The baseline plan is used to make commitments to subcontractors, to plan procurement and to prepare subcontract tender schedule and milestone information. To achieve these objectives, the start and finish dates should be reliable to within one week of accuracy.

Current schedule

The *current stage* functions in a similar way as the baseline, however it specifically recognises the need for change in the project plan to take into account new information which was not available when the baseline plan was made, including both changed project data and more detailed construction planning, such as look-ahead planning. The current plan is changed whenever new information becomes available. This new information can include information about resource availability, prerequisites of production, quantity changes or change of logic. However, even if there are changes, the original baseline constrains the finish dates in each location—necessary for the management of commitment to trade packages.

The location-based control model establishes the mapping between these two planning stages, using a new set of location-based quantities (*current quantities*) and a set of current stage tasks (*detail tasks*) to manage the changes involved in current stage planning. Detail tasks also consist of *detail activities* in each location.

Monitoring progress (statusing)

During the production phase, the baseline and current information is gradually augmented by progress information. This information highlights the deviations from the plan, eases the task of updating the current plan and is critical in the subsequent evaluation of the quality of original plans. In the location-based system, status information should be tracked by location. The most benefit is gained if progress is tracked for all components of the planning and control system. The basic components of the system (additional, optional components are introduced in Chapter 9) include:

• Actual quantities

- Actual resources
- Actual shift length and days off
- Actual begin and finish dates.

This information can be used to calculate the following important values:

- Actual resource consumption
- Actual production rate.

Monitoring of these items should be made on the most accurate planning level. Effectively this is the chosen location level of the detail task. The mapping between the detail and baseline tasks allows progress data to be compared with either of the two levels of planning.

Progress

The *progress stage* monitors the actual time performance of the project and therefore tracks data in the detail tasks. The progress of the production is measured by recording task start and finish times in each detail activity location. Actual production rates for detail tasks can be calculated from this and, if actual resources are known, the actual resource consumption rates (man hours per unit) can be calculated.

Forecasting

The current and progress data can be used to calculate forecasts to predict the total effect of schedule deviations and variations, and therefore reveal problems. Forecasts should assume that production will continue with the achieved production rate (rather than the planned) unless control actions are taken. It is inaccurate to forecast based on planned completion rates when they are different from those achieved. Forecasts empower management to react to problems early enough to take effective action and to provide the data required to support control action decisions.

Forecasting is a process which utilises the best currently available information. In the early stages of the project, the original plan can be used. The forecast is then updated when new information about quantities or schedule becomes available. During production, the actual production rates should be used as the basis for the forecast.

Control actions

Location-based control actions are the steps taken to recover from deviation in order to prevent interference or to prevent project delay. Control action planning resembles rescheduling detail level tasks. However, control actions are needed when someone else's work will be interfered with, therefore there is a concrete goal for control action planning: finding a feasible solution to prevent interference. The list of available control actions is usually shorter than those available in planning both the schedule and detail tasks, because the control action must be implemented in the near future. Moreover, people close to the production should be included in the decision-making process. Carrying out control actions requires that everyone commits to the decisions. The following actions are available to prevent interference:

- Changing the number of resources (the same productivity will be assumed)
- Changing shift length or working overtime (on weekends or holidays)
- Changing the location sequence
- Splitting a task (either to allow working in multiple locations at the same time or to allow a break with workers returning at an agreed date)
- Removing or switching technical dependency (this may cause interference in locations). This is most relevant in detail task planning
- Increasing productivity by reducing non value-adding activities (waiting, materials handling, rework, etc.)
- Shifting the start date of a successor task to make that task continuous.

With control actions, it is the forecast which is adjusted directly to correspond with planned control actions, rather than the plan. The plan is not changed because the fact that there was a deviation in the first place would then become hidden—possibly leading to a false sense of security. By updating the forecast instead of the plan, management accepts that there was a deviation but commits to action to restore the original plan. It is also desirable to maintain a log of the control actions taken. A forecast which has been adjusted with any control actions becomes the location-based look-ahead plan (Seppänen, submitted).

LOCATION-BASED REPORTING

Location-based planning and control is heavily influenced by graphical techniques of representation. In addition to the use of traditional Gantt charts, LBMS makes extensive use of flowline diagrams and control charts.

Gantt charts

Gantt charts are used in the traditional way and do not differ significantly from heavily location-coded charts produced by CPM systems. The only difference arises from the inherent property of being location-aware. This enables collapsing of locations into task summaries—or conversely the expansion of tasks into their locations.

Gantt charts provide little flexibility for visualisation of planning and control. They are therefore mainly used as a tool for planning (some location-based techniques can be easier in this mode, such as linking tasks) or for site distribution. Some believe that this familiar representation provides less on-site resistance to the new approach.

Flowline

The flowline representation is the primary communication method for LBMS. A single view is able to project a great deal of information about the plan for the work, particularly the continuity and planned breaks. Experienced users are able to *read* the plan in a similar way to reading a floor plan, and are able to interpret detail about the method of construction which is not possible from a Gantt chart.

Flowline is an even more effective representation tool when used in the control phase. Here the inclusion of lines for actual work highlight the actual circumstances through the entire history of the project. Forecasts predict the likely outcome given current rates of progress. In the control mode, an experienced user is able to rapidly read the history of the project, compare it to the plan and interpret the future consequences.

A single chart is able to provide more information than a set of progress charts and is able to provide immediate comparison against the baseline.

Control charts

While a flowline can be rapidly read by experienced users, the control chart is extremely effective for communicating dates, work sequences and progress performance for individual tasks. Those responsible for a given task are able to compare the planned dates with actual dates, and colour coding provides rapid feedback about project status. This form of communication is ideal for project site meetings, where responsible individuals can be rapidly held to account for their lack of progress by reading the colour coding of the tasks. However, a control chart should not be used as a controlling tool, as it is a static report. Controlling requires knowledge of production rates and available locations and thus flowline diagrams and forecasts should be used.

LOCATION-BASED QUALITY MANAGEMENT

One of the greatest challenges in project control which can be significantly influenced by the efficiency of the production system is the quality management system. This is not the quality specified, but the degree of match between the specified quality and that achieved. Quality problems can occur where materials are damaged, or where work is undertaken out of sequence.

Figure 11.1 Damaged bottom plates due to materials handling and work sequencing problems

One example is metal stud work, where if the floor plate is placed too far ahead of the studs, there is an opportunity for materials handling to damage the plate. Another example is painting being undertaken after carpets are installed, with paint damage occurring as a result. Projects are rife with such examples. The LBMS can address these problems in two ways.

First, work should planned to prevent quality problems arising from work sequencing or division. Proper sequencing of the work and avoiding inappropriate task division (such as separating bottom plate from studs) will ensure that the site team has every chance to get it right. It should be possible to step into a location after the completion of an task and find all predecessors complete and no commencement of successors. If the work has been properly planned, then no defects should be apparent due to interference.

Second, to ensure damaged work is not built in, inspections should be carried out. Handover inspections between all tasks will ensure that no sub-standard work is accepted or allowed to remain in place and be built into the project.

There is a further discussion of these issues, and in particular the use of inspections and handover, in Chapter 12.

LOCATION-BASED FINANCIAL CONTROL

Another important factor in driving site performance relates to the motivation of the sub-contractors to perform. Current CPM techniques usually mean that subcontractors have no motivation other than to complete all their work by the end date. How they get there is considered their own business. This, of course, can have severe impacts on other trades. If they are running slow (very common), they will hold up following trades. If they are running fast, they will interfere with preceding trades.

Location-based financial control is the mechanism to avoid these problems. This is a system of management through the timing of payments. Typically payments are currently made on the basis of a lump sum periodic (monthly) assessment of the total work done anywhere. This usually results in all the easy bits being done first to get as much money as possible upfront. Equally, it can be very difficult to get completion of the fiddly bits due to the cost to complete and the relative lack of reward. Both situations can be disastrous for a project and the LBMS can use the payment system to drive project performance through location-based payments.

Allowing multiple payments, made progressively and on completion of locations, is a very powerful motivating force for project performance. The LBMS allows just this situation, whereby payments are triggered by location completion *in accordance with the plan*. Thus work once completed and inspected may be paid for as long as it was undertaken in the correct sequence. This removes the current difficulties in assessing the value of progress payments in large areas, as completion of smaller locations is required.

Chapter 12 contains a further discussion of using payments to improve performance.

nD VISUALISATION

Location-based planning and control can be undertaken using Vico Software's Control software. However, location-based management systems are also able to take advantage of 3D modelling to produce 4D and 5D visualisations of the project.

If using Vico Office, the results of the location-based project plan may be exported from Control into Vico Office. There they may be matched to the estimating data and shown as a 5D visualisation.

The advantage of a 5D visualisation in the planning phase is that the planning team is able to view the construction of the project—including zooming into close detail to observe how the project construction sequence will work in practice. This is a very powerful method for visualising problems with the construction schedule. Estimating data also makes it possible to view the planned cash flow profile over time.

The real power of 5D visualisation is provided by the use of the Control software to monitor progress and to record resources used. If this data is output, then the visualisation is able to show the actual construction sequence, this may be compared to the planned sequence, and costs from the estimating data can be used to provide earned value analysis.

These techniques are extremely powerful for remedial and forensic project management for problem projects.

REFERENCES

Kelley, J.E. and Walker M.R. (1959). "Critical-Path Planning and Scheduling". *Proceedings of the Eastern Joint Computer Conference*. 160–173.

Seppänen, O. (submitted). *Empirical research on success of production control in building construction projects*. Unpublished PhD thesis submitted in fulfilment of the requirements of the degree of PhD at Helsinki University of Technology, Helsinki, Finland.

Chapter 12

Implementing LBMS

INTRODUCTION

If you have decided by this point that location-based management is the way to go, you will be wondering what to do next. This chapter deals with some of the issues involved in implementation. From the basic elements of engaging with management and the site team, through to technical issues about the way the planner and the site team must change their thinking.

It will come as a surprise to many in senior management that their decisions may not be followed through. In construction there are many players and all are required to work together as a team in order to develop a successful project. Similarly, these must all work together in order to change an entrenched management system. This is not easy to coordinate and the strategy to achieve this will depend on the context and the team, their experience and their adaptability. For many organisations, the first question should perhaps be: are you ready for this?

This question needs to be asked because an initial implementation is much more likely to fail than succeed. Such failure can arise from many sources, but the most likely is resistance to change. Resistance can vary from passive (failure to change actions) to active (sabotaging initiatives). If an organisation supports such behaviour through weak or mixed messages or even running parallel systems, the overall result is likely to be reversion to old practices. Care should be taken to properly prepare and manage the change in order to achieve success.

Even with a willing and supportive project team, there will still be barriers. Players will have been managing projects the old ways for many years. They will have entrenched practices and beliefs that deliver outcomes they are familiar with. Shifting to a lean production approach and using location-based methods with continuity will not only challenge these, it will require a new way of thinking. This is a lot to ask of one person, let alone a team. These problems can be loosely gathered under the heading *change management*.

There will also be *practical problems* to address. These are the problems of sourcing required data, modelling projects, recognising the role of continuity and buffers in planning, obtaining progress data and developing understanding of reports and forecasts. Similarly there will be challenges around the reliability of plans and reports—these will be very different from the familiar CPM methods. This will produce challenges in interpretation, communication and most importantly in contractual relationships.

One of the important decisions that must be made early is the extent of the implementation. A really useful technique for early adoption is to just quietly prepare the plans and report progress as if nothing has changed. This will give the site a chance to understand that changing technology is not a scary thing. The opposite approach is to undertake professional change management including the use of an industrial psychologist. This will ensure the best uptake of the production changes and the new way of thinking. Of course, designing an implementation plan is a very individual thing for most companies and contexts.

DEALING WITH CHANGE

Learning from CPM

Moving from an activity-based to a location-based management solution will change the working environment for many participants, both on and off site. Many of these will resist the change. There is nothing new in this, many of the early publications indicated there was resistance to CPM when it was first introduced, despite there being no computerised alternative at the time.

Arditi reviewed the uptake of CPM in 1983 and found that while it differed between countries, it followed an overall pattern. He found early rapid adoption followed by a plateau and then a decline. He also found that uptake followed a pattern similar to that outlined by Freeman (cited in Arditi, 1983) who described the diffusion period for a novelty in the plastics industry as:

1. To Germany and the US; 2–3 years.
2. To the UK and France; 5–7 years.
3. To Japan and Italy; 12–15 years.
4. To all other countries (except Canada, Switzerland, and Sweden); more than 20 years.

Arditi described the construction industry as very conservative and therefore likely to take longer to adopt new technologies. "Consequently, natural inertia and reactions to novelties should be expected from the people employed in this industry" (Arditi, 1983).

Writing from the standpoint of 20 years since adoption, Arditi noted that there were structural reasons for the plateau in industry acceptance. The resistance to CPM was ascribed to several causes, as these can now be seen to apply equally to the adoption of location-based methods. These were:

- "First, because the older generation of engineers are not familiar with network techniques, the necessity arises of forming a central planning unit staffed with planning experts." Arditi further noted that a matrix structure is needed which may challenge the authority of the project manager. The same problem exists today with the lack of knowledge of location-based methods. It is not reasonable to expect sites to possess sufficient knowledge to be able to rapidly adopt the new methods.
- "Secondly, project managers of the older generation have no knowledge of network planning. With years of site experience, however, they have been used to exert nearly unlimited authority on site." Arditi noted that the bright enthusiastic engineer introducing the techniques on site, usually without site experience, is resented by the experienced site staff. The parallels with location-based methods are significant. Here the experience is with activity-based methods. In fact, the site practices are generally, in commercial construction, techniques based on ignoring the schedules and directly managing the project. In this context, the introduction of location-based methods with tight control systems will naturally be resisted.

Arditi's solutions involved either training of site managers, or development of central planning departments with expertise. These solutions are equally applicable, and equally problematic today. While the former solution is accepted by sites, it tends to not be achievable. In contrast, the latter solution is achievable but not accepted.

There are two factors operating here. One is resistance to change, the other is expertise. These factors work in quite different ways. Arditi focused on expertise: training, education, talent and flow of knowledge. His view was that adequate knowledge dissemination would be sufficient. He tended to ignore the reluctance of the industry to adopt different methods—in particular those which reduce a manager's autonomy.

"The length of time a firm waits before using a new technique tends to be inversely proportional to its size" (Mansfield cited in Arditi, 1983). However, bigger firms are more likely to have access to specialised knowledge and therefore are more likely to be capable of early adoption. It is certainly true that the adoption of location-based management techniques has tended to start with the bigger, often multinational, firms rather than smaller operations. This supports the view that knowledge, and therefore training and education, plays an important role in acceptance of new techniques.

Such high-level views ignore the real practicalities of sites and their interpersonal factors. It is at least equally true that acceptance of new techniques requires an understanding of the relationship between the managers and the operatives and the dynamics between head office and the site. For example, in the early days of PERT, Fazar observed :

> A difficult problem was posed, early in the program, by the natural tendency of most participants to draw a dividing line or a dichotomy between management, as represented by the Plans and Programs Division, and technical effort under the responsibility of the Technical Division. Effective diagnosis of progress and integrated program management are always constrained by this dichotomy that people tend to draw (Fazar, 1962).

The problem is that these issues are not to be dealt with lightly. Upsetting the spirit of a site will very likely lead to problems on a project. As Fazar had also noted: "we must not upset the equilibrium by introducing PERT!"

Thus we may also see that another important component required to implement change is acceptance and engagement in the process by the site teams.

> ...An increase in job satisfaction was necessary for acceptance of new, more productive methods. ...the best way to introduce productivity change among foreman and craftsmen was to allow the workers to plan the changes in their own methods. In this way, any perceived loss of status could be avoided (Smith, 1981:51).

The alternative, loss of status, is not to be confronted—rather it must be carefully managed. Human nature dictates that problems will be dismissed and history rewritten. As Fazar also noted in 1962:

> The expression by one leading industrial manager was that no new system would get him to admit that he was not going to meet his existing due dates. He may not have known that hundreds of his original schedules had experienced rescheduling several times over and serious slippage (Fazar, 1962).

Those involved in the centralised provision of construction planning will quietly admit that nothing has changed in 45 years.

While the experience with adopting CPM has relevance to location-based management, the methodology relies heavily on lean principles, which has similarly had difficulty in engaging practically.

Learning from lean construction

Location-based management is essentially a lean production environment supported by a better way of organising and managing project data. Thus it is relevant to look to the experience in adopting lean construction as a guide for developing a successful location-based environment.

Lean construction has been found to experience significant resistance to change (Alarcón and Diethelm, 2001; Picchi and Granja, 2004; Ansell et al., 2006; Arbulu and Zabelle, 2006; Pavez and Alarcón, 2006; Neto and Alves, 2007). This is because lean project delivery implies a different way of doing business in construction (Arbulu and Zabelle, 2006). A number of different approaches to the problem have been identified, having many aspects in common.

Arbulu and Zabelle (2006) identify the project as the mechanism for implementation. They observe that a company-wide, or *shallow and wide* approach, communicated from the top-down without proper stakeholder engagement, supported by big announcements such as 'we are going lean' does not lead to success. The authors' experience is that a top management driven decision to implement LBMS has often resulted in poor implementation. Arbulu and Zabelle adopt Peter Drucker's view of entrepreneurship to argue that new (entrepreneurial) initiatives must be organised separately from the continuation of old or existing processes—as using existing units to carry new entrepreneurial projects will fail. Thus, individual projects (and, while they did not specifically say this, by extension new, separately established projects) should be used as the vehicle for change. Such an approach they term a *narrow and deep*—bottom up—approach.

The desire for a project-based approach is, perhaps, natural in a project-based industry desiring to change. Nevertheless, it is important when developing a strategy to understand whether it is targeted at the company or the project level. More importantly, it is not really possible to separate the two. A shallow and wide approach must use projects, a narrow and deep approach must engage with leadership. Alarcón and Diethelm (2001) summarised four components to consider, which may be characterised as targeting leadership. These were signals from upper management, commitment from site managers, early constitution of an improvement committee and leadership. However, Alarcón's later work is much more interesting, as it developed a well-accepted hierarchy of adoption.

Pavez and Alarcón (2006) observed that the problem was not only about the application of new tools and techniques, but more about organisational and human issues. Following the work of Picchi and Granja (2004), Pavez and Alarcón developed a scenario-based model for levels of implementation on projects (Table 12.1). This approach is more sensitive than Arbulu and Zabelle's two approaches, and embeds the role of projects. They talk of progressing in levels from fragmented, tools-based, through integrated job-based (tools and principles) to enterprise-wide, philosophical approaches. This important observation provides a key to unlocking strategies for location-based management, as Pavez and Alarcón struggled to explain the role of both technology (tools and techniques) and culture (organisations and management methodology).

In reality, location-based management requires two fundamental changes which can each be divided in much the way Pavez and Alarcón describe. These then form two axes of a continuum: technical advancement—the use of systems, technical support and adoption of a virtual environment; and methodological advancement—moving toward location-based and lean management methodologies (either or both). It is useful to view this diagrammatically (Figure 12.1).

Table 12.1 Lean implementation scenarios (Pavez and Alarcón, 2006)

Scenario	Description
Scenario 1: Fragmented tools applications	Is the most frequent pattern observed in construction companies, and means the fragmented application of lean tools, without a rigorous consideration of lean principles and concepts (lean thinking). The focus is put on lean tools application in specific project.
Scenario 2: Integrated job site application	Represents a major step towards a wider application of lean thinking on job sites: the systematic application of the five lean principles, combined with tools use, driven by a future state value stream mapping designed to improve flow. There do not yet appear to be implementations in construction with the amplitude of scenario 2. In this scenario the focus is put both on tools and principles.
Scenario 3: Lean enterprise application	Is the application of lean philosophy to the job as a part of a company-wide transformation. In this scenario the focus is put simultaneously on tools, principles, the project and the enterprise as a whole.

The advantage with this representation is that it makes clear that there are two simultaneous strategic changes being demanded of the organisation, whether this be the individual project or the whole company. Visualisation is important for understanding. It also helps for individuals to track their progress on charts, which display a company's progress toward a complete implementation. While Figure 12.1 illustrates the comparison between the adoption of technology and the advancement of methodology, it is also useful to extrapolate from Pavez and Alarcón (2006), see Table 12.1, to chart methodological developments toward both lean production and LBMS (Figure 12.2).

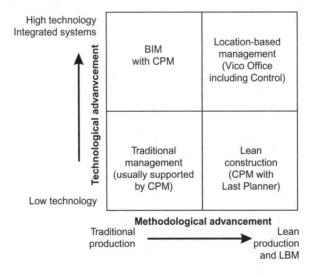

Figure 12.1 LBMS development grid

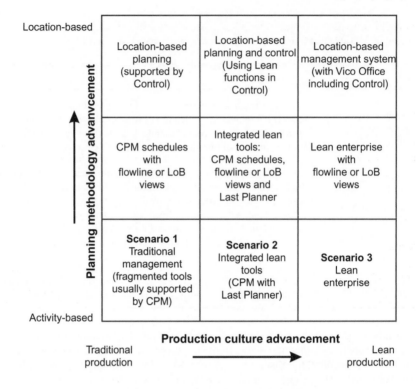

Figure 12.2 Expanded methodology development grid

The development of lean production in construction without the use of location-based techniques has been well covered in the literature[1]. The development of a location-based management system in construction has not yet been covered. The following solutions are intended to move an organisation towards the square in the top right-hand corner in Figure 12.2, but they take the approach suggested by Neto and Alves (2007), who argued for gradual and visible intervention, quoting Womack and Jones:

> Lean should be first implemented in activities that are 'important and visible', e.g. production, so that all people in an organisation can see the benefits achieved with lean.

Accordingly, two different approaches to implementation are discussed. These represent the ends of a continuum from very simple to very thorough.

A low impact solution

The practical on-site implementation of location-based scheduling can be treated very simply. Most site staff do not really care about the technology behind the planning, which is in many ways only the business of the planner. Therefore it is possible to continue operating on site as previously, merely producing reports as required.

---

[1] See IGLC conference publications which may be found at http://www.iglc.net/conferences

In other words, work as if nothing has changed except for some additional reports. The following is a guide:

- Plan the project using location-based principles.
- Plan for continuity of crews.
- Include large buffers to make allowance for the problems that will arise on site due to an 'as soon as possible' mentality.
- Produce Gantt charts for the site, showing when the work should be done but keep this at a high location summary level, such as per floor.
- Produce control charts, filtered for individual subcontractors or supervisors and showing start dates and end dates in each location at the relevant lowest level.
- Report progress with frequent inspections and monitoring of progress in the control phase. Highlight deviations from the plan by the use of colours on the control chart.
- Show simple flowline diagrams to illustrate deviations from plans and their reasons

This approach is recommended if no resources can be devoted to changing the site culture. No immediate benefit will arise from subcontractor pricing, because they will be unaware of the potential for better productivity. Furthermore, any promise of benefit would have to be highly qualified, as the site would very likely have a high degree of variation from the plan and earlier trades would interfere with later trades causing cascading delays. Even though past problems can be better explained, this approach does not prevent problems from recurring in the future.

Nevertheless, this approach would be an ideal first stage for early adopters. It works best where the project manager is involved in the process, especially if he/she is the planner as well.

There are a number of tricks that can be used to increase the effectiveness of location-based planning in this low-impact approach. These are intended to create the flow effects and continuity without dealing with training and culture change.

- Make the locations highly visible on the site.
- Place floor plans on the walls with the location (higher level: floors or zones) and sequence of work (lower level: zones or rooms) clearly visible at entrances to locations. An example is shown in Figure 12.3.
- Number work sequences. For example, if 18 hotel rooms are being constructed per floor, paint a number in the correct construction sequence in each room on each floor.
- Remind contractors, at regular site meetings, to follow the sequence of the control chart.
- Prefer to deliver materials to the locations according to the control chart in preference to where the work is being done.
- Do quality inspections in accordance with the control chart and highlight problems due to continuity and interference problems early.

While such an approach will always have problems, it will nevertheless be often easier than managing a project simply using a CPM schedule.

Andersson and Christensen (2007) report the empirical results of trialing location-based management using DynaProject. They reported three major benefits: an improved overview of the project schedule and better communication with the flowline view, improved resource management and avoidance of clashes, and improved project control as crew locations were known and could be monitored. Nevertheless, their's was a low impact approach, and among the difficulties they encountered was that the builders did not have adequate starting data—particularly in the form of a bill of quantities. There are ways to solve this problem.

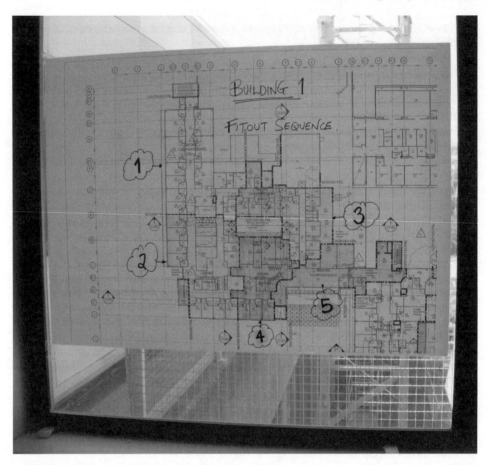

Figure 12.3 Typical site plan mounted on a window just inside the entrance to a floor showing zone sequence

Planning without a bill of quantities

Location-based planning can be used without a bill of quantities, in the same way as conventionally planned projects are often scheduled, by scheduling activities (or task locations) in days—or some other unit of time. However, using quantities brings a lot of extra power to the planner and greatly reduces the workload when combined with production rates.

The best way to do this is to approximate durations and crew sizes by location to simulate quantities. For each work type an item in the bill of quantities is created with duration in worker days as the quantity. This forces an assessment of the differences between locations. It is also possible to go into finer detail than often done in regular CPM, for example having fractional durations of, say, 0.5 shifts per apartment instead of whole days, or planning in worker hours. This is possible because all the information is contained in a single task. Using worker days allows the planner to resource load the location-based schedule with minimum effort. Durations of the CPM schedule can be given to subcontractors asking them to provide a crew size. Crew size times duration gives the number of worker days in a location.

It is quite easy to manipulate as many as 300 locations in the same task. Imagine doing this using ordinary CPM and linking all those locations to other tasks—possible to set up, but impossible to evaluate alternatives effectively or keep up to date, let alone to individually resource load every activity.

Using a bill of quantities without location information

While the best solution is to have a BOQ measured by location, having some quantities as basis for planning is still better than not having any quantities at all. However, the accuracy of flowline scheduling suffers somewhat, because the exact amount of work in each location is unknown. To get the best out of available quantities, they should be distributed to locations using one of the following options:

1. Distribute according to total area of locations—this option works for many work types if the locations are approximately similar.
2. Distribute according to some other numerical measure—for example, using the number of apartments per floor.
3. Distribute manually, row by row—this option requires a lot of work, possibly amounting to a partial measure, but is usually worthwhile.

The best strategy is some combination of the above methods. For example, some quantities can be quite accurately distributed by area using manual checking for obvious errors (for example, spaces which do not have any quantity).

Often the items in existing bills of quantities or cost estimates cannot really be used as a basis for planning. For example, painting and plaster work can have the quantity measured in building volume. In planning the work, we need the quantity of painted area, not the cubic metres of the building. In these cases, the quantities should be re-estimated on the basis of floor, ceiling and wall areas. Another common problem with BOQs is the use of *items* (work measured complete as a single item) which do not really say anything about how much work should be done. Such quantities should be replaced by real ones if known or alternatively have duration estimates approximated per item.

All work is done by subcontractors, why should a general contractor care about resources?

Many general contractors do not believe it is their responsibility to manage to the level of resources for tasks being serviced by subcontractors. This is a fallacy, as it is precisely this attitude which leads to a loss of control of production on the site. Rather than being a responsibility to be avoided, the size, mix and location of work crews should be planned by the contractor with the cooperation of subcontractors.

The general contractor needs a way to forecast problems. If it is known that to achieve a scheduled production rate, the subcontractor needs four workers (of average productivity), and the subcontractor provides two, it is safe to assume that there may be problems in the future (unless the workers are twice as productive as the average).

On the other hand, only a given number of workers can work in a single location. If the desired duration requires too many resources, either the duration must be increased or the resources must be split so that work is undertaken in more than one location at a time. This is a critical planning decision, because it may affect the continuity of other tasks and if it is not explicitly taken into account, cost and risk will increase.

Locating productivity data

It does not matter what planning system is used, consumption or productivity data is needed for effective management, so if you do not have it then it is time to start gathering it. Using the control mechanisms presented earlier in this book, it is quite easy to estimate productivity rates for individual works from past projects. These productivity rates can be used to estimate realistic durations for future tasks.

Meanwhile, it is possible to roughly estimate the productivity rates by estimating the duration and estimating the number of resources needed to achieve that duration. This gives weights for individual items in the bill of quantities (larger weight takes more time per unit relative to the other items). Then the consumption can be calculated backwards from this data distributing the worker hours in relation to quantity times the weight.

1. *Man hours needed = duration × number of resources × shift length.*
2. Calculate the weight for each item in BOQ (*defined weight × total quantity*).
3. Distribute the needed worker hours to items according to the weights from 2.
4. Calculate the resulting consumption for each item = *man hours ÷ quantity.*

This approach calculates an estimate of productivity which can be checked after actual data becomes available. Moreover, it distributes the total duration to locations in proportion to worker hours needed. By defining some initial values, it is easy to start collecting productivity information.

The general contractor can use the productivity information in:

• Assessing the price level of subcontractor bids
• Measuring changes in productivity after implementing new management systems
• Making realistic, feasible schedules
• Optimising production
• As a tool in subcontractor meetings (how many workers do we need for the next two months).

A thorough solution

At the other extreme it is desirable to take a more comprehensive approach to implementing a location-based management strategy on site. This discussion will not consider the off-site aspects to planning thoroughly, such as obtaining data from modelling software, but will address the steps which will build a supportive culture and drive productivity improvement.

The first thing to recognise is that culture change requires a deliberate strategy of training, consultation and engagement. The idea is to build a team which is working cooperatively towards achieving a successful outcome. In fact, one contractor decided that implementing the LBMS was to be set as "the project" for the project manager of each LBMS project, as management considered that achieving its aims would necessarily deliver performance for the project as measured by traditional criteria.

At the start of the process, the following questions should typically be answered and communicated to stakeholders:

• What are the specific business benefits to the organisation?
• What will be the barriers on this project?
• Who will take responsibility?
• Who will do the work of planning, scheduling, monitoring and reporting?

- What is in it for the site staff and will they commit?
- What is in it for the subcontractors and will they commit?
- Will senior management commit to the process?
- What performance indicators will be reported against?

The key issues are understanding and commitment. Both of these are hard to achieve.

Training

Training is critical to all parties in the process, although the level and type of training differs greatly depending on the role.

Planners and schedulers need the appropriate technology training to be able to drive the process. This is mainly training in the use of relevant software, such as Vico Software's Control. However training in the principles of lean production, their application to construction and the role for location-based management in achieving productivity improvement is highly desirable. They must also be fully able to monitor and report progress and to use the tools to control the project, including the planning of control actions.

Design managers need to understand the principles of lean construction and location-based management. They have a responsibility for ensuring that the LBS is considered in the design phase. They are also in a position to ensure that the required data is easily available. For example, they can plan for the use of model-based quantity take-off software such as Vico Software's Take-off Manager as a way of producing the required quantity data—important for resource-based task scheduling. Design managers often set up the procurement system through their decisions, and it helps if they understand the benefit of logistics driven by location-based management.

Project managers must understand the system as per the design manager. However, they must also understand the site management issues and therefore must understand how control phase monitoring, forecasting and reporting works. They are the principal communicators to all the other stakeholders, so it is very helpful if they have a clear picture of the entire process.

Site managers and project engineers need sufficient training to demystify the management system, particularly issues of timing and progress reporting. They occupy a critical locus in the entire process and any implementation will fail without their support. They also will have the most to lose in terms of their self-respect. The training must therefore focus on why the new processes will deliver better outcomes for them. In part, this must also focus on shifting emphasis from early and quick to planned and consistent.

Site supervisors, whether direct or subcontract, must understand the significance of timing, and be able to interpret the control charts. It is their responsibility to deliver the plan in practice, so they must understand the basic principles of continuity and working to planned start dates with planned resources. They must also understand the significance of control actions.

Even subcontractor managers or owners should receive basic training about the benefits, particularly where a long-term approach is being taken and cost savings are to be anticipated and exploited.

Team building

It is advisable therefore to establish a training program which suits the level of implementation and the roles of participants. This can be combined with a team-building approach, possibly with the participation of an industrial psychologist, to get the entire team focused on the goals and objectives of both the project and the management system. This will deliver the best long-term benefits.

One of the biggest problems with managing construction efficiently is the distrust between subcontractors and the contractor's team. This distrust is a subtle thing, rarely expressed directly, but manifest in the actions of the participants.

The contractor's team likes to own the schedule and demand performance, with little consideration for the impact on the subcontractors. They in turn anticipate being frustrated and manage their many contracts to maximise their own benefit with little regard to the project. The consequence can be viewed as a state of undeclared war.

The solution is to bring the players to a position of mutual trust. This requires transparency in both planning and action. It is not sufficient to plan to manage resources for continuity if the action is to manage them in the traditional way, as this causes unnecessary costs, usually borne by the subcontractors. They will anticipate this and price accordingly and behave accordingly—as it suits them. If the plan is transparent and there is communication and consultation during the progress of the works, then trust will emerge and system costs will reduce. The team will become more productive and all players will benefit.

There are team-building initiatives which deliver benefit without requiring location-based management. These typically concentrate on the dual aspects of team spirit and cooperation. In a scheduling sense, such methods require the crews to identify who is before them and who is after them and to work to ensure they pass work smoothly between the crews. Erroneously, this is often claimed to achieve the same results as LBMS, however in location-based management, it is not sufficient to rely on these aspects of team building. Rather, it is not only important to manage the interface between crews, but also to align the production to minimise disruption. In other words to work the schedule. Commitment to a work plan at such a detailed level requires a great deal more trust—in both the planning process and the team.

Site tricks and techniques

There are several techniques which can be employed on site to make communication and control easier. These range from the basic techniques discussed above (and repeated here) to sophisticated manipulation of contracts and payment systems.

- Plan the location breakdown structure for the project to match with the constructability on site. If using modelling software to define the locations, do not merely divide the building by wings and floors, but explore the best work sequence of construction zones, suites and rooms that will provide suitable control on the site. Where there are obvious units of completion, such as hotel rooms, apartments or lecture theatres, make these manageable in the location breakdown.
- Make the locations highly visible on the site. Crews should be aware that they are leaving or entering a location. They should be able to visualise ownership of the location.
- Place floor plans on the walls with the location (higher level) and sequence of work (lower level) clearly visible at entrances to higher level locations.
- Number work sequences. For example, if 18 hotel rooms are being constructed per floor, paint a number in the correct construction sequence on each room floor.
- Remind contractors, at regular site meetings, to follow the sequence of the control chart. The control chart can be used as a very effective way to get the team to focus on poor performance, as it highlights work which is behind schedule. A team approach will allow peer pressure to apply to those not performing.

- Maintain consistent report formats at site meetings. By attending multiple site meetings, participants will gradually begin to understand the information presented in the new formats. The format of print-outs should be decided as early as possible during the project. The same print-outs with the same locations and the same tasks should then be presented in all subsequent site meetings. The only things that should change are the colours and dates in cells. This will maximise the learning and sense of comfort with the system.
- In large projects, flowline figures tend to become quite complex without filtering. Subcontractors who are not used to reading flowline figures will resist heavily if they find that the diagrams are difficult to understand. Concentrate on showing problems and control actions—show problem task, predecessor and successor tasks. Avoid figures with too many lines and figures where lines are crossing.
- Flowline figures work in site meetings only if they are very clear and simple. A good basic rule is that each single A4-sized figure should not contain more than 20 locations and ten tasks. There should not be crossing lines or very complex work patterns because these will not be easily understood and will result in resistance. Ideal figures have locations sorted so that workflow proceeds from bottom left towards top right.
- Plan the logistics for the project such that materials are delivered to the locations just-in-time for the work being done. This has a double benefit: work is constrained to the scheduled sequence, and waste in materials damage and double-handling is avoided.
- Do quality inspections in accordance with the control chart and highlight problems due to continuity and interference problems early. Frequent inspections will ensure that crews do not work out of sequence.
- Establish a quality assurance system that includes carefully managed location ownership and transfer. A crew should not take over a location until it is ready, which means predecessors must be complete and the quality of work approved. In turn, the contractor should not take back the location until the crew has completed. Crews should only be allocated locations in accordance with the planned sequence.
- Contracts should be amended to support the location-based management system. In particular, payment systems should be matched to controlling the work. Traditional systems rely on monthly assessments of work completed. This is a major problem for project control, because subcontractors are encouraged to complete as much easy work as possible in order to get as much money as soon as possible. This usually involves a disconnect from the desired sequence of work and, in particular, does not require location completions. LBMS supports a payment system based on frequent payments for completed locations (if frequency is a problem, they can be aggregated into monthly claims). As the quality system records completion of a location a payment approval can be triggered. The only instance where a progressive work assessment is required is for large locations which are worked in the correct sequence and remain incomplete. In this case an assessment of work completed can be used. Work which is commenced out of sequence should not be paid for, especially if it is incomplete. If payments are only made in the correct sequence and on completion, subcontractors will rapidly fall into line with site management.
- Forecasts should be used to plan control actions. This ensures that the site team gets into the habit of identifying problems and taking planned corrective action. Communication of this to the team will ensure that all understand the situation and can take suitable action to both accommodate and rectify the problem (as appropriate).
- Visualisation of the model, using tools such as Vico Office, can be used to show crews (on site) the planned sequence of the work. For example, if a hotel is being constructed, a typical floor could be shown and the crews could see the fit-out of the floor proceeding from structure to final completion. Being able to see the importance of work sequencing is a powerful communication tool.

There are many methods which can be used, some generic and some site specific. Project managers should be encouraged to develop their own suite of tools and way of working. While these may appear very different from traditional methods, it will be found that there is little substantive difference in practice and, in time, the new methods will become just as familiar.

Managing the client

One of the most common arguments used against LBMS is that the client requires CPM— usually in the form of a specification of a certain software application. This is not really a barrier except perhaps on budget sensitive small projects.

Consultation by the authors with many clients, including government clients, confirms that these specifications are generally considered a minimum requirement. Thus provision of a higher standard, such as a location-based schedule, will exceed the requirement and will generally be approved.

Where it is not approved, there may be two ways to handle this. The preferred way is to provide the software to the client (or their agent) such that they can do their own analysis. An alternative is to export the data into a CPM tool. This will break the location-based logic and must necessarily involve fixed, not calculated, dates in the resulting schedule. The reason is that CPM does not include the continuity heuristics of the layered logic LBMS.

One of the critical concerns in the relationship with the client is the reliability of the schedule and assessment of contractor performance. A location-based schedule will provide much greater reliability and confidence for the client. It is in their best interests to accept such schedules. In fact, it is likely to be the contractor that will take some getting used to the accountability of location-based reporting. Currently, the best representation of progress is provided by earned value analysis (EVA). Yet this is relatively imprecise in comparison to location-based reporting.

Clients will generally embrace location-based management but, when they do not, it is then possible, if undesirable, to maintain dual systems. However, dual systems will always undermine the acceptance of LBMS on the site.

Empowerment and restraint

As with any significant culture change, the introduction of location-based management requires support from senior management. However, its penetration to the level of site supervisors means that it is vulnerable to attack from managers all the way down the line. The starting point should be strategic management. Ideally, the organisation's strategic vision and goals should be altered to reflect the intent to target production efficiency.

Senior corporate management must support the implementation. This is rarely a problem, because implementation, whether a trial or a more complete run-out, generally follows the presentation of a business case. Senior management recognises the business advantages of moving towards location-based management and enthusiastically supports the initiative. In most cases there is a single supporter for whom it 'just makes good sense' or who 'has seen it work before'. These become champions for the change.

Champions are important because senior management support is insufficient. Champions can take the support and make it clear at the lower levels where generally they will find increased scepticism—arising from increased impact on current practice. A champion can do much more than merely instruct lower levels of management, they can pass on their enthusiasm and their vision of how projects should be run.

Project managers must support the initiative for their projects. This is the most common blockage to change. Modern management styles place project managers in a position of significant power. No new systems will work without their support. Training and change management support is therefore highly desirable at this level. Similarly, projects with the levels of project director and site manager should consider the same level of intervention to ensure support, as each of these can block progress at different stages.

It does not matter how good the plans are or how much preparation has been done, if the site management team (the site manager and site supervisors) do not implement the plan. It is very common for participants at this level to refuse to change and to block implementation through passive or even active resistance. Unfortunately, senior management needs to watch for such resistance and be prepared to remove such individuals to other sites where LBMS is not being used.

A key stakeholder in ensuring successful implementation is the planning team. Generally contractors have a central planning manager, and on bigger projects a locally based site planner. All of these must be fully trained and aware of the strategic advantage to the organisation. Resistance here, particularly on site, will lead to all the wrong messages being given out and to failure of both planning and control of the work in accordance with LBMS principles. It is important to recognise that planners have generally invested years of their life in becoming proficient with current technology. They will sometimes be very reluctant to start using a new technology—particularly one which challenges their very ways of thinking.

When an organisation is considering LBMS as a way forward, they will generally ask their planner for an expert view of the technical solution. It is common to see resistance to change right from this stage, as the planner will test the software to see if it is as good as existing products for doing things the traditional way—a long list of 'shortcomings' will then result. Most of these shortcomings will actually arise from the location-based methodology, in particular the need for planning for continuity, and reflect the lack of knowledge of the planner rather than a shortcoming of the technology. Once again, because their support is so critical, training and support are required to help the planners understand. At the site level, it is sometimes necessary to replace the planner with someone able to understand the new production principles.

Contracts

Construction contracts are a complex area and beyond the scope of this book to cover in detail. However, current contracts often contain restrictive provisions, or fail to provide for the opportunities which a location-based management system provide. The following points should be addressed when considering contracts for LBMS.

- Clauses which relate to scheduling and the provision of construction models should reflect the technology to support a location-based system. At a minimum, a location-based software solution such as Vico's Control should be specified, rather than a basic CPM package. Preferably this should include a provision for exchange of digital models and schedule files, as well as printed reports or PDFs.
- An approved baseline schedule can be defined for the project. Typically this should be a risk adjusted schedule including buffers between trades for variation. The buffers should be owned by the contractor as a planned component of the production system.
- Subcontractors could be required to provide detailed current schedules that fit within the constraints of the approved baseline schedule.

- It is highly desirable that subcontractors adopt the same planning system and files as the head contractor. Contracts can request this of contractors and subcontract agreements can reflect this.
- Locations should be specifically defined in a contract (but not the detailed LBS) so that references to location will be specifically understood as applying to a location-based management system.
- Contractors should own their production efficiency. This opens scope for new rigour with regard to time-related claims. Contractors can demonstrate the impact of changes on their production efficiency, but clients in return can demonstrate deviations from the plan.
- The contract should clearly state that progress by location is a key component of the contract. It is insufficient to have merely start and finish targets, there should also be milestone dates based on locations. This can apply to the subcontracts as well, to ensure progress is linked to the location-based plan.
- Payment terms should be linked to completions of locations rather than percentage complete. This is the change most likely to drive changes in behaviour. At a minimum, monthly payments should be restricted to payment for aggregation of completed trades in locations according to the planned sequence. A full implementation of the LBMS would require micro-payments to be triggered by the completion of locations (as long as it is according to the planned sequence).
- Clauses targeting time-related claims should be changed to reflect the location-based methodology and the recognition of the production method as part of the contractor's entitlement. It is important that concepts of float are adjusted to reflect that planned continuity is more important than early start.

It is perfectly possible to run projects using traditional contracts, particularly subcontracts. Nevertheless, the system would benefit from optimising these critical components of the relationship between the parties.

Financial control systems

Payment systems have been discussed above, but it is worthwhile highlighting again the use of progress payments to drive behaviour and therefore performance.

Current contracts generally provide for valuation of the work-in-progress (WIP) at the end of the month for the work done to date. Typically, a quantity surveyor will assess the work done on site and calculate the payment due as a lump sum. This method, while very familiar, is vague and ineffective as a project control mechanism. Anyone who has had to conduct an assessment of work completed knows the difficulty of being accurate in such an activity, especially when work can be undertaken anywhere on the project. The assessor is required to rapidly develop an awareness of all the work-in-progress and to what extent. Frequently, a process of negotiation is required between the client and the contractor about the value of the work-in-progress.

If payments are restricted to an aggregation of completed trades in locations only according to the planned sequence of work, then the assessment task is greatly simplified. For each trade it is simply a matter of progressing through the sequence, recording 100% complete, until reaching the first incomplete location. At this point an assessment can be made for that location and all remaining locations can be ignored. The only exception would be where a revised work sequence had been approved due to special circumstances.

The most extreme method, as adopted by many Finnish projects, is to link all payments to completion of locations or milestones for trades as they progress. On a major project, this may lead to as many as 30 payments per month. Micro-payments are triggered by completions of locations, but only according to the planned sequence. Traditionalists would see this as a huge administrative task. But there is no reason why, with modern computer systems, such a process could not be merely the recording of completions (which is necessary for progress reporting anyway) which triggers a payment in the management system. The advantage is that both parties are motivated to have accurate completion data.

ENVIRONMENTAL FACTORS

There are factors which must be considered in the uptake of location-based planning arising from the cultural and technical context. Two such factors are the challenges to thinking that participants, particularly planners, need to consider when sifting from activity-based to location-based thinking. Another is the impact of a technological environment such as 'virtual construction'.

Location-based planning for CPM thinkers

The challenge of introducing a location-based management system is not just about dealing with change. The underlying epistemology of location-based planning is different from activity-based planning. This means that one of the most interesting challenges is for the planner to change the way they think about scheduling. In the early days, even with the best intent, mistakes will occur as a result of the application of CPM thinking in both the planning phase and the construction phase.

There are two major sources for this change in thinking. The first arises from the adoption of lean production principles, requiring that work only commence when it can adequately be completed—when all the precedents and prerequisites are completed. The second arises from the desire to protect production efficiency by planning and managing for continuity, requiring that a sequence of work be planned not to commence until the entire sequence can be completed without interruption. As with adopting lean construction, "it is necessary to change mental models and/or the way of thinking" (Pavez and Alarcón, 2006).

Applying lean thinking to construction is a significant but well-documented challenge. As previously mentioned, the International Group for Lean Construction is a large international researcher network[2] devoted to the field of lean construction and a great deal of effort has been devoted to developing understanding of lean principles in the industry.

In general, the following need to be at the forefront of thinking, especially for those used to activity-based thinking:

- Use quantity and resource-based scheduling
- Think of tasks as work crews moving through the project systematically
- Plan for continuity, not early starts
- Make tasks flow sequentially, unless deliberately planned otherwise
- Resist the temptation to start ahead of schedule or out of sequence
- If precedents are not complete, get it fixed rather than working around them
- Use location completions to drive quality
- Plan for optimum production rates (much shorter durations per activity)
- Use large buffers between trades as a risk management technique (optimum duration plus buffer equals traditional duration)

---

[2] Internaltional Group for Lean Construction (IGLC) http://www.iglc.net

• Slow and steady wins the race. Just as in a time trial, catching the preceding team increases disturbance and risk.

It is possible to look to integration of technology as the way to bridge the problems of implementation. The use of integrated suites, such as Vico Office, can make the process easier and deliver more robust outcomes.

PLANNING FOR CHANGE

There are special circumstances in which it is often not possible to prepare detailed plans. Such circumstances might be the fit-out of shops in a shopping centre development, or the detail of apartments which are sold progressively. In such circumstances, design information is only available late in the process and is reactive to client change. There are in fact many circumstances in which client changes may be expected right until very late in the production process.

Generally, such matters are either not planned until known, or form the content of major disputes between the parties. A location-based planning system can provide a mechanism for handling such uncertainty, either by removing its impacts or providing a clear mechanism for measuring its impact.

Where change may be anticipated, such as late letting or progressive sales, then it is sensible to use buffer mechanisms to shield the production from the change. Buffers may enable time for undertaking detailed planning to accommodate changes, along with required detail planning and re-sequencing that may be required.

Where change cannot be accommodated, the location-based management system allows for changed quantities and locations to be planned reactively and the consequences of production duration and efficiency can be measured.

Virtual construction

Adopting an approach such as Vico's *virtual construction* presents opportunities and challenges to the implementation of location-based management.

First, opportunities arise from the comprehensive environment that a location-based 3D model environment, such as Vico Office, can provide. This is hugely enabling for resource- and location-based scheduling. The data is rich and accurate, often providing data when the alternative is to make do without.

Second, the potential simplicity and elegance of a dedicated flowline schedule, and the associated concentration on production efficiency, can be lost in the greater environment—with its associated 'wow factor'. It is certainly true that 5D modelling is extremely powerful and can lead to both acceptance and effectiveness of LBMS. Nevertheless, it is easy to lose sight of lean production thinking, and instead to concentrate on modelling, scheduling and visualisation, without due regard for the practicalities of an efficient site.

RETURN ON INVESTMENT

It is difficult to place a specific financial measure upon the return on investment in location-based management. It is possible, however, to explore the costs and benefits in a broad sense and to draw conclusions about the net benefit. Cost and return depend the following issues:

- The level of implementation
- Contracts
- Level of knowledge and competence of the team/organisation
- The support technology available.

The costs include:

- Software
- Training
- Change management
- Team building
- Virtual construction services or capacity
- Scheduling services or capacity.

The benefits include:

- More accurate and reliable scheduling:
  - Harmony on site
  - Reduced waste—labour
  - Reduced waste—materials
  - Reduced waste—rework.
- Better data and information
- Early warning of problems
- Avoiding the end of project rush and spending money to finish on time, or handing over unfinished buildings
- Improving the productivity of management
- Confidence in senior management
- Client confidence in the project
- Improved Quality
- Improved Safety
- Reduced remedial works and defects lists
- Reduction in measuring and estimating costs
- Reduction in cascading delays.

Schedules can typically be compressed by 10% without adding further risk and, with the implementation of better control, another 10% can be taken from the schedule. However, it is not recommended to aim for 20% or more duration compression in the first project because it requires the effective implementation of location-based controlling methodologies, which are typically the most difficult to master. In addition, productivity benefits have been estimated to be 20% from implementing continuous workflow. An additional 20% can be saved if location-based controlling is fully implemented to prevent location congestion. By adding virtual construction tools to the mix, additional benefits of better design coordination and reduction of requests for information (RFIs) will further decrease delays and improve productivity.

A minimum implementation but including training and team development might be expected to cost around 0.15% to 0.3% and deliver benefits of around 2% (minimum strategic benefit) to 10% (harnessing the long-term benefits of productivity improvement). A thorough implementation, including a full virtual construction environment, could cost between 0.25% and 0.5% and deliver benefits of around 3% to 15%.

REFERENCES

Alarcón, L. F. and Diethelm, S. (2001). "Organizing to introduce Lean principles in construction companies". *Proceedings of the International Group for Lean Construction , 9th* Annual Conference. Singapore.

Andersson, N. and Christensen, K. (2007). "Practical implications of location-based scheduling". *CME25:Construction Management and Economics: past, present and future.* Hughes, W. (Ed.) Taylor and Francis Group, London.

Ansell, M., Holmes, M., Evans, R., Pasquire, C. and Price, A. (2006). "Lean Construction trial on a highways maintenance project". *Proceedings of the International Group for Lean Construction ,* 15th Annual Conference. Michigan, USA: 119–128.

Arbulu, R. and Zabelle, T. (2006). "Implementing Lean in construction: How to succeed". *Proceedings of the International Group for Lean Construction,* 14th Annual Conference. Santiago de Chile: 553–565.

Arditi, D. (1983). "Diffusion of network planning in construction". *Journal of the Construction Division.* American Society of Civil Engineers. **109**(1): 1–12.

Fazar, W. (1962). "The origin of PERT". *The Controller.* 1962(December): 598–621.

Neto, J.de.P.B. and Alves, T.de.C.L. (2007). "Strategic issues in Lean Construction implementation". *Proceedings of the International Group for Lean Construction ,* 15th Annual conference. Michigan, USA: 78–87.

Pavez, I. and Alarcón, L.F. (2006). "Qualifying people to support lean construction in contractor organisations". *Proceedings of the International Group for Lean Construction,* 14th Annual conference. Santiago de Chile: 513–524.

Picchi, F.A. and Granja, A.D. (2004). "Construction sites: Using lean principles to seek broader implementations". *Proceedings of the International Group for Lean Construction,* 12th Annual Conference. Copenhagen, Denmark.

Smith, D.A. (1981). "Productivity engineering is 'task management'". *Civil Engineering.* American Society of Civil Engineers. **81**(8): 49–51.

Chapter 13

Planning project types

INTRODUCTION

This chapter describes how to adapt the location-based management system for different project types and special cases, with an emphasis on planning and control. Projects can vary due to the type of building, special circumstances, contractual phases or construction stages. Earlier in the book, generic methodologies have been described, whereas this chapter considers how to handle common project circumstances.

There are many tools which are not suitable for all project types and some tools need to be adapted for specific circumstances. In the following discussion, it should be noted that optimal accuracy levels of pre-planning and control depend on suitable available information. Fast-track projects will have less information available on commencement, while routine production can be pre-planned very accurately. If the LBMS is implemented midway through a project, options for changing the schedule (re-planning) will be limited. However, whatever the plan, it may be successfully controlled using location-based control, whatever the stage of intervention.

To date, location-based management has been applied on some project types more than others and, as this chapter is based on real experience, some of the following sections are more detailed than others.

Guidelines for typical project types, phases, circumstances and stages

In the following sections major project types (residential, office, retail, health care, refurbishment), common project features (industrial projects, highly repetitive projects, multi-purpose projects, refurbishment, sport stadiums, linear projects, civil projects, maintenance projects and large or complex projects) as well as special interventions (late intervention or chaotic projects), contractual phases (owners schedule, bidding, pre-construction) and construction stages (earthworks and foundations, structure, façades, finishes and MEP, commissioning and handover) are assessed for the best way to use the LBMS for those projects or circumstances. Guidelines are provided for typical project types and the discussion is broken down to describe the special aspects, typical location breakdown structures and recommended ways to use the LBMS in planning and controlling different types of projects. Some project types, such as residential, offices and retail projects have so far seen more application of LBMS than other project types, enabling more complete recommendations to be made.

RESIDENTIAL

Description

Residential projects usually consist of highly repetitive apartment or housing units. They can include anything from detached housing, through apartment buildings, to entire residential complexes consisting of apartment buildings, condominiums and houses.

Residential projects are highly suitable for location-based planning because the sequence of trades is straightforward, quantities are easily available and changes do not have a great effect on the schedule because they are usually localised to individual units. The typical form of interference that takes place in residential construction involves trades running out of work when their technically mandatory predecessor is too slow. While interference can occur between trades which do not maintain technical dependency, it is rare because the sheer number of locations means they can easily change sequence by just skipping one apartment.

Location breakdown structure

The location breakdown structure for a residential project is normally straightforward. Individual buildings should form the highest level. If there are underground facilities such as parking or mechanical areas that are shared by multiple buildings, they should usually form their location at the highest hierarchy level. If some of the buildings have large floor areas and it is possible to raise the structure in two or more stages, these buildings can be split to two or more highest level locations. This allows duration compression because, by building the structure in stages, it is possible to commence work inside the building earlier. The next hierarchy level can be risers, where there is more than one riser in some of the buildings. The next levels are the floors, and the final level is the apartment level, which is usually accurate enough for controlling purposes.

It is beneficial to go to the apartment level of detail because generally only one trade should be allowed to work in an apartment at the same time. Additionally, it is easier to show the effects of variations on the schedule if changes can be localised to apartments. In addition to apartments, the lowest functional level often includes corridors or stairwell rooms and technical rooms in some floors. Sometimes there are special functions in the lowest floor which should be separated to different locations.

Condominiums or houses with apartments extending to all floors may not require the floor level because the interior work usually proceeds apartment by apartment—not floor by floor. If more accuracy is needed, a floor level can be added below the apartment level.

Starting data

Earthworks, foundations and structural quantities of buildings are key factors in defining the optimal sequence of buildings. Therefore, much care should be put into estimating their quantities. A typical mistake in planning a residential construction project is to fix the construction sequence before analysing alternative sequences for building the structure. In many residential projects, months of duration can be saved, without increasing resources, merely by optimising the sequence.

It is easier to get accurate starting data in residential work than in many other project types, because residential work is often quite standardised. In many residential complexes, there are a certain number of apartment types which share the size and basic finishes of other apartments of the same type. Therefore, it is sufficient to estimate quantities in one apartment and then copy the quantity to all similar apartments. Quantities do not change a lot during production because the function of the space stays the same—for residential use. Changes in floor covering materials or types of tiles used, or the addition or moving of some partition walls do not typically have substantial schedule effects. Also, the standard nature of production means that productivity rates can be drawn from similar projects and utilised directly while planning. For this project type, using template projects with standard crew sizes, productivity rates and logic can be very beneficial.

Pre-planning

Very accurate planning is already possible in the pre-planning stage due to the comparatively good starting data and the relatively small uncertainty associated with quantities and productivity rates. All the tools in the planning chapter can be used. High predictability allows that an accurate logistics plan and cash flow plan are possible in the early stages of the project.

One result of being able to have space for many crews, for a given task, in one building, is that it is usually possible to perfectly synchronise production rates for the finishes trades by using multiples of optimal crews. Normally, it is possible to do all the work of the same trade with the same resources but sometimes this is not desirable because of milestones and time constraints. Even though the predictability of productivity rates and quantities is high, some buffers will be needed because subcontractors often do not provide enough resources in the beginning of their contract. It is normally enough to have a space buffer of three empty apartments during the finishes phase. This allows the subcontractors the flexibility to work in parallel in three apartments with different crews. It is not necessary to plan the exact movements of crews in the pre-planning phase because this will change, depending on selected subcontractors and the circumstances on site when production begins.

One typical planning problem in residential buildings is caused by the comparatively slow production rate of the structure compared to the faster production rates of the finishing trades. This presents a problem because the structure cannot be accelerated without large cost increases. To achieve continuous flow of work, it is necessary to increase the scope of work for subcontract (tasks) or to delay the start of finishes work to achieve continuity.

For visualisation purposes, it is often beneficial to hide the apartment level when showing the baseline schedule. Otherwise the schedule may become difficult to read unless printed at a very large scale. Although all calculations are done at the apartment level of detail, the visualisation level can be rougher depending on need. Floor-level visualisation can be used to visualise the basic flow of the project and apartments can be introduced for more detailed look-ahead plans and for logistics planning purposes.

End user satisfaction, and thus planning for quality, is critical in residential construction. This also impacts heavily on viability for a contractor due to prolonged maintenance periods. In many circumstances, a project's profit can be subsequently expended on maintenance. As an example, concrete floor drying in wet spaces should be ensured before any floor covering work by having precedent logic links to the concrete floor finishing and building dry-in with enough lag. Quality checks in apartments should be pre-planned and controlled for each trade.

In addition to quality, procurement and deliveries are the largest risks in typical residential projects. The need times should be carefully analysed based on location-based schedules and enough time should be planned between procurement events.

Risk analysis does not add much value in this repetitive project type. From a well-planned synchronised location-based schedule, it is easy to see risky spots visually without the use of advanced simulation tools.

Implementation phase

Production control should occur at the apartment level of detail. Instead of tracking percentage completed, it is often sufficient to track started and completed locations because locations are small and many locations can be completed by a crew in one week. Also, because residential projects can have hundreds of locations, visualisation can be a challenge. For most of the time, print-outs can be simplified by showing the floor level of detail. However, for detailed controlling of work, control charts and flowline diagrams need to be printed at the apartment level. Frequently it is not feasible to show the entire project unless a large format, such as A0 size paper, is used. Generally speaking, overall progress and flow can be evaluated at the floor level and detail scheduling and production control can use look-ahead control charts and flowline figures for the next five to ten floors at the apartment level. The A0 print-out can be hung on the site trailer wall as an overview.

Production is usually frictionless because there are many locations in which crews can work. Even though two trades are not going to be in the same apartment at the same time, there is usually room for them to work around each other. There is some flexibility in logic, so it is easy to switch the sequence of tasks as a control action. Interference occurs mainly when preceding trades makes work ready for the next trade. Those trades which must be completed in a fixed sequence in each apartment should be the focus point of control. They can be identified by first building only mandatory technical dependencies which cannot be violated and evaluating free float based on those. Those trades which can be done in many alternative sequences are not so critical because they can easily be switched around with very small loss of productivity.

Detailed planning is important for quality, cost, logistics and safety issues. However, because the pre-planned schedules can already be very detailed, task scheduling or detail task planning is often limited to updating start dates and planning control actions if there are deviations. Exceptions include drywall and tiling, which are often scheduled as one baseline task. These should be divided into multiple sub-tasks in detail planning—based on subcontractor input.

It is important to tie the contractual milestones to these detail activities. Otherwise a subcontractor may do the easy work first and not complete the work required to release work to the next subcontractor. It is recommended that payments should only be triggered by completion of that work which releases ready work for the following trades.

Example

Figure 13.1 shows a location-based baseline schedule for a typical residential project at the floor level of detail. The location breakdown is composed of individual building projects that are subdivided into buildings and then to floors. The apartment level is not shown but all quantities related to finishes and MEP tasks have been estimated for each apartment and calculations are at that level.

Tasks selected to be shown in the flowline are those tasks which make work ready for succeeding tasks and which must occur in a fixed sequence. Other tasks are not shown because they can easily be adjusted to work around other crews. The scheduling process makes sure that there are enough buffers between trades that potentially cause conflict and that work is continuous for all these main tasks. The only overlapping tasks are tiling in bathrooms and priming and sealing which can be done in other rooms of an apartment. Many things have not been scheduled in detail because contractors have not been selected when doing the baseline. For example, plasterboard walls will be broken down to framing, board on first side, electrical piping and board on second side during the detail planning process. For baseline purposes and for planning procurement it is enough to know that the required production rate is on average 60 m^2 per day. This information, together with planned start and end dates of buildings can be used to specify required production rates and milestones in bidding documents. Subcontractors will then participate in more detailed planning of their work within the contractual production rate and milestone constraints. In this example, although the work was subcontracted, the general contractor was using location-based data to buy all boards related to suspended ceilings and plasterboard walls from the same supplier. Boards were lifted to floors by a tower crane once the structure of the floor was completed. This forced plasterboard walls to be installed before concrete floor finishing because otherwise stacks of boards would interfere with pouring the finishing surface.

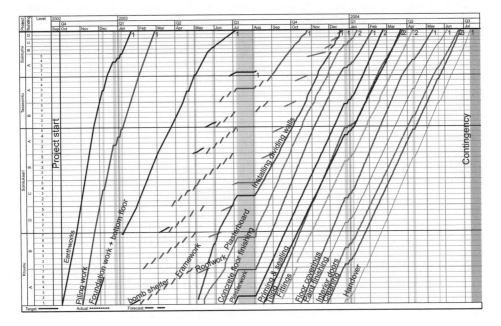

Figure 13.1 Schedule for a typical residential project

OFFICES

Description

Office buildings tend to have both repetitive and non-repetitive components. In most office projects, all the floors except the ground floor and the top floor tend to be typical (repeat). The ground floor may have a lobby and reception area and special-purpose spaces such as kitchens, restaurants or retail spaces. The top floor is often reserved for top management or entertaining guests and usually has better quality finishes and special spaces.

Some special challenges arising in office buildings include end user companies who often have special requirements for MEP systems. Modern office buildings have significant amounts of MEP systems in relatively confined spaces, so accurate location-based planning is very beneficial. Although there are some opportunities to switch the sequence, most areas (especially corridors) have very constrained logic.

Variations can be also have much greater impact in office than in residential buildings. In residential projects it is typical to have strict deadlines beyond which no changes can be made to apartments, whereas in office projects, companies are very dynamic in their needs and the layout of office space can change many times—even if the end user is known. Business concerns normally override production issues and changes get accepted late in the process. Variations typically affect location and quantity of interior walls, floor covering materials, suspended ceilings and electrical installations. In some cases, the end users may be allowed to add additional facilities such as restrooms or kitchens, and these will result in large changes in the schedule.

Typically waste, in the form of loss of productivity, manifests as the slowing down of work when multiple crews are working in the same location. This is caused by a perception of space arising because office walls are typically movable partitions and are therefore built after any floor covering work. However, in practice crews tend to slow each other down in unpredictable ways, so one of the main challenges of production control is to keep each work area for one crew only and resist the temptation to send everyone in at the same time. This is very different from residential construction, where apartments naturally separate crews. Productivity issues also occur because it is often possible to complete some part of the scope. These partial dependencies are very challenging to manage.

Location breakdown structure

The location breakdown structure for office buildings is usually straightforward. Individual buildings or structurally independent sections should be the highest hierarchy level.

However, when deciding independent sections, the effect of mechanical services zones should be considered. There is no point in creating a separate section if the structure cannot be completed independently from other sections, which would be the case were mechanical zones to be inconsistent with section boundaries. Otherwise the benefits of planning sections would be lost during the finishing phase, when a different area needs to be free of dust during testing of the mechanical equipment.

Floors form the next hierarchy level. Structure should be divided into pour areas if it is cast in place or can be handled using floors when it is steel or precast concrete. For finishes, floors should be divided to functional space groups inside each section.

Because the walls of offices are often completed quite late in the process (demountable partitions are placed on top of any floor coverings), many offices can be lumped into one space. The whole office area is often open for a long time. During this time, before individual office rooms exist, corridor work and work in rooms will interfere with each other. Therefore they can be considered part of the same location. Other common space groups include restrooms, lobbies and stairwells which should be separated because they have different finishes. If the floor area of a section is small, it is sufficient to have one location for each space category. Otherwise, the space categories need to be subdivided into more detailed levels. The main principles hold: one trade should be able to work productively in the area, the locations should correspond with physical reality and it should be easy to say if the location has been completed or not. Special locations, such as kitchens, restaurants and mechanical rooms, and all other special-purpose rooms, such as data centres, should naturally be separated to their own locations.

Pre-planning

Pre-planning for earthworks, structure, façade and roofing can be completed accurately before specific user needs become known, as they are not generally impacted by the fit-out. In contrast, quantities related to finishes and MEP remain generally inexact until the end user(s) needs are known. Nevertheless, a generic layout can be used to estimate quantities for early versions of the schedule by assuming standard finishes. This baseline finish can then be used to show the effects of changes to the client and to management—in terms of both time and cost. Approving the baseline schedule for finishes and MEP should be done at a later stage when there is more information about floor layouts and requirements for management and client reporting purposes. However, the main mechanical room work and finishes of any technical rooms should be planned in detail as early as possible because they define important milestones for MEP contractors and often require long lead time deliveries. Also, approximate production rates and workflow for interior works is needed for pull scheduling of procurement and design activities—especially when long lead time deliveries are involved. These may change when more information becomes available.

Optimising the sequence of earthworks and structural phases is critical in order to be able to commence finishes and MEP work as early as possible. Large office projects often have staged handovers with some buildings or parts of buildings handed over to the client before others. These constraints may prevent planners from selecting the optimal sequence. In office buildings, it is common to start interior construction from the top floor and continue down, finishing the ground floor last (this will be slower, but the idea is that quality will be improved). The ground floor can be used as a storage area and changes affecting the ground floor tend to be expensive. However, because the ground floor often has higher work content for MEP and finishes works, it is often beneficial to finish at least some spaces which have a low risk of change in the early stages of the project. Starting from the top always loses time, so any work that can be completed on the ground floor will reduce the rush at the end of the project. To prevent the risk of rework, early work should be limited to spaces with a low risk of changes. It is good practice to establish these starting data milestones for each space together with the client, so that the client understands what will happen if there is a late change for any location.

The network logic is generally more difficult in office buildings than in residential construction. The most difficult part is defining the logic between building services and construction work. Because there is little space, selecting the sequence is critical. However, production rates are easier to synchronise because there is a lot of repetition in the floors. An added challenge in office buildings is the frequently high pressure on time. There is a natural tendency to reduce buffers and overlap work in sections to shorten the project duration. This will increase risk in the production system and will necessitate good production control. It is important to use risk analysis and simulation to quantify this risk. For risk analysis and forecasting to work, logic links need to be added between all tasks that would interfere with each other were they undertaken in the same location. A typical mistake is to leave the logic out if it is not relevant for the planned schedule. When delays occur, the forecast and risk analysis will not observe the missing logic, resulting in overly optimistic outcomes. A better way to achieve this is to define the list of activities which occupy so much space that they will interfere with each other. Typical tasks include all floor covering work, painting, plasterwork, MEP overhead ducts and distribution, cabling and partition walls. Interference is not normally caused by tasks such as installation of doors and hardware, electrical fit-out in office rooms, lighting in office rooms and all work in restrooms (unless restrooms have been defined as separate locations). These non interference causing tasks must obey their technical predecessors but can be executed together with other tasks which also do not cause interference. For all tasks in the interference causing category, (dummy) logic links need to be defined to ensure that the tasks do not cause interference in a location. This often requires selecting logic links, which can later be switched around during the detail planning phase should the preceding subcontractor become slower than the successor.

Pull scheduling techniques are critical to scheduling the design and procurement tasks for office projects. Design is often the critical factor. Many office projects are executed using fast track procurement, with the design and production tasks heavily overlapping. Therefore, the production of design should be monitored in the same way as construction-related production. In particular, design should be divided into separate parts: design packages needed for procurement, and design packages needed for construction work (regardless of whether design management is the client's or the general contractor's responsibility). Additionally, design coordination should be tracked.

If 3D models are used, design coordination can be done relatively quickly. With traditional design methods, achieving coordination can take a long time because even when problems are fixed, they tend to move to other parts of the design as designers do their design changes in isolation. Starting construction with an uncoordinated design will result in loss of productivity and rework because of the inability to prefabricate MEP parts and the need to solve coordination issues on site.

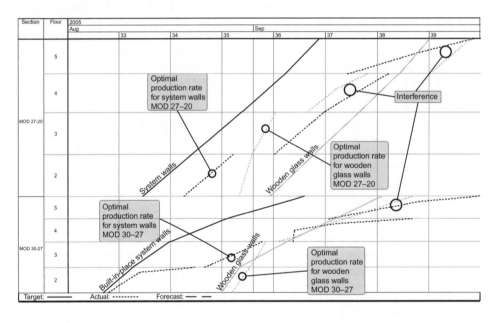

Figure 13.2 Slowing work in an office project where tasks interfere in the same locations

Implementation phase

There are usually more trades working in the same location on office projects than in most other project types, therefore controlling production rates becomes critical. Immediate reaction is required when production rates do not meet expectations. Although the sequence of trades may often be changed, having multiple trades working in the same area will slow work down and interrupt tasks unpredictably. Figure 13.2 shows an example from a real office project where there are no mandatory technical logic links between tasks occurring in the same location but slowdowns and interruptions occur when tasks try to enter the same location at the same time.

Sometimes all tasks slow down, sometimes some tasks take priority and others will be suspended—the production control system will fail if this behaviour is allowed to continue on the project. The best way is to calm the situation is to re-plan production to allow one subcontractor to work at their optimal production rate, while the others have production breaks. In this case, it is best to leave some buffer between the tasks so that frictionless production can be restored. Figure 13.3 shows an updated detail plan for the same example. After week 35, wooden glass walls will continue with their optimal production rate and system walls will have a work break. In addition to subcontractors saving production cost, they will finish production one week earlier than in Figure 13.2. Of course, real situations involve many more tasks so they are not always this clean. In the worst situations on this project, eight different tasks were working in the same location resulting in huge slowdowns.

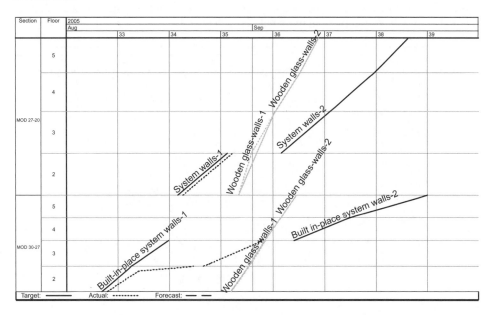

Figure 13.3 Re-planning work on the office project to resolve conflicts and speed production

Detail planning is often needed because the logic of the pre-planned schedule might not be correct. There is often a lot of flexibility in logic and many of the links are decisions, not mandatory technical dependencies. Mistakes in logic—often pointed out by the subcontractor in detail planning phase—may cause delays to other trades. The critical issue in detail planning is to prevent tasks which can cause interference to each other (typical examples are plasterwork or floor covering with almost any other task) going to the same locations at the same time even if there is no technical dependency. If a relationship has been planned to prevent tasks from happening together, it is possible to switch the sequence as a control action if the predecessor is slower than the successor.

Space constraints make planning material storage and deliveries critical. If material arrives too early on site, other trades will have to move around it thereby leaving those locations unfinished. Prerequisites for starting and continuing production are critical. For example, structure, roofing and façade systems often have complicated design and connections to other systems. Detailed planning at least should be undertaken for the most critical activities.

Example

Figure 13.4 shows part of the finishes and MEP detail schedule of a typical office building. This was an interesting project, because it contained buffers in the first section and did not experience significant production problems or delays there. Most of the work in the first section is not shown in the figure. However, the tight second section (MOD 27–20) had many problems because there the schedule did not have buffers. A delay by a subcontractor caused production to queue with slowdowns of 20–50% and unanticipated interruptions. Note, for example, that all the other tasks were suspended when the vinyl floor covering enters the space! The schedule did not have a technical dependency because the planner did not anticipate that predecessors could be delayed so much that they would actually occur together. Also note that the chaos ends when there is a break in work of two weeks. This project was one of the early case studies for the development of location-based control methods and will be described in more detail in Chapter 15.

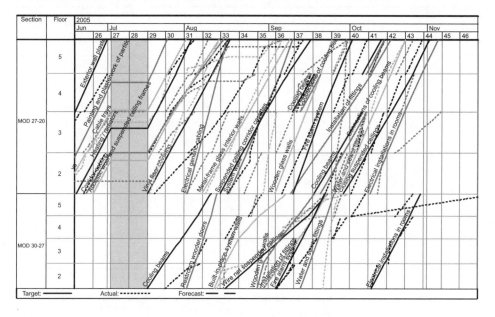

Figure 13.4 Problems arising due to inadequate buffers

RETAIL PROJECTS

Description

Retail projects are shopping centres or malls. They can have single end users (for example, a shopping chain) or multiple end users. They can equally be developed by an end user, or they can be developed by a property developer and then leased to tenants. Large retail chains usually want to have similar facilities in all locations and they have well-defined requirements for each building. However, centres developed for leasing will generally involve considerable uncertainty with respect to the final tenants and their locations, and those smaller players will not always know what they want, all of which can result in considerable variations or change orders.

Retail projects often have a combination of large spaces (halls) and small spaces (small retail areas). In large spaces, the sequence of trades is often not technically strict and there is room for many crews. The critical parts of the projects are typically earthworks, foundations, structure, air conditioning plant room and the small retail spaces. In many cases, end users are not known before commencing the project and the actual commercial process of renting the premises becomes a critical part of the project.

Strict deadlines are often a characteristic of retail projects. If the new shops are not completed and open for Christmas sales, the end users face enormous losses. However, it makes commercial sense to only open a new retail centre in certain time slots and end users need to know well beforehand when sales can begin. Therefore, any duration cuts need to be predictable and communicated very early in the process or they will not benefit the owner. Predictability in achieving the agreed end date is much more important. Therefore, duration compression and associated risks should be evaluated during the pre-planning process.

Interference is very rarely a problem, except in small retail spaces and mechanical rooms, because retail projects tend to be wide instead of high, so there is a lot of space. However, the critical areas require very effective location-based control. Large spaces often have flexible logic and crews can easily work around each other, so slowdowns and work stoppages typically experienced in office buildings do not occur in retail projects except with mandatory technical logic.

The main source of waste in retail projects is relocation cost and rework cost. The workers almost always have sufficient space to do work elsewhere so they do not have to mobilise and demobilise. The sequence of trades is more flexible so work can always be found when workers look for it. Instead, money is wasted because workers cannot finish locations and they move around looking for the next location to work in. It is easy for management to lose control and allow work to dynamically self-organise, often having to return to complete areas. This leads to excessive need for buffers and consequential bloated durations.

Location breakdown structure

Planning the optimal location breakdown structure plays a critical role in enabling duration cuts in retail projects. Location breakdown structures of wide buildings are generally much more difficult to design than high-rise buildings. Highest levels of location breakdown should be based on the principle of independence of structure. It is always beneficial to complete the structure of a section to roof level before moving on to the next section so that the roof and indoor work can start in that section. This is usually straightforward if the structure is steel or precast. If the structure is cast-in-place, the need to have multiple locations going on at the same time results in individual areas being completed to roof level more slowly. However, it is also possible to define structural sections which can include multiple pour areas for cast-in-place structures.

Finishes are generally difficult to organise into areas in wide buildings, especially when the lay-outs of retail spaces are not known during pre-planning. Often, structural sections can be used to get proper links to dry-in of the section. However, this may cause difficulties with control because large open spaces spanning multiple sections are generally regarded as a single location by subcontractors and it will be difficult to get accurate completion rates by section. When retail spaces become known, each space should form its own location. At this stage, the project becomes easier to manage using locations.

Pre-planning

It is frequently not possible to plan very accurate schedules during the pre-planning phase because there remains a lot of uncertainty. How much uncertainty depends on how well the tenancy needs are known and whether all the spaces have been leased yet. If there is no information about an end user, types of finishes have to be guessed based on basic assumptions about generic tenants. Locations belonging to unknown tenants should be scheduled to be last in any location sequence, so that there is time to react if the needs do not match expectations.

Retail projects are typically wide (as distinct from high, sometimes called wide-rise) and have multiple sections. Therefore, optimisation of earthworks and structure can result in significant time savings for the project. This can be evaluated by building the basic tasks first and then changing Layer 3 logic for all tasks to try out different scenarios. Important factors to consider are earthworks and structural quantities, expected quantity of finishes and MEP, including the location of mechanical rooms, as these will often become a critical part of the project. It is important to be able to start sections with lots of MEP and finishes as early as possible and finish with those sections where indoor quantities are smaller.

In some cases, there is almost no information available about tenancies when pre-planning begins. Scheduling the required dates for information becomes critical. A good trick is to assume that some percentage of retail spaces will become available for construction at certain points of time (typically three or four stages). This will result in every indoor task which requires tenancy information to be split into as many parts as there are stages. The method for determining how tasks should be split depends on the owner. Typically, owners want to let spaces evenly throughout the building instead of starting the commercial process in some predefined location. In this case, all the tasks will have all locations, only the estimated total quantity is split according to rental stage proportions.

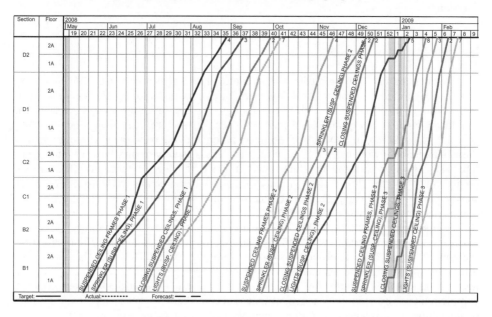

Figure 13.5 Scheduling retail tenancies—Hartela project

Figure 13.5 illustrates this concept. Suspended ceilings and associated MEP installations are shown in a large retail project which had no rented spaces at the time the schedule was planned and analysed. The schedule uses the structural sections. There are three stages of tenancy information: in the first phase 50% of information will be available, 30% will be available in the second phase and 20% will be available in phase three. These percentages were also incorporated into the agreement with the owner. If enough spaces had not been rented by the specified time, the contractor was allowed to build the remaining missing rental area with standard finishes. If all rental areas were standard, then the schedule would be perfectly accurate. In reality, some retail spaces will be released earlier, so work can start earlier with a slower production rate. Also quantities can have large changes—for example, some spaces might not require suspended ceilings at all and some spaces will have special and time-consuming suspended ceiling installations.

Much of the work can be done without tenant information. For example, the main MEP ducts and conduits, main fire protection, corridor walls and so on can be built. Tasks should be separated clearly so that they either have scope which can be completed before end user information or that they are dependent on end user information.

In retail projects, logistics, quality and cost tools can be used in pre-planning. However, because there is high uncertainty, it should be accepted that the cash flow and deliveries will change, often substantially, during implementation. The pre-planned schedule cannot be predictable while uncertainties remain in the starting data.

Risk analysis is very useful in retail projects because major risks are related to readiness of design and tenant starting data. Resource risks together with risks related to design and renting processes are often the most critical risks, because space is not usually an issue. Risk analysis can be used to evaluate different scenarios. For example, what happens if most of the spaces will not be rented until the end of the project?

Implementation phase

Using and planning detail tasks is very important for retail projects because changes are common. Pre-planned schedules are usually inaccurate because of the lack of tenant information. Therefore, detail tasks should be used to actually control the work. Quantities of work for each known retail space are estimated and they are added as new locations. When information about new retail spaces becomes available, the locations can be added to existing detail tasks to preserve continuous flow for each crew and to prevent crew clashes. The estimated quantities of non-rented spaces should be decreased based on the proportion of rented space, using the original assumptions about general finishes. In this way, approximate resource forecasts can be kept up-to-date equally for the non-let parts of the work and accurate resource forecasts will be available for all known retail spaces.

Large open areas generally do not need the additional location detail during implementation. Interference with multiple tasks working in one large location can be prevented by detailed weekly work planning, either by using floor plans or a 3D model. The idea is to show the actual areas where crews are going to work each day. If a 3D model of the building is available, good practice will show the actually finished production and will indicate next weeks production with different colours—allowing visual identification of conflicts.

However, because many locations are large, degrees of completion are often needed in the monitoring process. In practice, some small part of the work in a large location will often remain unfinished with the result that production stagnates at 95% complete. In this case, an new detail task should be added for the 5% work required to complete the baseline task in question and the project rescheduled. This should explicitly show which part of the scope is unfinished and also record the reason, by providing correct logic to the new detail task.

   To decrease the waste of moving around, logistics can be planned so that each structural section of large open areas has its own equipment and material storage. This way, only the workers have to move and little time is wasted by relocating to other locations. The available space will be getting more and more constrained as more rented spaces become available, so logistics planning of the later stages of project becomes even more important.

   Retail projects tend to be more resource than space-constrained, therefore it is very important to do resource forecasts for all critical resources and subcontractors. Often, a lack of starting data tends to affect the electrical trade the most, resulting in the last few months of the project having requirement for large numbers of electricians. Nasty surprises in the last few months of a project can be prevented by being able to communicate the resource forecasts to subcontractors.

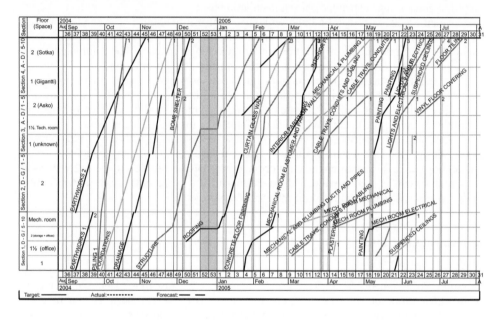

Figure 13.6 Typical retail project pre-planned schedule

Example

Figure 13.6 shows a schedule for a typical retail project involving a rough level of detail. In this case, most of the tenants were known with only one location not being allocated (Section 3, Floor 1). The pre-planned schedule assumed that information for the empty space would be received prior to starting interior work in the location (before February 2005). Because the MEP design had not been completed and subcontractors had not been selected, MEP tasks were poorly detailed and had large compensating buffers. In this project, all users except one fitted neatly into one floor in one section so a separate finishes location breakdown structure was not needed. The second section did not have any indoor work, this was because the tenant Asko spanned over two structural sections. All quantities for Asko were allocated to Section 3, Floor 2 and left out of Section 2. This makes sense because the work for Asko could start only after both Sections 2 and 3 were completed. The sequence of sections was optimised so that Section 3 (with bomb shelter) would be built late to enable an early start for the indoor work.

HEALTH CARE BUILDINGS

Description

Medical buildings and hospitals have a special feature: they require a lot of MEP systems. Coordination of MEP design is one of the major challenges of these projects. The function of spaces is relatively uniform, so accurate schedules can be planned before the start of construction. Quality is very important in hospital projects, because of the critical importance of hospitals, so there are often more inspections required which can increase the durations of tasks significantly.

Location breakdown structure

Similar to most other building types, the highest levels of a location breakdown can be defined based on the structurally independent sections and floors. Finishes and MEP can be handled at the section and floor level or divided into functional areas. Hospital projects typically have a separate breakdown for interior, exterior and structural tasks.

Pre-planning

Health care buildings have a lot of MEP in what are frequently small locations. Because of the lack of space and congested systems, the dependency relationships are rigid and a certain sequence has to be followed. This makes health care buildings ideal for location-based planning and control. Examining resource constraints and running resource risk analysis are especially important for medical buildings, because multi-skilled MEP trades dominate.

Production phase

Tight technical constraints usually mean that medical buildings require good production control to prevent interference. On the other hand, many contractors are able to be multi-skilled, which allows them to work on other tasks in the event that one task gets interrupted. Thus medical projects can be run with relatively small buffers providing multi-skilling is an option (this depends on the industrial relations context and local practices). However, resource management becomes very important because of the importance of the MEP contractors. If subcontractors provide too few resources, the cascading delay chains are difficult to stop. Hospital projects typically exceed their planned duration because of the need for frequent inspections and cascading delay chains. This can be prevented with strong production control, following the planned sequence and directly managing subcontracted resources. Hospital projects can benefit from location-based payment systems because completely finishing locations and handing them over to the general contractor increases quality levels and decreases the need for rework. Most advanced hospital owners are researching cost reimbursable procedures where main subcontractors are paid directly based on their cost, while savings in worker hours are shared between the subcontractor and the owner.

Example

Figure 13.7 shows an example of a schedule for finishes and MEP of a hospital building. The building had four quadrants and a centre area. These quadrants were divided into three floors and roof. Work on the first and second floors could be started before work on the third floor. Because of this constraint, locations were sorted so that the third floor of each quadrant was on top. Without sorting, the work would appear discontinuous as the third floor is completed last in sequence. The project achieved the same total duration with the resource-loaded quantity-based schedule, with almost all trades continuous—in comparison with a previous CPM schedule which had lots of starts and stops in it. All resource constraints communicated by the subcontractors were taken into account. The production system cost was calculated for both the CPM schedule and the location-based schedule and it was determined that the location-based schedule resulted in 20% lower production system cost.

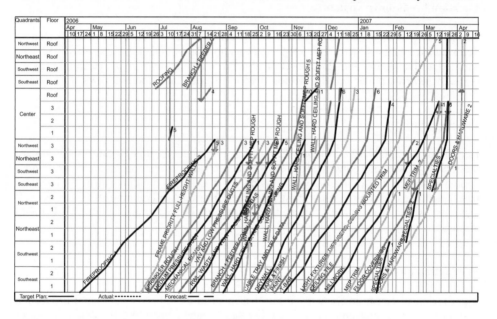

Figure 13.7 Typical hospital project pre-planned schedule

COMMON PROJECT TYPE FEATURES

Various different project types are discussed in this section. These are brief discussions of specific components which are a feature of that type, rather than a complete analysis as above.

Industrial projects

Industrial buildings often have very little construction work except from earthworks, foundations, structure and roofing. Inside the building, the main tasks include slab-on-grade and MEP work. There are some minor office or social spaces which require partition walls but these are usually very limited in scope. Selecting a suitable location breakdown structure becomes a challenge because it is very difficult to map all work in a large open factory floor to one location structure. Often grid lines form the locations on the shop floor and special functions, such as restrooms or laboratories, get their own locations. Guidelines from other project types apply for smaller locations. Guidelines for large halls can be used for the factory floor.

Highly repetitive projects

Some projects require building exactly the same location over and over again. For example a large casino project may include thousands of identical hotel rooms. In this case, the number of units can be used as the quantity of each trade rather than forming a location level on the LBS. This requires estimating how many worker hours of work each trade has in each unit. The location breakdown structure can then be entirely based on sections and floors. Production control reduces to finding out how many rooms have been completed in each location. As a supplement, room-based checklists should be posted on the door of each room to show which tasks have been completed in the room. The control chart would show that an area has uncompleted rooms, and checklists enable workers and management to quickly identify which rooms have not been finished. This approach does not burden the planning and controlling system with thousands of locations but achieves all the same benefits.

Multi-purpose projects

Multi-purpose projects can include multiple project types in the same project. A Finnish case study had three office projects, two underground bus stations, a shopping centre, metro station, three apartment buildings and two levels of retail spaces in the same project. These huge projects can be split to multiple location-based schedules depending on resource use. If different subcontractors are being used in different parts of project, the parts which do not share resources can be handled as separate sub-projects. Even though all parts of the project do not need to be in the same location-based plan, control information needs to be aggregated to be able to report planned and actual completion rates for the total project. If all parts of project are interconnected, different functional parts of buildings should be divided to their own high level locations to be able to easily filter data.

Refurbishment

Refurbishment projects often have many special constraints. The client may require that part of the building stays operational during the project. This requires staged handovers and starting some locations at a much later time (after some other locations have been handed over). These issues make it difficult to plan flow. Location-based schedules should include tasks to illustrate when the client is using each location. This can be used to visually check which locations are operated by the client at any given time.

Refurbishment projects often have a high degree of uncertainty in terms of their task quantities. As-built documentation is not usually available for old buildings. Therefore, all quantities related to demolition of existing structures are largely guesses. A common risk in many refurbishment projects is the presence of asbestos which requires a lengthy demolition process. Demolition related activities therefore need to have large buffers. The effects of quantity uncertainty can be examined by using simulation.

Sport stadiums

Sport stadiums have different requirements for the location breakdown structure for the structural and interior phases. The structure is usually built in bays, while finishes are built in functional areas which can overlap multiple bays. This special feature requires a finishes branch to the location breakdown structure, separated from the structure branch. It is necessary to link tasks in these branches together by using Layer 5 logic, because structural locations are separated from finishes locations. In other respects, normal LBMS can be used.

Large open locations

Location-based planning and control is typically challenging to implement for very large open spaces. Logic is difficult to assign because some parts of a large location can have different logic from other parts. For example, mechanical ducts may be above cable trays in some places, but some cable trays may not be located in the same position as ducts so they can start earlier. In some parts of the location, the logic can go in reverse and cable trays go in first. Typically, in each large location, only a part of each task will be dependent on other tasks. There are a few exceptions, such as slab-on-grade and underground utilities, which have clear technical dependencies. The typical way to handle this in CPM schedules is to lump all overhead MEP into one task and not even try to model their relationships. However, this does not allow detection of production rate deviations or interference. In location-based management there are two basic approaches for handling large open spaces.

First, it is possible to try to divide each large location into smaller virtual locations and try to define logic in each one of the small locations. This can be very difficult to implement because locations should be physical and easily understood. It may be feasible to 'split' the hall based on column lines. This approach has the benefit of being easy to understand and easy to track progress. However, this usually requires that the same task occupies multiple areas at once, so it may be impossible to visualise production flow. For example slab-on-grade can be poured in many locations at the same time. Overhead MEP is assembled system by system instead of location by location. Therefore, some tasks may reserve multiple locations at once and some will not finish locations but will visit every location multiple times. This approach may result in too many detail tasks to be useful. It has been tried in a case study but it was not possible to implement the work according to the plan— the case study was a failure in this respect.

Second, it is possible to only divide a large space up when most trades can finish a location before moving to the next location, and otherwise accept that multiple locations may be underway at any time. Tasks which may cause interference to each other should be spaced out using start-to-start and finish-to-finish relationships with a delay. Controlling is based on production rate control to achieve the planned production rate for each task and on start date control, not letting a subcontractor in before the allocated start time. Because subcontractors are spaced out temporally, they should be able to handle interference at the weekly planning level. 3D models, or marking weekly plans onto the floor plans, will normally be enough to prevent interference. The number of dependencies between the tasks defines the required lag. For example, slab-on-grade will cause interference to all tasks so the lag must be longer. The lag can be shorter if there is only minor overlap. After the plan has been finished, a location-based resource graph can be used to evaluate the area each worker has available, an alternative used successfully in many projects. Results show that interference can easily be prevented on site this way. Controlling the production rates, and having subcontractors spaced out five days or more, has been found sufficient to prevent clashes. However, this approach should be used only with single function open spaces.

Linear projects

Linear projects are a special type of project which involve many aspects beyond mere location-based planning. Accordingly Chapter 14 expands on this brief discussion with much more detail relating to the planning and control of linear projects, particularly involving mass haul. There are, however, some features of linear projects which may be summarised here.

Linear projects usually involve continuous production activities. In this aspect, they are closer to factory production than the typical construction project. It is no surprise therefore that location-based methods and in particular location-based charting methods have survived as a primary form of communication for these projects. Indeed, this is why the location-based scheduling method is often termed 'linear scheduling'.

Methods for scheduling such projects ordinarily require dedicated tools as discussed in Chapter 14, however they can also be scheduled using basic techniques. The important issue is to convert the continuous nature of the production into discrete locations by aggregating to a summary level. Each aggregated unit of length becomes the smallest unit in the LBS. This may be illustrated by examining the case of a tunnelling project, scheduled using standard location-based methods.

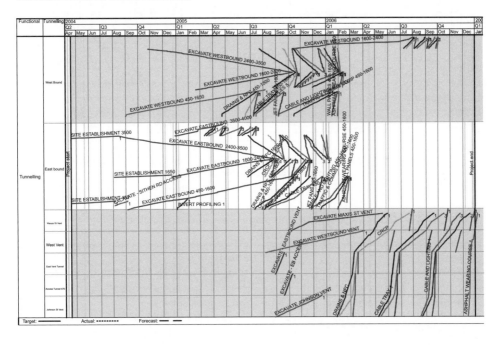

Figure 13.8 A flowline for a tunnelling project using conventional flowline methods

An example LBMS approach to planning tunnelling

To demonstrate standard LBMS approaches for tunnelling projects, a suitable distance increment was selected within which the location-based quantities may be collected. Typically for tunnelling, this should relate to a logical repetitive element of the project such as cross access tunnels. In many projects these are at intervals of 100 yards or 120 metres.

Figure 13.8 illustrates a complete tunnelling schedule, excluding building construction. The initial long sloped lines represent the excavation using a road header, and the production rate is dictated by the production capacity of the machine. The following trades represent the fit-out works in the tunnel. This schedule remains to be optimised, as evidenced by the considerable location buffers between tasks.

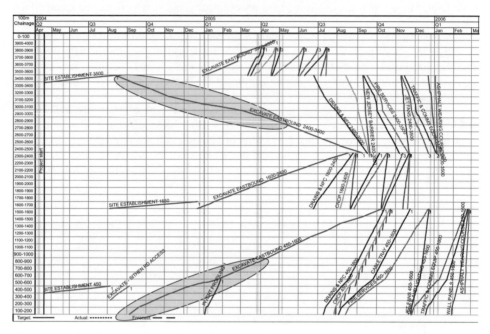

Figure 13.9 A flowline for a tunnelling section showing production rate changes due to changes in material type and tunnel size

Table 13.1 The BOQ for a section of tunnel showing differing types of excavation and differing quantities

Tunnelling:		East bound						
100m Chainage:		100-200	200-300	300-400	400-500	500-600	600-700	700-800
Item	Consumption person hours/units							
Eastbound Exit to Hwy Fan Niches	0.0217	55	110	176	275	286		
Eastbound Exit to Hwy 2 Lane (Bottom)	0.0217	4067.2	6560	4067.2				
Eastbound Exit to Hwy 2 Lane (Top)	0.0217			2492.8	6560	4723.2		
Eastbound (Ch 1630 to CH 1550) Fan Niches	0.0237							
Eastbound (Ch 1630 to CH 1550) BreakDown	0.0206							
Eastbound Portal to Ch 1550) Fan Niche	0.0237							
Eastbound Portal to Ch 1550) Breakdown	0.0206						3805	
Eastbound Portal to Ch 1550) 2 Lane	0.0237				533	5330	2665	

The advantage of scheduling this way over CPM methods is the clarity of the representation of the work fronts, and the matching of the trend line to actual production. This is seen more clearly if one section of the tunnelling project (not shaded in Figure 13.8) is examined more closely. Figure 13.9 illustrates thus section of tunnel. Here it can be seen that the tunnelling task varies in slope as the road header moves into the tunnel. This is caused by variations in the extent of tunnelling and the material being excavated. For example, hard rock takes longer to cut than soft rock. Similarly, two lanes are slower than a single lane. This effect is highlighted by the shaded areas on the diagram.

Each of the task lines can consist of multiple bill of quantity items, each involving different production properties. For example, the tunneling tasks highlighted in Figure 13.9 include special items such as fan rebates, emergency parking bays, etc. These are best illustrated in a table of one small part of the tunnelling task's bill of quantities (Table 13.1).

Civil engineering projects

Civil engineering projects typically have few tasks and many locations. Linear projects are one example of such projects, but work on mining projects, airports, civil infrastructure also has these properties.

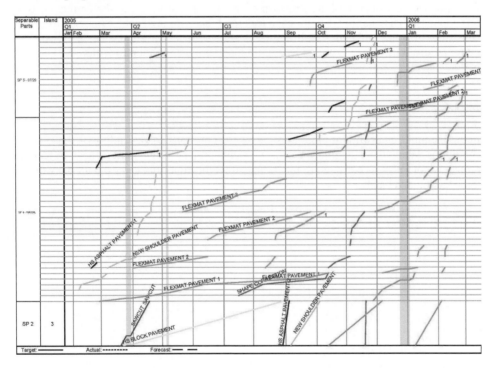

Figure 13.10 Example of LBMS applied to runway expansion for the Airbus A380

These projects clearly represent resource management problems. How can contractors manage the crews undertaking the usually highly repetitive tasks in a large number of locations? The problem ceases to be one of specifically managing production in individual locations (although this remains important) but specifically managing the multiple crews which may undertake the work. The projects also present very different demands on a LBS, from grids, spots to networks. All of these can be managed using a carefully designed LBS. The advantage that LBMS brings to such projects is extreme speed in both schedule development (due to the typically very small number of tasks and the ability to map location quantity data to the LBS) and enormous flexibility in making location changes in both quantity and sequence. It may also be noted that civil engineering projects typically are rich in starting data—in stark contrast to commercial building projects.

This problem is illustrated here using two specific projects. The first is the reconstruction of an airport's runways to make provision for the Airbus A380. The second presents the renewal of pipework in a city's water infrastructure network.

Airport works

In this example, a grid of runway paths causes a network of nodes which are the islands between the runways. The project is defined as the reduction in size of these islands by widening the runways. The project is therefore not linear (as might at first be assumed by examining the linear nature of runways) but nodal.

For this demonstration, it is assumed that the demands of daily production will require flexibility in the sequence of construction of the nodes. Location-based planning enables this with ease. The quantity data for each task (and there are few) can be mapped to a LBS consisting of zones and nodes. Splitting tasks allows for increasing resources by working multiple locations (space is not an issue) to reduce project duration to within the contract period.

Pipework renewal

In this example, a network of pipes must be patched or replaced, depending on the existing conditions. While there are many aspects to such a project which appear linear, in fact the project is not linear as crews go to individual street locations and carry out all the work in that location in a batch.

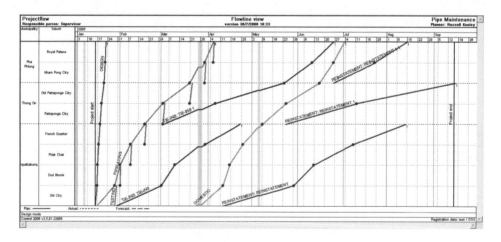

Figure 13.11 Example of pipework maintenance and renewal

In such an exercise, the main need is to assess required resources given the production rates assumed for optimal crews. Data management is the key to rapid scheduling of such work. Table 13.2 shows a section of data for a pipework maintenance and renewal project.

The data is in metres of pipe and numbers of special installations per street location. The data is organised by suburb and municipality—which provide a logical LBS.

Table 13.2 Table of maintenance data for pipework refurbishment

Location	Suburb	Municipality	Pipe size	Length to renew	# of Level 2	# of Level 1
Chau Duoc Av	Old City	Bumpattabumpah	100	101	1	1
Governor Henri	Old City	Bumpattabumpah	100	382	1	1
Canal Rd	Old City	Bumpattabumpah	200	78	1	1
Chiung Chiong hwy	Old City	Bumpattabumpah	100	241	1	1
Thanon Klangpaang	Dud Bhonk	Bumpattabumpah	100	133	2	1
Phlat Tiht Rd	Dud Bhonk	Bumpattabumpah	100	122	2	1
Airport Drive	Phlat Chat	Bumpattabumpah	100	153	3	1
Phlat Chat Way	Phlat Chat	Bumpattabumpah	100	174	3	1
Thanon Pong	Phlat Chat	Bumpattabumpah	100	180	3	1
Boulevards Aux Arbres	French Quarter	Bumpattabumpah	100	294	4	1
Bum Sloh Duk Lu	French Quarter	Bumpattabumpah	150	143	4	1
Chuloc	French Quarter	Bumpattabumpah	100	171	4	1
Kai Truc Lu	Pattaponga City	Thong On	100	170	1	2
Thong On Av	Pattaponga City	Thong On	150	324	1	2
Bamboo Rd	Pattaponga City	Thong On	100	89	1	2
Kru Kut Rd	Pattaponga City	Thong On	100	157	1	2
Bansak Rd	Old Pattaponga	Thong On	100	213	2	2
Wat Ratchasek Rd	Old Pattaponga	Thong On	100	282	2	2
Yu Nong Dr	Old Pattaponga	Thong On	100	302	2	2
Pattaponga Rd	Old Pattaponga	Thong On	100	488	2	2
Voi Da Donglu	Nham Pong	Pha Phlung	100	217	1	3
College Rd	Nham Pong	Pha Phlung	100	137	1	3
Thanon Pong Rd	Nham Pong	Pha Phlung	100	177	1	3
Achutack Rd	Royal Palace	Pha Phlung	100	194	2	3
Pha Phlung Rd	Royal Palace	Pha Phlung	100	220	2	3

In such a project, it is possible to aggregate all the data into suburbs, treating each street location as a BOQ item. This reduces the data workloade normously. The resultant LBS with the required tasks is shown in Figure 13.11.

Maintenance projects

Maintenance projects present a particular problem for the planner as they generally do not need to follow logical sequences and, for large projects, it can become extremely difficult to track the work. Maintenance projects usually only require work to be undertaken where there are defects. Consider a 500 room hotel, a defects list may have several trades in any number of locations. There may also be defined logical sequences between certain trades where these exist.

The best way to manage such projects is to create a LBS for the entire project down to the lowest level of detail, such as rooms, and then to load the data into each location. The resultant flowline and control chart views would be extremely difficult to read were all (often predominantly empty of work) locations shown. However, if the view is restricted to show only those locations where work is required, then it becomes a simple matter to plan the flow of work through the building as shown in Figure 13.12.

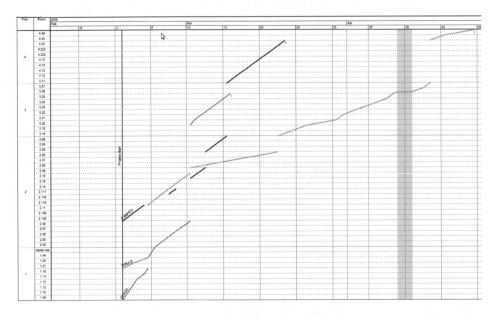

Figure 13.12 Example of hotel room maintenance only showing required rooms

Large or complex projects

Large and complex projects are characterised by being mixed use in most cases. Therefore they take on many of the properties of other project types for sections of the project. Care must be taken in designing the LBS to ensure that the greatest opportunities which the project offers are utilised. However, care must also be taken to not allow one component to dominate the project to the disadvantage of the remainder.

Typically the needs of structure will determine the overall form of the LBS. The most important thing is to enable rapid and efficient construction. Thus, for example, it is usually better to hold to LBMS principles with structure, vertically dividing the building where appropriate.

For example, a multi-use building with a 70 level residential tower of 50 levels on top of a larger 15 level commercial complex, all on top of a six level car park, should be divided vertically rather than horizontally. Thus the tower sections—even in the commercial zone and the car park—should be separated. This will allow the higher levels of the residential tower to get underway while waiting for the commercial floors to be completed. Building entire floor plates progressively upward will significantly delay the overall project.

GUIDELINES FOR SPECIAL INTERVENTIONS

In the following discussion, the special characteristics of particular interventions in the planning process are discussed with emphasis on the impact on planning and control.

LBMS implementation during production

While it is desirable to commence implementation of the LBMS early on in the process, it is still possible to implement even after the baseline has been previously approved in a non-location based system. This requires creating a location-based plan based on the activity-based plan, a step which requires some effort. The baseline can be replicated in location-based system by grouping each distinct task type from the schedule to one quantity item and using days as quantities with a consumption rate of 8 hours per day. Locations can be defined based on the task names in the activity-based schedule. This allows the baseline to be an exact copy of the original baseline. Copying the schedule is then simply a matter of assigning start and finish dates to locations which match the activity-based schedule. Typically, schedules replicated from an activity-based schedule will not have continuous flow and will have a lot of overlapping locations being worked by the same crew. This is because CPM durations are usually 'windows of opportunity' and do not reflect the actual scope of work.

Figure 13.13 shows a typical CPM schedule changed to a location-based schedule without any optimisation. It can be seen that every floor has the same set of tasks and the same pattern occurs as early as possible on each floor and just the start dates are different. This will compress the duration for finishing each floor but creates huge inefficiencies in the form of discontinuity and waste. This same pattern is observed in the case study documented in Chapter 17, page 526.

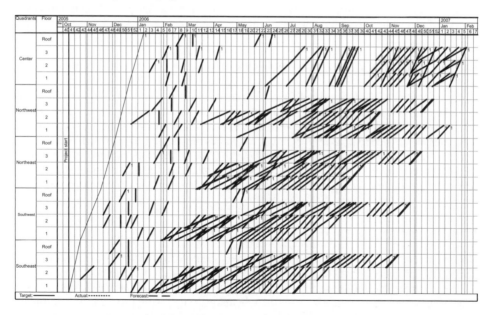

Figure 13.13 Example of converting a typical CPM schedule to the LBMS

Once the schedule has been replicated, the LBMS can then be used in detail planning and production control. To get the maximum benefit, it is best to calculate quantities and resource load for all detail tasks—even though the baseline tasks are not based on quantities or resources. This will allow planning for continuous flow at the detail task level while reporting against the baseline schedule. It is also possible to just control production of the original baseline schedule. However, because there is no continuous flow or buffers, this will easily lead to serious deviations and constant alarms. Controlling a non-continuous project on a weekly basis will show that there is a need for better planning!

Chaos projects

Chaos projects are those chaotic projects where the flow of work has been lost and there is no opportunity to take a 'time-out' to restore flow. When projects reach this stage the focus is finishing on time with an acceptable level of quality, rather than achieving continuous flow and productivity. For chaos projects, just using the control chart is recommended. In addition to showing production, it is critical to show all inspections which are required to pass quality work to the next trade. In chaos projects it becomes extremely important to completely finish locations before letting other trades in. Otherwise, the chaos will worsen and become unmanageable—small amounts of work remain everywhere. In chaos control charts, only grey, white and green are used:

- Grey means that location is unavailable for work because a predecessor has not finished
- White means that location is free for the trade
- Green means that all work in that location related to a task has been finished.

Production control works by ensuring that each subcontractor works only in white locations until they are ready and turn green. The control chart is updated daily to reflect progress and crews self-direct their work by looking at white locations for their task in the control chart.

GUIDELINES FOR CONTRACT PHASES

Different contract phases require different levels of detail. In commercial construction, the first schedules are typically created by the owner or owner's consultant in very early stages of the project life-cycle. Contractors get involved comparatively late in the process and their schedules are created in phases. Typically, first comes the bidding schedule which is used to estimate the project and to convince the owner that the contractors can deliver the project most productively and with little risk of delay. Once the general contractor has been selected, the schedule is planned by the team who are actually going to manage the project.

Owner's schedule

Before the selection of any contractors, the owner evaluates the feasibility of the project and designs the project to sufficient detail for contractors to bid on the project. Usually, the owner is responsible for the schematic design, which allows a very broad budget cost and target schedule to be established. After the schematic design, work proceeds to preliminary design. In some areas of the world, and in some contract forms, contractors will already be involved at this stage. However, most commonly the owner will also develop contract documents, including dimensions, and send these to the contractor for bidding.

Location-based management can support the owner, mainly through a location-based design schedule. Planning continuous flow for designers, and taking into account the constraints of gatekeepers, may expedite the pre-bidding process. Another possibility which has been explored by developer-contractors in Finland in their own production, is to create databases of worker hour consumption rates for building a typical space. By using location-based management, standardised logic and consumption rate databases, they have been able to develop schedules based on rough worker hour estimates and locations very quickly in the schematic design phase. Implementation of this requires the following information:

- Rough location breakdown structure
- Tentative lay-out of functional spaces
- Type of structure (cast-in-place, precast, steel, mixed)
- Type of façade.

Man hours are calculated as proportions of floor areas (for structure) or as proportions of the areas of different types of spaces (for example total office floor area, total corridor floor area) and based on surface area for the façade. Companies have found that they are able to develop databases which give good approximate durations for different lay-outs and structural methods in very little time.

This approach can also be used by owners who are not contractors by hiring a consultant. The location-based management approach of using approximate worker hours helps in checking the feasibility of the duration the owner has in mind. This analysis also helps in setting contractual milestones which are very important for evaluating progress of the project from the owner's point of view. Showing the location-based analysis to contractors as part of the bidding documents may influence them to implement location-based methods and thus achieve better productivity during construction. The goal of increasing productivity in construction can also be helped by tying payments to the completion of locations and therefore forcing contractors to finish a location completely before moving to the next location.

Bidding

In contract forms which require bidding, schedules may be created to estimate the risk level of the project—to identify opportunities for duration reduction and to estimate the overall resource needs for the project. Location-based schedules can be used during the bidding phase to optimise the total duration and to check the validity of estimated quantities through the schedule. A typical problem is that the location-based bill of quantities may not be available in this phase. However, approximately distributing total quantities across locations is usually sufficient detail for bidding. The goal of the schedule is not to be exact but to give a good overview of the possible ways to implement the project and to understanding the risk levels. Therefore, tasks which do not have a major impact on implementation can be left out of the bidding schedule. Location detail does not need to be high, for example sections and floors are enough for normal purposes. Selecting tasks to include should be based on availability of resources and availability of space. All tasks which cause interference to other tasks by working in the same location should be considered. The main things to optimise and decide are:

- Overall sequence—which parts of the building will be built in parallel and which will be built sequentially?
- Critical resource needs—the number of cranes, number of parallel pour areas, where to overlap finishes work in the different phases?
- Overall rhythm of project—bottleneck trades, overall strategy for crew continuity?
- Total duration and risk level—is there room for buffers, what are the resource needs?
- Cash flow—what are the demands on finance, the proposed payment schedule (unless specified in bidding documents)?

Risk analysis results and the bidding schedule can be submitted as part of the bid to demonstrate the feasibility of the schedule.

Contractor pre-construction

Before construction starts, a contractor should re-evaluate the project. This time, the team which is actually going to build the project is involved in the planning. The end result is going to be an owner-approved baseline schedule, which forms a contractual document to be used to report progress and show the effect of changes. The schedule will also be used to pull schedule the owner's decisions, design and procurement, so the start dates for tasks need to be accurate to within one week. It will be difficult to start tasks before the pre-construction scheduled dates because all other processes will have been planned based on the baseline schedule information.

The initial baseline schedule should show the overall strategy for construction. The critical decision points are:

- Location breakdown structure and strategy for evolving it
- Sequence of locations
- Production rate targets and continuity for each subcontractor
- Buffers
- Logic
- Decision about subcontracting or direct labour.

The detail of the baseline schedule should depend on available information. If the subcontractors have already been selected for some part of the scope, or if there is a decision to do some part of the scope with directly employed resources, the team responsible for the actual work should participate in the scheduling. In this case, this part of the scope can already be scheduled in detail during the baseline planning phase. If subcontractors have not been selected, it is enough to have an overall production rate requirement and schedule at a rougher level of detail (for example, schedule plasterboard walls as one task, instead of splitting into sub-tasks). Tasks where subcontractors have not been selected should be allocated more buffers to enable flexibility once the subcontractor is selected.

The baseline schedule should use both quantities and consumption rate information. However, the scheduler needs to evaluate all durations which are derived from the quantities and consumption rates. If the duration does not seem to be correct, the reason may be an incorrect quantity, or incorrect resources or an incorrect consumption rate.

For success of the location-based management system, it is very important that production rates are synchronised and known during the pre-planning phase. The overall production rate for the baseline needs to use assumptions about subcontractor resources before the subcontractors are selected. For some tasks, it is not possible to utilise large numbers of resources due to space reasons. For other tasks, it is logistically impossible to supply materials beyond some limit of production rate. Frequently, for some tasks, the subcontractor will be small and unable to mobilise sufficient resources for the task. Additionally, any increase in resource amounts on site will affect the need for supervision. Therefore, there are natural realistic limits for production rates even when subcontractors have not been selected. In finding the bottleneck trades, these realistic limitations need to be considered for each trade. Structure can typically only be accelerated to a point. In the finishes phase, the erection of the structure sets limitations upon the practical production rate. Otherwise, there is enough space in a large building to have many crews working in parallel. Therefore, in the finishes phase, it is usually logistics and resource availability that can turn a task into a bottleneck.

These realistic production rates should be used in a call for bids to ensure that contractors can achieve the required production rate. Note that if the production rate requirements are too high, the smaller subcontractors may be unable to bid and this will lead to an increase in resulting bid prices. For simple work, where quantities can be easily calculated, production rate requirements may be communicated in actual physical quantities per day. When doing this, remember that if the quantities change, this kind of contract will automatically result in a time extension for the subcontractor. For more complex work, it is easier to use location-based milestones. Actual crew sizes based on estimated consumption rates should never be used in bid documents because they will be different depending on the skill level of the workers. Crew sizes are used in pre-planning phase only to evaluate the risk and feasibility of the schedule.

GUIDELINES FOR CONSTRUCTION STAGES

Different construction stages require different levels of detail. During production, detail schedules are planned and updated weekly by the production team. The handover or commissioning phase requires a different way of scheduling because tasks are not known before either quality inspections or punch lists have been completed.

Earthworks and foundations

Earthwork tasks are dominated by machinery rather than manpower. Quantities and difficulty are often highly uncertain. For example, the ratio of soil excavation to rock blasting depends on the level of bedrock. This also affects the need for piles. Although assumptions may be made about the soil conditions, there remains high uncertainty with the selection of consumption rates and quantities. Therefore, earthwork tasks require some buffer before foundations to avoid nasty surprises. Earthwork tasks are typically weather prone which increases their risk. Because the same machines are used for many tasks (for example, excavation and fill) and many tasks are very short in duration and will be done when possible (backfilling of foundation walls), it is usually possible to ensure continuous use of excavators without too much pre-planning. Scheduling problems related to earthworks typically concern the overall production rate, increased quantities and inadequate information regarding bedrock (leading to additional rock blasting or the need for longer piles than expected). Machines are typically in short supply and it may be impossible to mobilise additional resources in a short time, so deviations should be detected early to be able to react. Earthworks also have complex dependencies to site logistics.

The foundations stage typically includes cast-in-place activities with formwork, rebar and concreting of footings or basement walls. Because there are many work areas, these activities can be carried out in parallel and it is easy to plan continuous flow. For this reason, location-based master schedules typically show just foundations as a task. The sub-tasks formwork, rebar and concreting can be planned close to implementation during the detail planning phase. Foundations are typically pretty standard and do not have major productivity risks but if the design is fast-tracked and overlapping with construction, there may be a risk of not having the design ready in time.

During production, there have been difficulties in getting reliable progress information from earthworks. This arises because of quantity uncertainties. In many projects, it has been very difficult to estimate the percentage of work done and remaining. Detail tasks should be used to update quantities when more information is received. In the foundations stage, progress data is more straightforward. Detail tasks can be used to explode the foundations task to trade sub-tasks (formwork, rebar, pouring) to be better able to track production and find the bottleneck trade in case of deviations.

Figure 13.14 shows a simple foundations phase schedule. There are three sections in the project: A, B and C. The basement level has been separated to its own section with location B, C, Parking hall, A and 4 Basement. The basement letter locations are directly below the corresponding sections A, B and C. The foundation schedule uses just formwork because that task provides the quantity which defines duration. The wall elements, columns and slabs are predecessors to foundation wall formwork work on the first floor. Section B has a bomb shelter which must be completed before any foundation walls. In the implementation phase, the foundations task was broken down to multiple detail tasks: pile footing formwork and rebar, basement wall elements and their joint pours, fills, hollow core slabs and their joint pours.

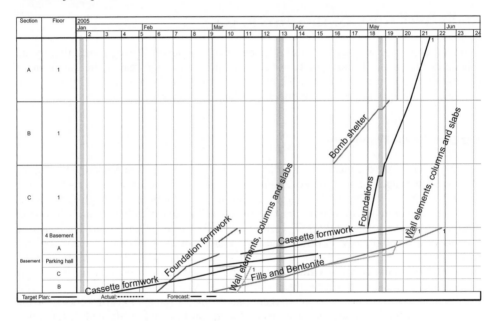

Figure 13.14 A simple foundation stage schedule

Structure

Structural trades are typically on the critical path of any project. Therefore, most effort has traditionally been spent on scheduling structure in great detail. From the point of view of location-based management, structure is a special case because it can suffer from interference only on the ground floor. The tasks of structure typically have no buffers and it is extremely difficult to catch up production once there are delays. The structure typically suffers from slowdowns on the early floors but achieves much higher production rate once it has incurred sufficient repetition. Guidelines for structure depend on whether the structure is cast in situ, precast concrete or a steel structure. Structure is often a mix of these three basic categories.

Cast in situ structures

Cast in situ structure is very different from other main structural types. It requires multiple skills—at least form, reinforce, and pour (FRP). Each one of the trades can have multiple tasks. For example the formwork crew can be associated with the following tasks:

- Formwork set of columns and upturn beams
- Formwork strip of columns and upturn beams
- Formwork set and embeds of suspended deck and beams (including horizontal formwork, edge form and construction joints)
- Formwork strip of suspended deck and beams
- Formwork set core wall lead side
- Formwork set core wall double up side
- Formwork strip core walls.

Because these activities do not tie up the crane, except for logistics and moving materials, resources can be added quite freely compared with other structure types. For very high buildings, cranes or pumps become a constraining factor because concrete must be poured using a crane or pump. Typical constraining factors are resource availability, material availability (how many sets of forms) and space availability. The type of formwork system affects many things in optimisation because forms can be prefabricated or built in place. Prefabricated forms tend to take less worker hours to install but may require the crane for lifting and supporting. Forms built in place are more labour intensive but it is easier to add resources to increase production rate. Typically, highly skilled carpenters are used in formwork and less skilled labour is used to strip the forms.

Crews often specialise for a specific limited function even though they could do a greater range of available work. For example, the formwork of slabs often requires more worker hours than general formwork, so there may be a dedicated crew for slab formwork and they will work continuously through the project just doing slab formwork. Another crew may be dedicated to core walls and yet another crew to columns. Resource constraints often make one resource a bottleneck, which prevents further speeding up of the overall building of the structure. Resource availability and continuity can be evaluated by using a combination of resource graphs and flowline figures.

Limitations in material availability applies especially to formwork. It is important to choose how many sets of forms will be used during the project. Formwork is committed between the tasks setting and stripping, such constraints can be modelled by adding Layer 5 links from stripping to setting formwork tasks.

Space availability is a function of many planning decisions. Each pour area can accommodate only a limited number of productive workers at a time. Increasing the number of pour areas makes it easier to get continuous flow for crews and thereby increases productivity. On the other hand, the costs of concrete pours are decreased if concrete is placed in large pours. As a pour must be completely finished within a day, the pour area has a maximum size dictated by the maximum concreting volume for one day. This can be increased by having more concrete pumps or by working longer hours. In projects with large floor areas, it is normally possible to plan continuous flow for all trades by adding multiple pour areas on every floor. In high-rise projects which have relatively small floor areas, continuous workflow is difficult to ensure and there will be waiting times in the work cycle. These can sometimes be solved by work-time arrangements, for example by having some crews work four days a week but with longer hours each day (subject to local industrial agreements).

Location-based management can be used to model cast in situ structures very quickly with a high level of detail. The most powerful tools are the layered logic dependencies which can automate much of the structural cycle logic. This makes it very rapid to test different alternatives. Optimisation of the structure schedule is at its most powerful when pour areas are planned using a 3D model and when alternative scenarios with different location structures and quantities are analysed.

In pre-planning, flowline figures are most useful for visualising empty areas where no work is occurring and for optimising crew sizes for those crews which work continuously. In the production phase, forecasts can be used to find which trade is holding up progress. Because the structural schedule is typically very detailed already at the baseline level, new tasks are rarely added during production. Detail tasks are updated if there are changes in quantities or to agreed start dates or production rates.

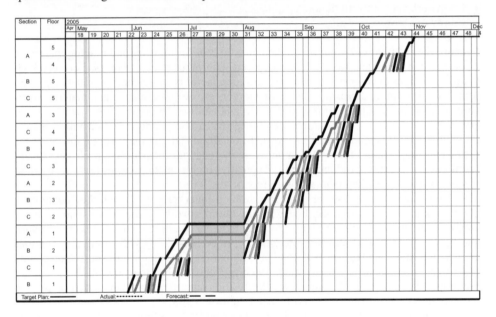

Figure 13.15 A typical cast in situ structural schedule for three connected apartment buildings

In many cases, the crews working on structure can also be utilised in other less critical tasks. If this is the case, the continuity requirement does not need to be met in structure because the crew can proceed to workable backlog tasks when it does not currently have work in the structure. Figure 13.15 shows a typical cast in situ structural schedule for three connected apartment buildings. There is a parking hall structure which can be worked on by the crews while they are not working on critical structural tasks. Therefore, small breaks and discontinuities can be easily handled on site by utilising the workable backlog. In the figure, all sections and floors have the tasks in the same sequence: Element installation of exterior walls, formwork of slabs, bottom side rebar, MEP in slabs, upper side rebar, floor heating pipes and pour. The scheduling challenge in this project is that section B is available for starting first and the interior work needs to start in section C. Therefore, the sequence shifts so that section C is finished first. Work is almost continuous for precast wall elements and for slab formwork.

Precast concrete structures

Precast structures can be divided into two types of work: dry installation and wet installation. Dry installation requires the crane to support the element which is being installed. Thus resources cannot be increased unless another crane is added as well. Wet installation is related to cast in situ areas to fill in the holes where precast slabs do not fit and to fill in the joints between precast slabs. Continuity is often ensured by working four-day weeks for dry installation crew and doing wet installation on the fifth day of the week. Because wet installation crews do not have continuous work for structure, something else can be planned for them—for example, pours on the roof, pours inside the building or related to the foundations, or slab-on-grade. However, in some buildings, there is a large quantity of cast in situ work which thus has a longer duration than one day per week.

The number of cranes determines the duration of the precast concrete structure, so in a baseline schedule it is common practice to plan just the installation of elements using one location-based task. Cast in situ areas can be added in the detail planning phase. Many precast concrete case study projects preferred not to go to this level of detail at all, instead just planning and controlling the installation of precast elements and handling wet installation at the weekly planning level. This approach does not cause any production problems. However, when this is done, most projects separate columns and beams to one task and slabs to another.

The location breakdown structure for a precast structure depends on the location of movement joints. To get continuous flow for finishes, it is best to try to finish sections of buildings from ground to roof level, as fast as possible, before starting other sections. Because continuity can be achieved very easily for precast structure, the total area of installation horizontally can be much smaller than with cast in situ structures.

Precast structures are easier to perform productively and have many fewer decision points than cast in situ structures. Their major risk lies in the material deliveries and design. Typically, the design for a floor needs to be finished at least six weeks before fabrication. Start times, and when materials are needed, for each floor and area must be established for both the design and fabrication of elements. Design and fabrication can be tracked in the control chart, together with production. This makes it possible to have design and production forecasts based on the actual production rates. These forecasts give early warning about forthcoming delays in design or element deliveries.

Quantities are easy to measure because installation is affected only by the number of elements of different types. Cast in situ areas are somewhat more time-consuming to measure by location due to the complexity of the work required on site.

If a structure has been planned with just one baseline schedule task, it makes sense to split it into multiple detail tasks for monitoring purposes. Actual production rates tend to fluctuate daily if the only baseline task is monitored by calculating the number of installed elements in each location, because normally the work is organised so that precast slabs are installed on separate days. Because they have a much higher production rate than other elements, weekly production rate is much too dependent on how many 'slab days' occur in the period. This has a large effect on forecasts. For this reason, tracking should at least separate slabs from other work stages.

Steel structures

Steel structures have similar characteristics to precast structures. The number of cranes defines the duration. Continuity is easy to achieve because the crew is creating space for itself—there is always another location to move to. The main difference with precast concrete is that slabs are often poured once the frame is erected, or precast slabs may be used. Cast in situ work also includes filling steel columns with concrete.

Similar to precast concrete structures, it is enough for pre-planning purposes to plan one continuous line for steel elements. Quantities are straightforward to measure because installation is affected only by the number of columns, beams and trusses. These need to be separated, mainly based on connection type (bolting or welding) and size.

The same principles apply in the production phase as with precast concrete structures. All element types which have different production rates, and take one day or more to install, should be separated as different detail tasks to make forecasts more stable.

Façade

The façade and skin-related work occurs on elevations. They are dependent on the structure and have important links to dry-in of the building and the corresponding start of finishes work. However, they work with different location logic. The structure is raised by floor and finishes are built by functional area. Façades are typically completed one elevation at a time. Depending on the type of façade and the construction method, floor information may or may not be relevant. For example, a curtain wall can be installed as one entity for an elevation and it does not make sense to plan or control it by floor. A brick masonry façade can be done in various ways, depending on the type of scaffolding and, in this case, division to both elevation and floor may be appropriate. It makes sense to go for a more detailed breakdown, especially in cases where the façade includes a series of interrelated tasks.

Subcontracted finishes and MEP

Finishes and mechanical, electrical and plumbing services (MEP) should be scheduled as one entity because they are tightly interrelated. The scheduling of MEP tasks in detail is often omitted by general contractors in current schedules because there is little information about their dependencies, quantities or production rates. Often there are just a few items at a very rough level of detail—for example, overhead MEP. In many cases, the location breakdown structure decided for the finishes trades is not suitable for the MEP trades. Because it is generally easier to adjust the workflow of finishes, MEP should actually play a major part in defining the location breakdown structure for the fit-out stage. While the location breakdown structure of the structure and façade are heavily constrained by technical considerations, such as joints and access by the crane, the interior location breakdown structure can be more freely defined. Typical CPM schedules stay at the building and floor level of detail but location-based management benefits from more detailed locations.

Detailed locations should be decided in cooperation with the general contractor and the main MEP contractors in order to get everyone's input. The best way is to draw the boundaries of locations onto floor plans in a joint meeting. Areas with same tasks and same logic can generally be contained in one location. The key question to ask is whether it is possible to do all work of a task at one time in this location. If the answer is no, then either a new task or a new location needs to be added. If the answer is no for only a particular task, then it is logical to create a new task, otherwise a new location should be considered instead. If a location has unique tasks which do not exist anywhere else, then these unique tasks can occur inside the larger location. It is common practice to include restrooms in an apartment location or in an office lobby or corridor location, because the tasks can be used to distinguish which work is related to the restroom. For example, other parts of the location may not include floor tiling, wall tiling or sewer fittings. All technical rooms, such as the main switchboard and main mechanical room should be separated to their own locations. This is to ensure that the walls of technical rooms are built on time and MEP contractors know when they can enter the space. In cases where locations can be created in many different ways, it is best to look at MEP system effect areas—for example, group switchboards have an effect area which can be used to define the logical locations from the electrical contractor's point of view.

After the location breakdown has been decided, quantities are needed for the locations. The quantity take-off for construction work can be easily measured based on any selected location break down from the drawings or a 3D model of the building. For MEP work, it has been more difficult to get quantities by location, because quantities are measured for each system which may not be the same as locations. In the authors' experience, it is easier to get estimated worker hours per location than it is to get actual physical quantities. These worker hours can then be used as quantities by location (with a consumption rate of 1) to develop a resource loaded location-based schedule. However, caution should be used when utilising these worker hour estimates. The following problems commonly occur in practical implementations:

- Subcontractors understand the location breakdown differently
- Subcontractors include waste and risk in their estimates (based on historical data) or use standard hours (which are based on piece rates)
- When reporting progress, subcontractors report used worker hours instead of the physical rate of completion.

The following are easy ways to mitigate the risk of getting the wrong information.

Subcontractors understand the location breakdown differently

Sometimes the drawings include references to locations. These may have different boundaries in different versions of drawings or with different designers. More confusion is caused by the fact that some locations may have many names. The best way to prevent misunderstandings is to issue a drawing set where location boundaries have been drawn in CAD software with clearly visible names. Because location boundaries can be different on different floors, the locations should be marked on every floor (even if they are similar).

Subcontractors include waste and risk in their estimates

Subcontractors tend to believe that the worker hour information will be used against them. They may be reluctant to give this information and even if they give it, they tend to inflate the hours based on the worst case scenario. This behaviour is understandable because, by giving out worker hours, subcontractors feel that they are exposing their cost structure and thus profitability. The authors have been able to get reliable worker hours from subcontractors after explaining to them how the information will be used. The idea is to level the resources of subcontractors and to identify any resource bottlenecks. A resource-based schedule can also be used to show the effects of variations or change orders to the general contractor and finally to the owner. Typical problems which subcontractors may experience in terms of productivity, can be decreased by planning for continuous work and for one trade in one location at one time. Accurate resource need forecasts are very important for MEP contractors. Without reliable worker hour information they are very difficult to create. There are easy ways to ensure that worker hours are correct.

First, showing the resource graph of the finished schedule to subcontractors will often initiate a change of worker hours. If worker hours have been inflated, the resource graph will show many more resources than the subcontractor is actually going to mobilise. Showing the resource graph often initiates many cycles of adjusting worker hours and changing logic so that resource use is more balanced. In the end, while the baseline is being approved, all main subcontractors should also commit to mobilising resources roughly as indicated in the resource graph. The actual resource needs will be forecast and discussed in site meetings for the look-ahead period of 4–6 weeks.

Second, during production it is easy to calculate the approximate actual worker hours used and then to compare that to the value of worker hours produced. If there is a difference, it should be analysed and the remaining worker hours corrected if the same pattern is likely to continue in the future. In a Finnish project, the electrical trade had estimated 115% of actual worker hours, the plumbing trade 144%, the fire protection trade 227% and the mechanical trade a remarkable 346%. This calculation was carried out after only two months of production and all the remaining worker hours were corrected, depending on the reason for the overrun. The fire protection trade had increased prefabrication, which explained their overrun. Plumbing and mechanical trades had very good crews on site but the subcontractor admitted that there was some padding in their estimate: estimates had been based on piece rate hourly assumptions. However, good salaries for plumbers are explained by production rates that are much higher than in standard piece rates! All worker hours were corrected and the accuracy of resource forecasts improved dramatically to benefit both the general contractor and the subcontractors.

When reporting progress, subcontractors tend to report spent worker hours instead of the physical completion rate

Because quantities are reported in worker hours, subcontractors often mistakenly report the actual hours of work done instead of the physical completion rate. The need to report physical completion rates must be emphasised at the beginning of the project. To make sure that contractors have understood this, the general contractor should check the progress data for the first month—location by location. If the completion rate is below 100% and the location is actually completed or completion rate rises to over 100% (especially if the location is not completed), it is likely that the subcontractor is reporting hours. For some tasks, such as electrical cabling, it is difficult to estimate the physical work done, so subcontractors prefer to report hours. This is OK for the beginning of production in a location, but at some point the estimate has to turn around and estimate how much work is remaining. This is critical for forecasting progress.

In addition to having quantities by location and task and location breakdown structure, logic and constraints are needed to build the schedule. These are best derived in a joint workshop with all the main contractors participating. This logic meeting can have the same structure as otherwise in normal CPM logic meetings, but special care should be taken to investigate the logic of different functional space groups. For example, the logic of corridors and the logic of office rooms is likely to be very different. A logic network can be developed for each identified location group. Often, location groups have some minor links to other groups (for example, the main switchboard is needed before some cabling can be undertaken, or lobby floor tiling should not be done before the heavy construction work of adjacent retail spaces is complete). However, mostly the links relate to structure, façade and internally within the location group. Logic diagrams should be approved by all participants because they form a decision about how the project will be implemented. Any change in logic needs to be approved by all team members before it can be initiated in the field. In addition to mandatory technical logic, tasks should be identified which will interfere with each other if they occur in the same location at the same time. Optimal schedules keep these tasks away from each other.

Schedule constraints can relate to resource constraints, long lead time items, end user requirements and design. Resources should be discussed with each subcontractor to see how many workers they have been assuming they need in the project, what the main risks are and any other constraints they see in the project. MEP tasks especially contain many long lead time deliveries such as lifts, switchboards and mechanical plant room equipment. Long lead time items cannot typically be pulled to site by production requirements, so must instead rely on push production. These expected delivery times should be taken into account in the schedule.

The location-based schedule for finishes and MEP can be planned and optimised with this starting data. Methods for achieving continuity, and buffering against variability, are the most critical features of location-based planning. Risk analysis is very useful to evaluate the correct size of buffers. In the determination of buffer sizes, quantity risks should be evaluated in addition to the resource and production rate risks. For example, end user changes will affect different tasks in different ways. In office buildings, suspended ceilings or lights are normally not changed much but in retail buildings, every user might have a different type of suspended ceiling and lighting system. User changes also affect the start dates for tasks because the user information may be received too late to start work on time. Risk analysis should be used to calculate the last responsible times for any owner decisions, by location and by information type (for example, information about whether there will be a suspended ceiling in the space is needed before information about the exact type of suspended ceiling).

Monitoring of finishes and MEP is very time-consuming unless subcontractors assess and report their progress. A practical way to implement self-reporting is to print-out a filtered control chart showing only the tasks for that subcontractor and to ask them to return the correctly filled information before the weekly subcontractor meeting. Every week, the project engineer or site manager should blind check some of the actual data on site to make sure that the reporting is correct. Special care should be taken to ensure that the physical degrees of completion are reported. Normally, subcontractors tend to have errors in their actual data at the start of the project, but after some rounds of corrections they will learn the process. More time needs to be budgeted for double checking at the start of trade.

Subcontractor meetings are critical for controlling MEP and finishes because there are many subcontractors involved and production is not as clear as in previous construction phases. Instead of focusing on the progress of past weeks, meetings should focus on current deviations from the plans, upcoming problems, prerequisites of production, control actions and the resource needs for the next 4–6 weeks. Tools offered by location-based management which can support this are control charts, flowline figures with both actuals and forecasts, control action planning and resource forecasts. A typical problem in many case projects has been that control charts are well implemented and used in subcontractor meetings but the forecasting and control action parts are left out. This means that the project team is only reacting to problems which already have occurred and is focused on mitigating the consequences, instead of trying to proactively prevent problems from happening. For proactive control, forecasting and control actions are essential.

Commissioning and handover stage

Flowline does not plan very well for the handover stage, where the work to be done is only defined once the results of self-commissioning inspections and client inspections are known. It is impossible to define beforehand which errors will be found. In the baseline schedule, this can be taken into account by having five location-based tasks going through all locations: self-commissioning checks, fixing errors round 1, client inspections, Fixing errors round 2 and commissioning. Checks, commissioning and inspections can be very short duration tasks in each location, while a lot of time needs to be reserved for fixing the errors.

After self-commissioning checks have been finished in each location, the location will have a scope of work for fixing errors. These can be planned as detail tasks for the fixing errors round 1 baseline task. Each detail task should contain the work of one subcontractor only. This can then be controlled using a control chart which has all the fixes and checks necessary to complete the building. The goal of scheduling for the handover stage should be to finish all the work of a subcontractor in a location before handing over to the next subcontractor. Continuity can also be planned, by trying to make detail tasks of a subcontractor continuous. However, this can be difficult, because fixing errors will start in the first completed locations before any errors are identified in later locations.

Chapter 14

Planning and control of linear projects

INTRODUCTION

Linear projects are a particular type of civil engineering project which involve long runs of continuous sequences. Typical examples are road and rail projects, but they may also include tunnelling or mining projects such as open cut or strip mining. Linear projects often use line-of-balance techniques for both analysis and representation, so location-based management has traditionally been considered especially suitable for linear projects. However, in contrast with earlier literature, this book regards linear projects merely as a special case of location-based planning rather than as a separate type of project needing a separate planning technique.

For location-based planning, the location breakdown structure of linear projects, and the linear parts of some other projects, are different from other construction projects. The nature of the project also enables some of the calculations to utilise the actual location measured in metres or other distance unit (such as chains) from a starting point. In other words, unlike other project types, the locations in linear projects can generally be translated to distance travelled along a line—thus linear. This chapter extends the location-based planning system to make the most of the linear structure when planning such projects. Most of the basic logic stays the same, however there are changes to the dependency calculations, the planning methodologies are different, and industry prefers a different form of visualisation.

A linear project's location breakdown structure (LBS) can utilise the linear relationship of the locations to automate creation, indeed the lowest level of the location breakdown structure must be allowed to have continuous parts. In normal projects the locations are logical unless defined into a location group (such as vertical risers, or vertical pour sequences). In contrast, instead of having each location define a discrete area, the linear parts of a project can be handled continuously and cumulatively (by distance) and therefore the calculations can be reduced to a metre level of accuracy. This chapter generalises the LBS of previous sections to be able to also use continuous locations defined in metres or any other distance unit.

The best schedule visualisation for a linear project is the time-distance (T–D) diagram. Unlike the flowline diagram, this diagram has location placed on the horizontal axis and time on the vertical axis. The reason for the reversal is simple: the horizontal representation of the LBS enables meaningful comparisons between the schedule and other maps, charts or diagrams. As distance is normally placed horizontally in engineering charts—for example, mass haul diagrams or longitudinal sections—these can be printed using the same scale as the T–D chart or may even be combined with the schedule. This allows visualisation of production in the same way as with flowline diagrams but also enables direct comparison with other information using the same visualisation logic. Such a comparison between design drawings and the schedule is not possible in commercial construction.

Planning methodologies should change because of the fact that resources are nearly always 'multi-skilled'. For example, the same excavator can be used in many different tasks in a road building project. Because production rates are a factor of machine size and capacity and there are fewer quantity types (types of work to be carried out) than in commercial construction projects, production rates may be used instead of consumption rates to calculate durations. Resources also have relatively high hourly rates and relocation times are much greater than in building projects, thus optimising the production system cost is even more critical to linear projects.

Mass haul optimisation

Mass haul optimisation is a major concern in infrastructure projects and huge cost benefits can be achieved if haul distances and stockpiling can be minimised. Therefore, this chapter places a major emphasis on projects involving mass haul. Integrating the schedule with a mass haul plan is a critical factor for economic success of any such project. Some of the difficulties associated with scheduling mass haul include the fact that cuttings are executed in the reverse sequence to the corresponding material needs for embankments. For example, topsoil must be removed first, despite being required for the filling of slopes and other processes which occur last in the construction sequence. Additional challenges include that the same work chain includes both cutting and embankment (where the masses are being hauled). Therefore, a delay in one location of the road line can affect multiple other locations. This has an important effect on the way procurement is planned in linear projects involving mass haul.

The same resources may be used for multiple tasks during linear projects, therefore subcontracts for linear projects, especially road or rail projects, are often defined by location—not by trade. There are very few specialist subcontractors which follow the flow logic of the simple flowline as used in commercial construction, rather they involve a complex interaction of sequential cuts and sequential fills, broken by physical needs (arising from the land and road profile) and also by intersections and traffic management requirements. When subcontracts are defined by area, buffers are not required in the same sense as in commercial projects. However, linear project subcontractors often need to be separated from each other by constraining the mass hauls to occur within their own mass economy area (a locally defined region with balanced work). This is the only way to localise the schedule deviations which occur in mass haul projects. Thus mass haul planning becomes critical for defining contract boundaries and the schedule.

Tunnelling and other infrastructure projects

Tunnel construction is a special form of linear project that shares some of the characteristics of both linear projects (involving mass haul) and the cycle planning of commercial project structures. Similarities to mass haul projects arise from the fact that masses from tunnelling project need to be hauled and utilised somewhere. On the other hand, tunnelling is very similar to the pour cycle of an in situ vertical structure—just as it is not possible to build vertically out of sequence, it is not possible to tunnel out of sequence. Activities of different subcontractors need to occur in close sequence before the drilling machine can proceed to the next location in the tunnel. Tunnelling uses multiple ends or maintenance tunnels to balance expensive drilling resource use while other trades are exploding, clearing and reinforcing the end of the previous tunnel.

Many other infrastructure projects share characteristics of both linear and building projects. For example, ports, airports and mining projects have significant mass haul components but they are not necessarily linear. These projects can use discrete locations in a coordinate system. Standard location-based planning systems can be used for scheduling, but mass haul can be combined by using the methods which are outlined in this chapter.

Control methods

Location-based control methods similar to those developed for building construction can be used in linear projects. Linear projects pose a special challenge when getting accurate information about the work which has been completed on site. They may involve travelling long distances for inspections, which will usually make it impossible to directly observe the amount of work done. Rather, delivery notes from truck drivers need to be relied on to get an understanding of the actual progress. Additionally, work stages are rarely as clear cut as in commercial projects—it is therefore often difficult to say where one structure ends and another begins. These challenges result in higher uncertainty about the validity of progress data. Forecasting is also difficult in projects involving mass haul. Mass haul project fills are linked to the cuts which supply materials to them. Any schedule forecast for fills requires information about the mass haul plan and progress of the related cuts—as fills cannot proceed without associated material from cuts. Furthermore, there are also more control actions available—such as changing the mass haul plan or even changing the design to deliberately affect quantities (such as raising or lowering the alignment), instead of merely changing production rate by adjusting resources or the length or number of shifts.

Contents of this chapter are in large part based on results from the master's thesis of Juuso Mäkinen, CEO of DynaRoad Oy, the developer of DynaRoad software (Mäkinen 2007). This chapter describes methods and practices developed on projects in Finland and Australia. The graphics are generally produced by DynaRoad version 4.2[1], the linear project scheduling and mass haul optimisation software.

PLANNING LINEAR PROJECTS

Linear project location breakdown structures

Linear projects require that the lowest level of the location breakdown structure must be allowed to have continuous parts. Continuous parts are defined by starting and ending stations (for example, 0–500). All higher levels of the LBS are discrete and follow the logic defined in previous chapters, as for any other LBS. Multiple road lines having different stations can be modelled by introducing additional higher level discrete locations (for example, one higher level location for each road line or intersection having different stations). Linear projects have location breakdown structures, Figure 14.1 illustrates sample structures for both road and tunnel projects. The important distinction from general location breakdown structures are the branches labelled 'continuous' and with a range definition (Shaded in the figure). These equate to a packet of many locations, all with linear properties.

As shown in these examples (Figure 14.1), the lowest level locations do not need to be continuous across higher hierarchy levels, but can be discrete, and continuous logic can be combined in the same project. To enable distance calculations, which are critical for mass haul and relocation time optimisation, the various locations can be given coordinates and possible routes that can be used for equipment to move (such as haul paths). This also facilitates calculating layered logic dependencies.

There are two types of special locations in linear projects: usage breaks and temporary locations. Usage breaks are locations which prevent equipment or mass haul from travelling through the location for the period of the break. Usage breaks can be temporary. For example, river can form a usage break until a bridge has been built. A mountain will form a

---

[1] Details of DynaRoad software may be found at www.dynaroad.com

usage break until a tunnel has been completed. Temporary locations are site roads, detours and other temporary structures. They become available as a haul route once a task has been completed (for example, when activities related to building a road are finished). They cease to be available once the road masses have been excavated. Usage breaks and temporary locations are critical for schedule optimisation of most linear projects.

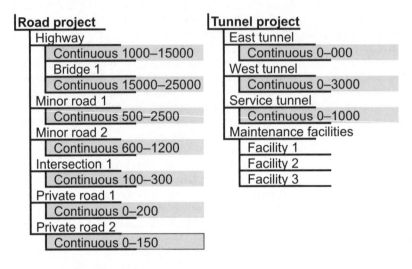

Figure 14.1 Sample ocation breakdown structures for linear projects

Linear projects such as dual highway roads and combined road and rail projects create special needs for the LBS. It is necessary to make early decisions regarding the treatment of the multiple road or rail lines. If dual lanes exist for a single road alignment, multiple lines can be treated as a single hierarchy on the LBS—like a wide single lane road. However, apart from completely greenfield projects, such a structure is rarely practical. It is more common for traffic management to interfere in the construction of the road, necessitating the treatment of the upstream lanes to be capable of management separately from the down-stream lanes. In such a case, multiple road lines must be created, not withstanding their closeness to each other, with intersections at logical points of connections. This provides maximum flexibility in planning complex schedules for traffic management.

Quantities in a linear project

Just as with commercial projects, there are quantities for linear projects. These can relate to the construction work required along the line or the masses requiring to be moved. When concerned with mass haul, quantities relate to the cuts and fills of different material types (cuts of rock, soil, silt, sand, etc.; fills of bulk fill, crushed rock, sand, etc.). Quantities for linear parts of a project can be defined using station intervals. For example, many road design software packages export a bill of quantities (BOQ) for each 20 metres of each road line. Each quantity point consists of starting station, ending station, and the total quantity within the interval. Within each interval, the quantity may assume an even distribution. It is straightforward to calculate the date of reaching any subsequent station along a continuous location (road line) by using a constant production rate.

Table 14.1 Typical starting data for road projects with a twenty metre station interval.

Station	LandCut1 BCM	LandCut3 BCM	RockCut3 BCM	Landfill CCM	Slopes CCM
0	216		0	867	53
20	128		0	816	34
40	176		0	657	33
60	224		0	660	35
80	248		0	552	36
100	264		0	533	35
120	320		0	344	147
140	0		0	101	69
160	176		0	34	51
180	184		0	33	43
200	3528		0		41
220	2224		0		69
240	4328		0		66
260	3008		0		66
280	280		0	1	86

Unlike normal construction, tasks which involve mass haul are usually not clearly defined. The problem is that materials sourced from excavations may be used in many places as fills. Indeed material may be purchased or disposed of off-site. Therefore, when defining quantities in mass haul projects, it is important to consider the suitability of material for other structures. Each suitability class should have its own quantity for each range of stations. Table 14.1 shows typical starting data for location-based planning of road projects with 20 metre station intervals. For planning purposes, it is cumbersome to schedule each 20 metre section, so it is typical that quantities be aggregated into larger location groups, for example 300 metres, and that this is set as the planning level of accuracy. Accordingly, each 300 metre section has a centre of mass which can be used for haul distance calculations.

Graphically, the starting data can be shown in a mass haul diagram. These show the station of the project on the horizontal axis and quantities as boxes above or below the road line. Cuttings are above the road line, and materials excavated first are closest to the centre of the chart. Similarly, fills or embankments are below the road line, and fills placed first are nearest the road line. The box size is related to its volume. Mass hauls can be shown with arrows. Figure 14.2 shows a mass haul diagram with hauls displayed that are over 1000 m^3. The small black boxes at the top of the figure represent disposal areas (for example, landfill). Each work site box has a length of 300 metres, which is the chosen accuracy level for planning in this sample project.

Another graphical presentation of mass haul is the mass balance curve, which shows the cumulative cut and fill from start (lowest chainage) of the project. Figure 14.3 shows a sample mass balance curve.

A special characteristic of many linear projects is that while the total volume of any given excavation can be accurately measured, the relative quantity of different kinds of work (type of material being excavated) has high uncertainty as it is not possible to accurately assess the exact material under the ground before excavation. For example, it is usually not feasible to determine where bedrock begins and soil excavation ends for all locations of the project. Other typical inaccuracies relate to the quantities required for soil reinforcement. Even with accurate soil surveys, it is not possible to know these quantities exactly. By spending more money in soil studies, it is possible to get a more accurate picture, yet the process cannot be totally accurate. Because these uncertainties are commonplace when scheduling a linear project, location-based techniques become even more important.

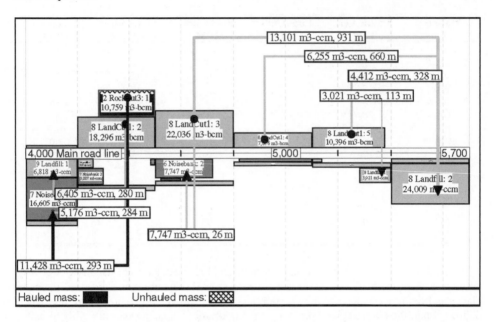

Figure 14.2 Mass haul diagram showing hauls over 1,000 m³

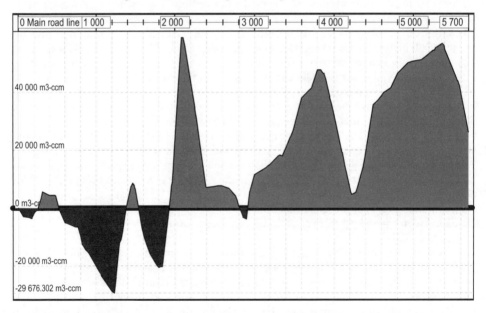

Figure 14.3 Mass balance curve calculated from start of project (chainage = 0)

Mass haul tasks in linear projects

There are typically far fewer task types in linear projects compared with commercial projects. The process of task definition should be based on the nature of the work. For example, all soil excavation of different suitability classes should be lumped to a single task if the same resources will used for the excavation and it is not known in which sequence they will be excavated. In this case, for mass haul calculating purposes, it is assumed that each material class is evenly distributed over the total duration of the task.

In some cases, excavation or fill can occur in layers, which means that a large area will be unavailable for succeeding tasks until the final layer is completed. Work continues from one end of the task to another and then goes through the same stations again. For these tasks, the number of layers should be selected and controlled.

Some tasks can prevent mass haul passing over that location while the work is in progress. For example, soil stabilisation will prevent mass haul through the area while it is being done. A large rock cutting in difficult terrain will prevent mass haul through it until it has been blasted away. Often, contractors will not be able to haul across final topping layers once placed.

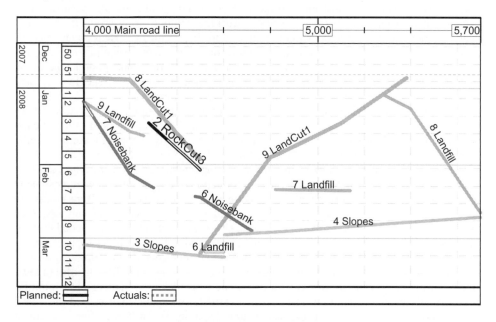

Figure 14.4 Example T–D diagram

Visualisation

Much of additional graphical information available from a linear project—such as longitudinal sections or mass haul diagrams—is presented by having location on the horizontal axis. Thus, it is beneficial to show location on horizontal axis for continuous locations. This allows the schedule to be printed at the same scale as other print-outs, allowing easy comparison of information. These graphs are called time-distance diagrams (T–D charts). Figure 14.4 shows an example time-distance diagram for a road project. Another benefit of using a time-distance diagram, is that haul distances and possible hauls can be read in the horizontal direction (from a cut to a fill at the same time).

There are usually fewer tasks for each individual location, making the charts easier to interpret than flowline diagrams for building construction.

Resources, crews and durations

In a linear project, most of the work is performed using machines, not labour. Therefore it is the machine which determines the production rate and selecting the size of the machine has a critical effect on production rate. Typical resource consumption rates, as used in normal LBMS, typically cannot be used to calculate durations because they are too variable according to the equipment selected. On the other hand, a single resource can be used for multiple types of work. In linear projects, it is more economical to have production rate as a property of both the resource type and the different kinds of work. For example, an excavator of 40 tons will have a different production rate from a 60 ton machine, but also might have a different production rate for removing topsoil than for doing soil excavation or loading rock into a truck.

In commercial construction, space constraints play a larger role than resource constraints. However, in most linear project types such as road and rail, space is not an issue—it is possible to have many resources working without losing productivity due to space restrictions. Resource constraints play a more important role. Heavy machinery is expensive and difficult to move around. It is often not feasible to move equipment to another site when there is insufficient work available for that plant for a short period. Therefore, optimisation to ensure work continuity for each machine is a major cost issue. In tunnelling projects, both space constraints and resource continuity are important.

In projects involving mass haul, resource utilisation is more difficult to optimise because it is a function of other resources. For example, a bulldozer could theoretically achieve a very high production rate but only if there is a steady supply of material coming from soil cutting operations. One land cutting is frequently not enough to achieve full resource utilisation for a bulldozer simultaneously making an embankment. This results in waiting times at the embankment. Therefore, production rates of chains of resources—land cutting, haul equipment and embankment resources, have to be balanced[2]. Big savings can be achieved by optimising production rates—either by having two land cuttings, each sending material to a single embankment or by selecting a smaller bulldozer. Schedule optimisation of this kind is easy to do locally for single cutting but to achieve global savings, mass haul distances and resource utilisation needs to be considered as a whole.

Figure 14.5 shows a simple example where material required to fill a landfill (a location requiring fill) is coming from two cuttings. The schedule has been planned to minimise mass haul distances and duration. The landfill is started simultaneously from two ends, each corresponding to a land cut starting at the same location. When the smaller land cut of 10,000 bcm (bench cubic metres) has been completed, resources transfer to increase the production rate of the second land cut. At the same time, the slope of the embankment changes because another bulldozer is shifted to the other end of the landfill. If the mass haul plan had only one of the lancets supplying the fill, the duration of the fill would be double and an additional mass of 10,000 ccm (compacted cubic metres) would have to be hauled from another source (possibly purchased off-site). Note that in the traditional scheduling process, only cuts are scheduled. As shown by this example, this will result in inaccurate plans which will make it difficult to determine the start dates of successors accurately.

---

[2] This type of production balancing is very different from that used in line-of-balance—where production rates of following trades should be balanced. Here, it is necessary to balance both the cut and the fill operations, and following trades are less important, or not considered in balancing production.

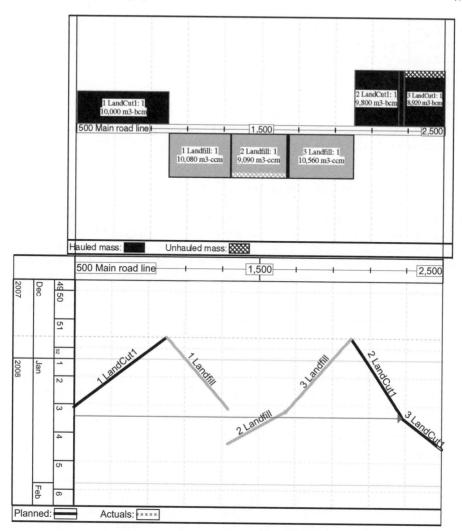

Figure 14.5 Combined mass diagram and schedule

Resource loading of mass haul

Resource loading the mass haul cannot follow the standard rules of production and consumption rates, because mass haul always occurs between two tasks: between cuts and fills. If the distance is very short, the same resource may be used in cutting, moving the mass and then filling. A scraper is a good example of this type of resource. The number of resources required in a haul operation is a function of the machine capacity, the available haul route and the distance. Because resource numbers are a function of both the schedule and the mass haul plan, it becomes computationally excessive to use any resource levelling heuristics in connection with haul resources. Any excessive resource use would delay some work sites, which would change available haul destinations, which in turn would change haul resource requirements possibly causing new resource peaks. Attempts to resolve this have thus far failed to produce heuristics which would be accurate and able to work on large enough projects in a reasonable time. At the moment, the only way known by the authors is to do manual adjustments to either the schedule or the mass haul plan to remove excessive resource use.

Logic links

All location-based layered logic can also be used for continuous locations in linear projects. However some changes are required, as described below.

Layer 1 logic: logic within the same location

In continuous locations, Layer 1 logic works at the station level of accuracy. When a task is calculated, the start and finish time for each centre of mass (for example each 300 metre section) is calculated. If the predecessor and successor have a conflict (for example a F–S relationship with the predecessor's centre of mass not yet finished when the successor starts), then the calculation is taken to metre level of accuracy by assuming linear production rates within the centre of mass. This way it is possible to calculate when the predecessor and successor reach any station and therefore calculate logic at the metre level of accuracy.

Layer 2 logic 2: logic on higher hierarchy level

Higher hierarchy levels for continuous locations are always discrete locations, so this logic layer is used to create links for all road lines or tunnels. It works in exactly the same way as described earlier in Chapter5.

Layer 3 logic: internal links

All of Layer 3 logic can be modelled in the continuous environment. It does not make sense to make tasks start 'as soon as possible' (when faster than the predecessor) because the level of accuracy is one metre—in this case, faster tasks would complete one metre of work in the first location and then wait before doing the next metre. Instead, a new parameter for discontinuous work is needed. This parameter is the minimum distance within which it is rational to complete work continuously (the work should be treated as a single work site). For example, setting a value of 100 metres would delay starts until at least 100 metres of work could be executed continuously, but would allow a break after this minimum threshold, continuing until the work site is completed.

It usually does not make sense to jump between different locations within a work area, because moving machinery costs money. Therefore, it is enough to specify for each task the starting point and direction of work. If there is a specific reason for a change in the sequence, a task may be split to model different workflows.

Layer 4 logic: location lags

Location lags are just as important in scheduling linear projects, but location lags for continuous locations are specified in metres (the distance two activities need to be separated by). For example, a predecessor may need to be finished 100 metres before a successor comes in.

This is useful for ensuring that there is enough physical distance between preceding and succeeding operations. This is also very powerful in tunnelling, where it is a critical planning decision to decide how many metres to drill before blasting. Instead of having to change locations every time, only this parameter of a link needs to be changed.

Layer 5 logic: random CPM links

Layer 5 logic is used to link different road lines or tunnels with different station systems together. For example, work located on a main road line can be a predecessor to work located at an intersection.

Schedule calculations

The schedule calculations of Chapter 5 can be directly used, but each station needs to be handled as a separate location. Continuity calculations must take into account the continuity threshold—the minimum number of metres which must be produced continuously. To find the most constraining point and to define the time of other stations based on that point, it is enough to proceed in segments of threshold in the direction of production, because each task has a starting point and direction.

PLANNING METHODOLOGIES FOR LINEAR PROJECTS INVOLVING MASS HAUL

Almost all linear projects are heavily resource constrained and the main objective of a plan is to maximise the utilisation rate of expensive equipment. In projects involving mass haul, the minimisation of both haul distances and stockpiling are equally important. Schedule optimisation becomes even more critical and complex in linear projects, despite the lower number of tasks to be scheduled, because there are large direct cost effects in the schedule (as opposed to mostly indirect production system costs in commercial projects) and there are many complex interdependencies caused by the mass haul.

Planning constraints

Linear projects can typically start in multiple locations at the same time. However, every linear project has constraining factors which are known when creating the schedule and mass haul plan. In a typical linear project, constraints can be caused by:

- Traffic
- Bridges
- Rivers
- Hard terrain
- Tunnels
- Opening direction
- Environmental concerns.

These constraints can prevent hauls through an area, increase haul costs because hauls need to be made through other traffic, increase haul distances or force a certain sequence of work.

For example, work cannot occur in an area before traffic has been directed to some other route. If a bridge is needed to haul rock over a river, the rock blasting operations and embankment cannot be started before the bridge is finished. The constraints should be figured out before planning starts because constraints make a large impact on the optimal outcome.

Mass haul considerations

In linear projects involving mass haul, the schedule is of critical importance. Many road projects have 70% of their costs coming from work related to mass haul. Optimising the schedule and mass haul plan together provides an opportunity for significant direct cost savings. To optimise the schedule for mass haul, you need to consider some of the following issues.

Avoiding stockpiles

Stockpiles, which increase costs, will be required if there are not resources available to work at both the cuts and fills when cutting direct to fill. It is possible to haul mass from a cutting to an embankment only if resources are working at both ends. If the embankment must be completed after cutting, there will be a time delay between the cut and fill operations, making it impossible to haul from a cut to a fill without the use of stockpiles.

Slopes are filled near the end of a project. To decrease stockpiling requirements, it makes sense to delay the start of land cuttings which are not above rock cuttings so that material can be hauled directly from cut to fill. However, stockpiling is often necessary when all land cuttings are on top of rock cuttings, because it is usually not possible to delay rock cuts if their material is required earlier in embankments.

Optimising resources

The schedule should be optimised in combination with the mass haul plan to maximise the utilisation of heavy equipment and to minimise haul distances and thus the need for transportation equipment. The schedule is also critical in circumstances where hauls between two locations can be made only after some other work is completed in between (for example, a bridge or large rock cutting). Note that haul distances resulting from the schedule should be compared to the theoretical minimum haul distances, calculated with linear optimisation methods to see how much cost the schedule is adding to the mass haul costs.

Crushing rock for structural materials

Rock crushing is a critical consideration in many mass haul projects. The crushed material is generally required, late in the process, as sub-course and base-course material. In this case, it is often best to delay the land cutting on top of bedrock to be able to utilise the material in slopes which are also filled near the end of the project. While delaying crushing has a cost benefit because of a decreased need for stockpiling, the project management can be easily tempted into wasting the rock for other purposes in the mean time. Also, delaying the crushing operation increases risk, because if it turns out that the selected rock is unsuitable for crushing, all other rock material may have already been used in other structures.

Uncertain soil materials

For those masses whose suitability is uncertain (for example, masses affected by weather), the schedule should have an embankment and disposal area close to the cutting and available at the same time. Material suitability is evaluated separately for every truckload in real time and then will be taken to either the fill or else the disposal area. If an embankment is not available, all material must be disposed—this will often waste suitable materials.

Temporary structures

Mass hauls always require a route. Building temporary site roads costs money and their locations need to be carefully planned before the start of construction. A good linear schedule minimises the need for temporary roads by using the built structures as site roads. If temporary roads need to be built, the optimal time to dismantle them should be selected based on the possibility of reusing their materials.

Some embankments may require temporary surcharges to settle or compact the structures. The required duration of the temporary surcharge can be modelled by a lag in the Layer 1 dependency. If decreasing the settling time results in large benefits for the project schedule, it is often possible to accelerate compaction by building vertical drainage or placing heavier surcharges—for example, in the form of crushed material stockpiles. In this case, the material used for the surcharge can be optimised.

Optimising work packages

Selecting optimal work packages is often not a straightforward task in a linear project (except tunnelling). Subcontract packages should be optimised based on minimised interdependencies with other packages. Packages often become interdependent if material excavated by one subcontractor is needed for fills done by another subcontractor. Dependencies will result also if many subcontractors share the same crushing plant, disposal area or stockpile or require material from the same borrow pit. Resolving these interdependencies requires on-time information about the start and finish dates of dependent work sites and daily production rates. However, the project becomes much easier to manage if these interdependencies are reduced by optimising work packages so that they have internal mass balance. In effect, hauls outside of the package are minimised. Such areas have mass balance. Each subcontractor works in his own *mass economy area*. There should not be any obstacles to mass hauls inside this mass economy area.

Figure 14.6 shows a mass haul plan for a project divided into contract packages. There are three contract packages in this mass haul plan—from station 0 to station 1,800, from station 1,800 to 3,800, and 3,800 to 5,000. It can be seen from the figure that very few hauls cross over the contract boundary. Only the noise bank from station 3,800 to 4,300 is shared by multiple contracts. Also, the large rock cutting in station 2,000 is shared by all packages (hauls are not shown in the figure) because it is being used to crush aggregates that are needed for the sub-base and base-courses.

In addition to the mass haul, complex dependencies between contractors can be caused by bridges, earthworks in bridge areas, or temporary roads required to avoid public traffic.

In some cases, not all work in a location will done by the same subcontractor. A typical example is rock excavation. The rock is blasted by one subcontractor, the rock is hauled by another subcontractor and yet another contractor may be working on receiving the materials. In such cases, the production easily fails if one of the contractors cannot achieve the required production rate. This can easily lead to idle time for resources and subsequent reduction of the resources on site—with return delays or compensation claims for lost time. Even if the contractor who caused the problem corrects his production rate, it is likely that the other subcontractors will already have reduced their production to match the bottleneck. In all cases, the general contractor pays, either in time or money.

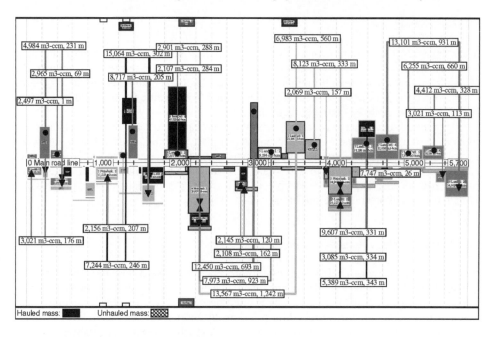

Figure 14.6 Mass diagram showing hauls

Optimising flow

Continuous flow of heavy construction equipment resources is even more critical than in building construction projects, because they are very expensive. A good way to approach scheduling is to plan for the flow of a particular individual resource throughout the project. It helps to start scheduling from major cuttings and then to schedule fills based on the shortest haul distances. Figure 14.7 shows a schedule with cuttings scheduled first and then fills scheduled to minimise associated haul distances. Cuts have been delayed so that the work crews can work continuously taking into account that mass should also be hauled over the bridge.

Optimal starting points for work chains are locations where the direction of mass haul shifts. Work chains are built from that location towards the next shifting point. The required number of work chains is determined by schedule constraints. It is often necessary to start multiple cuttings at the same time in order to achieve the overall production rate requirement. Often, it is not possible to start from an optimal starting point because there is no access to that location. Alternatively, sub-optimal starting locations may have to be chosen because of schedule milestones that are unrelated to the mass haul work—for example, building site roads or preparing work for large bridges.

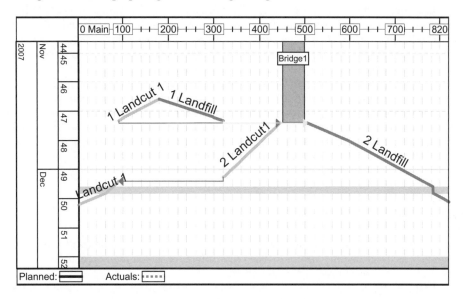

Figure 14.7 A schedule with cuts and fills scheduled to minimise associated haul distances

Optimising design

Large linear projects are often implemented as design-build contracts. In this case, the general contractor can achieve large cost savings by optimising the design based on the mass haul and schedule requirements. The design is often constrained by a required alignment but there are usually tolerances within which both the vertical and horizontal alignment can be adjusted. Such adjustments can easily affect the quantities of cut and fill, shifting the balance. The critical issue is not so much to have balance (design is ordinarily adjusted for this) but to ensure that there is optimal balance of materials once scheduled. Soil reinforcement and noise protection methods can also often be adjusted by the contractor, so long as they adhere to the client's quality requirements. These have a major effect on the project mass balance (for example, use of soil replacement instead of stabilisation) and the schedule. An optimised mass haul plan will reveal any surplus or deficit, or excessively long haul distances, and this information may be used to guide design decisions.

It is possible that an alignment with an otherwise balanced mass plan is sub-optimal when scheduled. Adjusting the alignment to deliberately unbalance the mass haul plan may allow an improved overall plan.

Managing uncertainty

Mass haul material quantities are highly uncertain in linear projects involving identification of in-ground materials. Site testing is inaccurate and expensive, so it is rarely possible to know exactly how much of different materials will be excavated, even though it may be possible to estimate accurately the total volume of excavation. The mass haul plan and corresponding schedule are heavily dependent on actual quantities, therefore sensitivity analysis should be undertaken by using different quantities in areas of high uncertainty and risk.

In earlier chapters, the role of buffers has been emphasised in mitigating risk. In linear projects, buffers can similarly be used when subcontractors have direct schedule dependencies in relation to each other. However, it is more usual that subcontractors have tight dependencies caused by mass haul and must work with balanced production rates. In such cases, an ordinary schedule buffer cannot be used. Stockpiles take their place as buffers, effectively buffering the source and haul resources from any production rate problem at the destination. Stockpiles can be used to reduce the interdependency between two contractors—the dependency changes from having to work at exactly the same pace to having to cut before filling. However, stockpiling creates an additional direct cost related to receiving and loading materials at the stockpile, as well as occupying physical space. Therefore, stockpiling presents a similar dilemma in linear projects as buffers do in commercial building construction—they increase direct costs but decrease the risk of waiting time.

Procurement and design schedule

The procurement and design schedules for linear projects can be handled with similar pull techniques as described in Chapter 6. The only difference is that many procurement packages are location-based instead of trade-based.

In the design schedule, design which results in better information about quantities is critical. For example, soil reinforcement design reveals the quantity of rock required for soil replacement. Disposal area design shows how much material is needed to prevent disposed material from spilling over. Other important considerations include the last day when the designer needs to start work to finish prior to work being scheduled to start. It is often beneficial to start the final detailed design of some areas as late as possible, to be able to change the structures to accommodate possible late changes in the mass balance—for example, by changing noise protection to accommodate a larger quantity of soil cutting than anticipated.

Linear project cycle planning

Tunnelling projects are examples of linear projects where cycle planning methods can be applied. In principle, tunnelling is similar to in situ concrete structures in commercial construction: finishing a set of tightly interrelated work stages releases the next work area for construction. Because there are multiple subcontractors working in a cycle, continuity of resource use can only be achieved by having multiple ends from which to bore the tunnel. These ends can be maintenance tunnels, underground parking areas or other tunnels. Critical planning variables include the length of a work segment (analogous to size of pours in a structure), number of ends (analogous to number of pours) and size and number of crews to balance work. Similar methodologies to those described in Chapter 7 for cycle planning of buildings can be used. A special factor in linear cycle planning is that locations can be adjusted simply by changing the mandatory lag in metres between work phases—quantities do not need to be recalculated for different work segments.

MONITORING LINEAR PROJECTS

In earlier practice, most monitoring of linear project schedules has concentrated on getting information about the production rate (at the cutting end) and the haul distance. The actual use of hauled materials and the destination of hauls has not been tracked. However, all this information exists in the delivery dockets of haulage contractors. They have generally not been tracked systematically due to the large amount of data required. In a location-based linear scheduling system, this information can be utilised effectively and thus it becomes worth devoting additional effort to monitoring the actual mass hauls. If mass haul data is to be correctly monitored, the following process should be followed.

In the proposed process, delivery information is collected on Fridays, entered to the progress database on Monday morning and reports are distributed (at the latest) on Tuesday morning to all parties. By doing this, the information is on average only one week behind— much earlier than with most existing processes.

Delivery information typically has problems in allocating information correctly at interface points of structures. For example, the border between the respective masses of an interchange and a main road line is not possible to accurately identify on site. This may result in different actual quantities to the interchange and the main road line when compared to the original plan. Other challenges include interfaces between various structures within the same location. The mass haul contractor cannot identify where soil replacement ends and rock embankment starts. The same problem occurs when embankment and noise banks are made of the same material. These deviations from the plan need to be checked as soon as it becomes clear that actual quantities are not matching the planned quantities.

Accurate actual information enables useful reports to be generated for various needs. Examples of important reports for decision making during the construction phase are:

- Mass completion report
- Control chart
- Mass use compared to plans
- Average haul quantities by contractor and soil type
- Actual work in time-distance diagram
- Forecast work in time-distance diagram.

Mass completion report

The mass completion report (Table 14.2) shows the actual masses compared to the planned masses for each mass economy area. This information gives a good overview of progress of the project compared to the plan.

Mass use compared to plans

Actual mass use compared to the plan can also be shown by use of a mass haul diagram. In the diagram, the coloured parts of boxes (solid) indicate completion and white ones (empty) have not started. Boxes filled with cross-hatch are work sites which do not have a mass haul plan. Actual hauls can be shown together with the planned ones to detect deviations. Figure 14.8 shows a sample mass haul diagram with actual information.

Table 14.2 Mass completion report

Source	Subcontractor 1			Project		
	Planned	Actuals	Unfulfilled	Planned	Actuals	Unfulfilled
RockCut1 (m³-bcm)				45,720		45,720
RockCut3 (m³-bcm)	10,759		10,759	34,773		34,773
LandCut1 (m³-bcm)	59,414	11,000	48,414	137,631	11,000	126,631
LandCut3 (m³-bcm)				37,322		37,322

Destination	Subcontractor 1			Project		
	Planned	Actuals	Unfulfilled	Planned	Actuals	Unfulfilled
Crushing (t)						
Base (m³-ccm)	6,770		6,770	19,780		19,780
Landfill (m³-ccm)	39,113	8,000	31,113	53,788	8,000	45,788
Noise bank (m³-ccm)	7,911	3,000	4,911	110,173	3,000	107,173
Rock Disposal (m³-ccm)				26,038		26,038
Rockfill > 600 (m³-ccm)				35,708		35,708
Slopes (m³-ccm)	9,305		9,305	29,359		29,359
Sub-base (m³-ccm)	4,993		4,993	15,035		15,035

Linear control chart

In the linear schedule adaptation of a control chart, location on the line is shown on the horizontal axis and work types are shown on the vertical axis. This works well in linear projects because usually there are just between 10 and 20 tasks. Having distance on the horizontal axis is useful because the control chart can be shown together with other information using distance—such as time-chainage diagram, mass haul diagram or vertical section of road line. Similar conventions may be used to other building construction: green work sites are completed, blue ones are in schedule, yellow ones have started but are delayed and red ones are late and not started. Blank (white) work sites are planned to start later. Figure 14.9 shows a sample linear project control chart.

Haul quantities by contractor and soil type

It is useful to list hauls, by both the contractor and soil type, to see if the subcontractors are following the mass haul plan. Even small deviations in the mass haul plan can be costly and immediate reaction is required if there are deviations.

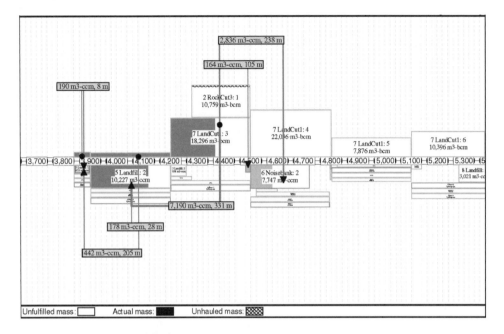

Figure 14.8 A sample mass haul diagram with actual data

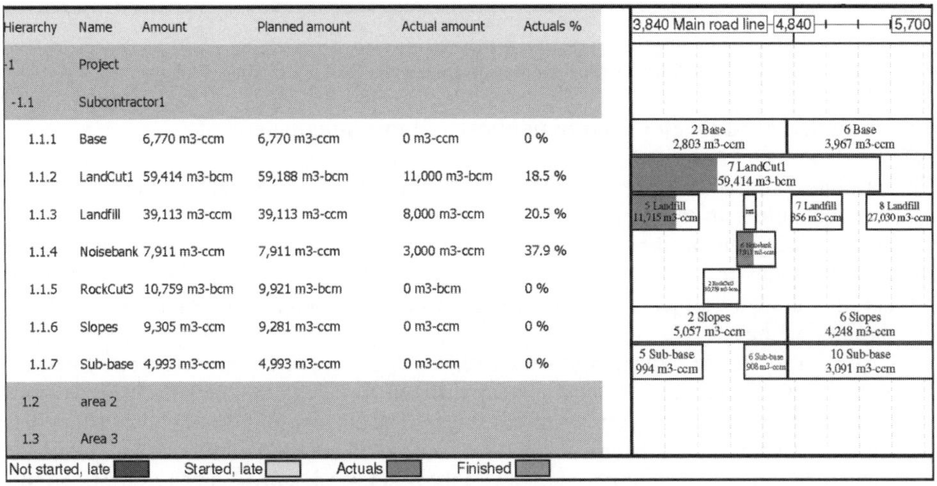

Hierarchy	Name	Amount	Planned amount	Actual amount	Actuals %
-1	Project				
-1.1	Subcontractor1				
1.1.1	Base	6,770 m3-ccm	6,770 m3-ccm	0 m3-ccm	0 %
1.1.2	LandCut1	59,414 m3-bcm	59,188 m3-bcm	11,000 m3-bcm	18.5 %
1.1.3	Landfill	39,113 m3-ccm	39,113 m3-ccm	8,000 m3-ccm	20.5 %
1.1.4	Noisebank	7,911 m3-ccm	7,911 m3-ccm	3,000 m3-ccm	37.9 %
1.1.5	RockCut3	10,759 m3-bcm	9,921 m3-bcm	0 m3-bcm	0 %
1.1.6	Slopes	9,305 m3-ccm	9,281 m3-ccm	0 m3-ccm	0 %
1.1.7	Sub-base	4,993 m3-ccm	4,993 m3-ccm	0 m3-ccm	0 %
1.2	area 2				
1.3	Area 3				

Not started, late ▓ Started, late ☐ Actuals ▒ Finished ☐

Figure 14.9 A sample project control chart

Actual work shown on the time-distance diagram

Actual work can be plotted on to a time-distance diagram, in the same way commercial projects use dotted actual lines in the flowline diagrams (see Chapter 8). In mass haul projects, the corresponding cuts and fill lines are always connected because a delay in the cut will delay the fill. Figure 14.10, for example, shows a typical time-distance diagram with actual information displayed. The complex relationship between cuts and fills and progress makes forecasting of progress difficult.

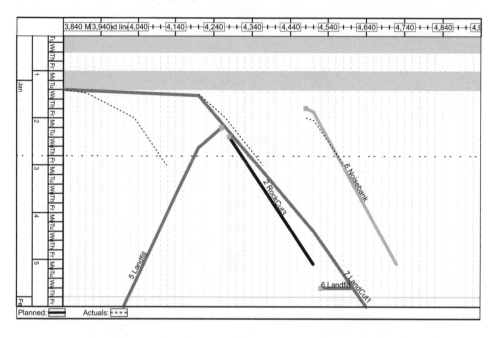

Figure 14.10 A typical time-distance diagram with actual information displayed

Updating the schedule and mass haul plans and micromanagement

The baseline comparison systems which apply to controlling the model for linear mass haul projects are similar to those which apply for commercial construction (Chapter 8). The current detailed plans need to be compared against the original baseline plans in order to raise alarms about any deviations and to prevent problems from shifting to the end of the project. In mass haul projects, the situation is made more complicated by the necessity of also baselining the mass haul plan.

The four stages of information are the same: baseline, current, progress and forecast. For the mass haul plan, the forecast is very difficult to calculate—thus it currently only works on the baseline, current and progress stages. The current mass haul plan is generated by removing all masses that have already been hauled and creating new mass haul plans for the remainder. This can then be compared with the baseline plan to detect any variations and to forecast costs for the remainder of the project. Any control actions related to the mass haul are modelled against the current mass haul plan directly.

More detail is sometimes added to a schedule during implementation. For example, a rock blasting operation could be divided into drilling and blasting tasks to be able to track the work more effectively. Drainage, bridges and soil reinforcement are often planned at a higher level of detail. The principles are the same as with commercial construction, except that some of these work stages result in material requirements or material supply which will affect the current mass haul plan.

Controlling production rate by use of forecasts

Schedule forecasts can be calculated based on actual production rates just as they can for commercial construction projects. However, the interdependencies of work sites caused by mass haul need to be taken into account. It would be computationally very expensive to forecast mass hauls. Therefore, forecasting can only be done based on actual production rates and then the forecasting rules from Chapter 8 can be used without modification.

Control actions in linear projects

Control actions in linear projects can be divided to three categories:

- Control actions related to resources
- Control actions related to mass haul
- Control actions related to design.

Control actions related to resources

Control actions related to resources are the same as for commercial construction: adding or decreasing production by changing resources (production rate) or the hours worked. These control actions can be modelled to directly affect the forecast by adjusting its slope. This only works for cuts, because embankments are forecast based on the progress of cuts which supply materials to them. The rock cut in the previous example might be accelerated by working longer hours and on weekends. This would also affect the slope of the embankment line.

Control actions related to mass haul

Control actions related to mass haul adjust the mass haul plan to change the schedule. The effects of delays on the schedule, or ways to better optimise mass hauls for the remaining masses can be modelled using control actions. These are modelled directly on the current mass haul plan. This mass haul plan is then used in forecasting schedule progress. In the previous example, another supply of material could be found for the embankment. This would remove the link from rock cutting to the embankment.

Control actions related to design

In design and construct contracts, the general contractor can control the design within specific design parameters set by the client. The design schedule often overlaps the construction schedule, so it is possible to change the design as a control action in response to deviations on site. These control actions affect the quantity and quality of masses directly and can be used to control the current mass haul plan and schedule. Examples of available control actions include changing soil reinforcement methods, replacing noise banks with sound barriers, changing the vertical or horizontal alignment or structure of the road. Risks can also be diminished by concentrating design on areas with the biggest uncertainty and impact on the mass haul plan and schedule.

VISUALISATION

One of the distinctive features of linear projects is that, while they are extremely repetitive, it remains important to know where in the project work is being undertaken. For this reason it is very helpful to be able to visualise both mass haul and the construction schedule in 2D or 3D.

Figure 14.11 illustrates progress of mass haul superimposed onto a map image of a road project. As the project progresses, the location work progress can be easily visualised, with the work in the right image showing work started inside the ramps.

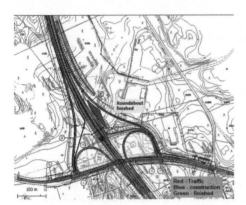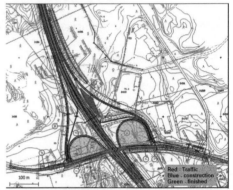

Figure 14.11 Two sample map-views showing project progress in DynaRoad 5.

Figure 14.12 provides another example of the way visualisation can assist in managing complex projects like a freeway interchange. In this example, it is possible to track the very detailed progression of individual scheduled works, not just the mass haul.

SUBCONTRACT AGREEMENTS

To be able to use the methods described above, subcontract agreements need to motivate the subcontractor to behave in the best interests of the project. Each contractor will then maximise their profit. Subcontract agreements should be created so that subcontractor and project goals are similar. As an example of a poor subcontract agreement, consider where a subcontractor is paid based on haul distances without an adequate mass haul plan. In this case, it is in the subcontractor's interest to maximise haul distances and stockpile as much as possible. Obviously this is sub-optimal from the point of view of production system cost and the general contractor will suffer.

DISCUSSION

There is a great deal that can be said about the planning and scheduling of linear projects, and this book cannot do this topic justice. This chapter is just a taste of the power of location-based management of this type of project. Tools such as DynaRoad allow rapid and efficient planning and control of complex mass haul and construction work of linear projects, while more familiar location-based tools, such as Vico Software's Control can also be used to schedule projects without mass haul consideration.

Linear scheduling techniques, which are in fact location-based methods, have a wide acceptance in current practice. Therefore the issues discussed here will be reasonably familiar to many. However the practical application of the LBMS to linear projects can bring better planning and control to linear projects, reducing both cost and risk.

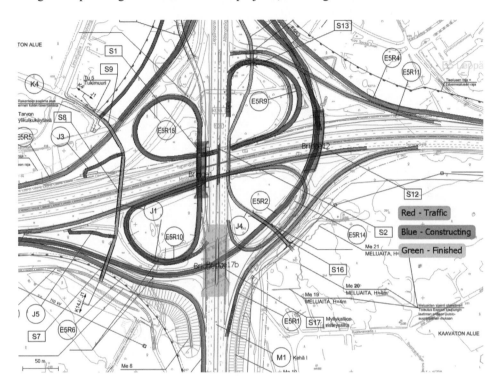

Figure 14.12 Another sample map-view showing progress of a complex interchange in DynaRoad 5.

REFERENCES

Mäkinen, Juuso (2007), *Pääurakoitsijan massatalouden hallinta suunnittelua sisältävissä hankkeissa* (Cut and fill management by main contractor in design build projects). Master's thesis submitted in fulfilment of the requirements of the degree of MSc (Civil engineering) at Helsinki University of Technology, Helsinki, Finland.

Section Five

Case studies

SECTION FIVE—CASE STUDIES

Section Five contains case studies of projects which have been instrumental in the development of the location-based management system in use. The first two case studies were selected because they demonstrate the integration of the LBMS with BIM. The multiple small case studies have been selected because they highlight learning stages of the LBMS.

- Chapter 15 presents case study 1: Opus Business Park—Stage 3.
  Opus Business Park is a 14,500 m^2 office building in eastern Helsinki. The general contractor of the project, NCC Construction Ltd, is one of the largest Nordic construction firms. The Finnish subsidiary of the company uses LBMS in all its projects. Opus was its first case study for the location-based controlling system and also incorporated location-based contracts. The LBMS provided good results in this project. In this project, the baseline schedule was reliable to within two weeks, resulting in less re-planning and re-negotiation of dates. Benefits included handing the building over to the client two months earlier than planned. The project showed that subcontractors and other stakeholders of the project can be taught to understand flowline and control diagrams.
- Chapter 16 presents case study 2: St Joseph's NE Tower Addition.
  The St Joseph Northeast Tower addition is a 9,476 M^2 addition to an existing hospital in Eureka, California. St Joseph Health System is an innovative hospital owner that has been at the front line of implementing 3D model-based trade coordination and change management processes. This project was selected as its first pilot project for implementing the location-based management system using a 3D model to generate the quantities. This case study presents a typical way to start using location-based management for projects which have already commenced using CPM methods. As the first step, the CPM schedule is analysed to find improvement opportunities. Second, the schedule is optimised within contractual constraints. The third step is to start tracking production and to implement the location-based controlling system.
- Chapter 17 presents multiple case study components.
  All of these small cases have been influential in the development of the location-based planning system in some way. The projects include:
  - Kamppi Centre, Helsinki: the largest city-centre project in the history of Finland. There were three underground bus stations, an underground parking hall and underground retail areas and restaurants. Above ground, the centre had 35,000 m^2 of retail space, 99 apartments and 12,000 m^2 of office space.
  - Camino Medical Center is a US$100 million medical office building with surgery centre and urgent care clinic. The total area of the project was 22,500 m^2.
  - The Form 302 residential development consisted of four buildings on and around a podium with two levels of underground parking. The buildings consisted of a fourteen level tower, an eight level building, and two four level walk-ups.
  - The Parramatta office project was a simple office building refurbishment consisting of eight floors to be gutted and refurbished.
  - Mission Hospital is a health care project from St Joseph Health Systems in southern California. The New Acute Care Tower is an addition to the existing hospital on the campus. It is four stories over a basement for a total of 8,750 m2.
  - Skanssi retail centre is planned for 96 small or medium retailers and a supermarket. With a total of 128,000 m^2 it has retail spaces on two floors and 2,400 car spaces including two parking structures and parking on the roof.
  - The final section presents three case studies from Seppänen's PhD research into how location-based planning and controlling tools are implemented on site and how those tools and processes can be improved to achieve better results in production control.

Chapter 15

Case study 1: Opus Business Park

DESCRIPTION OF THE PROJECT

Opus Business Park is a 14,500 m^2 office building in eastern Helsinki. The general contractor of the project, NCC Construction Ltd, is one of the largest Nordic construction firms. The Finnish subsidiary of the company uses LBMS in all its projects. Before this case study, the emphasis had been on implementing the planning components of the system. Opus was its first case study for the location-based controlling system and also incorporated location-based contracts. The developer of the project was part of NCC Construction, and thus an internal client. This meant that time to market was important—earlier completion would provide earlier rental income. A normal contract was signed between the construction and the property development organisations. It was a target price contract with a guaranteed maximum price.

Figure 15.1 Opus 3: Opus Business Park—composite image

Opus 3, commenced in 2004, is the second part of a development of three office buildings. The first part, Opus 2 had already been completed in 2002. The Opus 3 project comprised two sections, arranged in a 'V' and which could be built independently of each other (a section is shown in Figure 15.1), and a parking hall below the main building. The structure was precast concrete. Identical office sections were connected by a diagonal corridor area with a glass façade (Figure 15.2). This diagonal area was built together with the second section. Both office sections had six floors. There were mechanical rooms on the roof of each section. Because the offices had not yet been rented at the start of the project and because companies have varying space requirements, the fit-out design incorporated flexible walls which could be moved within certain tolerances. The ground floor had special spaces, such as an auditorium, a dentists clinic with special requirements, and a lobby. Other floors were more or less repetitive construction although floor area reduced on higher floors and the top floor had a sauna section (almost a mandatory requirement in Finland).

Figure 15.2 Opus 3: Opus Business Park—joining corridor façade

AVAILABLE STARTING DATA

NCC Construction Ltd was the first company to implement location-based management systems widely in its organisation. For two years before this project it had been familiar with the required starting data. In this project, accurate location-based quantities, based on the project's location breakdown structure, were available very early on to support the planning process. Quantities were available for both the construction and MEP trades. However, MEP quantities were rough estimates based on project characteristics and size. NCC also has an internal database of productivity rates for both directly employed and subcontractor labour which could be used for scheduling purposes. Consumption rates are collected from all projects and compared to a Finnish productivity database which uses information provided by all major construction companies in Finland. The main principle of scheduling is that subcontracted work should be planned as well as if it were done with direct labour because otherwise effective control becomes much more difficult. Also, it is possible to evaluate when work should be done using subcontractors and alternatively to assess when using direct labour might be more efficient.

There were a few external scheduling constraints. The parking hall below Section 1 had to be finished by the end of October 2004. The area under Section 2 was used as a temporary parking area so excavation of that area could not begin until the parking hall was completed. This constraint meant that the highest level of the LBS had to be sequenced such that the parking area was built first, then Section 1 and finally Section 2 once the parking area was commissioned.

The total project duration, based on a previously built project of the same type and size, was originally planned to be 21 months. The start date was May 17th 2004, and this would have resulted in an end date of February 2006. This date was used as the basis of the contract between the construction and the property development organisations. There was no other time information available before planning the first location-based schedule.

Methods of LBMS that were used in this project included:

- All components of basic location-based planning (Chapter 5)
- All components of basic location-based controlling (Chapter 8)
- Procurement planning (Chapters 6 and 9)
- Production system risk (Chapters 6 and 9).

Not all methods described in this book were available in 2004. For example, logic Layers 4 and 5 logic and location-based definitions of float had not been invented and production system risk methods were only in the early stages of development. Many of the methodologies in Chapters 7 and 10 were developed based on this, and many other, case studies—so the information was not available at the time to help planning. Therefore, the case study describes a critical stage in the development of these methods.

SCHEDULING PROCESS

Contracts with subcontractors had not been established when scheduling started. The baseline schedule was used in the call for tenders to specify the start and finish dates, the required production rates and intermediate milestones. The input from subcontractors was used to plan detail tasks during implementation.

Location breakdown structure

During the pre-planning phase, there was not enough detail available about the actual spaces on each floor because the design had movable walls and the quantities were only approximations. Therefore, sections were used as the highest level hierarchy and floors as the lowest level hierarchy.

In hindsight, it would have been better to divide the floors to at least three or four areas or zones for both finishes and MEP trades—for example, east offices, west offices and corridors and restrooms). The chosen locations were quite large, so the used finish-to-start links caused some implicit buffers. In reality, many trades were able to work in the same location at the same time without losing productivity. Planning a more detailed LBS would have made the schedule more accurate and control more effective. However, this information could also have been added during the detail planning phase (during construction).

Tasks, resources and quantities

There was good starting data, therefore all tasks had quantity and productivity information. This was used to calculate the initial durations for each task in each location. Multiple quantity items were included in the same task if they could be done by the same subcontractor and had the same external dependencies.

For example, all tasks related to structure were lumped to the same task. This was good enough for modelling the structure because it was precast concrete and all major work stages (except the small in-place poured concrete strips) were done by the same crew, which completely finished one floor before moving to the next floor. This gave a realistic estimate of duration for structure and all finishes tasks could be tied to it with Layer 1 links.

All tasks related to the roof were similarly lumped together. During production, this was revealed to be a mistake, because it consisted of the work of multiple subcontractors and only a part of work had dependencies to finishes and other work. For example, to be safe to install plasterboard walls, the roof should be waterproof. Before starting the finishing of external walls, eaves also needed to be completed. Linking all roof work to beginning of plasterboard with a Layer 2 dependency created an implicit buffer which, although it was a mistake, proved to be beneficial in production.

MEP tasks were planned together with NCC's MEP specialists. Tasks were integrated into construction work but quantities and durations were guesses and estimates during the pre-planning phase due to a lack of suitable data.

Tasks were often lumped together to slow down overly fast tasks. The information about the scope of the subcontract was thus included in the quantities of the task. This information was used in the call for tenders to ensure that production could proceed according to plan.

Schedule optimisation

Overall sequence

The overall sequence could not be optimised because of the parking area constraint. This was an unfortunate constraint because otherwise it would have been possible to shorten the duration without increasing risk by building the larger section first. The smaller section had a bomb shelter below it (a mandatory requirement in Finnish buildings) which delayed the start of erecting the structure. The duration could have been decreased by two months without increasing any resources if it had only been possible to change the sequence and build the other section first.

Splitting versus continuous work

Two alternatives were explored. The first had almost totally continuous work through the buildings. The second had structure going continuously but all finishes trades had a break of about one month between the two sections. Figures 15.3 and 15.4 show the difference. The names of only some of the tasks are shown in the figures to maintain clarity. By splitting the work, it was possible to achieve a two month reduction of the total duration. There are also other differences between the versions. The first version assumes the same resources in both sections. Because the second section is larger, this results in a change of slope in tasks for the second section. In the second version, more resources are used in the second section. Also, the logic was re-examined for the second version and unnecessary links were removed. On the other hand, the production rate of structure of the first version was deemed too high in Finnish winter conditions and was corrected for the second version. Altogether, the second version was less risky for the structure but more risky for finishes. However, because there were large benefits related to cutting the duration by two months, the second option was ultimately selected. The idea was to control the risk by taking the work break and the increased resources of the second section into account in the call for tenders and in subcontract negotiations.

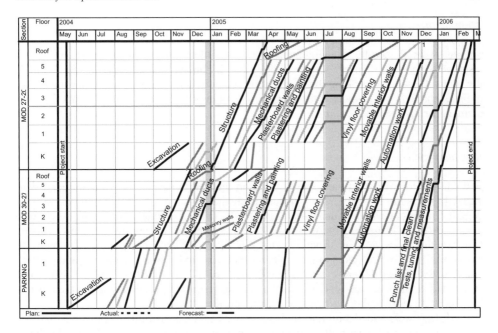

Figure 15.3 Scheduling with a continuous construction sequence without splitting

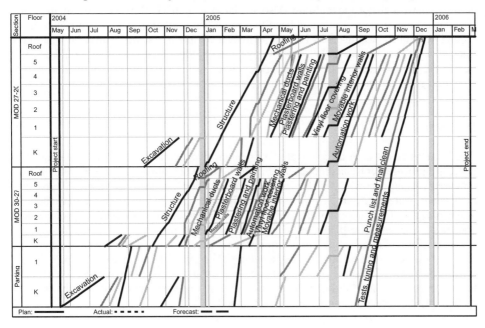

Figure 15.4 Scheduling with a split construction sequence

Aligning of work

All work was aligned by selecting appropriate multiples for optimal crews. Therefore, all tasks were planned to proceed with their optimum rhythm without arbitrary manipulation of durations. Generally speaking the first section required less resources and, after the break, the subcontractor was expected to add more crews in order to achieve the production rate planned for the second section.

Risk analysis

A crude version of the risk analysis simulation incorporating just duration and start date variability was run to analyse the optimised schedule. As a result of this analysis, buffers were inserted between those tasks found most risky in the schedule. For example, there was a big uncertainty factor with the structure which was to be erected during the Finnish winter. Cold weather stops work on structure. This was taken into account by allowing more time before finishes work started. For example, buffers were added between roofing and plasterboard walls. Other big risk factors were the (specialised) system interior walls and suspended ceilings because end users were not known during pre-planning and there was a big risk of quantity and design changes during project.

Buffers were added until risk analysis showed a high probability of finishing on time, two months ahead of the original contract. Major risk of interference was identified to be in the structural and roofing phase, installation of air handling units in mechanical rooms and automation work inside the building. Also, the latter section's buffers were too small during the finishes phase, so it was recognised that much better control would be needed in the second section. It was decided that performance of subcontractors would be accurately tracked in the first section so that any resource and production rate problems could be fixed before the critical second section. The risk related to the structure was mitigated by the structural subcontractor agreeing to work longer hours on days of good weather.

Procurement

The location-based plan affected procurement in two main ways. First, it was used to develop a procurement schedule for procurement-related activities. The project had its own procurement engineer who gave his input after production had been optimised. All procurement events were achievable by the latest start indicated by the pull procurement schedule. Therefore, the production plan did not need to be changed to accommodate procurement constraints. Second, the plan was used as starting data for invitations for tender. Each subcontractor had at least one interim milestone—usually this was tied to finishing of the first section according to the master schedule. The general contractor also promised a starting week for subcontractors. Any change in the starting week would cause a renegotiation of milestones and the finish date. Therefore, it was very important that the original master schedule could be implemented as planned.

CONTROLLING PROCESS

Updating of baseline

The baseline schedule was updated only once during the project. This update was only done to correct the MEP schedule after feedback from MEP subcontractors during the early stage of the project and before many finishes subcontractors were chosen, so it did not affect procurement. Disregarding this one update, the original baseline schedule could be used for comparing progress and current plans. This baseline schedule stayed accurate to within two weeks throughout the entire project.

Use of detail tasks

To decrease the workload during the finishes stage, planning of the detail tasks had already started during the foundation and structure phase. Every week, one baseline schedule task with enough available information was exploded into initial detail tasks by the project engineer. These detail schedules were then evaluated by the subcontractor responsible for the work. Detail schedules were approved by both the GC and the subcontractors. Significantly, a detail schedule could not be approved until all prerequisite work had detail schedules approved. Thus their was step-wise progression in the planning.

Detail schedules were updated when there was a change of circumstances. For example, a change of quantities, a procurement deviation or unforeseen surprises could trigger a detail schedule updating process. Detail schedules were updated weekly, so that the next few weeks were always accurate.

Getting progress data

Progress data was gathered weekly from the site. A site tour through all locations took two hours. Actual progress in each location was manually marked on the control charts and then entered into the computer system after the tour. A computer software package[1] generated the forecasts and visualised them together with the plan and actual data on flowline diagrams. This information was used to identify production problems. Problems were discussed in site meetings and control actions were planned to get production back on track.

Subcontractor communication

Every week the subcontractors were given a flowline diagram and a control chart of last week's progress. Sample site charts in use are shown in Figure 15.5. Any future production problems were indicated. If the flowline forecast for a subcontractor shifted because a predecessor was too slow, the resultant peer pressure was effective in getting production back on track. Forecasts were also used to warn subcontractors about upcoming penalties if milestones or the contract end date was threatened. Peer pressure, visualising bad performance and showing future problems were very effective controlling tools.

It is interesting to note that in this project, Gantt charts were generally not used in communicating with subcontractors. Everyone related to the project had to learn to

---

[1] The software was DynaProject from DSS—an early forerunner of Vico Software's Control software.

understand flowline diagrams and control charts. After a few weeks of complaints, everyone agreed that this was actually a much better way of working!

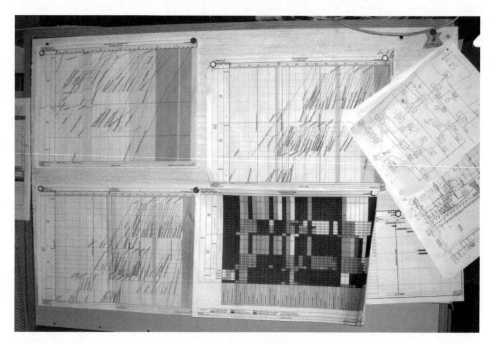

Figure 15.5 Site schedules and control charts

Client communication

Schedule problems, highlighted by forecasts and alarms, were communicated to the client weekly. The overall planned degree of completion of baseline was also compared weekly to the total degree of completion. The planned degree of completion was calculated from the baseline task by calculating how many worker hours should have been completed at the control date. The total degree of completion was calculated from progress data—how many current worker hours were completed. This calculation method took into account quantity changes. After a delay in structure of the first section, the project was on average behind the master schedule by 2–6%. However, because of buffers, the delays did not affect the finish date although they did cause some problems with work continuity for some trades.

This approach was different to the usual way of communicating progress, where a project was always said to be on schedule even though there were delays. The more honest approach was well received by the client and also prompted real planning for recovering the delay instead of pushing the problem into the future. After the case study, NCC has won many competitive bids through more transparent client reporting, so it was able to turn this finding into a competitive advantage.

Examples

This section describes some specific examples. The example tasks have been selected because of their criticality during the risk analysis or because they actually encountered production problems.

Structure

In the baseline, structure was represented by one continuous line, with a planned holiday break at Christmas. The quantities were larger in the second section, which would have caused a change of slope. The production rate was planned to increase in the second section to accommodate the difference. The baseline, and immediate predecessor and successor flowlines, are shown in Figure 15.6. Predecessor progress data as available just before the detail schedule of structure was planned is also shown. All predecessors have finished on time and there is one week of unused buffer between the tasks.

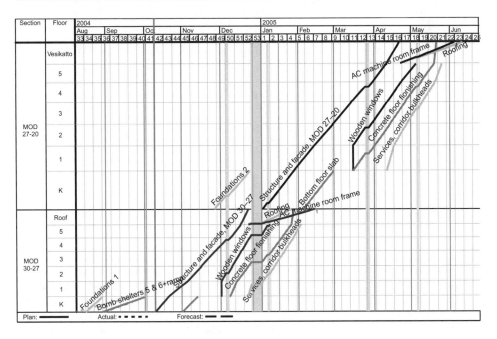

Figure 15.6 Baseline schedule for structural phase

In practice, it is often difficult for a trade to start earlier when a preceding buffer is not used because the subcontractor has been promised a start date and has reserved resources and equipment to mobilise on that date, and in this case the concrete precast factory has delivery dates set according to the original baseline. Therefore, the unused buffer was wasted in this example. Optimally, the additional time should have been used to ensure that all prerequisites were available for production. However, it turned out later that design was partially incomplete, which caused production problems on the first two floors. Figure 15.7 shows the original detail plan for the structure of the first section. Detail tasks for the second section were planned later, after observing the actual production rates from the first section. Detail tasks included *columns, beams and walls, slabs, joint reinforcement and concreting, façade elements* and *cast-in-place parts*. The tasks *columns, beams and walls, slabs* and *façade elements* used the element installation crew. Tasks *joint reinforcement and concreting* and *cast-in-place parts* had different resources which only came on site as required. The detail schedule had accurate element quantities in each location, whereas the baseline had estimates. In cold months, one day of bad weather a week was anticipated by planning dummy holidays (horizontal line segments in the figure on Fridays). This ensured that the production rate on good weather days was high enough. In the current schedule, there were no buffers. They were not needed between sub-tasks done by the same crew and for the other tasks the contractor had other work available in finishing the parking hall (less critical workable backlog) and could flexibly apply resources as required.

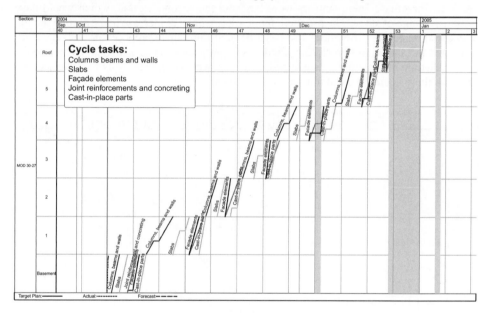

Figure 15.7 Original planned detail tasks for the structure of the first section

The first three weeks of the project suffered production problems. There were work breaks because of incomplete design. This was not noticed early enough because the subcontractor foreman was on vacation. The lack of supervision resulted in the production rate being too low. However, because of the design errors, the general contractor had to take responsibility for the delay. Therefore, the detail schedule could be changed. Figure 15.8 shows the production problems and the updated detail task schedule. New information had become available about the structure for the main mechanical room, which forced a sequence change for the roof level.

The production problems were corrected after a few floors. Subsequently, the production rates were as originally expected. However, the delay in the early floors could not be caught up. Therefore, when planning the detail tasks of the critical second section, a decision was made to start the ground floor using a mobile crane and a new crew and switch to use the tower crane and the original crew when the first section was finished. This allowed most of the delay to be caught up before end of the second section.

In the end, structure was delayed one week from the original baseline. Because of the buffers, finishes of the second section were not affected. In the first section, there was a delay to the roofing task which eventually made a change necessary to the detail tasks of plasterboard walls. In this case, the baseline schedule was detailed enough. The planned production rate could be achieved consistently. However, the prerequisite screening process was not adequately handled by the general contractor, believing it to be the subcontractor's problem. If this had been done by location, production problems could have been avoided.

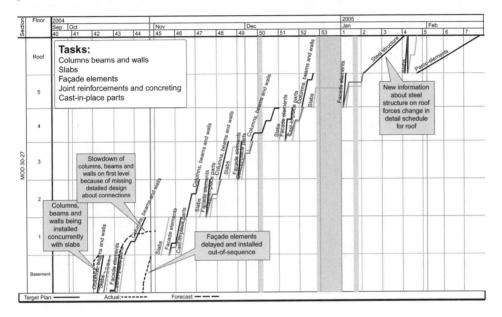

Figure 15.8 Progress delays and updated detail task schedule for the structure of the first section

Roofing

In the baseline schedule, roofing consisted of just one flowline even though many subcontractors work on the roof and different roofing stages have different external dependencies. The link to finishes was solved by having a start-to-start Layer 2 link at the section level of accuracy. The baseline ignored the fact that the roof was actually on many levels—for example, a roofing zone was on Floor 4 because the floor area decreased when going up the building. Because structure of the first section was delayed, detail tasks were examined very carefully to find opportunities for getting back to the original schedule.

Roof was planned for one section at a time so that the second section could benefit from learning of the first section. The LBS branch of the roof was subdivided into seven roof zones. It consisted of nine detail tasks done by four subcontractors. Steel structure of air conditioning machine room was done by one subcontractor. Masonry related to air conditioning machine room was done by another subcontractor. Initial pours on the roof were done by the same subcontractor that poured the finishes concrete inside. All other work was done by a roofing subcontractor. The detail schedule tried to maximise work continuity for subcontractors. For the steel structure and masonry subcontractors this meant a continuous line. Eaves were done by NCC's directly employed carpenters, who had work also inside the building. For the concrete pouring contractor, this meant trying to find dates on which both inside and roof pours could be done on the same day. For the roofing contractor, the sequence of sub-tasks did not matter because they were completed by the same staff. Therefore sub-tasks could be done in any sequence, as long as just one task was being worked at a time. Figure 15.9 shows the initial detail schedule for roofing. Black lines are for contractors trying to achieve continuous work. Concrete pouring activities are light grey and roofing contractor activities are darker grey.

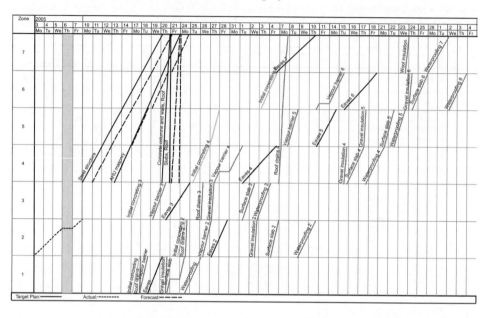

Figure 15.9 Initial detail schedule for roofing

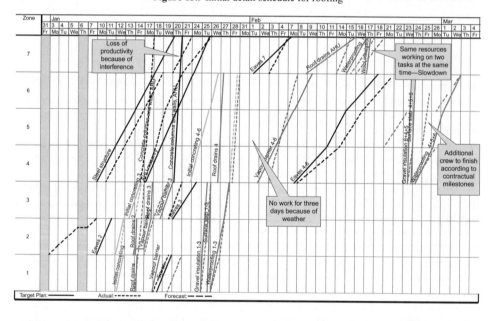

Figure 15.10 Revised final detail schedule for roofing with progress information

The roof schedule was updated during production to reflect actual commitments by subcontractors. The overall duration stayed the same but the sequence of sub-tasks was juggled around to improve productivity. This resulted in more continuous runs of similar work. When a weather delay event occurred, all detail tasks were updated to start on the first day of good weather. In the first schedule, there were dummy holidays each Friday to account for bad weather. Figure 15.10 shows the final task schedule with progress information. The actual finish date was one day earlier than in the original detail schedule. Also, the first three zones were waterproof well ahead of schedule. There were two periods when bad weather caused delays. The first period was from 21st to 24th of January. Only steel structure and AHU masonry were progressing on those days and they were slower than planned. Another period of bad weather was from 31st January to 2nd February when no work occurred on the roof. Resources were exactly as planned in the original schedule, except that in the final week another crew from the roofing contractor came to work on gravel insulation and waterproofing. This can be seen as two lines being worked on in parallel.

Compared to the original baseline, the roofing task did not succeed well. Roof work started two weeks later than planned and finished one month later because the subcontractor could not work efficiently with the resources required by the master plan. The master schedule also failed to take into account the hand-offs between different subcontractors working on the roof and did not have enough allowance to cover the bad weather that occurred on five days. The next example illustrates how the schedule was restored during succeeding detail tasks.

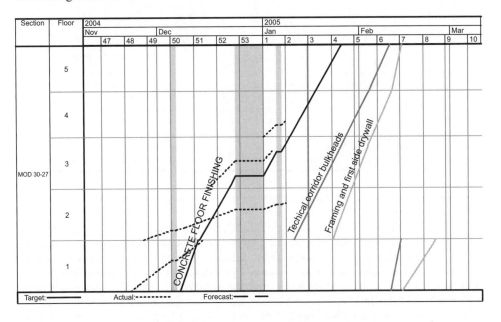

Figure 15.11 Planned detail schedule for plasterboard walls and corridor MEP, with progress information

Plasterboard walls and corridor MEP bulkheads

A subcontractor from Estonia was selected to do both plasterboard walls and corridor MEP bulkheads. At the time of planning detail schedules, it was known that roofing would be delayed. As a control action, frames and first board of plasterboard walls were planned to be built together with bulkheads. Installation of the second board waited for the roof to be finished. This decreased the risk of rework caused by wet conditions but allowed work to commence earlier. The detail schedule and actual data of preceding tasks are shown in Figure 15.11.

It turned out that the subcontractor was heavily involved in the Kamppi Centre, the largest construction project in Finland ongoing at the time. Therefore, there were constant resource problems because the larger project was given priority by the subcontractor. The same crew was also installing bulkheads and plasterboard walls. When the subcontractor was shown the production rate problem in one sub-task, all resources were merely transferred to that sub-task while the other sub-tasks were suspended. This was because additional resources were not available to cover all sub-tasks which caused delays in the first section. Figure 15.12 shows actual progress of the first section.

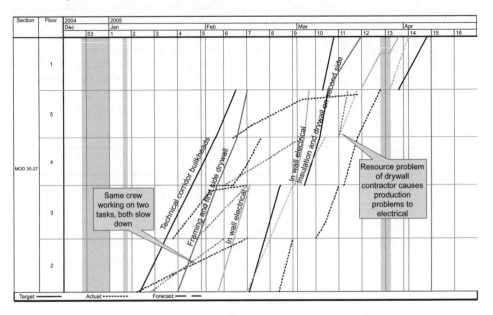

Figure 15.12 Actual progress for the detail schedule of plasterboard walls and corridor MEP—first section

Although the project team assumed that all problems would be somehow solved in the second section, the final outcome was that the subcontract finished three weeks late. The next trade was the plaster render and painting subcontractor. Interference between the plasterboard walls and plaster render was removed by rendering the exterior concrete walls first, thus removing the technical dependency to plasterboard walls. When the crew was finished with this task in the building, they could start again from the ground floor (Figure 15.13). This is a good example of how production continuity can be preserved for each subcontractor by evaluating different scheduling options during the detail scheduling phase. By the end of the plaster render and painting trade, the production was again following the master schedule.

DISCUSSION

Some interesting special issues came up during construction. Part of each floor was used as a storage area in the first section. This meant that floor covering work could not be finished on the floor of the first section—about 100 m^2 was always left unfinished. Instead of marking the location unfinished on each floor, the unfinished quantity was transferred to vinyl floor covering in the second section on the same floor. The floor covering crew would come back to finish the work when the second section was being covered. By doing this, the real production rate of floor covering could be estimated without sacrificing total accuracy of the system.

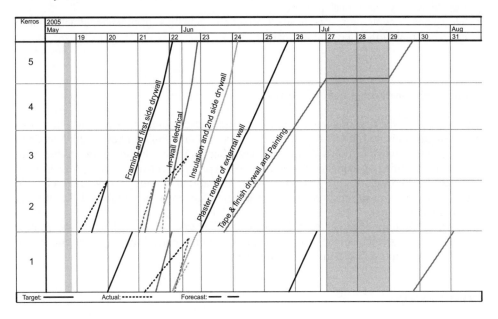

Figure 15.13 Drywall detail schedule to maintain continuity

Lessons learned

Despite the success of the LBMS system overall, there were many places where control could have been improved. The LBS was not considered suitable for the tasks final cleaning and punch-list work (rectification of faults), which caused the last month of the project to be managed with activity-based systems. Specifically, the LBS was flipped around so that finishing up was done by floor and not by section. This was caused by air handling units having different operation areas than the LBS. In this case, a new LBS branch for finishing up should have been created at the room level of detail. Detail tasks could have been planned for each room which had punch-list work and continuous work planned for each of the crews. Controlling this stage with activity-based systems at the floor and task level of detail led to unnecessary confusion and loss of productivity during the last month.

The original location breakdown structure was not sufficiently detailed. For controlling purposes, percentages of completion had to be used for MEP and finishes work. The LBS should have had at least one more hierarchy level below floors. This was actually done for the first floor during the detail planning phase but the other floors would also have benefited from a more detailed LBS.

The original baseline schedule had an inappropriate level of detail for many of the critical tasks.

- The level of detail in the original baseline was inappropriate for many critical tasks, including roofing work

- The chosen LBS was not relevant to final cleaning and punch-list work which caused the final month of the project to be managed with activity-based systems.
- A lack of adequate processes for prerequisite screening caused many of the production problems
- The procurement control failed for cooling slabs, because MEP procurement was still using old activity-based push processes
- The general contractor was hesitant to claim penalties from subcontractors who exceeded their milestone or finish dates.

Influence in the development of LBMS

This was one of the first projects where the location-based controlling system (Chapter 8) was fully implemented. It was one of the case studies in the PhD studies of Olli Seppänen and helped to define the proposed methodology. Many of the production problems were not accurately identified by the schedule forecast, which prompted changes in how forecasts are calculated and how alarms are generated. Also, the habit of updating the detailed schedule based on changes in production caused problems with implementation and the original commitment was lost.

These findings together with findings of other case studies, changed the control system to have greater emphasis on managing commitment, ultimately resulting in the three-level commitment system described in Chapter 10.

CONCLUSIONS

The project was handed over on time, two months before the original contract. The original baseline was updated only once and its dates were accurate until the end of project. This was achieved through control of both production rates and a detail task planning process where solutions were found for each work package to get close to the original baseline without causing interference to other trades. In most cases, the master schedule dates could be achieved within a week of accuracy. Subcontractor contracts did not need to be renegotiated even though dates were given in the contract instead of just contract durations.

In conclusion, the LBMS provided good results in this project. The project manager was positively surprised by the lack of production problems. In normal projects, schedules were typically a month behind the original baseline at some point and only crashed to completion during the final months of the project. In this project, the baseline schedule was reliable within two weeks. This resulted in less re-planning and renegotiation of dates with subcontractors. Concrete benefits included handing the building over to the client two months earlier than would have been likely using traditional processes.

The project showed that subcontractors and other stakeholders of the project can be taught to understand flowline and control diagrams. Gantt charts were not used for communication; everyone had to use flowline diagrams exclusively. Client reporting was done based on LBMS results.

Chapter 16

Case study 2: St Joseph's NE Tower addition

DESCRIPTION OF THE PROJECT

The project is the St Joseph Northeast Tower addition. This is a 9,476 M² addition to an existing hospital in Eureka, California. St Joseph Health System is an innovative hospital owner based in Orange County, California. It has been at the front line of implementing 3D model-based trade coordination and change management processes. This project was selected as its first pilot project for implementing the location-based management system (LBMS) using a 3D model to generate the quantities. Skanska (USA), who had no prior experience or knowledge of the LBMS, was the general contractor for the project.

The Northeast addition included:

- An emergency response (ER) unit (twenty rooms)
- An information technology (IT) area,
- A Diagnostic Imaging Department,
- A Surgery Department
- An intensive care unit (ICU) department (12-Beds)
- A 40-bed telemetry unit
- Shelled space for future expansion
- The renovation and expansion of 1,115 M² of existing space.

Figure 16.1 A 3D rendering of St Joseph's Hospital

AVAILABLE STARTING DATA

A 3D construction model was available and provided the necessary starting data (Figure 16.1). It is possible to quickly estimate quantities according to any decided location break-down structure from 3D construction models. Therefore, quantities could be calculated based on the location for any elements which were included in the 3D model. It was decided that quantities which related to construction trades would be taken from the model.

However, in this case the modelled MEP systems did not include sufficient detail and also the MEP design had changed. Thus, these quantities were not taken from the model. For MEP tasks, the durations were estimated and worker hours by location were requested from subcontractors.

The project used a CPM schedule as the contractual schedule. The location-based schedule was used for schedule analysis of the CPM schedule, to find opportunities for acceleration and for production control during implementation. The total project duration had been agreed based on the CPM schedule. Because many subcontracts had been purchased in accordance with the CPM schedule, there were constraints on available options for optimisation of the schedule. In the schedule analysis phase, the location-based schedule was an exact copy of the CPM schedule. During the optimisation phase, the main milestones were preserved (such as the start and finish dates of foundations) but continuity was optimised within each phase.

LBMS methods that were used in this project included:

- All components of basic location-based planning (Chapter 5)
- An initial implementation of the monitoring features of basic location-based controlling (Chapter 8).

SCHEDULE ANALYSIS

Schedule analysis was undertaken by entering the CPM schedule into the location-based format. This required the identification of many unique activities and locations. Because CPM schedules ordinarily have locations mixed in task names or in a work breakdown structure (WBS), this work is normally a manual process. Importing directly from CPM planning software to location-based planning software will not achieve the right outcome as locations are not handled in a consistent way between CPM software packages or their users. Figure 16.2 illustrates the resulting location-based schedule analysis and findings.

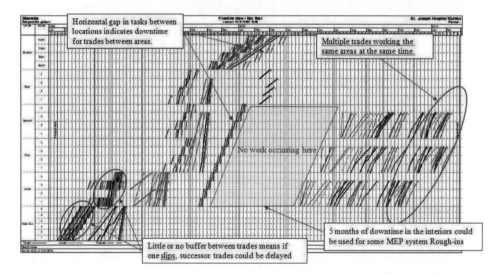

Figure 16.2 Original CPM schedule recast as a location-based flowline schedule

The CPM schedule planned in Primavera by Skanska was well planned and had a high level of detail. However it required many pages to display and there is no point reproducing it here. The schedule had over 1160 activities for the hospital addition alone. However, because activities are repeating over multiple locations, this translated into only 201 location-based tasks. This is more detail than normally seen in location-based schedules where the typical number of tasks is 100–150. Even though the CPM schedule was well planned and had deliberately considered resource flow for most tasks, it was possible to find improvement opportunities and conflicts in the schedule. The following opportunities were identified:

- Many tasks had been planned in the same location at the same time (location congestion).
- Discontinuous work was required for many trades.
- Empty locations, indicating opportunities for better synchronisation of the schedule.
- There were no buffers at the end of the project
- Durations were the same in all locations for any given interior trade, despite locations not being identical. This indicates that activity durations were not quantity driven.

Skanska used these schedule analysis findings to update its contractual schedule.

It should be noted that almost all CPM schedules converted to flowline look like Figure 16.2, because it is impossible to plan continuity using the standard CPM algorithm.

SCHEDULE OPTIMISATION

The next phase of implementation of the LBMS was to generate an optimised location-based schedule as a proof of concept. The goal of the location-based schedule was to identify opportunities for duration compression without increasing risk. The 3D model-based quantities were now used for construction trades. MEP trades were resource-loaded based on MEP contractor estimate of worker hours in each location. The durations of individual tasks and locations were based on productivity rates published in the RS Means database (US productivity rates) and the measured quantities. The total durations for each major construction phase were planned to be approximately the same as in the contractual CPM schedule in the first draft schedule. This was achieved by selecting crew sizes to match the CPM schedule milestones while maximising continuity. Then opportunities for schedule compression were investigated.

Location breakdown structure

The location breakdown structure was defined based on Skanska's CPM schedule. In this, there were three hierarchy levels: floors, construction phase and zones. Floors were divided into structural phase, exterior phase and interior finishes phase. There were four structural areas on each floor. The exterior phase was planned by elevation (north, south, east and west). Additional detail was added to finishes which originally had three areas on each floor. These were divided further to north and south, and MEP rooms were separated from corridors. Zones were communicated to participants by showing them clearly on the floor plans (Figure 16.3).

Figure 16.3 Zones used in the LBS shown on the floor plan

Tasks, resources and quantities

The same tasks were mostly used as in the schedule analysis, however with some changes. For example, the formwork of multiple foundation types was combined to one formwork task because all foundation types were to be done with a single crew. The same was done to the related rebar and concrete pouring tasks for foundation elements. This was necessary to show the correct task flow through the building.

The original CPM schedule did not show resource information. In the optimisation, the tasks were resource-loaded based on the RS Means productivity data for tasks with quantities. For MEP tasks, worker hours in each location were estimated by MEP contractors.

Schedule optimisation

Overall sequence

The overall sequence was not optimised. The sequence in the CPM schedule was used. Also, safety constraints meant it was not possible to start interior work before the whole of the structure had been completed and fireproofed.

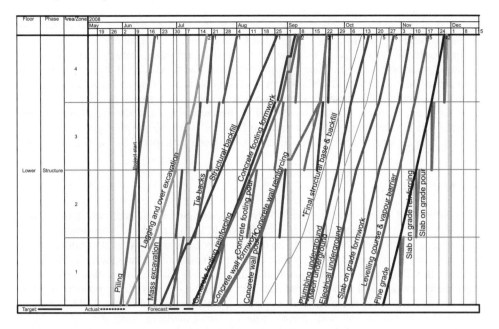

Figure 16.4 Foundations phase tasks

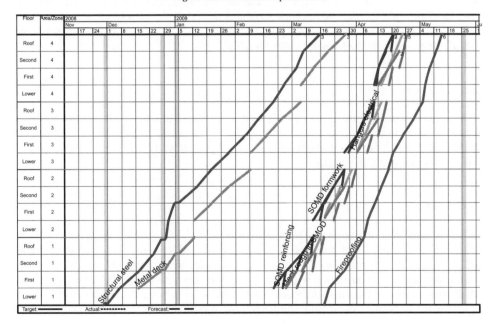

Figure 16.5 Structural phase tasks

Foundations phase

Foundation tasks were divided into four areas (Figure 16.4). Most of the tasks were planned to be performed continuously from area 1 to area 4. Most of the tasks had quantities provided by the model, however some changes to the level of detail were required. For example, formwork of footings was created to include all the formwork which would be undertaken by the same crew at the same time—pit slabs, pit walls, continuous footings, pad footings, grade beams and tie beams. In the original CPM schedule, all these activities had been separated, which would not allow production rate monitoring and would result in an implicit assumption that separate crews would be mobilised for each activity. Production rates for the tasks were aligned by changing the crew sizes and the crews were cross-checked with the site manager. Crew sizes matched well with the site manager expectations. In some cases, a larger crew was required than had been anticipated by the site manager.

Structural phase

The building had a steel structure. Erection would be completed one quadrant at a time from the bottom to the top. The slab metal decking would follow right after the structural steel but, because of safety reasons, the slab on the metal deck could only start after both adjacent quadrants had been erected (for example, quadrant 1 could start after quadrant 3 had been erected). In the schedule, this was handled by using Layer 5 logic. In addition to structural steel, metal decking and slab on metal deck, fireproofing was considered part of the structural scope. All tasks in the structural scope had quantities derived from the 3D model and structural steel pieces and total lineal footage were calculated from the model. Metal decking and slab on metal deck used the area of deck (formwork used deck edges). Fireproofing was measured by looking at the total surface of all steel elements. All production rates were aligned by changing the crew sizes and validating the crews with the superintendent. Figure 16.5 shows the structural phase schedule. Note that floors have been sorted so that all the floors of a quadrant are shown together.

Exterior phase

The exterior phase (Figure 16.6) was planned using elevations for the locations. Roofing was also added to the exterior scope. In this phase, it was not feasible to synchronise all stages because of a slow production rate for cement plaster walls and soffits. All skin types were quantified using the model.

Interior phase

Unsurprisingly, the interior phase (MEP, rough-in and finishes—Figure 16.7) presented the greatest opportunities for improvement. The location-based schedule was divided into more areas on each level than in the CPM schedule. All trades were optimised to have the same production rate and to have only one crew in any location at one time. Information for the MEP trades was received from subcontractors who were excited at the opportunity to be able to work without location congestion and task interruptions (starts and stops).

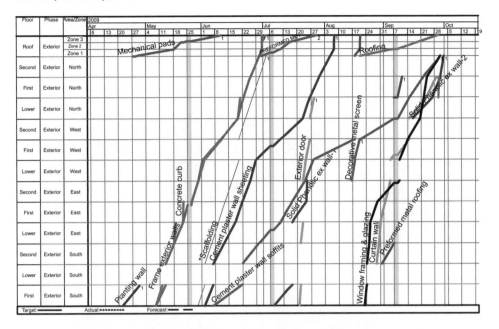

Figure 16.6 Exterior phase tasks

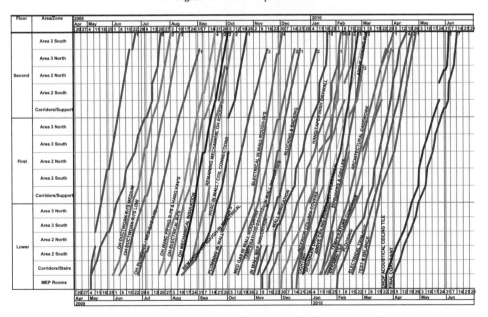

Figure 16.7 Interior phase tasks

4D SIMULATION

3D models had been used to generate quantities and locations, so the location-based schedule could be automatically linked back to the model using Vico Software's

Constructor, and to 5D Presenter to show a 4D simulation of the schedule. This was useful for making sure that all the logic related to the structure and exterior was correct. In these phases, floating objects indicate if the logic is right. Figures 16.8 to 16.10 show progressive snapshots of the 4D simulation.

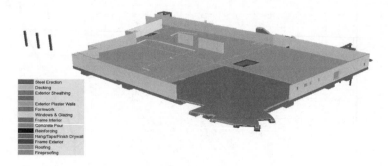

Figure 16.8 4D simulation sequence image 1

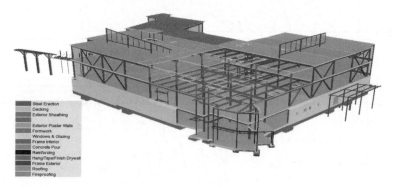

Figure 16.9 4D simulation sequence image 2

CONTROLLING PROCESS

The use of location-based controlling tools is still very much in early stages in this project because the decision to implement was made after the CPM baseline had been approved. At the time of writing, progress data is being collected weekly but information is not yet used in a systematic way.

Updating of baseline tasks

After completing the first optimised schedule, the findings were presented to the project's main subcontractors. They understood the potential of the system and re-estimated their tasks by calculating worker hours in each location. At the time of writing, this information is being updated to the schedule. Updates are also required because of delays in the foundation phase which requires planning an accelerated schedule or showing the total impact on completion time.

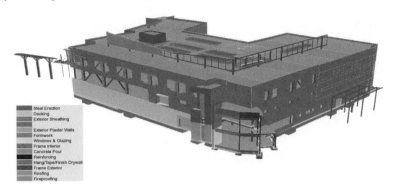

Figure 16.10 4D simulation sequence image 3

Use of detail tasks

Detail task planning has not been implemented at the moment. Presently, monitoring is going on by tracking production of baseline tasks.

Getting progress data

A project engineer gathers project data weekly from the site. The progress information is entered weekly to control charts in Vico Software's Control 2009 software package. This takes just a small amount of time each week.

Subcontractor communication

The project is still in the foundation stage and the location-based controlling system is running in parallel with the contractual Primavera schedule (CPM). Subcontractors have demonstrated an interest in optimising their work continuity and getting access to information but subcontractor communication based on location-based information has not yet started on the project.

Client communication

The implementation of the location-based management system was initiated by the hospital owner, St Joseph Health System, so there is a high level of interest from the owner to improve the client reporting based on the location-based management system. Monitoring data quality and monitoring frequency is improving in the project, so completion rates based on accurate quantities could be reported compared to plans. Look-ahead planning based on resources has not yet been implemented, so forecasts may not be reliable at the moment. The next step for implementation is to ensure the reliability of forecasts, so that the total effects of production deviations can be communicated and control actions can be planned.

DISCUSSION

This case study presents a typical way to start using location-based management for projects which have already commenced using CPM methods. As the first step, the CPM schedule is analysed to find improvement opportunities. Second, the schedule is optimised within contractual constraints. The third step is to start tracking production and to implement the location-based controlling system.

Chapter 17

Multiple case study components

INTRODUCTION

The following case studies have been selected because they have all been influential in some way in the development of the location-based management system through a case-study approach. Many of these have helped refine the location-based planning theory, others have been instrumental in the development of the theory of location-based control. The projects span the world, from Finland, through the USA to Australia. They are only a selection and there are many others that could have been chosen. They are also limited to that component which they influenced, and therefore do not seek to present a full case study of the project. It is hoped that they will help understanding, by looking at the practical problems which any planning and control system must deal with on real projects.

These cases also highlight that the LBMS is not a theoretical construct, but rather an applied system built upon real-world experience and practical application. In this, the system is new and the learning continuous. The authors are always looking for new projects to try out the LBMS and to further develop the theory and knowledge of the system.

The cases are presented roughly in chronological sequence.

CASE STUDY: KAMPPI CENTRE (2002-2005)

Figure 17.1 Construction of the Kamppi Centre building complex, Helsinki, Finland

Description of the project

The Kamppi Centre building complex is located in the centre of Helsinki, Finland. The total project budget was €500 million and it was the largest city centre project in the history of Finland. The building complex was designed for multiple purposes. There were three underground bus stations—two for passenger buses and one for freight buses, an underground parking hall and underground retail areas and restaurants. Above ground, the centre had 35,000 m² of retail space, 99 apartments and 12,000 m² of office space. The total gross area of the project was 13, 000 m².

The total project duration was four years. During the peak of construction, the project had 800 subcontractors, up to 2,000 workers on site simultaneously and involved 5,000 construction workers in total.

LBMS implementation in the project

Location-based management was adopted as the sole schedule planning and control system. The implementation started after the extensive earthworks phase and lasted until the hand-over of the building. The following parts of the LBMS were used in the project:

- All components of basic location-based planning (however, only Layers 1–3 of layered logic existed at the time, and automatic CPM-like setting of tasks as early as possible was not yet designed).
- Control charts and visualising progress in flowline charts
- All monitoring was done at the baseline level (the current stage of information had not yet been designed).

One full-time person was tracking production and one full-time person was analysing the information and generating reports. The approach was one of macromanagement and concentrated in having the right amount of resources in aggregate working in the project instead of trying to track every work crew for the project (an approach which developed later). Subcontractors were managed with control charts, progress information in flowline views and numerical information about quantities and production rates. Separate views were generated for each contractor. Because of the full-time dedication of two team members to the effort, the information in the LBMS was typically the best available. Any contractor or superintendent could come to the LBMS nerve centre and get answers right away to almost any schedule-related question. The key to the success of the project was that the superintendents and subcontractors learned to use the LBMS as their first and most reliable source of information.

Limitations of the LBMS implementation included the fact that the MEP trades were coordinated and managed using a separate schedule. Productivity benefits were not sought in this project. The key benefits were schedule compression, schedule reliability (reducing risk) and the ability to better manage a complex project of this size.

LBMS results

The case study was the first where the LBMS was used to aggressively manage a project. The results were spectacular. The contractor was able to handover the complex large-scale project six months ahead of schedule, saving millions in overhead costs. This confirmed that the LBMS can compress large-scale schedules by 10% or more. This result was achieved by using the LBMS as a macromanagement tool even without trying to achieve productivity benefits by micromanaging crews and their continuity.

LBMS learning

This has been the biggest project implemented using the LBMS. Although the overall results were great for all participants, the project helped the authors to find many critical development needs in the system. Based on the case study results and observed shortcomings of the system, the following learning was achieved:

- The cast-in-place structure could not be properly planned using Layers 1–3 and splitting–layered logic was developed as a work-around to simulate Layers 4 and 5.
- The need to integrate MEP trades better to the schedule—research was started to get better data for MEP.
- The forecast was not used as a management tool because of shortcomings in both the planning and the control systems. This triggered Olli Seppänen (author) to undertake PhD research to improve the forecast.
- Quantities of a large project keep changing and it is difficult to keep up with changes using manual quantity take-offs. This started research efforts toward integrated 3D model-based quantities to assist location-based schedules.
- It is not feasible to rely on just one level of planning—the reports often tended to look overly optimistic because the plan was being updated. This triggered the development of the current stage of information in the location-based controlling system.

After Layers 4 and 5 and the production system cost and risk approach had been developed, a follow-up Master's thesis was commissioned to test how much more the schedule duration could have been compressed if the schedule optimisation could have been undertaken using Layer 4 logic without having to split all structural tasks (a work around). In the same research, the productivity effects were researched for both structural and interior trades by cost loading the production system with costs.

The results were different for structural and interior trades. For structural work, it was possible to level resources using Layer 4 logic. This decreased the theoretical production system cost by 16%. Of this production system cost decrease, 55% was caused by a decrease of waiting time of the formwork and reinforcement tasks. For the balance of 45% of the production system cost saving, the cause was the decrease in total duration. The optimised plan had much less work breaks. However, the risk analysis results showed that although it was more probable to achieve milestones with the optimised schedule, the risk of cost overruns was larger in optimised schedule.

This shows that, to achieve the optimised plan, production must be controlled so that the variability is smaller than in a non-optimal schedule. Planning a better plan has no productivity benefits given the same variability. Using layered logic allows significant cost and time benefits which can be realised in practice only by reducing the variability of production.

In the case of the interior schedule, the results showed a production system cost saving of 5%. Of this production system cost decrease, 95% was explained by the optimisation of total construction duration and 5% in the reduction in demobilisation costs. It should be noted that interior finishes were well optimised already in the original schedule because they typically do not need Layer 4 and 5 dependencies. In a risk simulation, it was noted that the benefits of optimisation were greater for the construction phase with an optimised schedule despite the same level of variability. The optimised schedule achieved a 6% saving in production system cost despite the variability. This research showed that because of the highly constrained nature of structural work, the benefits of optimisation are lost if variability is not reduced. However, in interior finishes, there is more flexibility and the optimised plan is able to perform better even with the same level of variability (Ojala, 2007).

CAMINO MEDICAL CENTER

Description of the project

Camino Medical Center was one of the first projects in the US which was analysed using the LBMS. The Camino Medical Group is a Sutter Health affiliate. The Camino project is located in Mountain View, California. It is a US$100 million medical office building with surgery centre and urgent care clinic. The total area of the project was 22,500 m^2.

LBMS implementation in the project

The LBMS was used to analyse an existing CPM schedule and to find opportunities for improvement. The scope of analysis included MEP, drywall and interior finishes. Quantities for each task were measured from a 3D model. Productivity rates related to quantity items were requested from subcontractors. This information was combined with the CPM durations to find required crew sizes for each location. Then another location-based schedule was planned and optimised from scratch with the same quantity and productivity assumptions trying to achieve the same duration with more continuous use of resources.

The two schedules were compared and contrasted to find out which one was better. Both schedules were cost loaded using the production system cost as per the LBMS to find how much improvement lay in the productivity which could be achieved using the LBMS.

LBMS results

Resource use of the CPM schedule was found to be extremely uneven, with many starts and stops for all subcontractors. The durations in each location did not have any relationship with the quantities in that location, which therefore made an implicit assumption that the site workers would be able to change the crew size for each location. In contrast, the location-based schedule could be planned to achieve level (even) resource use while achieving the same duration. The production system cost analysis indicated that the same duration could be achieved with 20% lower labour cost using LBMS and the conservative assumption that, for each (de)mobilisation, just two hours of productive time would be lost.

Figure 17.2 shows an example of the differences between the original CPM schedule and the location-based schedule for the drywall contractor.

LBMS learning

The main learning derived from these (and many other) CPM schedule analyses was that a huge improvement in terms of productivity and correctness of schedule can be achieved by using LBMS instead of a CPM schedule.

The case also illustrated an implementation challenge in the US market—much more time is devoted to planning than to actually ensuring that the plans are implemented. Even though the theoretical productivity savings can be demonstrated to be 20% in a project, to achieve them the plans must be followed.

The biggest implementation challenge in the US was found to be the difficulty of getting location-based quantities for projects. As a result, research was commenced to implement model-based quantity take-off solutions and, in the mean time, the first implementations of LBMS were started using location-based worker hour information instead of using real quantities as a work-around.

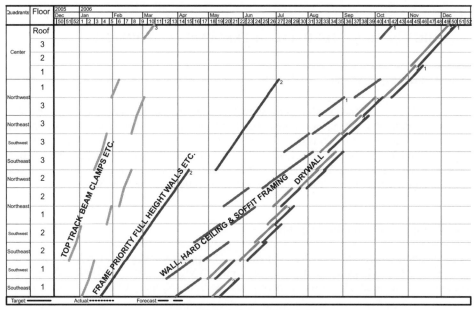

CPM schedule remapped

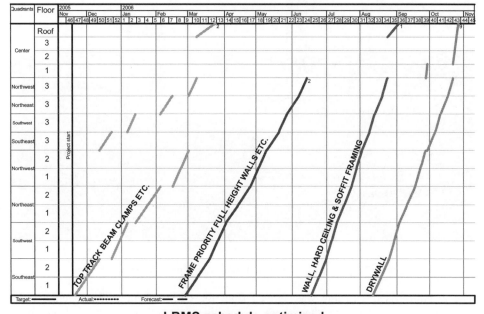

LBMS schedule optimised

Figure 17.2 Camino Medical Center, CPM for drywall—remapped to flowline view

The case study was used as a teaching assignment at Stanford University where the students had to optimise the location-based schedule based on resource constraints of subcontractors.

VICTORIA PARK RESIDENTIAL DEVELOPMENT—FORM 302

Description of the project

The Form 302 residential development (Figure 17.3) was a single stage in a large-scale residential development. The project consisted of four buildings on and around a podium with two levels of underground parking. The buildings consisted of a fourteen level tower, an eight level building, and two four level walk-ups between the two higher bookmark buildings. The builder was the Walter Construction Group, which was trialling location-based scheduling on the project (although it was not known by that name at the time).

LBMS implementation in the project

Implementation consisted solely of scheduling the project using location-based software (DynaProject from DSS solutions—the forerunner of Vico Software's Control). The project commenced late in 2003 and was contemporary with the Kamppi Centre commencement.

Figure 17.3 Victoria Park residential development—Form 302—in construction

The location-based schedule was derived from the CPM schedule (Primavera P3) by working in days. Therefore, there were no quantities (modelled or measured) nor were production consumption rates applied. In order to be able to accelerate and decelerate the production to align the tasks, crews were designated with a production factor of 0.10, and multiples of crews were allocated (with 10 being equivalent to the original CPM assumption).

The project LBS was broken down into buildings (four plus the podium), levels (up to eighteen, including basements, plant room and roof), pours and apartments (289). There were 40 structural and site tasks. Structure was never, however, the main focus of the effort. Attention was mainly devoted to planning for more efficient fit-out works—as experience indicated that this was where most cost and time problems had occurred on previous stages. There were seven major summary tasks (initial work, bathrooms, kitchens, finishing, corridors, laundry and final fit-out) of units, consisting of a total of 70 tasks at that level of the LBS. In total, the model represented approximately 22,630 activities. This was significantly more than previous project models.

LBMS results

The schedule was the first completed which attempted to model tasks down to represent individual crews in a concept termed micromanagement (Kenley, 2004). Previous schedules, as per the Kamppi Centre, were modelled in aggregate. Similarly, this project was the first to model such a large project down to the level of apartments. With such fine task detail, there were many more tasks than were normal at that time, this stretched the processing power of computers then available and led to significant improvements in the software to handle large models.

A comprehensive schedule for the project was constructed, but never practically implemented. A concurrent system was being employed on the site to build team spirit and to ensure pass-on of work between trades. This admirable system (Crow and Barda, 2004) adopted a methodology founded in CPM thinking. While there is no reason this should have presented a barrier to the implementation of the LBMS, in practice it was found that a focus on passing work along the parade of trades (a lean concept) prevented a focus on establishing workflow within trades—a key requirement of the LBMS. Site inspections showed that no practical sequence of work was ever achieved on the project and the productivity gains were therefore not realised (Kenley and Seppänen, 2006).

LBMS learning

The most significant learning was that the scheduling system was in fact location-based. This term had not been used prior to this and was published for the first time at the lean construction conference in Denmark (Kenley, 2004). The concept of *micromanagement* was also first proposed in that same paper. Micromanagement is a methodology of working within the LBMS which focuses deliberately on tracking the physical presence of crews, and therefore must define the work at the level of a task for each crew (where a crew indicates same work, even if there are multiples). The most important element of a location-based schedule is that it provides information about the location of work and work crews. From this comes the ability to manage the work to ensure workflow, work reliability, avoidance of interference, improved quality and reduced rework. For micromanagement, rather than working in aggregate, the following guidelines should be applied:

- All activities must belong to a location.
- A location may display in a hierarchy, but only where there is a physical reality. For example, apartments logically exist within floors and floors exist within buildings, so it is

logical that work done in an apartment is also displayed within a floor and in turn in the building.

- Location should be logical. For example, excavation should not 'appear' in the upper floors. Thus activities should be constrained within their logical place. Thus a 'site' location or a 'project' place can be used to isolate work into a logical location.
- The indication of a line (for an activity) in a location should correspond to a physical presence of a work crew. This enabled rapid identification of conformance.
- The boundaries of the diagrammatic representation should represent the physical boundaries to visually identify interference. For example, the architectural documentation standards which indicate the top of finished floor as being the commencement of a new level should correspond to the graphical representation in the schedule. This ensures common annotation, and also ensures that floors and their associated temporary support structures exist within the physical interference space.
- Gangs may be multiplied, with computer directed location optimisation
- Multiplied gangs should be represented as multiple parallel lines, with the slope representing the productivity of a single gang. This allows manipulation and identification of the physical location. This level of detail is not required for macromanagement.
- The schedule should clearly indicate who should be where and when.
- Deviation from the schedule should be highlighted at the location level of accuracy, allowing rapid identification of required control actions.
- Control actions should allow acceleration to restore schedule, delay acceptance and resultant activity delay, and earned value interpretation. Micromanagement recognises the immediate impact of a delay (in the absence of buffers) and manages that impact on following trades. This requires careful procurement and subcontractor management.
- Both resource flow and workflow are explicitly addressed, as work crews are ensured available locations to move to, as well as readiness at that new location for their work.

While not all of these principles have been strictly applied in the LBMS, micromanagement remains a useful and effective way of using the LBMS.

PARRAMATTA OFFICE BUILDING REFURBISHMENT

Description of the project

The Parramatta office project was a simple office building refurbishment consisting of eight floors to be gutted and refurbished. The project was won by Walter Construction Group late in 2004 on the basis of a CPM schedule for the project which presented a typical floor construction sequence to be repeated on each floor commencing approximately every two weeks. One floor was set to be completed on a fixed date at the end of this series.

LBMS implementation in the project

In this project, the site team gathered for a detailed planning session, at the end of which a location-based representation for the project was developed and contrasted with a flowline view of the original CPM schedule. The process was that the CPM schedule was mapped to the flowline as a series of tasks and the original durations (in days) were used. The same crew multiplier method developed for Form 302 was used for this project.

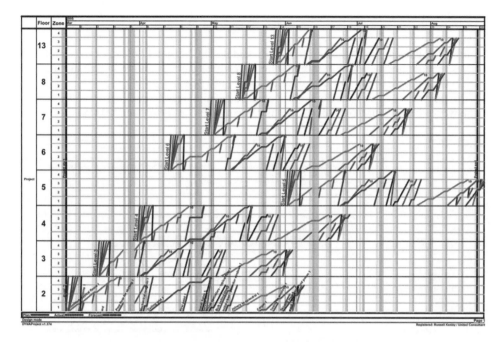

Figure 17.4 Parramatta office refurbishment—flowline of CPM showing discontinuity

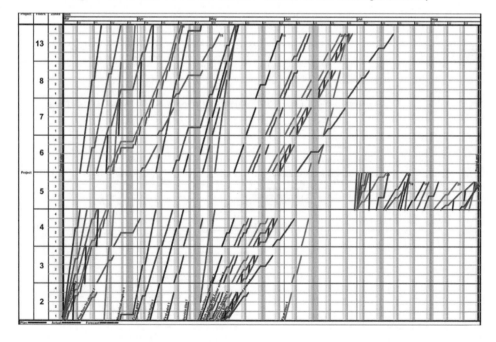

Figure 17.5 Parramatta office refurbishment—flowline of LBMS showing continuity

LBMS results

Figure 17.4 shows the CPM version of the schedule and Figure 17.5 shows the optimised LBMS version. It is clear from this that running a typical floor cycle through each floor of a project does not assist the crews to develop continuity (Kenley, 2006).

Examination of this schedule clearly indicates both overlap and discontinuity for the same work on different floors. This does not occur when optimised for LBMS.

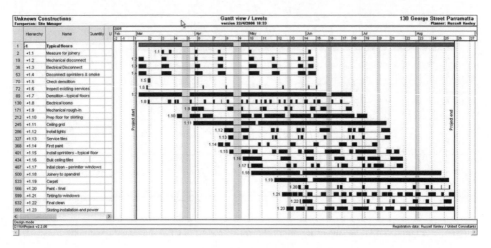

Figure 17.6 Parramatta office refurbishment—Gantt of CPM showing discontinuity

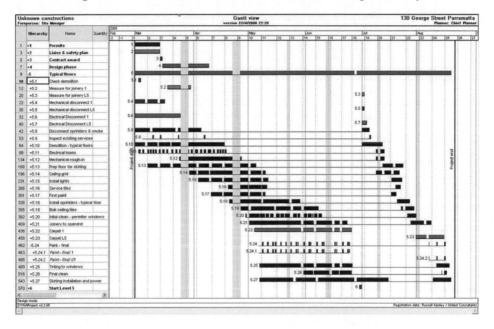

Figure 17.7 Parramatta office refurbishment—Gantt of LBMS showing continuity

The contrast is perhaps clearer in the compressed Gantt view. Figure 17.6 shows the non-optimised schedule with a single line for each task and with the line being solid only when work is planned to be continuous. Figure 17.7 shows the contrasting improvement in the optimised Gantt chart. This highlights the stop-start nature of the CPM version compared to the continuous work of the LBMS version. From this you can see that 4 weeks of work can be distributed over 13 weeks.

LBMS learning

This project demonstrated the speed with which the team could be brought on-side with the location-based management system and the effectiveness of re-planning a CPM schedule as a LBMS schedule. This project also demonstrated the significance of the schedule to the perception of chaos in construction. It is a frequently made claim, particularly in the lean literature, that construction projects are inherently chaotic and are unpredictable. In contrast, this project's schedule analysis indicates that CPM tends to cause this perception through chaotic discontinuities and overlaps—which must be dynamically managed on site. This demonstrated the significance of the LBMS to reducing uncertainty in project control.

SKANSSI RETAIL CENTRE

Description of the project

Skanssi retail centre is a retail project planned for 96 small or medium retailers and a super-market. With a total of 128,000 m^2 it has retail spaces on two floors and 2,400 car spaces including two parking structures and parking on the roof. The total budget of the project was €100 million with the total duration from January 2007 to opening in April 2009. The general contractor was Hartela.

LBMS implementation in the project

The LBMS was implemented from the beginning of the project and was the only scheduling and production control tool on the project. The implementation used 3D model-based quantities for all trades except MEP. The MEP contractors provided worker hours for each task in each location of the building. Because the general contractor and the MEP contractor were owned by the same company, this project presented a unique opportunity to benefit from productivity improvements for MEP contractors.

The following components of the LBMS were used in the project:

- Model-based quantities
- All components of basic location-based planning
- Risk analysis of the plan
- Control charts, visualising progress, forecasts and alarms
- Improvements in subcontractor meetings and production control processes based on empirical research results
- Owner reporting using completion rates.

Figure 17.8 Skanssi retail centre—in construction

LBMS results

At the beginning of the project, the general contractor and the MEP contractor were also the owner-developer of the building. The total duration of the project was optimised using risk analysis results. There are only a few commercially viable opening dates for large retail centres during any year. The probability of achieving these dates was evaluated by the use of risk simulation. The schedule was optimised to achieve the minimum duration within acceptable levels of risk.

The project was sold to venture capitalist companies during construction. The main risks identified by the schedule simulation were excessive resource needs of the electrical contractor in the final stages of the project and the effect of delayed design information on resource loading of MEP contractors. The risk analysis results were used to define latest need dates for end user design information. Balancing the resource use for the electrical contractor was implemented through changes in the design, aimed at enabling standardised solutions so that the majority of electrical cabling could be completed before identifying end users. Additionally, all dependencies were redefined multiple times in workshops with all the main subcontractors participating. The understanding of the criticality of the electrical trade was taken as the focus point of control and the project finished with less rush at the end of the project than is typical for large retail projects.

At the time of writing, the project is in its final stages and will finish on time. The peak number of electricians will stay under the most pessimistic scenario due to the recognition of its importance for the success of the project and the consequential design changes and control actions which were taken on the project.

All the main subcontractors were excited by being able to participate in the scheduling process and being able to get resource-loaded realistic schedules as the end result. They were also positively surprised by the fact that the location-based schedule was able to prove the importance of tenant information and affected the commercial contract between the general contract and the owner. The general contractor learned better ways to cooperate with the subcontractors and realised the importance of balancing resource use. The risk analysis results were validated in practice and resulted in decisions to decrease the risk levels. Control charts were adopted as the main production control tool and were used as communication tools in subcontractor meetings.

However, the control methods were still push methods and the use of the schedule forecast was not fully implemented. Subcontractors continued to move between locations without finishing them in sequence and slowdowns were evident in project data. The tendency to start too early with low productivity was demonstrated also in this project.

LBMS learning

This project demonstrated the danger of using control charts as the sole production control tool. Control charts provide only a snapshot of current progress and fail to show trends—these are better visualised in flowline diagrams. Using control charts as a production control tool is likely to lead to push control because the control chart only shows deviations from plans. Subcontractor meetings tend to focus on red squares (late starts). Because yellow squares (ongoing, but delayed) are deemed less dangerous than red squares, the controlling tends to focus on starting in all locations which are red, which is likely to make matters worse. This is also a factor contributing to push controlling being commonly observed in current practice. Production management should use the schedule forecasts of the flowline when controlling and the role of the control chart should be limited to monitoring and communicating the status of production compared to plans.

Although many responses to the resource problem of electrical work were adaptive, counterproductive actions were also taken. In order to balance the peak at the end, all tasks were started early. However, because of a lack of design detail and completed and preceding work, started locations could not be finished—instead all locations were completed only to 50% level. This caused messy flowline diagrams which were difficult to read, contributing to inadequate use of forecasts during the project. Analysis showed that locations which started too early suffered from lower productivity and did not help with the overall resource peaks, because they included only a small percentage of total worker hours. This shows that all production management decisions benefit from analysis using location-based tools.

MISSION HOSPITAL

Description of the project

Mission Hospital is a health care project, with a relatively small footprint, from St Joseph Health Systems, in Southern California.

The New Acute Care Tower is an addition to the existing hospital on the campus. It is four stories over a basement for a total of 8,750 m2. The basement includes a central plant building, tunnel to the existing tower, and a shelled space. The first floor includes the imaging department and a chapel. The second floor contains more shelled space. The third floor houses the ICU. And the fourth floor is dedicated to medical offices. The addition is connected to the current building with a bridge and as a special remark it is completed with a chapel on the southwest corner and a linear accelerator on the southeast corner.

LBMS implementation in the project

LBMS was implemented from the commencement of construction of the finishes stage only. The implementation used worker days which were calculated from a Primavera activity-based schedule. The following parts of the LBMS were used for that stage of the project:

* All components of basic location-based planning
* Control charts
* Owner reporting using completion rates.

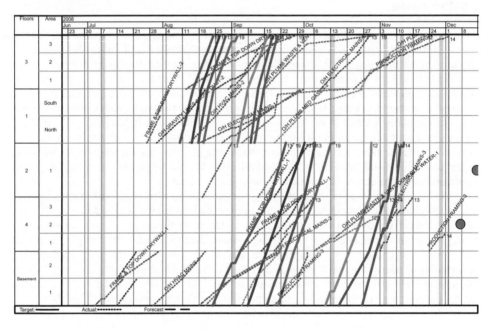

Figure 17.9 Actual versus planned production for Mission Hospital: note the early starts in locations, the cascading delays caused by location congestion and a general trend toward slower than planned production

LBMS results

This project was unique because the owner, St Joseph Health Systems, requested the general contractor to use location-based management for client reporting. The starting point of the plan was a resource-based Primavera schedule. This schedule was transformed into a location-based format. The location-based schedule followed the same logic and had the same durations as the Primavera schedule but used more locations and was planned for continuous workflow.

Control charts were used to monitor production on a monthly basis. The general contractor generated a monthly client report showing the total schedule forecast and completion rates for each major work stage. Monthly progress data was sent for analysis by Vico Software to explain production problems and to forecast upcoming issues.

After four monthly updates, the results showed major deviations from the planned schedule. Tasks had tended to start early, with major slowdowns, a lot of evidence for congested locations and disruptions to workflow was evident. Achieved productivity for many subcontractors was much lower than planned. Deviations resulted partly from lower than planned resources on site, not following the planned sequence and out-of-sequence work. The results were very similar to those observed in Finnish empirical research.

Location-based progress data was able to explain many of the deviations and to forecast upcoming problems. However, at the time of writing, it had not so far been used to actually control the project. Control methods are push methods and the schedule forecasts have not been used for production control. Because forecasts were shown to accurately predict problems each month, there have been discussions about switching to weekly updates and starting to use the LBMS to control the project.

LBMS learning

This case study illustrates an easy way to get started with location-based management. The LBMS can be implemented by using worker days instead of actual quantities in locations. However, the results show that the problems of production control found in Finnish case studies also generalise to US hospital construction. Plans are not being followed and production control operates using push mechanisms. One lesson from this case study is that location-based planning cannot be implemented alone. If location-based controlling is not implemented, the benefits of continuous workflow remain theoretical. However, this approach is minimally disruptive to the site and can be used to demonstrate the potential of location-based management for production improvement by correctly identifying production problems.

EMPIRICAL RESEARCH ON LOCATION-BASED PRODUCTION CONTROL

Description of the project

One of the authors (Olli Seppänen) used three case studies in his PhD research (Seppänen, submitted) to empirically discover how location-based planning and controlling tools are implemented on site and whether those tools and processes can be improved to achieve better results in production control. The research was motivated by previous analysis of various location-based schedules which were not implemented as planned, and instead suffered very low reliability during production. The research had a passiv, observatory approach to compare with earlier contradictory findings: implementations where the author (or other researchers) had been participating in the decision making process had been very successful whereas implementation's by construction companies without the same level of support had less successful results. The goal was to find how LBMS was currently being used and to use these results to explain how it should be used.

Three case studies were used from three different companies. Case studies included a retail mall expansion of 6,000 m², a 10,638 m² new retail centre and a 14,528 m² office building (the case study of Chapter 15). All buildings had a precast structure but were otherwise very different in terms of function (single-user store, multi-user retail centre and office building) as well as tightness of schedule.

LBMS implementation in the project

Location-based management was implemented within each case study as the sole schedule planning and control system. The implementation started at the beginning of the project and continued until the handover of each building. The following parts of LBMS were used in the project:

- All components of basic location-based planning (only Layers 1 to 3 of layered logic existed at the time)
- Control charts, visualising progress, early versions of forecasts and alarms
- Detail planning during production.

In addition to LBMS data, the research data included subcontractor and owner' meeting memos, together with direct observations about how the LBMS data was actually used by management during the project.

LBMS results

All projects finished on time, with in one case (Opus, Chapter 15) the LBMS resulting in a cut in duration of two months. All project teams were satisfied with the LBMS results. However, the analysis of production problems on site and the reliability of the baseline, detailed and weekly schedules, indicated that the schedules were very unreliable and did not achieve the productivity benefits of continuous production. Discontinuities and slowdowns were common in all projects.

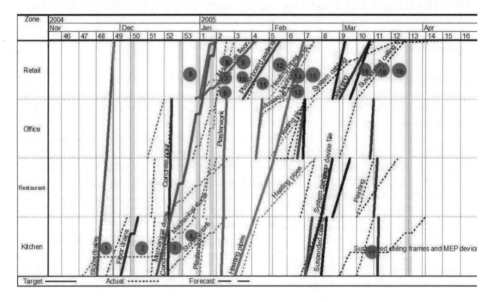

Figure 17.10 Problems identified on one project

Cascading delay chains started with the beginning of the finishes and MEP phase and continued until the end of the project. The only reason these cascading delays did not affect the final handover were the large end-of-project buffers reserved only for commissioning activities. In reality, production and commissioning overlapped in each project with a corresponding rush to finish. The total effect of cascading delays meant that an additional buffer of 10–20% was needed at the end of the project to avoid loss of productivity during production. The causes for these cascading delays included the lack of analysis of production management decisions, unsystematic production control processes, concentration on the past instead of the future in subcontractor meetings, and denial of problems. Additionally, control methodologies were push control, emphasising the start of work in locations rather than ensuring completion of work. When a subcontractor delay was noticed, an immediate reaction was to request more resources even if the locations were congested. These decisions actually contributed to increasing the problems. Figure 17.10 shows an example of a cascading delay chain in four small retail spaces (Seppänen, submitted). Descriptions of the numbered production problems are available from the authors and on publication of Seppänen's thesis.

The LBMS was able to predict 40% of production problems before they occurred in the two case studies where dependencies were correctly used. However, the production team rarely took action to prevent the forecast from turning into reality. In the rare cases where action was taken, production problems could be prevented with control actions but only if the problem was known two weeks or more in advance. Alarms given one week before were typically impossible to successfully circumvent.

LBMS learning

The research results were used to develop better forecasting methods by examining why an alarm was not generated in 60% of the cases. The improved forecast was able to predict problems in 90% of the cases and predict 57% of the problems more than two weeks before they occurred. Additional information requirements include commitments of subcontractors to resources for the next two weeks (knowledge of additional mobilisations and demobilisations), knowledge of decisions about starting and continuing dates and knowledge of production management prioritisation decisions. It is possible to obtain all this information by implementing a systematic production control process and incorporating these items into subcontractor meeting agendas. The results of this research form the bulk of the controlling theory and methodologies in Chapters 8 and 10. Follow-up research was started to test these processes in the field and try to change push controlling to pull controlling.

REFERENCES

Ojala, J. (2007). *Vaativan erityiskohteen aikataulusuunnittelu*. (Schedule planning of a challenging special project.) Master's thesis, Construction Economics and Management, Department of Civil and Environmental Engineering, Helsinki University of Technology, Espoo, Finland.

Crow, T. and Barda, P. (2004). "Project strategic planning: A prerequisite to lean construction". *12th Annual Lean Construction Conference.* Copenhagen.

Kenley, R. (2004). "Project Micromanagement: Practical Site Planning And Management Of Work Flow". *12th Annual Lean Construction Conference.* Copenhagen.

Kenley, R. (2006). "Location-Based Management". *AUBEA 2006*, Proceedings of AUBEA, UTS Sydney, 13p.

Kenley, R. and Seppänen, O. (2006). "Location-based management system for construction: principles and underlying logic". *CIB W55-W65*. Rome, CIB.

Seppänen, O. (submitted). "Empirical research on success of production control in building construction projects". Unpublished PhD thesis submitted in fulfilment of the requirements of the degree of PhD at Helsinki University of Technology, Helsinki, Finland.

Authors index

A

B

C

D

General index

A

B

C

M

N